AF352879

Natural and Synthetic Bioactives for Skin Health, Disease and Management

Natural and Synthetic Bioactives for Skin Health, Disease and Management

Editors

Jean Christopher Chamcheu
Anthony L. Walker
Felicite Noubissi-Kamdem

MDPI • Basel • Beijing • Wuhan • Barcelona • Belgrade • Manchester • Tokyo • Cluj • Tianjin

Editors

Jean Christopher Chamcheu
University of Louisiana at
Monroe
Monroe, LA, USA

Anthony L. Walker
University of Louisiana at
Monroe
Monroe, LA, USA

Felicite Noubissi-Kamdem
Jackson State University
Jackson, MS, USA

Editorial Office
MDPI
St. Alban-Anlage 66
4052 Basel, Switzerland

This is a reprint of articles from the Special Issue published online in the open access journal *Nutrients* (ISSN 2072-6643) (available at: https://www.mdpi.com/journal/nutrients/special_issues/ Skin_Health_Management).

For citation purposes, cite each article independently as indicated on the article page online and as indicated below:

LastName, A.A.; LastName, B.B.; LastName, C.C. Article Title. *Journal Name* **Year**, *Volume Number*, Page Range.

ISBN 978-3-0365-6067-0 (Hbk)
ISBN 978-3-0365-6068-7 (PDF)

Contents

About the Editors

Jean Christopher Chamcheu

Jean Christopher Chamcheu, PhD, is an assistant professor of pharmacology (Tenure-track) at the School of Basic Pharmaceutical and Toxicological Science, College of Pharmacy (COP), at the University of Louisiana at Monroe (ULM), USA. He received graduate training in dermatology and venereology at Uppsala University, Sweden, and undertook post-doctoral and staff scientist training at the Department of Dermatology of the University of Wisconsin-Madison through the cutaneous and cancer biology and chemoprevention program. Since 2017, as a ULM COP faculty member, he has established a research program on collaborative cutaneous biology, molecular dermatology, and experimental therapeutics. With expertise in the tissue engineering of normal and diseased human skin, murine models of disease, and new pharmacological and oncolytic viral therapies, Dr. Chamcheu is currently researching the interplay between disease targets, immunity, cancer, and interventions based on dietary bioactive ingredients, as well as synthetic scaffolds and oncolytic viruses, in the development, prevention, and treatment of chronic inflammatory skin diseases and cancer to promote health. He has authored over 41 peer-reviewed publications and book chapters, edited books, and supervised and trained several students, scientists, and fellows. He has received several awards from the American Skin Association, SID, PC, IJMS, LBRN, and the Louisiana Board of Regents.

Anthony L. Walker

Anthony L. Walker, PharmD, is a clinical associate professor in the College of Pharmacy at the University of Louisiana at Monroe (ULM), USA. He is a licensed pharmacist, laboratory coordinator, and pharmaceutical care lab manager of the Integrated Laboratory Sequences (ILS) courses at ULM. He teaches clinical pharmacy practice in the College of Pharmacy. His general interests include community health volunteer services, providing clinical training to student pharmacists, and mentoring undergraduate, graduate, and professional students. His research interests include clinical pharmacy practice education for student pharmacists; student pharmacists' attitudes towards, and perceptions of, pharmacy practice in the Integrated Laboratory Sequences courses; cost savings through the utilization of in-house prepared mock medications for teaching and learning; and extemporaneous compounding for the delivery of natural and synthetic products in skin disease research.

Felicite Noubissi-Kamdem

Felicite Noubissi-Kamdem, PhD, is an associate professor of cancer biology in the Department of Biology at Jackson State University. She teaches genetics and general biology to undergraduate and graduate students. The mission of her research program is to investigate the cellular and molecular mechanisms underlying tumor development and resistance to drugs. Dr. Noubissi graduated from Jawaharlal Nehru University and the Center for Cellular and Molecular Biology in India with a PhD in Genetics and Molecular Biology. She went on to complete postdoctoral training in cancer biology at the University of Wisconsin, Madison, where she studied the regulation of mRNA turnover and its role in cancer development. This research led to the discovery of an RNA-binding protein, insulin-like growth factor 2 mRNA binding protein (IGF2BP1), which appears to play a critical role in cancer development and metastasis. These findings were published in Nature. Dr. Noubissi's work also demonstrated a mechanism of crosstalk between the Wnt and

Hedgehog (Hh) signaling pathways, which is mediated by IGF2BP1 and plays an important role in cancer development and resistance to drugs. Aspects of her current research include the targeting of IGF2BP1 to sensitize cancer cells to chemotherapeutics and the identification of the mechanisms driving tumor heterogeneity and metastasis. She previously demonstrated that breast tumor cells fuse spontaneously with other cells to form heterogeneous hybrids exhibiting an increased metastatic potential. Her current work aims to delineate mechanisms of cancer cell fusion. Dr. Noubissi' s research has been published in many high-impact journals and supported by a Career Development Award and the Freinkel Diversity Fellowship from the Society for Investigative Dermatology, an R21 and an R03 from the NIH, and a U54 from the NIHMD.

Preface to "Natural and Synthetic Bioactives for Skin Health, Disease and Management"

Among the most important challenges facing the pursuit of global health are diseases that present on the skin and their psychosocial effects. The skin represents the outermost, most exposed part of the human body and largest organ in the integumentary system, which is designed to rapidly prevent and/or treat ailments affecting systemic or local dermal areas and provide the rapid transformation and maintenance of cutaneous homeostasis, thus reducing the harmful effects of genetic and environmental risk factors affecting the skin. However, in today's world, our efforts to treat or prevent skin ailments are failing at an alarming rate, primarily due to the skyrocketing emergence of skin diseases. Hence, the discovery, development, and employment of effective, economical, and practical natural and synthetic products as solutions to these ongoing and emerging challenges, which have profound psychosocial impacts on the affected individuals, are essential. Recent decades have witnessed a resurgence in the application of natural and synthetic products, as sources for the therapeutic treatment of numerous human skin diseases, due to advances in research on natural product chemistry and medicinal chemistry, pharmacological testing, and the identification and development of natural phytochemical compounds, synthetically derived pharmaceuticals, and formulations derived from natural botanical extracts.

This Special Issue comprises state-of-the-art scientific literature reviews, original research papers, and communications that address the existing gaps in scientific knowledge and further advance our understanding of the roles of biologically active, plant-derived natural and synthetic products in the promotion of skin health and the prevention and management of skin diseases. These publications also highlight the molecular pharmacological basis of certain promising strategies that have exhibited a capacity to reduce the burdens of various human skin diseases and conditions, such as dermatitis, cutaneous T-cell lymphomas, melanoma, allergies, hyperpigmentation, inflammation, and more.

Of the fifty-five articles submitted, nineteen were accepted for publication after peer review, reflecting an acceptance rate of 35 percent. The published articles cover a range of topics and applications that are central to the promotion of skin health and the prevention and management of skin diseases. While this Special Issue is intended to inform its readers of the current state of research on bioactive natural and synthetic products in the contexts of skin health, disease, and treatment, it is our hope that the various articles published herein will shed light on new and valuable updates in this field and expand the current readership.

Although the closing date for submissions to this Special Issue has now passed, further in-depth research on the development, technologies, and clinical applications of natural and synthetic products is still required. Therefore, the investigation of this topic will continue in the new Special Issue entitled "Bioactive Natural and Synthetic Products in Human Health and Diseases: Basic, Pre-clinical and Clinical Studies."

This pre-printed book contains sections on the topic of skin inflammation, including atopic dermatitis, ageing, pigmentation, and melanoma, investigated through the use of skin cell cultures or mouse models in studies examining the effects of diverse phytochemicals and their extracts. The authors analyze the objective parameters of key skin disease markers, such as the gene and/or protein expression, and provide direct measurements of the anti-inflammatory actions of the investigated materials to validate any claims of their therapeutic potential.

Finally, we would like to express our gratitude to the staff at MDPI Books, the editorial team to nutritional choices. Bioactive natural products and their synthetic derivatives have garnered tremendous interest in health promotion, disease prevention and management since the ancient medieval era. These biologically active natural phytonutrients, extracts and their synthetic derivatives have numerous potential health benefits that are poised within their regulatory properties on critical physiological processes of life, such as transcriptional and translational programs, metabolism, differentiation and growth. These natural products have demonstrated efficacy in protecting the skin against premature aging, environmental health-related risks and infection. They also ameliorate skin conditions associated with inflammatory diseases such as atopic dermatitis, psoriasis, diabetic and chronic wounds in addition to skin cancers. Appropriate in vitro and preclinical model systems of human diseases have been employed to provide a mechanistic understanding of disease pathogenesis as well as proof of concepts targeting molecular biomarkers that may be useful for therapeutic benefits.

This Special Issue is a snippet of published original research, communications and quality reviews of the scientific literature that further increase our understanding of the role of natural dietary and synthetic bioactive compounds for skin health promotion, disease prevention and management. The findings not only highlight those natural phytonutrients, extracts and synthetic scaffolds that target diverse disease molecular markers, but delineate promising strategies to reduce the burden of cutaneous human diseases and beyond.

The first paper by Michalak et al. (2021) is a literature review of the effects of bioactive micro- and macronutrients such as vitamins, minerals, fatty acids, polyphenols and carotenoids on skin heath conditions, and how healthy dietary habits could serve as useful approaches in anti-aging interventions. It highlights the effects of these compounds on skin parameters, including elasticity, firmness, wrinkles, senile dryness, hydration and color, as well as their role in the processes of skin aging [7].

The second review by Kaur et al. (2021) assessed numerous skin fungal infections, impinging on their possible treatments and the effective utilization of plant extract and oil-embedded polysaccharide-based nanohydrogels. They discuss the advantages of the use of plant extract and oil-based nanohydrogels as emerging technologies in the treatment of various skin fungal infections over the use of standard drugs that require a long treatment time with harmful side effects and often low efficacy. Plant extract-based nanohydrogels can be applied to infected wound sites as they have great penetration power, are directly absorbed through the skin, reach the hypodermis or subcutaneous layer and effectively suppress fungal infections [8]. Nutraceuticals have been proposed to have beneficial effects on many other chronic inflammatory diseases of the skin as well.

Trzeciak et al. (2020) reported that the skin barrier defects in cutaneous T-cell lymphomas (CTCLs) mirrored those observed in atopic dermatitis (AD) with regard to their expression levels of genes and proteins of the cutaneous cornified envelope (CE) in CTCL, AD and healthy skin. They identified changes in their gene and protein expressions as well as their associations with the pathogenesis of CTCL and AD. These authors identified decreased mRNA levels of FLG, FLG2, LOR, CRNN and SPRR3v1, while increased mRNA levels of RPTN, HRNR and SPRR1Av1 were observed in lesioned and non-lesioned AD skin compared to healthy control samples. In contrast, they observed increased mRNA levels of FLG, FLG2, CRNN and SPRR3v1 as well as decreased mRNA levels of RPTN, HRNR and SPRR1Av1 in CTCL skin when compared to lesioned AD skin. The study suggested that in addition to AD, these markers, FLG, FLG2, RPTN, HRNR and SPRR1A, also play a key role in skin barrier dysfunction in CTCL and thus could be considered as consistent differential diagnostic biomarkers for both AD and CTCL, thus opening up opportunities for further studies on a larger study group [9].

In a study, Adamczyk-Grochala et al. (2020) investigated the biological activity of water and ethanolic extracts of modified microalgal clones within the context of their applications in cosmeceutical and regenerative medicine. They used normal human fibroblasts and keratinocytes, and found that treatment with modified extracts of the microalga Planktochlorella nurekis variant attenuated the development of oxidative-stress-induced senescence in these human skin cells. The authors concluded that these effects may be related to increased nitric oxide, niacin and biotin levels in cells compared to an unmodified microalgal clone. Their data suggested that selected microalgal extracts of Planktochlorella nurekis could be used in an anti-skin aging regimen [10].

Another study by Gęgotek et al. (2019) used proteomics profiling of 3D keratinocytes cultures exposed to UVA or UVB radiation to identify a synergistic cytoprotective effect of rutin and ascorbic acid [11]. Because of their antioxidant and anti-inflammatory properties, combinations of ascorbic acid and rutin are often used in oral preparations and thus can be used to protect against the effects of cutaneous UV radiation from sunlight. In particular, the authors observed alterations in the expression of proteins involved in the antioxidative response, DNA repair, inflammation, apoptosis and protein biosynthesis compared to when compounds are used individually [11].

Another study by Choi et al. (2020) investigated the effect of konjac (Amorphophallus konjac) glucomannan (KGM) on damaged skin. KGM is widely used as a dietary supplement with anti-inflammatory and immunosuppressive effects, in addition to skin regenerative potential in patch or sheet form. They reported that KGM alleviates and repairs/regenerates UVB-induced skin damage by increasing the proportion of proliferative young cell populations in UVB-exposed senescent human epidermal primary melanocytes and UVB-damaged human fibroblasts in a dose-dependent manner. This was the case as they found that the mRNA and protein levels of age- and pigmentation-related factors decreased depending on the rate of new cells generated. They identified KGM as a highly effective natural agent for maintaining skin homeostasis via the promotion of a reconstituted dermal matrix against UVB-induced acute senescence or skin damage [12].

Quah et al. (2020) investigated the anti-allergic, antioxidant and anti-inflammatory activities of the ethanolic extract of Cornus officinalis (COFE) as a potential treatment of AD. Here, using RBL-2H3 cells sensitized with the dinitrophenyl-immunoglobulin E (IgE-DNP) antibody after stimulation with dinitrophenyl-human serum albumin (DNP-HSA), they observed that COFE suppressed the release of β-hexosaminidase in a concentration-dependent manner. The extract was also observed to significantly inhibit both lipopolysaccharide (LPS)-induced nitric oxide (NO) production and the expression of iNOS and pro-inflammatory cytokines (IL-1β, IL-6 and TNF-α) in RAW 264.7 cells in addition to TNF-α-induced apoptosis in HaCaT cells. Collectively, this study suggests that COFE might inhibit allergic responses, oxidative stress and inflammatory responses by a mechanism that disrupts the binding of IgE to human high-affinity IgE receptors (FcεRI), and that these inhibitory effects might be achieved by COFE's compounds loganin, cornuside and naringenin 7-O-β-D-glucoside [13].

In another study by Kim et al. (2020), the usefulness of propolis, a resinous substance generated by bees, in treating UV-induced photoaging in human skin was introduced. Propolis treatment was also found to suppress UV-induced matrix metalloproteinase (MMP)-1 production and expression in human dermal fibroblasts as well as block collagen degradation in human skin tissues, suggesting that the anti-skin-aging activity of propolis can be mimicked in clinically relevant conditions. Their investigations revealed that propolis, especially its main components caffeic acid phenethyl ester, quercetin and apigenin, shows anti-skin-aging effects through the direct suppression of phosphoinositide 3-kinase (PI3K) activity. The content of active compounds was quantified, and among the compounds identified from the propolis extract, caffeic acid phenethyl ester, quercetin and apigenin were shown to attenuate PI3K activity. These results demonstrate that propolis shows anti-skin-aging effects through the direct inhibition of PI3K activity [14].

In another study, Hyuna Lee and Eunmi Park (2021) showed that Perilla leaf and callus extracts exert a protective effect against UVB-induced damage on skin and/or keratinocytes by a mechanism that involves an increased DNA repair response and G1/S cell cycle arrest [15].

Kim et al. (2021), in a randomized, double-blind and placebo-controlled study revealed the anti-inflammatory and anti-pathogenic bacterial activities of a dietary Lactobacillus plantarum, CJLP55, a natural product supplement, on clinical improvement, skin sebum, hydration and urine bacterial EV phylum flora in patients with acne vulgaris [16].

Furukawa et al. (2021) used neonatal normal human epidermal keratinocytes (NHEK-Neo) and interleukin-10 knockout (IL-10 KO) mice to show that eggshell membrane (ESM) has a beneficial effect on maintaining skin health and improving the skin aging process. The molecular mechanism was associated with this, which upregulates calcium signaling and results in an increased expression of keratinocyte differentiation markers (including keratin 1, filaggrin and involucrin) as well as growth factors, such as transforming growth factor β1, platelet-derived growth factor-β and connective tissue growth factor. They also showed that ESM changed keratinocytes' morphology and suppressed skin thinning [17].

Oh et al. (2021) showed that ethanolic extracts of Lithospermum erythrorhizon (LE) roots improve symptoms associated with AD (atopic dermatitis) in an NC/Nga AD mouse model. They found that LE restored defects in skin barrier function and the balance of T helper (Th) 1/Th2 AD-related cytokine and chemokine expression in the dorsal skin, serum and splenocytes of the NC/Nga mouse. This improvement of AD-related symptoms correlated with a reduction in the serum levels of IL-4, IgE and histamine via the regulation of the IgG1/IgG2a ratio, as well as the restoration of the Th1/Th2 immune balance in NC/Nga Mice. LE also decreased the levels of AD-related cytokines and chemokines in NC/Nga mouse serum and regulated the balance of cytokine and chemokine secretion in splenocytes in addition to their expression in the dorsal skin [18].

Olivera-Castillo et al. (2020) evaluated the bioactive-enriched components found in the sea cucumber (*Isostichopus badionotus*) body wall that promote anti-inflammatory activity and found that fucosylated chondroitin sulfate, the primary glycosaminoglycan, reduced the expression of critical genes, such as NF-kB, TNFa, iNOS and COX-2, and further lessened the inflammation and tissue damage caused by 12-O-tetradecanoylphorbol-13-acetate (TPA) in a mouse ear inflammation model [19].

Lee et al. (2019) demonstrated that triacylglycerol (TAG) metabolism is related to the acyl-ceramide (Cer) synthesis and corneocyte lipid envelope (CLE) formation involved in maintaining the epidermal barrier. Their investigation demonstrated that dietary borage oil-enhanced TAG content, the gene expression of TAG metabolism, acyl-Cer synthesis and CLE formation in the epidermis of essential fatty-acid-deficient guinea pigs [20].

A high melanin content has been shown to significantly reduce the efficacy of radiotherapy and chemotherapy in patients with metastatic melanotic melanomas, probably due to the radioprotective and free radical scavenging properties of melanin. Therefore, inhibiting melanogenesis has been suggested as a single, combination or adjuvant approach in the treatment of melanotic melanomas.

This Special Issue has also highlighted the utility of natural products tested and has shown their inhibitory properties in melanogenesis as well as their contribution to the treatment of melanotic melanoma or pigmentation disorders affecting the skin. These natural products include extracts of lotus seedpod (LSE), sorghum bicolor, black cumin seed and lactobacillus-helveticus-fermented milk whey (LHMW).

Hsu et al. (2020) showed that lotus seedpod extract (LSE), or its main active compound epigallocatechin (EGC), downregulates the signaling pathway, leading to melanin production in the α-MSH-induced murine melanoma cell line B16F0 by inhibiting the expression of MC1R, the expression of tyrosinase and the expression of microphtalmia-associated transcription factor (MITF). These authors also detected a reduction in the activity of p38 and protein kinase A (PKA) in the cells treated with LSE. These results were supported by their in vivo studies that showed a similar downregulation of melanin synthesis in mice ears treated with LSE or ECG and exposed to UVB. They suggest that the mechanism by which LSE inhibits melanin synthesis involves the inhibition of p38 and PKA signaling pathways [21].

Recently, Han et al. (2020) demonstrated the anti-melanogenic role of ethanolic extract of sorghum bicolor (ESB) in B16/F10 cells treated with the melanin synthesis inducer 3-isobutyl-1-methylxanthine (IBMX). The authors showed that IBMX-induced melanogenesis was drastically reduced in the presence of ESB by a mechanism involving the inhibition of MITF, tyrosinase and tyrosinase-related protein 1. ESB contains phenolic compounds such as 9-hydroxyoctadecadienoic acid (9-HODE), 1,3-O-dicaeoylglycerol and tricin, and exhibits a high antioxidant property. It could be used as a skin-whitening material in cosmetics or in the management of melanocytic melanomas [22].

Similarly, Ikarashi et al. (2020) showed that lactobacillus-helveticus-fermented milk whey (LHMW) suppressed melanin synthesis in B16 mouse melanoma treated with α-MSH by inhibiting MITF, tyrosinase and tyrosinase-related protein 1 at the RNA and protein levels [23].

Li et al. (2020) found that the black cumin seed extract Thymocid also inhibits melanin synthesis in B16F10 cells by a mechanism that plausibly involves the downregulation of MITF and tyrosinase activity. In addition, Thymocid treatment reduced collagenase and elastase activities as well as protein glycation and collagen cross-linking. This portrays Thymocid as a potential anti-aging compound [24].

Overall, the 18 papers published in this Special Issue highlight that an adequate dietary intake of natural products or their application is essential for maintaining normal skin homeostasis, thereby being crucial for human health. In summary, the daily dietary intake of natural products such as fruits and vegetables is a cost-effective and healthy way to strengthen human skin, significantly reducing the risk of skin disorders and improving general well-being. These natural dietary products are potential drugs for the management of premature aging and hyperpigmentation as they can serve as whitening agents or as chemopreventive and chemotherapeutics, as well as an adjuvant in the treatment of melanotic melanomas or other skin disorders involving melanogenesis and other chronic skin disorders. However, additional studies using more physiologically relevant models are warranted to determine their efficacy and potential clinical applications. Validating the beneficial effects of these natural and synthetic bioactives would improve the management of skin health and reduce the burden associated with the plethora of emerging skin diseases.

Author Contributions: J.C.C. and F.K.N. cowrote the original draft preparation. J.C.C., A.L.W. and F.K.N. contributed in writing—reviewing and editing the final manuscript. All authors have read and agreed to the published version of the manuscript.

Funding: This research received no external funding.

Acknowledgments: As Guest Editors of this Special Issue, we would like to acknowledge all the authors that have provided outstanding contributions to the Special Issue "Natural and Synthetic Bioactives for Skin Health, Disease and Management," and are very grateful to all reviewers for their constructive comments. This Special Issue has highlighted the need for further research into the mechanisms involved in cutaneous disorders as well as the need for the development of novel therapeutics to treat the broad range of illnesses under the umbrella of dermatoses. The research in Chamcheu and Noubissi's research laboratories are supported in part by the following grants: an NIH NIGMS project grant P20GM103424, a Louisiana board of regents support fund grant LEQSF (2021-24)-RD-A-22 (to JCC), an NIH/NCI grant R03 CA223099, an NIH/NIHMD grant 1U54MD015929-01 and the Society for Investigative Dermatology (SID) Freinkel Diversity Fellowship (to FKN).

Conflicts of Interest: The authors declare no conflict of interest.

References

1. Chamcheu, J.C.; Roy, T.; Uddin, M.B.; Banang-Mbeumi, S.; Chamcheu, R.N.; Walker, A.L.; Liu, Y.Y.; Huang, S. Role and Therapeutic Targeting of the PI3K/Akt/mTOR Signaling Pathway in Skin Cancer: A Review of Current Status and Future Trends on Natural and Synthetic Agents Therapy. *Cells* **2019**, *8*, 803. [CrossRef] [PubMed]
2. Watt, F.M. Mammalian skin cell biology: At the interface between laboratory and clinic. *Science* **2014**, *346*, 937–940. [CrossRef] [PubMed]
3. Bergqvist, C.; Ezzedine, K. Vitiligo: A Review. *Dermatology* **2020**, *236*, 571–592. [CrossRef] [PubMed]
4. Park, S.Y.; Jin, M.L.; Kim, Y.H.; Kim, Y.; Lee, S.J. Aromatic-turmerone inhibits alpha-MSH and IBMX-induced melanogenesis by inactivating CREB and MITF signaling pathways. *Arch. Dermatol. Res.* **2011**, *303*, 737–744. [CrossRef]
5. Zhou, X.; McCallum, N.C.; Hu, Z.; Cao, W.; Gnanasekaran, K.; Feng, Y.; Stoddart, J.F.; Wang, Z.; Gianneschi, N.C. Artificial Allomelanin Nanoparticles. *ACS Nano* **2019**, *13*, 10980–10990. [CrossRef]

6.   Brozyna, A.A.; Jozwicki, W.; Carlson, J.A.; Slominski, A.T. Melanogenesis affects overall and disease-free survival in patients with stage III and IV melanoma. *Hum. Pathol.* **2013**, *44*, 2071–2074. [CrossRef] [PubMed]

7.   Michalak, M.; Pierzak, M.; Krecisz, B.; Suliga, E. Bioactive Compounds for Skin Health: A Review. *Nutrients* **2021**, *13*, 203. [CrossRef] [PubMed]

8.   Kaur, N.; Bains, A.; Kaushik, R.; Dhull, S.B.; Melinda, F.; Chawla, P. A Review on Antifungal Efficiency of Plant Extracts Entrenched Polysaccharide-Based Nanohydrogels. *Nutrients* **2021**, *13*, 2055. [CrossRef] [PubMed]

9.   Trzeciak, M.; Olszewska, B.; Sakowicz-Burkiewicz, M.; Sokolowska-Wojdylo, M.; Jankau, J.; Nowicki, R.J.; Pawelczyk, T. Expression Profiles of Genes Encoding Cornified Envelope Proteins in Atopic Dermatitis and Cutaneous T-Cell Lymphomas. *Nutrients* **2020**, *12*, 862. [CrossRef] [PubMed]

10.   Adamczyk-Grochala, J.; Wnuk, M.; Duda, M.; Zuczek, J.; Lewinska, A. Treatment with Modified Extracts of the Microalga Planktochlorella nurekis Attenuates the Development of Stress-Induced Senescence in Human Skin Cells. *Nutrients* **2020**, *12*, 1005. [CrossRef] [PubMed]

11.   Gegotek, A.; Jarocka-Karpowicz, I.; Skrzydlewska, E. Synergistic Cytoprotective Effects of Rutin and Ascorbic Acid on the Proteomic Profile of 3D-Cultured Keratinocytes Exposed to UVA or UVB Radiation. *Nutrients* **2019**, *11*, 2672. [CrossRef]

12.   Choi, K.H.; Kim, S.T.; Bin, B.H.; Park, P.J. Effect of Konjac Glucomannan (KGM) on the Reconstitution of the Dermal Environment against UVB-Induced Condition. *Nutrients* **2020**, *12*, 2779. [CrossRef]

13.   Quah, Y.; Lee, S.J.; Lee, E.B.; Birhanu, B.T.; Ali, M.S.; Abbas, M.A.; Boby, N.; Im, Z.E.; Park, S.C. Cornus officinalis Ethanolic Extract with Potential Anti-Allergic, Anti-Inflammatory, and Antioxidant Activities. *Nutrients* **2020**, *12*, 3317. [CrossRef] [PubMed]

14.   Kim, D.H.; Auh, J.H.; Oh, J.; Hong, S.; Choi, S.; Shin, E.J.; Woo, S.O.; Lim, T.G.; Byun, S. Propolis Suppresses UV-Induced Photoaging in Human Skin through Directly Targeting Phosphoinositide 3-Kinase. *Nutrients* **2020**, *12*, 3790. [CrossRef]

15.   Lee, H.; Park, E. Perilla frutescens Extracts Enhance DNA Repair Response in UVB Damaged HaCaT Cells. *Nutrients* **2021**, *13*, 1263. [CrossRef] [PubMed]

16.   Kim, M.J.; Kim, K.P.; Choi, E.; Yim, J.H.; Choi, C.; Yun, H.S.; Ahn, H.Y.; Oh, J.Y.; Cho, Y. Effects of Lactobacillus plantarum CJLP55 on Clinical Improvement, Skin Condition and Urine Bacterial Extracellular Vesicles in Patients with Acne Vulgaris: A Randomized, Double-Blind, Placebo-Controlled Study. *Nutrients* **2021**, *13*, 1368. [CrossRef] [PubMed]

17.   Furukawa, K.; Kono, M.; Kataoka, T.; Hasebe, Y.; Jia, H.; Kato, H. Effects of Eggshell Membrane on Keratinocyte Differentiation and Skin Aging In Vitro and In Vivo. *Nutrients* **2021**, *13*, 2144. [CrossRef]

18.   Oh, J.S.; Lee, S.J.; Choung, S.Y. Lithospermum erythrorhizon Alleviates Atopic Dermatitis-like Skin Lesions by Restoring Immune Balance and Skin Barrier Function in 2.4-Dinitrochlorobenzene-Induced NC/Nga Mice. *Nutrients* **2021**, *13*, 3209. [CrossRef] [PubMed]

19.   Olivera-Castillo, L.; Grant, G.; Kantun-Moreno, N.; Barrera-Perez, H.A.; Montero, J.; Olvera-Novoa, M.A.; Carrillo-Cocom, L.M.; Acevedo, J.J.; Puerto-Castillo, C.; May Solis, V.; et al. A Glycosaminoglycan-Rich Fraction from Sea Cucumber Isostichopus badionotus Has Potent Anti-Inflammatory Properties In Vitro and In Vivo. *Nutrients* **2020**, *12*, 1698. [CrossRef] [PubMed]

20.   Lee, J.Y.; Liu, K.H.; Cho, Y.; Kim, K.P. Enhanced Triacylglycerol Content and Gene Expression for Triacylglycerol Metabolism, Acyl-Ceramide Synthesis, and Corneocyte Lipid Formation in the Epidermis of Borage Oil Fed Guinea Pigs. *Nutrients* **2019**, *11*, 2818. [CrossRef] [PubMed]

21.   Hsu, J.Y.; Lin, H.H.; Li, T.S.; Tseng, C.Y.; Wong, Y.; Chen, J.H. Anti-Melanogenesis Effects of Lotus Seedpod In Vitro and In Vivo. *Nutrients* **2020**, *12*, 3535. [CrossRef] [PubMed]

22.   Han, H.J.; Park, S.K.; Kang, J.Y.; Kim, J.M.; Yoo, S.K.; Heo, H.J. Anti-Melanogenic Effect of Ethanolic Extract of Sorghum bicolor on IBMX-Induced Melanogenesis in B16/F10 Melanoma Cells. *Nutrients* **2020**, *12*, 832. [CrossRef] [PubMed]

23.   Ikarashi, N.; Fukuda, N.; Ochiai, M.; Sasaki, M.; Kon, R.; Sakai, H.; Hatanaka, M.; Kamei, J. Lactobacillus helveticus-Fermented Milk Whey Suppresses Melanin Production by Inhibiting Tyrosinase through Decreasing MITF Expression. *Nutrients* **2020**, *12*, 2082. [CrossRef]

24.   Li, H.; DaSilva, N.A.; Liu, W.; Xu, J.; Dombi, G.W.; Dain, J.A.; Li, D.; Chamcheu, J.C.; Seeram, N.P.; Ma, H. Thymocid((R)), a Standardized Black Cumin (*Nigella sativa*) Seed Extract, Modulates Collagen Cross-Linking, Collagenase and Elastase Activities, and Melanogenesis in Murine B16F10 Melanoma Cells. *Nutrients* **2020**, *12*, 2146. [CrossRef] [PubMed]

Article

Lithospermum erythrorhizon Alleviates Atopic Dermatitis-like Skin Lesions by Restoring Immune Balance and Skin Barrier Function in 2.4-Dinitrochlorobenzene-Induced NC/Nga Mice

Jin-Su Oh [1], Sang-Jun Lee [2] and Se-Young Choung [1,3,*]

[1] Department of Life and Nanopharmaceutical Sciences, Graduate School, Kyung Hee University, 26, Kyungheedae-ro, Dongdaemun-gu, Seoul 02447, Korea; ok9638@naver.com
[2] Holistic Bio Co., Ltd., Seongnam 13494, Korea; leesjun2006@gmail.com
[3] Department of Preventive Pharmacy and Toxicology, College of Pharmacy, Kyung Hee University, 26, Kyungheedae-ro, Dongdaemun-gu, Seoul 02447, Korea
* Correspondence: sychoung@khu.ac.kr

Abstract: The incidence of atopic dermatitis (AD), a disease characterized by an abnormal immune balance and skin barrier function, has increased rapidly in developed countries. This study investigated the anti-atopic effect of *Lithospermum erythrorhizon* (LE) using NC/Nga mice induced by 2,4-dinitrochlorobenzene. LE reduced AD clinical symptoms, including inflammatory cell infiltration, epidermal thickness, ear thickness, and scratching behavior, in the mice. Additionally, LE reduced serum IgE and histamine levels, and restored the T helper (Th) 1/Th2 immune balance through regulation of the IgG1/IgG2a ratio. LE also reduced the levels of AD-related cytokines and chemokines, including interleukin (IL)-1β, IL-4, IL-6, tumor necrosis factor-α (TNF-α), thymic stromal lymphopoietin, thymus and activation-regulated chemokine, macrophage-derived chemokine, regulated on activation, normal T cell expressed and secreted, and monocyte chemoattractant protein-1 in the serum. Moreover, LE modulated AD-related cytokines and chemokines expressed and secreted by Th1, Th2, Th17, and Th22 cells in the dorsal skin and splenocytes. Furthermore, LE restored skin barrier function by increasing pro-filaggrin gene expression and levels of skin barrier-related proteins filaggrin, involucrin, loricrin, occludin, and zonula occludens-1. These results suggest that LE is a potential therapeutic agent that can alleviate AD by modulating Th1/Th2 immune balance and restoring skin barrier function.

Keywords: atopic dermatitis; *Lithospermum erythrorhizon*; NC/Nga; Th1; Th2; Th17; Th22; immune balance; skin barrier function

Citation: Oh, J.-S.; Lee, S.-J.; Choung, S.-Y. *Lithospermum erythrorhizon* Alleviates Atopic Dermatitis-like Skin Lesions by Restoring Immune Balance and Skin Barrier Function in 2.4-Dinitrochlorobenzene-Induced NC/Nga Mice. *Nutrients* **2021**, *13*, 3209. https://doi.org/10.3390/nu13093209

Academic Editors: Jean Christopher Chamcheu, Anthony L. Walker and Felicite Noubissi-Kamdem

Received: 11 August 2021
Accepted: 14 September 2021
Published: 15 September 2021

Publisher's Note: MDPI stays neutral with regard to jurisdictional claims in published maps and institutional affiliations.

1. Introduction

Atopic dermatitis (AD) is most common in early childhood, however it is also present in adults [1]. AD is accompanied by persistent itching and clinical symptoms such as eczema, erythema, dryness, lichenification, and abrasions [2–6]. AD is an early stage of the "atopic march" that leads to asthma and rhinitis [7]. The chronicization of AD progresses through damage to the skin barrier due to scratching caused by itchiness [8]. An important feature of AD is the differentiation of biased T helper (Th) 2 cells [9]; the increase in levels of Th2-mediated cytokines and chemokines causes a Th1/Th2 immune imbalance and impairs skin barrier function [6,10–12].

Th2 cell activation increases the expression and release of cytokines such as interleukin (IL)-4, IL-5, IL-13, and IL-31 [9]. Additionally, IL-4 has an autocrine function that continuously activates Th2 cells and inhibits Th1 cell activation [13]. Thus, Th1-mediated cytokines, including IL-12 and interferon-γ (IFN-γ), are downregulated resulting in Th1/Th2 immune imbalance. The normal immune system maintains the Th1/Th2 balance with antagonism between IL-4 and IFN-γ [14]. IL-4 and IL-13 promote isotype conversion in B cells that increases the production of immunoglobulins (Ig) E and G1 [6,15]. IgE is detected for AD diagnosis as it is observed at a high concentration in the serum of patients with AD [16,17]. Histamine, a biological response modulator, is released from degranulated mast cells and functions together with IgE to cause itching [5,18]. IL-31 impairs the skin barrier function by causing itching and inhibiting apoptosis in eosinophils [19–21].

AD skin lesions express and secrete cytokines IL-25, IL-33, and thymic stromal lymphopoietin (TSLP) [22]. IL-25, IL-33, and TSLP can stimulate Th2 cells directly or indirectly by stimulating dendritic cells (DCs), mast cells, and eosinophils [23]. TSLP activates DCs to induce naive T cell proliferation and primes Th2 cells for differentiation, which is followed by the secretion of high levels of IL-4, IL-5, IL-13, and tumor necrosis factor-α (TNF-α) [24]. Activation of mast cells and eosinophils releases AD-related inflammatory factors such as IL-4, IL-5, IL-6, IL-31, and monocyte chemoattractant protein-1 (MCP-1) [25–27]. MCP-1 is reported to be involved in the activation of Th2 cells in mice [28]. Chemokines such as thymus and activation-regulating chemokines (TARC) and macrophage-derived chemokines (MDC) are expressed and secreted in keratinocytes that bind to the C-C motif chemokine receptor 4 (CCR4) of Th2 cells to induce Th2 cell activation [29]. Moreover, regulated on activation, normal T cell expressed and secreted (RANTES) is involved in the degranulation and infiltration of eosinophils [30]. The Th2-mediated cytokine IL-5 induces chronicization of AD through the continuous survival and differentiation of eosinophils [9,31].

The skin barrier blocks the influx of external antigens and allergens; therefore, the maintenance of a healthy skin barrier function is vital for moisture homeostasis and environmental protection [32,33]. Specifically, filaggrin (FLG), involucrin (IVL), and loricrin (LOR) are the key skin barrier proteins components [34]. FLG has a vital role in skin barrier maintenance and constitutes the outermost barrier through the aggregation of the keratinocyte matrix [35]. FLG exists as pro-filaggrin (pro-FLG) and is converted to FLG through dephosphorylation and proteolysis by serine proteases [36]. FLG then participates in epidermal differentiation; following its decomposition into free amino acids, FLG is decomposed into components of natural moisturizing factors (NMF), such as sodium pyrrolidone carboxylic acid and urocanic acid [37]. IVL and LOR constitute, in part, the outer wall of keratinocytes and promote the final differentiation of the epidermis [35]. IVL forms a scaffold in which other proteins are cross-linked [38], while LOR is an insoluble protein in the cornified cell envelope and accounts for 80% of the total protein [39]. LOR binds to IVL and acts as the major protein in the outer skin barrier [40,41]. Tight junction (TJ) proteins are located under the epidermis and contribute to the selective permeability of various substances, including cytokines and hormones. TJ proteins bind to plasma proteins inside the cell, thus forming a loop structure outside the cell that connects with adjacent cells [32]. TJs are formed and interact with proteins such as occludin (OCC) and zonula occludens-1 (ZO-1) [42]. OCC promotes cell adhesion of neighboring cells [43], while ZO-1 binds to several TJ components and interacts with signaling proteins such as heterotrimeric G-proteins [44].

Defects in the skin barrier function are associated with Th2-mediated cytokines, including IL-4, IL-13, and IL-31 [45]. Meanwhile, recent studies have also shown that Th17 cells and Th22-mediated cytokines IL-17 and IL-22 impair the skin barrier function [46]. Th2-mediated cytokines can further activate Th17 and Th22 cells to promote defects in skin barrier function [47].

AD is traditionally treated with topical corticosteroids (TCS); however, long-term use causes topical and systemic side effects [48]. Topical calcineurin inhibitors have been approved to treat AD as replacements for corticosteroids [49], though side effects such as skin malignancies, lymphoma, and leukemia have been reported [50]. Therefore, research on AD therapeutic agents with fewer side effects is required to improve patient treatment and quality of life.

In NC/Nga mice, skin lesions similar to AD appear naturally in a typical environment and are widely used to study mechanisms related to AD [3,51]. 2,4-Dinitrochlorobenzene (DNCB) quickly penetrates the epidermis and causes an increase in IgE, which can lead to hypersensitivity of the skin [52]. Repetitive DNCB application under specific pathogen-free (SPF) conditions causes AD-like skin lesions in the skin of NC/Nga mice [35,51].

Lithospermum erythrorhizon (LE) has traditionally been used as a natural preparation in various East Asian countries, including Korea, China, and Japan [53]. Active ingredients of LE include naphthoquinones β, β-dimethyl acryl shikonin, lithospermic acid, and acetyl shikonin, which reportedly have antioxidant and anti-inflammatory effects [54]. However, studies on the improvement of AD-related clinical symptoms, Th1/Th2 immune balance, and skin barrier function recovery using LE have not been conducted. Therefore, this study focused on whether LE restores AD-related clinical symptoms, Th1/Th2 immune balance, and skin barrier function in a DNCB-induced NC/Nga mouse model.

2. Materials and Methods

2.1. Preparation of the LE Extract

Dried roots of LE were collected from Jindo, Jeollanam-do Province, Korea in October, 2017. A voucher specimen (17-10-009) was deposited in the herbarium of Holistic Bio Co., Ltd. (Seongnam, Gyeonggi-do, Korea). LE was prepared as a dried powder from the diluted ethanol extract, which was provided by Nutrex Co., Ltd. (Seongnam, Gyeonggi-do, Korea). For the preparation of the LR extract, 100 g dried LR was extracted with 300 mL 70% ethanol at 85–90 °C for 2 h with stirring. The residual extract was extracted again under the same conditions. These two extracts were combined and lyophilized to yield 33 g LR extract, which contained 0.23% lithospermic acid.

2.2. High-Performance Liquid Chromatography (HPLC) Analysis of LE

Approximately 5.0 g of LE extract was weighed, transferred to a 50 mL measuring flask, dissolved in 50% methanol (1:1 deionized water-MeOH), and passed through a 0.45 μm membrane filter to prepare the test solution. Chromatographic separation was performed on an Agilent 1100 separation module equipped with a C18 column (4.6 mm × 250 mm; 5 μm, CAPCELL PAK, Shiseido, Tokyo, Japan). The pump was connected to two mobile phases: A, 0.1% formic acid, and B, acetonitrile in H_2O (*v/v*), and the elution flow rate was 1.0 mL/min. The mobile phase was consecutively programmed in linear gradients as follows: 0–20 min, 95% A, 5% B; 20–25 min, 80% A, 20% B; 25–50 min, 70% A, 30% B; 50–75 min 0% A, 100% B; 70–75 min 95% A, 5% B. The UV detector was monitored at a wavelength of 320 nm. The injection volume was 10 μL for each sample solution. The column temperature was maintained at 30 °C.

2.3. Animals

Four-week-old male NC/Nga mice were provided by Shizuoka Laboratory Center Inc. (Shizuoka, Japan). Mice were housed at 23 ± 3 °C and 55% ± 5% humidity in individually ventilated cages under SPF conditions with 12 h light and dark cycle. The mice were provided with feed (Catalog number 5L79, Central Laboratory Animal, Seoul, Korea) and water ad libitum. Animals were anesthetized with isoflurane, blood was drawn from the inferior vena cava, and then they were euthanized using CO_2. Blood was stored at 25 ± 5 °C for 1 h and subsequently centrifuged at 3000× *g*, 4 °C for 15 min to collect serum samples. Serum samples were stored at −80 °C until use. All experimental procedures were performed according to the protocol approved by the Kyung Hee University Animal Care and Use Committee guidelines (approval number KHSASP-20-252).

2.4. Induction of AD-like Skin Lesions and LE Treatment

AD-like skin lesions were induced by topical application of DNCB (Sigma-Aldrich, St. Louis, MO, USA) to NC/Nga mice as previously described [4,35]. Briefly, after acclimatization for 1 week, the hair on the dorsal side of the NC/Nga mice was removed using an electric shaver. Mice were randomly divided into normal (naïve control), DNCB (negative control), prednisolone (PD; positive control; Sigma-Aldrich, St. Louis, MO, USA), and 50, 100, and 200 mg/kg LE groups, with six mice assigned to each group. To induce AD-like skin lesions, 1% DNCB was dissolved in a mixture of acetone and ethanol (2:3 *v/v*) and applied twice every other day to the shaved dorsal flank (200 μL) and right ear (100 μL). After sensitization, 0.4% DNCB dissolved in a mixture of acetone and olive oil (3:1 *v/v*) was applied to the dorsal skin (150 μL) and right ear (50 μL) three times a week for 14 weeks. After 9 weeks of induction, mice in the normal and DNCB groups were orally administered 0.5% carboxymethyl cellulose (0.5% CMC) for 4 weeks. PD (3 mg/kg prednisolone) and LE (50, 100, and 200 mg/kg) were orally administered daily for 4 weeks. CMC was used to dissolve PD and LE and was administered to normal and DNCB groups.

2.5. Dermatitis Score and Ear Thickness

The dermatitis score was recorded three times per week as previously described [55]. The score grades were 0 (none), 1 (mild), 2 (moderate), or 3 (severe) and were measured for each of the five symptoms tested (e.g., erythema, dryness, maceration, abrasion, and lichenification). The total dermatitis score was quantified as the sum of all individual scores for the five symptoms. The ear thickness was gauged on the right ear of each mouse three times a week using a thickness gauge (Mitutoyo Corporation, Tokyo, Japan).

2.6. Scratching Behavior

Scratching behavior was recorded three times per week [56]. After vehicle administration, the mice were acclimated to an acrylic cage for 1 h and the scratching movements around the neck, ears, and dorsal flank skin with hind paws were measured and recorded for 30 min. The scores ranged from 0 to 4 (0 points (none), 2 points (less than 1.5 s), and 4 points (more than 1.5 s)). The total score for scratching behavior was presented as the sum of the individual measurements.

2.7. Histological Analysis

The dorsal skin was cut and fixed in 10% formalin, then sliced to a thickness of 4 μm. Tissue sections were stained with hematoxylin and eosin (H&E) and toluidine blue (TB). After staining, images were taken with an optical microscope (400×, DP Controller Software; Olympus Optical, Tokyo, Japan). The epidermis thickness and the number of infiltrated inflammatory cells (e.g., mast cells and eosinophils) were measured at six sites per mouse using Image J software (National Institute of Health, Starkville, MD, USA).

2.8. Serum Immunoglobulin and Histamine Assay

The levels of IgE, histamine, IgG1, and IgG2a in the serum were measured using a mouse enzyme-linked immunosorbent assay (ELISA) kit according to the manufacturer's instructions (IgE, Shibayagi, Gunma, Japan; IgG1 and IgG2a, Enzo Life Sciences, Farmingdale, NY, USA; histamine, Elabscience, Huston, ID, USA).

2.9. Serum Cytokines and Chemokines Assay

The serum cytokine and chemokine levels were measured using a mouse ELISA kit, according to the manufacturer's instructions (IL-4 and IL-6, Enzo Life Sciences, Farmingdale, NY, USA; TSLP, MCP-1, Elabscience, Huston, ID, USA; IL-1β, TARC, MDC, and RANTES, R&D Systems Inc., Minneapolis, MN, USA).

2.10. Isolation of Splenocytes and Analysis of Cytokines and Chemokines

Splenocytes were isolated in a sterile environment. Splenocytes were crushed with a sterile syringe plunger and collected using a cell strainer (BD Biosciences, Franklin Lakes, NJ, USA). Subsequently, after treatment with red blood cell lysis buffer, splenocytes were washed three times with RPMI-1640 (Gibco, Carlsbad, NY, USA) supplemented with 10% FBS. The isolated splenocytes were treated with 5 μg/mL concanavalin A (Con-A) (Sigma-Aldrich, St. Louis, MO, USA) and incubated in 24-well plates for 72 h at a concentration of 1×10^6 cells/well at 37 °C, 5% CO_2. After incubation, supernatants were collected and splenocytes were homogenized in lysis buffer containing cOmplete™ Protease Inhibitor

Cocktail tablets (Roche Diagnostics, Indianapolis, IN, USA). The lysate was centrifuged at $10,000\times g$ for 10 min at 4 °C. The collected splenocyte supernatants and lysates were frozen at -80 °C for subsequent cytokine analysis. The levels of IL-5, IL-12, IL-13, IL-17, IL-22, IL-25, IL-31, IL-33, TNF-α, and IFN-γ cytokines in the splenocyte supernatant were measured using an ELISA kit according to the manufacturer's instructions (Elabscience, Houston, ID, USA). The lysate protein concentration was measured using a Pierce™ BCA Protein Assay Kit (Thermo Fisher Scientific, Rockford, IL, USA). The levels of cytokines and chemokines in the supernatants were normalized to the protein concentration of the lysate.

2.11. RNA Extraction and Quantitative Real-Time Polymerase Chain Reaction (RT-qPCR)

Total RNA was extracted from mouse dorsal skin samples using the Easy-Red Total RNA Extraction Kit. Chloroform was added, and the mixture was stored at room temperature for 20 min. The supernatant was collected by centrifugation at $10,000\times g$ for 15 min at 4 °C. The cells were treated with an equal volume of isopropanol as the supernatant and incubated overnight for 24 h. The mixture was then centrifuged for 15 min at $10,000\times g$ and 4 °C and washed with 75% ethanol. After RNA was dissolved in DEPC water, cDNA was synthesized using a cDNA synthesis kit(Takara Korea Biomedical, Inc., Shiga, Japan). Quantitative real-time polymerase chain reaction (RT-qPCR) was performed on an ABI StepOnePlus™ real-time PCR system (Applied Biosystems, Waltham, MA, USA) using the synthesized cDNA as a template and SYBR Premix EX Taq (TaKaRa Bio, Shiga, Japan). The primer sequences are listed in Table 1. Gene expression levels were normalized to GAPDH using the $2^{-\Delta\Delta Ct}$ method for the cycle threshold (Ct) value.

Table 1. Primer sequences (in vivo).

Gene	Forward ($5'$–$3'$)	Reverse ($5'$–$3'$)
IL-1β (m)	TGT GTT TTC CTC CTT GCC TCT GAT	TGC TGC CTA ATG TCC CCT TGA AT
IL-4 (m)	ACG GAG ATG GAT GTG CCA AAC	AGC ACC TTG GAA GCC CTA CAG A
IL-5 (m)	TCA GCT GTG TCT GGG CCA CT	TT ATG AGT AGG GAC AGG AAG CCT CA
IL-6 (m)	CCA CTT CAC AAG TCG GAG GCT TA	GCA AGT GCA TCA TCG TTG TTC ATA C
IL-12 (m)	TGA ACT GGC GTT GGA AGC	GCG GGT CTG GTT TGA TGA
IL-13 (m)	CAA TTG CAA TGC CAT CTA CAG GAC	CGA AAC AGT TGC TTT GTG TAG CTG A
IL-17 (m)	AAG GCA GCA GCG ATC ATC C	GGA ACG GTT GAG GTA GTC TGA G
IL-22 (m)	CAG CTC CTG TCA CAT CAG CGG T	AGG TCC AGT TCC CCA ATC GCC T
IL-25 (m)	CTC AAC AGC AGG GCC ACT C	GTC TGT AGG CTG ACG CAG TGT G
IL-31 (m)	ATA CAG CTG CCG TGT TTC AG	AGC CAT CTT ATC ACC CAA GAA
IL-33 (m)	GAT GAG ATG TCT CGG CTG CTT G	AGC CGT TAC GGA TAT GGT GGT C
IFN-γ (m)	CGG CAC AGT CAT TGA AAG CCT A	GGC ACC ACT AGT TGG TTG TCT TTG
TNF-α (m)	TAC TGA ACT TCG GGG TGA TTG GTC	CAG CCT TGT CCC TTG AAG AGA ACC
TSLP (m)	TGC AAG TAC TAG TAC CGG ATG GGC	GGA CTT CTT GTG CCA TTT CCT GAG
TARC (m)	TGA GGT CAC TTC AGA TGC TGC	ACC AAT CTG ATG GCC TTC TTC
MDC (m)	CAG GCA GGT CTG GGT GAA	TAA AGG TGG CGT CGT TGG
RANTES (m)	GGA GTA TTT CTA CAC CAG CAG CAA	GGC TAG GAC TAG AGC AAG CAA TGA C
CCR4 (m)	TCT ACA GCG GCA TCT TCT TCA T	CAG TAC GTG TGG TGG TGC TCT G
Pro-filaggrin (m)	GAA TCC ATA TTT ACA GCA AAG CAC CTT G	GGT ATG TCC AAT GTG ATT GCA CGA TTG
GAPDH (m)	ACT TTG TCA AGC TCA TTT CC	TGC AGC GAA CTT TAT TGA TG

2.12. Western Blotting

Dorsal skin tissues were frozen in liquid nitrogen, crushed using a pestle, and subsequently homogenized with lysis buffer containing cOmplete™ Protease Inhibitor Cocktail tablets (Roche Diagnostics, Indianapolis, IN, USA). The lysates were sonicated and centrifuged at $10,000\times g$ for 15 min at 4 °C. The protein concentration in the supernatant was quantified using the Pierce™ BCA protein assay kit (Thermo Fisher Scientific, Rockford, IL, USA). After quantitation, equal amounts of protein were loaded into a 12% sodium dodecyl sulfate-polyacrylamide gel (SDS-PAGE; Bio-Rad, CA, USA) for electrophoresis and then transferred to a polyvinylidene fluoride (PVDF) membrane. The membrane was blocked with 5% skim milk in Tris-buffered saline with 0.5% Tween-20 (TBST) and incubated with 1:1000 primary antibody overnight at 4 °C. The following day, the membranes were treated with a horseradish peroxidase-conjugated (HRP) secondary antibody at a dilution of 1:5000 (GeneTex, Inc., Irvine, CA, USA) for 2 h and were visualized using a ChemiDoc™XRS + System (Bio-Rad, Richmond, CA, USA). The expression level of each protein was analyzed using Image Lab statistical software (Bio-Rad, Richmond, CA, USA) and normalized to β-actin. The primary antibodies used for Western blotting were as follows: filaggrin (FLG), occluding (OCC), and loricrin (LOR) (GeneTex, Inc., Irvine, CA, USA), involucrin (IVL), β-actin (Santa Cruz, CA, USA), and zonula occludens-1 (ZO-1) (Abcam, Cambridge, MA, USA).

2.13. Statistical Analysis

Data are presented as means $\pm$ standard deviation (SD). Statistical analysis was performed using one-way analysis of variance (ANOVA) and Tukey's honestly significant difference test. Statistically significant differences were evaluated using SPSS (SPSS Inc., Chicago, IL, USA).

3. Results

3.1. Identification and Qualification of Lithospermic Acid in the LE Extracts

HPLC analysis at 320 nm was used to determine the components of LE, and the established standard chromatogram is shown in Figure 1. Lithospermic acid is highlighted, which is a major constituent of LE, and was detected in high concentrations in our LE samples. The retention time of lithospermic acid was 41.118 ± 0.21 min. The content analysis indicated that LE contained 1.34 ± 0.009 mg/g of lithospermic acid.

3.2. LE Attenuates DNCB-Induced AD-like Symptoms and Scratching Behavior

The schedule for DNCB induction and LE oral administration is presented in Figure 2A. Figure 2B shows representative images of the six test groups, including the normal, DNCB, PD, and 50, 100, and 200 mg/kg LE groups, on the last day of the 14th week of the experiment. Repetitive application of DNCB aggravated the clinical symptoms associated with AD, including erythema, maceration, lichenification, abrasion, dryness, and scratching behavior, over the first nine weeks. However, administration of LE and PD reduced these lesions, and the dermatitis score, ear thickness, and scratching behavior were significantly higher in the DNCB group compared to the normal group; however, LE administration significantly reduced these parameters in a dose-dependent manner (Figure 2C). The dermatitis score and scratching behavior showed a similar efficacy between the 50 mg/kg LE group and the positive control PD group, and ear thickness measurements showed a similar effect between the 100 mg/kg LE group and the PD group. Taken together, LE attenuated the dermatitis score, ear thickness, and scratching behavior in the AD mouse model.

3.3. LE Reduces Epidermal Thickening as Well as Eosinophil and Mast Cell Infiltration

To evaluate how inflammatory responses contribute to AD symptoms, we measured epidermis thickness and amount of eosinophil and mast cell infiltration, both of which were significantly higher in the DNCB group compared to the normal group. However, LE treatment reduced the epidermal thickness and the number of infiltrated inflammatory cells in a dose-dependent manner, showing a similar efficacy between the 100 mg/kg LE group and the PD group (Figure 3).

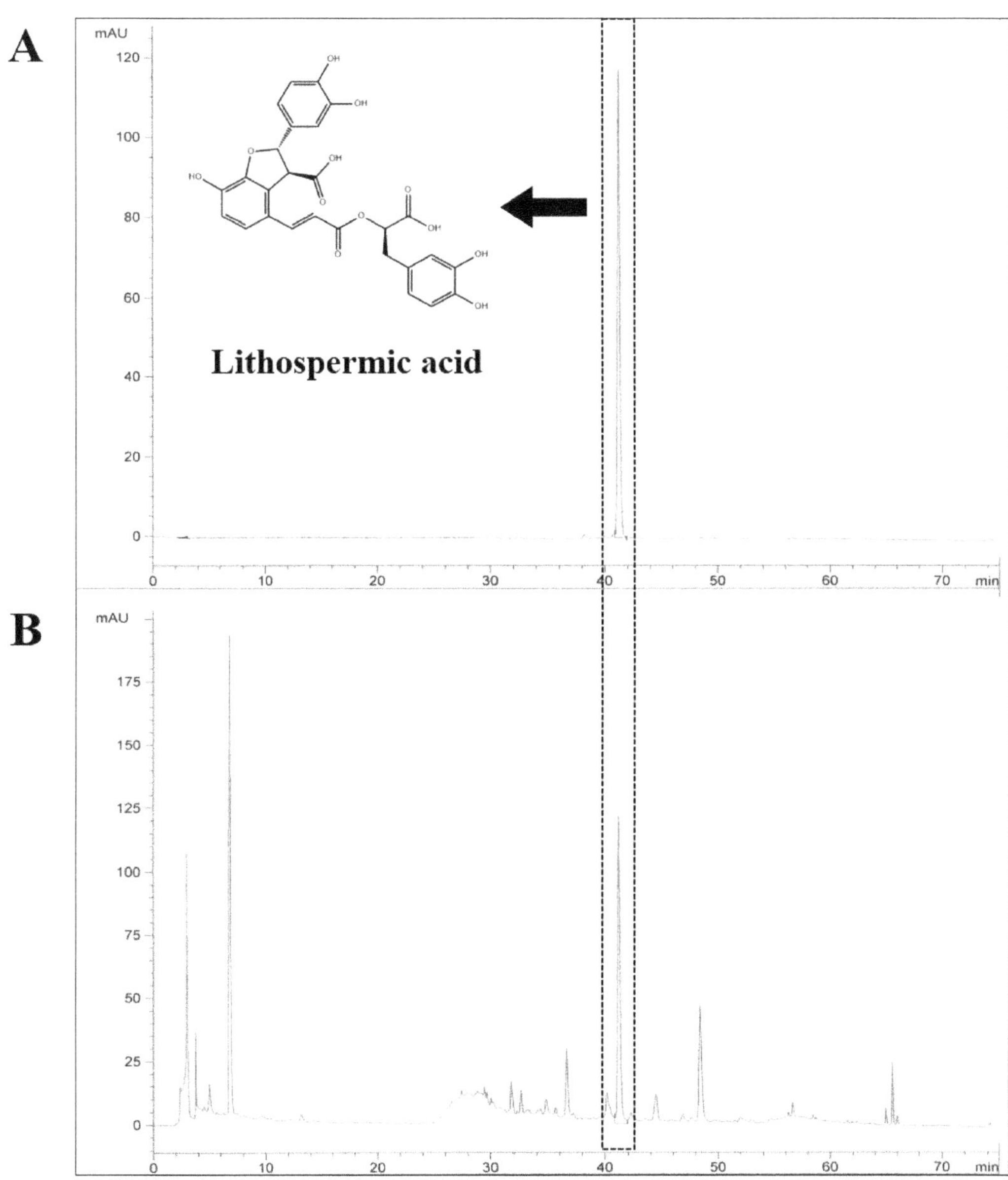

Figure 1. Representative HPLC chromatogram of LE. HPLC chromatograms of (**A**) lithospermic acid and (**B**) the LE extract. The arrows represent lithospermic acid. The retention time of lithospermic acid was 41.118 ± 0.21 min.

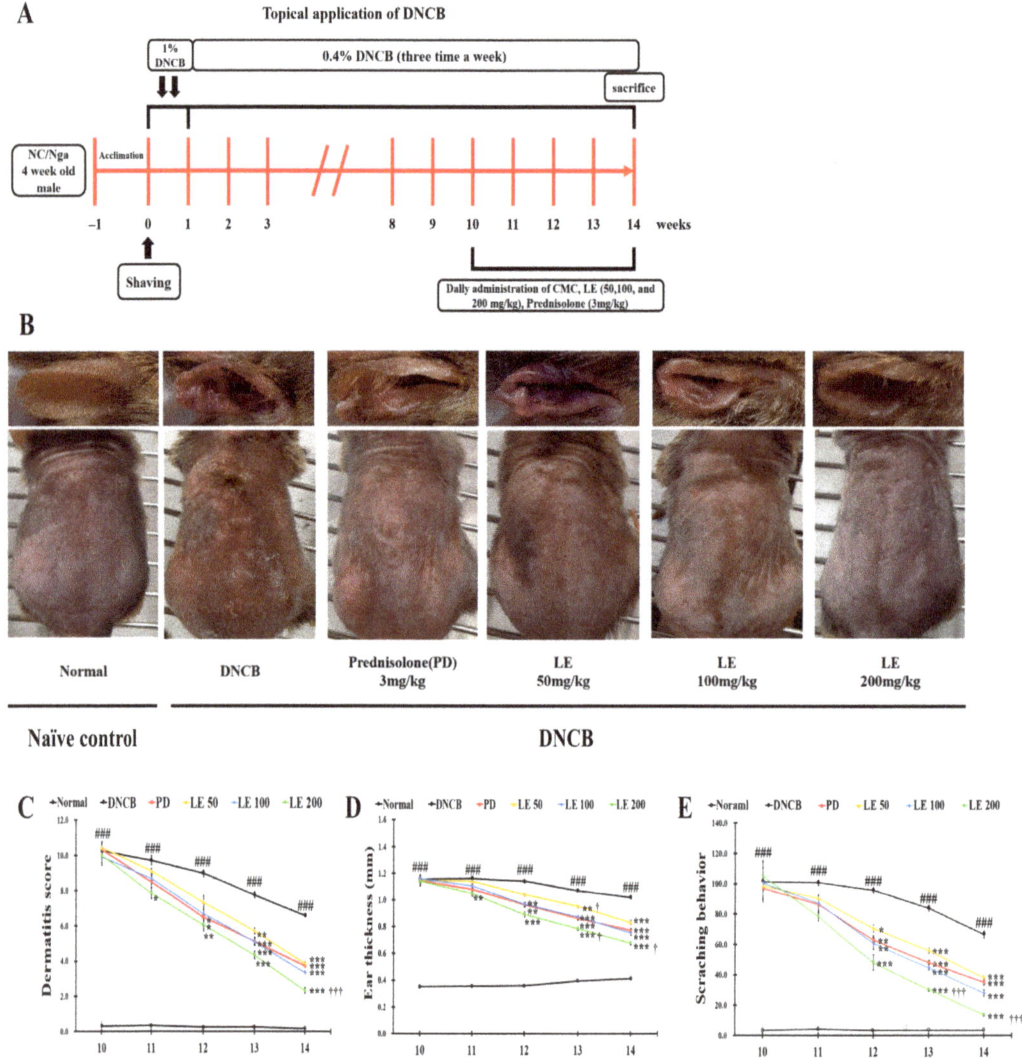

Figure 2. Experimental procedure and effects of LE on the clinical features of AD-like symptoms induced by DNCB in NC/Nga mice. (**A**) Schematic diagram of the experimental procedure for AD lesion induction and LE treatment. (**B**) On the last day of the 14th week of the experiment, the right ear and dorsal skin of mice representing each group were shown. (**C–E**) The clinical features of NC/Nga mice (**C**) dermatitis score, (**D**) ear thickness, and (**E**) scratching behavior were evaluated three times a week in the term of the administration of CMC, PD, and LE. The results were expressed as means ± SD ($n = 6$). ### $p < 0.001$ vs. normal (DNCB untreated group), * $p < 0.05$, ** $p < 0.01$, and *** $p < 0.001$ vs. DNCB (negative control; DNCB treated group), PD (positive control; prednisolone 3 mg/kg) treatment group, LE (LE 50, 100 or 200 mg/kg) treatment group, and † $p < 0.05$, ††† $p < 0.001$ vs. PD (positive control; prednisolone 3 mg/kg) treatment group.

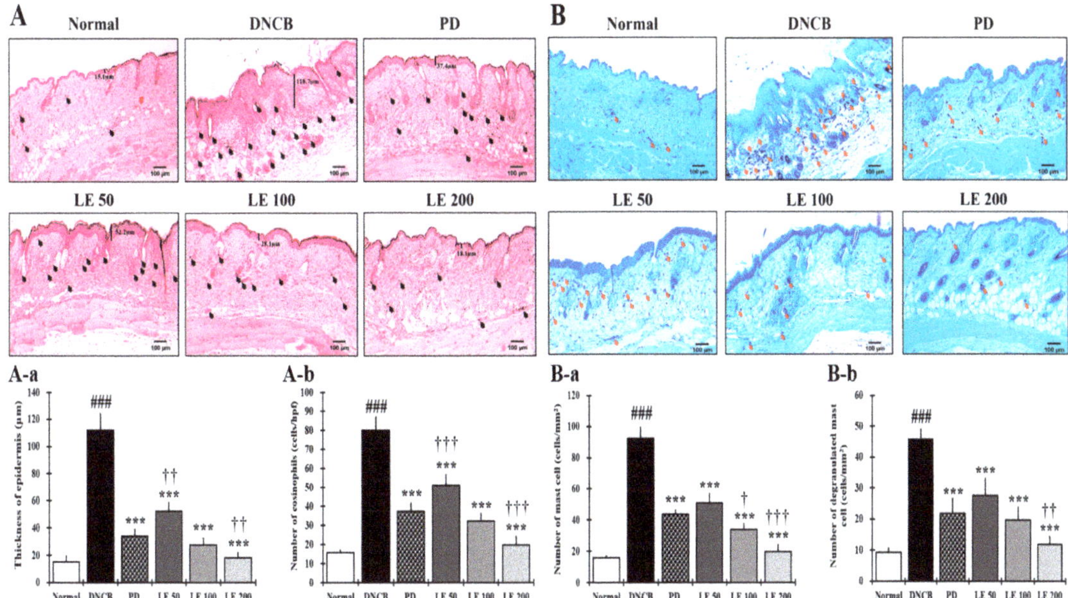

Figure 3. Effect of LE on DNCB-induced histological features of AD-like symptoms in NC/Nga mice. (**A**,**B**) Representative histological images of dorsal skin from mice treated with DNCB (400×, scale bar = 100 μm). (**A**) H&E staining of dorsal skin. The epidermis thickness is indicated by bars and black arrows indicate the infiltration of eosinophils into the skin. (**A-a**) The thickness of the epidermis was measured and averaged from the dorsal skin lesions. (**A-b**) The number of eosinophils infiltrated the dorsal skin. (**B**) Toluidine blue staining of dorsal skin. The red arrows indicate the penetration of mast cells into the skin. (**B-a**) Mast cells present in dorsal skin lesions. (**B-b**) The number of mast cells that infiltrated the dorsal skin lesion. All data were collected from six random dorsal skin sites for each mouse. The results are expressed as means ± SD (n = 6). ### $p < 0.001$ vs. normal (DNCB untreated group), *** $p < 0.001$ vs. DNCB (negative control; DNCB treated group), PD (positive control; prednisolone 3 mg/kg) treatment group, LE (LE 50, 100 or 200 mg/kg) treatment group, and † $p < 0.05$, †† $p < 0.01$, and ††† $p < 0.001$ vs. PD (positive control; prednisolone 3 mg/kg) treatment group.

3.4. LE Decreases IL-4, IgE, and Histamine Serum Levels

IL-4, IgE, and histamine serum levels were investigated to evaluate the effect of LE on AD itching. We found that the IL-4, IgE, and histamine levels were significantly higher in the DNCB group compared to the normal group. However, LE treatment reduced these levels in a dose-dependent manner, with similar efficacy in reducing IL-4 levels observed between the 100 mg/kg LE and PD groups. However, the LE group showed a higher efficacy in reducing IgE and histamine levels than the PD group (Figure 4A–C). These results show that LE reduced cytokines, IgE, and histamine in serum related to itching.

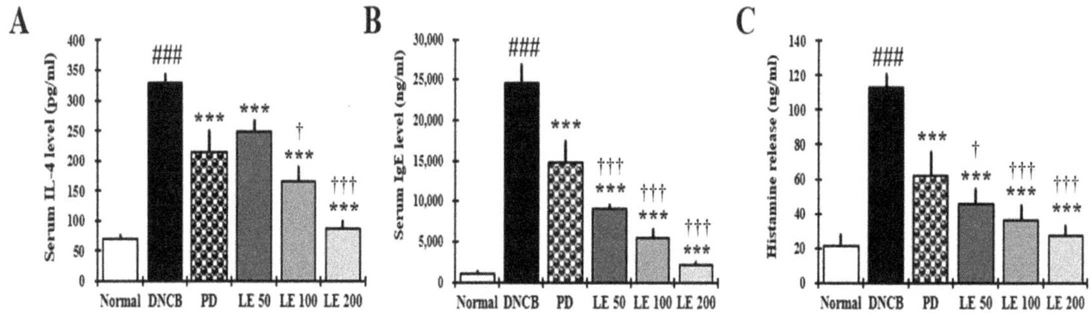

Figure 4. Effect of LE on IL-4, IgE, and histamine levels in the serum of NC/Nga mice. The levels of (**A**) IL-4, (**B**) IgE, and (**C**) histamine in the serum. Serum was collected on the last day of the experiment and measured using ELISA. The results were expressed as means $\pm$ SD ($n = 6$). ### $p < 0.001$ vs. normal (DNCB untreated group), *** $p < 0.001$ vs. DNCB (negative control; DNCB treated group), PD (positive control; prednisolone 3 mg/kg) treatment group, LE (LE 50, 100 or 200 mg/kg) treatment group, and † $p < 0.05$, ††† $p < 0.001$ vs. PD (positive control; prednisolone 3 mg/kg) treatment group.

3.5. LE Restores Th1/Th2 Immune Balance by Regulating Serum IgG1 and IgG2a Levels in NC/Nga Mice

The restoration of Th1/Th2 balance was investigated by measuring the levels of IgG1 and IgG2a in response to LE treatment. Serum IgG1 levels were higher in the DNCB group compared to the normal group; however, LE treatment reduced the IgG1 levels in a dose-dependent manner, with a similar efficacy observed between the 100 mg/kg LE and PD groups. Moreover, the IgG2a levels were higher in the DNCB group than in the normal group. Meanwhile, compared to the DNCB group, the LE group increased IgG2a levels in a dose-dependent manner, whereas IgG2a levels decreased in the PD group. These results showed LE treatment restored the IgG1/IgG2a ratio in a dose-dependent manner, with a similar efficacy between the 100 mg/kg LE and PD groups (Figure 5A–C).

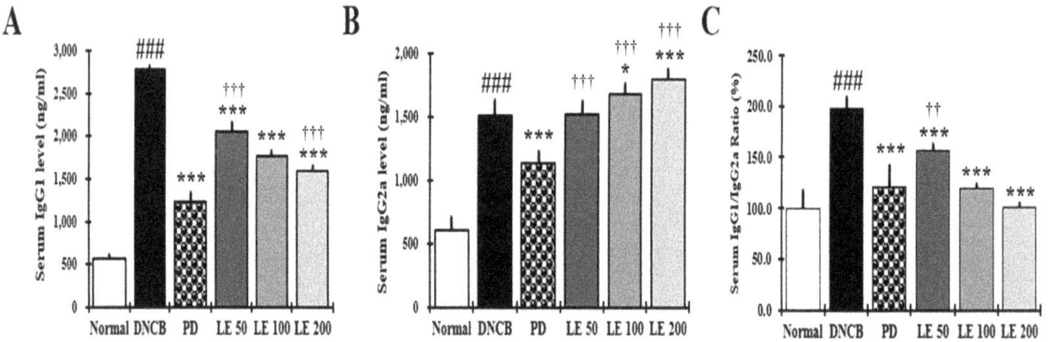

Figure 5. Effect of LE on the levels of IgG1, IgG2a, and the IgG1/IgG2a ratio in the serum of NC/Nga mice. The levels of (**A**) IgG1, (**B**) IgG2a, and (**C**) IgG1/IgG2a ratio (%) in the serum. Serum was collected on the last day of the experiment and measured using ELISA. The results were expressed as means $\pm$ SD ($n = 6$). ### $p < 0.001$ vs. normal (DNCB untreated group), * $p < 0.05$, *** $p < 0.001$ vs. DNCB (negative control; DNCB treated group), PD (positive control; prednisolone 3 mg/kg) treatment group, LE (LE 50, 100, or 200 mg/kg) treatment group, and †† $p < 0.01$, ††† $p < 0.001$ vs. PD (positive control; prednisolone 3 mg/kg) treatment group.

3.6. LE Decreases the Levels of AD-Related Cytokines and Chemokines in NC/Nga Mouse Serum

To evaluate the effects of LE on immune responses in the AD mouse model, we measured the levels of IL-1β, IL-6, TNF-α, TSLP, TARC, MDC, RANTES, and MCP-1. The levels of cytokines and chemokines in the serum of NC/Nga mice were significantly higher in the DNCB group than in the normal group. However, LE treatment decreased the levels of AD-related cytokines and chemokines in the serum in a dose-dependent manner and showed a similar efficacy between the 100 mg/kg LE and PD groups (Figure 6A–C). These results suggest that LE alleviated AD by reducing the levels of AD-related cytokines and chemokines in the serum.

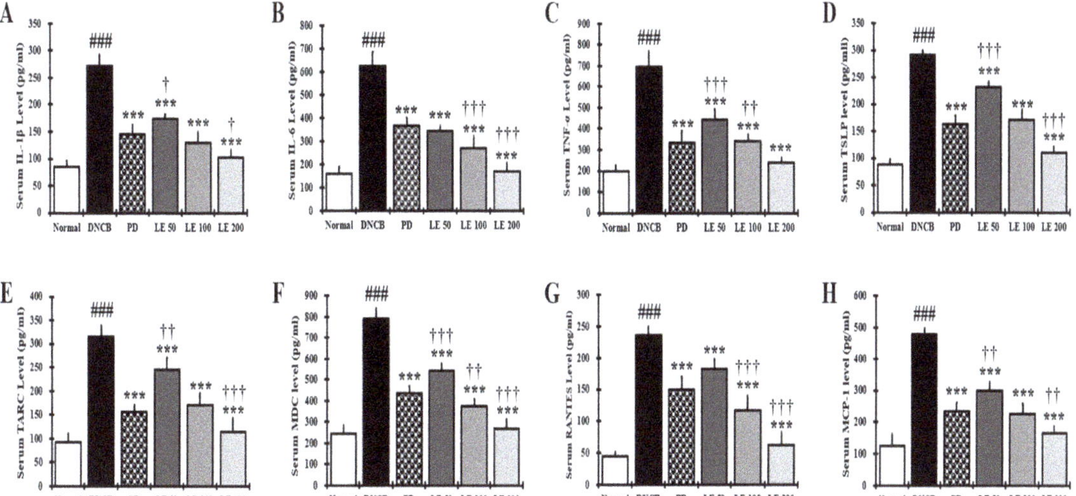

Figure 6. Effect of LE on AD-related cytokines and chemokines in the serum of NC/Nga mice. (**A–H**) The levels of (**A**) IL-1β, (**B**) IL-6 (**C**) TNF-α, (**D**) TSLP, (**E**) TARC (**F**) MDC, (**G**) RANTES, and (**H**) MCP-1 in the serum. Serum was collected on the last day of the experiment and measured using ELISA. The results were expressed as means ± SD (*n* = 6). ### *p* < 0.001 vs. normal (DNCB untreated group), *** *p* < 0.001 vs. DNCB (negative control; DNCB treated group), PD (positive control; prednisolone 3 mg/kg) treatment group, LE (LE 50, 100 or 200 mg/kg) treatment group, and † *p* < 0.05, †† *p* < 0.01, and ††† *p* < 0.001 vs. PD (positive control; prednisolone 3 mg/kg) treatment group.

3.7. LE Regulates the Balance of Cytokines and Chemokines Secretion in Splenocytes

We investigated the levels of cytokines and chemokines secreted from splenocytes to confirm the anti-atopic effect of LE. The levels of the Th2-mediated cytokines IL-4, IL-5, IL-13, and IL-31, and the cytokines IL-25, IL-33, and TSLP, which indirectly activate Th2, significantly increased in the DNCB group compared to the normal group. In addition, AD-related cytokines and chemokines, namely, IL-1β, IL-6, IL-17, IL-22, TNF-α, and MCP-1, were significantly higher in the DNCB group compared to the normal group. However, LE treatment reduced the AD-related cytokine and chemokine levels in a dose-dependent manner and showed a similar efficacy between the 100 mg/kg LE and PD groups (Figure 7A–E). Additionally, the levels of Th1-mediated cytokines IL-12 and IFN-γ were significantly reduced in the DNCB group compared to the normal group. Meanwhile, LE treatment restored these cytokine levels in a dose-dependent manner. Interestingly, the levels of Th1-mediated cytokines decreased in the PD group (Figure 7F).

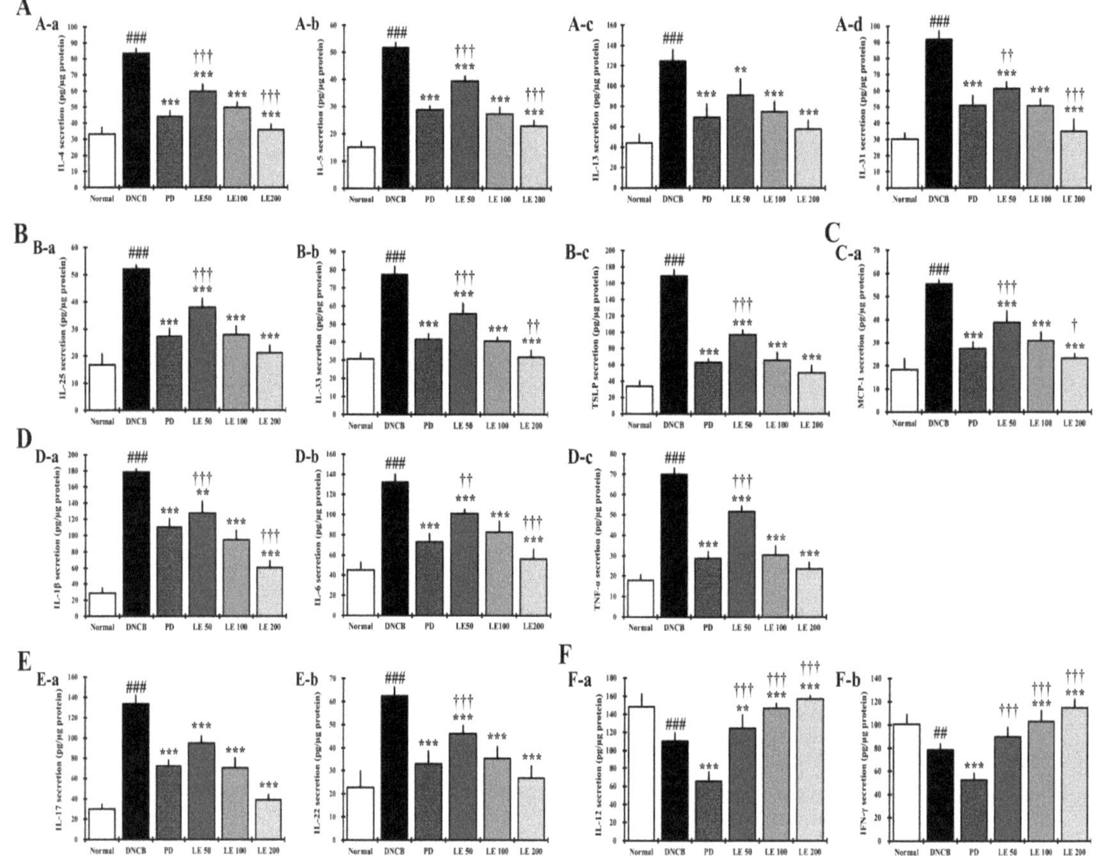

Figure 7. Effect of LE on AD-related cytokine and chemokine secretion in NC/Nga mouse splenocytes. (**A**) The levels of (**A-a**) IL-4, (**A-b**) IL-5, (**A-c**) IL-13, and (**A-d**) IL-31 in the supernatant of splenocytes. (**B**) The levels of (**B-a**) IL-25, (**B-b**) IL-33, and (**B-c**) TSLP in the supernatant of splenocytes. (**C**) The levels of (**C-a**) MCP-1 in the supernatant of splenocytes. (**D**) The levels of (**D-a**) IL-1β, (**D-b**) IL-6, and (**D-c**) TNF-α in the supernatant of splenocytes. (**E**) The levels of (**E-a**) IL-17 and (**E-b**) IL-22 in the supernatant of splenocytes. (**F**) The levels of Th1-mediated cytokines (**F-a**) IL-12 and (**F-b**) IFN-γ. Splenocytes from NC/Nga mice were stimulated with Con-A for 72 h, and then the supernatant was measured using ELISA. Cytokines and chemokines were normalized to the protein concentration of the lysate. The results were expressed as means ± SD (n = 6). ## $p < 0.01$, ### $p < 0.001$ vs. normal (DNCB untreated group), ** $p < 0.01$, *** $p < 0.001$ vs. DNCB (negative control; DNCB treated group), PD (positive control; prednisolone 3 mg/kg) treatment group, LE (LE 50, 100 or 200 mg/kg) treatment group, and † $p < 0.05$, †† $p < 0.01$, and ††† $p < 0.001$ vs. PD (positive control; prednisolone 3 mg/kg) treatment group.

3.8. LE Inhibits the Gene Expression of Cytokines, Chemokines, and CCR4 Involved in Th2 Activation in the Dorsal Skin

The gene expression levels of AD-related cytokines and chemokines that directly activate Th2 cells were investigated in the dorsal skin of NC/Nga mice. The expression of IL-25, IL-33, TSLP, RANTES, TARC, MDC, and CCR4 was significantly higher in the DNCB group than in the normal group. However, LE treatment reduced gene expression in a dose-dependent manner and showed a similar efficacy between the 100 mg/kg LE and PD groups (Figure 8).

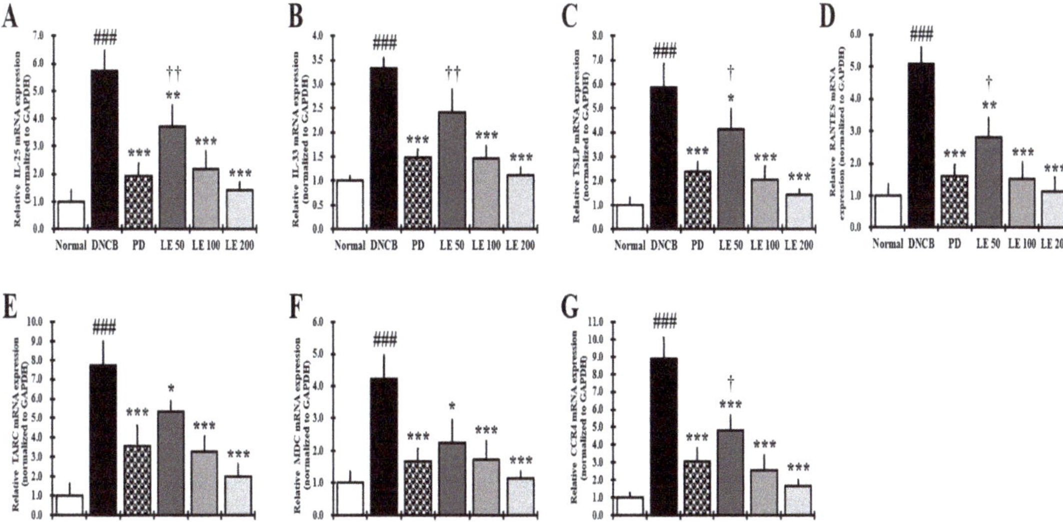

Figure 8. Effect of LE on gene expression of cytokines, chemokines, and CCR4 that activate Th2 cells in dorsal skin of NC/Nga mice. Gene expression of (**A**) IL-25, (**B**) IL-33, (**C**) TSLP, (**D**) RANTES, (**E**) TARC, (**F**) MDC, and (**G**) CCR4. Total RNA was extracted from the dorsal skin of NC/Nga mice and normalized to GAPDH. The results were expressed as means ± SD (n = 6). ### $p < 0.001$ vs. normal (DNCB untreated group), * $p < 0.05$, ** $p < 0.01$, and *** $p < 0.001$ vs. DNCB (negative control; DNCB treated group), PD (positive control; prednisolone 3 mg/kg) treatment group, LE (LE 50, 100 or 200 mg/kg) treatment group, and † $p < 0.05$, †† $p < 0.01$ vs. PD (positive control; prednisolone 3 mg/kg) treatment group.

3.9. LE Restores the Balance of AD-Related Cytokine and Chemokine Gene Expression in the Dorsal Skin of the NC/Nga Mouse

We investigated the levels of cytokines and chemokines expressed in the dorsal skin to confirm the anti-atopic effect of LE. Gene expression of the Th2-mediated cytokines IL-4, IL-5, IL-13, and IL-31 was significantly higher in the DNCB group compared to the normal group. In addition, gene expression of the AD-related cytokines IL-1β, IL-6, TNF-α, IL-17, and IL-22 was significantly increased in the DNCB group compared to the normal group. LE treatment reduced the expression of AD-related cytokines and chemokines, including Th2-mediated cytokines, in a dose-dependent manner, and showed a similar efficacy between the 100 mg/kg LE and PD groups (Figure 9A–C).

Expression of Th1-mediated cytokine (IL-12 and IFN-γ) genes was significantly reduced in the DNCB group compared to the normal group, while LE restored their expression in a dose-dependent manner. We also found that the PD group showed reduced expression of Th1-mediated cytokines compared to that in the DNCB group (Figure 9D).

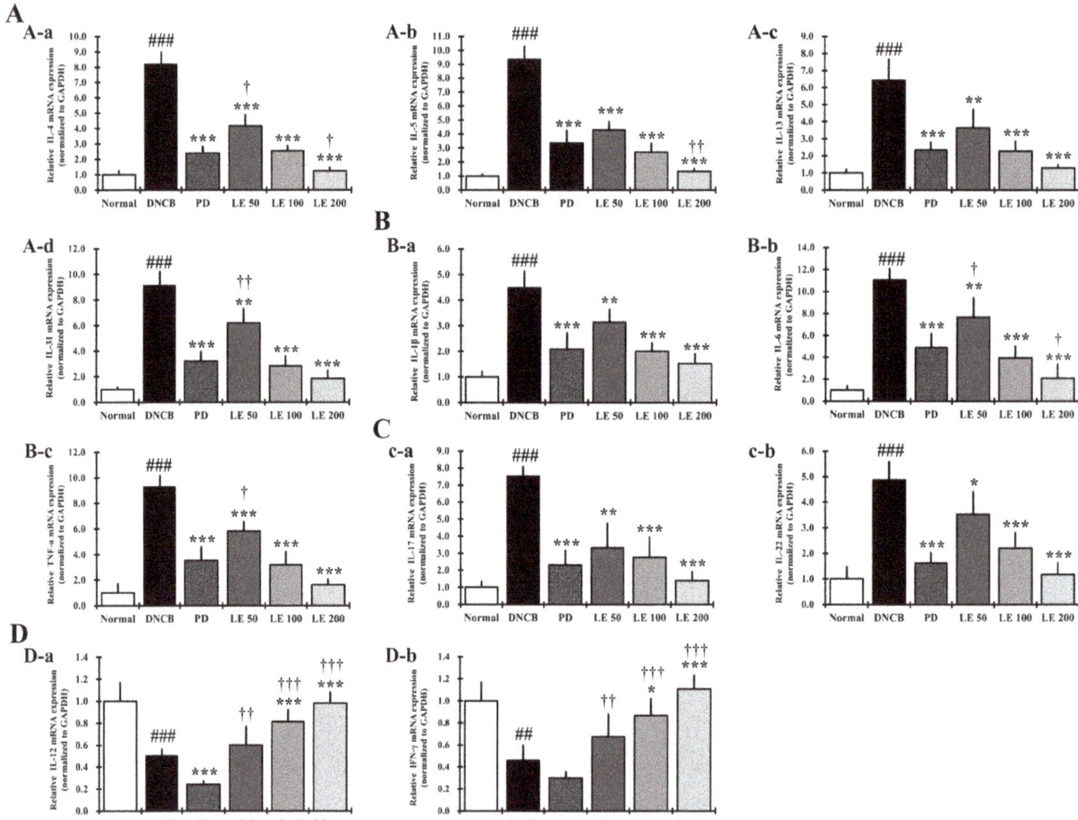

Figure 9. Gene expression of AD-related cytokines and chemokines in response to LE treatment of the dorsal skin of NC/Nga mice. (**A**) Gene expression levels of (**A-a**) IL-4, (**A-b**) IL-5, (**A-c**) IL-13, and (**A-d**) IL-31. (**B**) Gene expression levels of (**B-a**) IL-1β, (**B-b**) IL-6, and (**B-c**) TNF-α. (**C**) Gene expression levels of (**c-a**) IL-17 and (**c-b**) IL-22. (**D**) Gene expression levels of (**D-a**) IL-12 and (**D-b**) IFN-γ. Total RNA was extracted from the dorsal skin of NC/Nga mice and normalized to GAPDH. The results were expressed as means ± SD ($n = 6$). ## $p < 0.01$, ### $p < 0.001$ vs. normal (DNCB untreated group), * $p < 0.05$, ** $p < 0.01$, and *** $p < 0.001$ vs. DNCB (negative control; DNCB treated group), PD (positive control; prednisolone 3 mg/kg) treatment group, LE (LE 50, 100 or 200 mg/kg) treatment group, and † $p < 0.05$, †† $p < 0.01$, and ††† $p < 0.001$ vs. PD (positive control; prednisolone 3 mg/kg) treatment group.

3.10. LE Restores Defects in Skin Barrier Function Caused by AD

In the dorsal skin of NC/Nga mice, pro-FLG gene expression and the abundance of proteins responsible for skin barrier function were investigated. The expression of pro-FLG was significantly reduced in the DNCB group compared to the normal group; however, LE treatment restored its expression in a dose-dependent manner and showed a similar efficacy between the 100 mg/kg LE and PD treatment groups (Figure 10B). The abundance skin barrier proteins FLG, IVL, and LOR, and the TJ proteins OCC and ZO-1, was significantly reduced in the DNCB group compared to the normal group, while LE treatment restored the expression of skin barrier and TJ-related proteins in a dose-dependent manner. The skin barrier proteins FLG, IVL, and LOR were present at similar levels when samples were treated with 100 mg/kg LE and PD (Figure 10A,C–E). However, the expression of the TJ proteins OCC and ZO-1 was not recovered in the PD group compared to the DNCB group (Figure 10A,F,G).

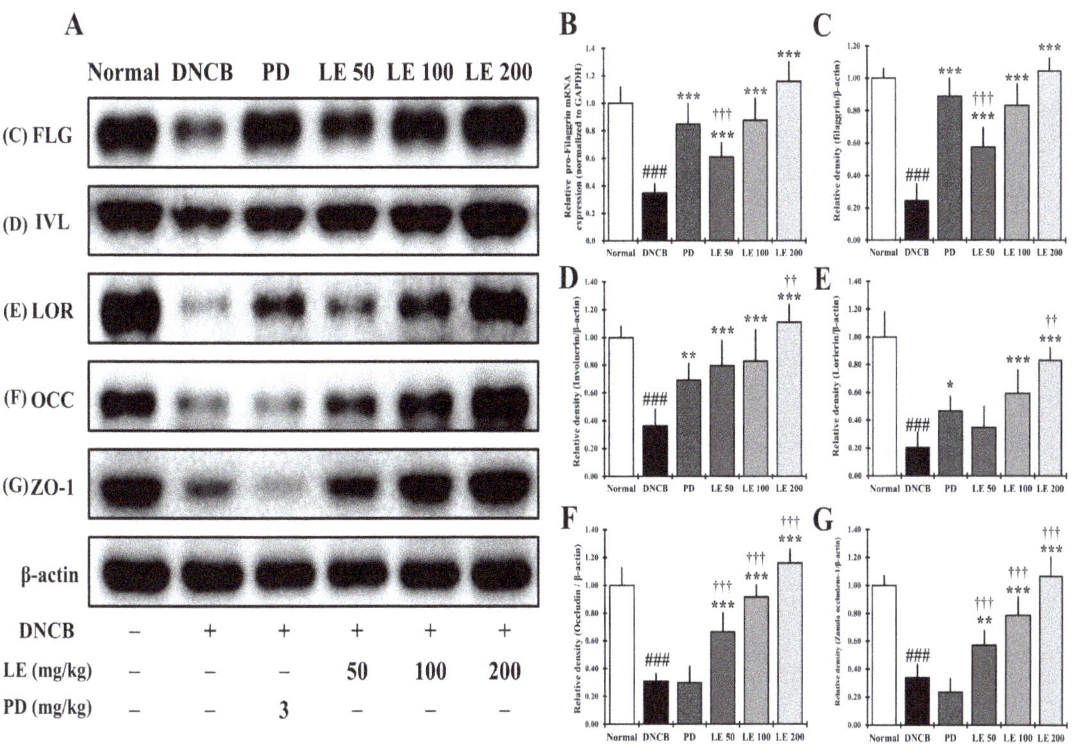

Figure 10. Restoration effect of LE on the skin barrier proteins in the dorsal skin of NC/Nga mice. (**A**) Abundance of proteins related to skin barrier function in the dorsal skin of NC/Nga mice. (**B**) Expression of pro-FLG, which was normalized to GAPDH. (**C–G**) Abundance of (**C**) FLG (34 kDa), (**D**) IVL (68 kDa), (**E**) LOR (26 kDa), (**F**) OCC (59 kDa), and (**G**) ZO-1 (187 kDa), protein levels quantified by band density and normalized to ß-actin (43 kDa). The results were expressed as means ± SD (n = 6). [###] $p < 0.001$ vs. normal (DNCB untreated group), [*] $p < 0.05$, [**] $p < 0.01$, and [***] $p < 0.001$ vs. DNCB (negative control; DNCB treated group), PD (positive control; prednisolone 3 mg/kg) treatment group, LE (LE 50, 100 or 200 mg/kg) treatment group, and [††] $p < 0.01$, [†††] $p < 0.001$ vs. PD (positive control; prednisolone 3 mg/kg) treatment group. FLG, filaggrin; LOR, loricrin; IVL, involucrin; ZO-1, zonula occludens-1; OCC, occludin; pro-FLG, pro-filaggrin.

4. Discussion

AD induces biased differentiation of Th2 cells, resulting in Th1/Th2 immune imbalance, followed by increased itchiness and complex pathophysiological changes that lead to the skin barrier function damage [10,35]. In the present study, the efficacy of LE extract against AD was investigated using an in vivo animal model, focusing on whether LE could improve AD-related clinical symptoms, Th1/Th2 immune balance, and restoration of skin barrier function. IgE secretion is increased in B cells by the Th2-mediated cytokine IL-4 [57], and the degranulation of mast cells by IgE increases itching through the release of histamine [27,58]. Continuous scratching can increase erythema, abrasions, and erosions, along with lichenification, which thickens the skin [59–61]. In addition, damage to the skin barrier caused by this scratching can further cause loss of moisture, leading to dry skin [62]. LE treatment led to improved AD-related clinical symptoms and reduced infiltration of inflammatory cells in a dose-dependent manner (Figure 2B–E).

The serum secretion levels of IL-4, IgE, and histamine decreased in a dose-dependent manner following LE treatment compared to the DNCB group (Figure 4). Therefore, LE appears to reduce infiltration and degranulation of mast cells, thereby reducing histamine release, as well as scratching, ultimately leading to improved clinical symptoms, including erythema, erosion, and abrasion. In addition, IL-4 reduced the expression of skin barrier proteins [45], further indicating that the dry skin symptoms were improved through the restoration of the skin barrier protein. LE showed a greater inhibitory effect on IL-4, IgE, and histamine than PD, which is used as a clinical treatment for AD (Figure 4).

Eosinophils increase in both the blood and skin of AD patients and are reportedly involved in chronic AD while playing an important role in the Th2 immune response [63]. LE treatment reduced the number of infiltrating eosinophils in a dose-dependent manner (Figure 3A(A-a,A-b)) and lichenification was recovered by LE, thus demonstrating its ability to improve a clinical symptom of chronic AD. The recovery of lichenification was consistent with a decrease in the number of infiltrated eosinophils and epidermal thickness.

In the keratinocytes of AD patients, the expression and secretion of cytokines IL-25, IL-33, and TSLP, and the chemokines TARC, MDC, and RANTES, which are directly or indirectly related to the activation of Th2 cells, are increased [64]. In addition, expression of CCR4, a receptor for TARC and MDC, increases on the surface of Th2 cells, thereby increasing Th2 cell invasion and activity [29,65]. LE dose-dependently reduced the levels of TSLP, TARC, MDC, and RANTES in the serum and gene expression levels of AD-related IL-25, IL-33, TSLP, TARC, MDC, and RANTES, including CCR4, in the dorsal skin of NC/Nga mice (Figures 6E–H and 8). In addition, LE dose-dependently decreased the levels of cytokines and chemokines, such as IL-25, IL-33, and TSLP, secreted by splenocytes in the AD mouse model (Figure 7B). These results suggest that LE reduces the expression and secretion of cytokines and chemokines that directly or indirectly activate Th2 cells in the serum, splenocytes, and dorsal skin. Inhibition of CCR4 expression reduced the binding of TARC and MDC, suggesting that the activity of Th2 cells was suppressed by LE treatment. The decreased expression and secretion of RANTES in the dorsal skin indicated that LE inhibited the progression to chronic AD. These results are consistent with the reduction in infiltrated eosinophils and restoration of lichenification (Figures 2B and 3A(A-b)).

IL-25, IL-33, and TSLP are involved in the activation of several immune cells, including mast cells and eosinophils, and induce the secretion of Th2-mediated cytokines [66]. Activation of mast cells and eosinophils is involved in allergic responses, in which gene expression and secretion of AD-related cytokines and chemokines such as IL-4, IL-5, IL-6, IL-13, IL-31, and MCP-1 increase [27,67,68]. We found that LE had a dose-dependent effect on the serum levels of MCP-1 as well as MCP-1 and Th2-mediated cytokines IL-4, IL-5, IL-13, and IL-31 secreted by splenocytes (Figures 6H and 7A,C). Moreover, LE treatment downregulated the expression of IL-4, IL-5, IL-13, and IL-31 in dorsal skin in a dose-dependent manner (Figure 9A). By suppressing the activity of immune cells, LE could resolve the immune imbalance in the AD mouse model via suppression of Th2 mediated cytokines IL-4 and IL-13, which continuously activate Th2 cells in an autocrine manner [69].

IL-5 and IL-31 have been reported as cytokines that delay the increase and apoptosis of eosinophils, while MCP-1, which is primarily expressed and secreted by eosinophils, is involved in the invasion of Th2 cells [70,71]. LE reduced both the gene expression and secreted levels of IL-5, IL-31, and MCP-1 (Figures 6H, 7A,C and 9A), indicating that LE could reduce the number of infiltrated eosinophils, consistent with improved clinical symptoms and lichenification.

IL-12 and IFN-γ are required for the differentiation of Th1 cells [72]. IL-12 promotes the differentiation of Th1 cells and IFN-γ secreted from Th1 cells suppresses the secretion of IL-4, a Th2-mediated cytokine, to suppress the production of IgE and IgG1 and promote the production of IgG2a in B cells [4,73]. Therefore, the IgG1/IgG2a ratio is used as a representative marker of Th1/Th2 balance [4]. In the splenocytes and dorsal skin of NC/Ng mice, secretion and gene expression of IL-12 and IFN-γ were decreased in the DNCB group compared to the normal group and recovered in a dose-dependent manner by LE treatment (Figures 7F and 9D). In the LE group, IgG1 decreased and IgG2a was restored, resulting in a dose-dependent recovery of the Th1/Th2 ratio (Figure 5). In contrast, PD was shown to restore the Th1/Th2 ratio by decreasing the levels of both Th1 and Th2 cells associated immunoglobulins. Taken together, our results indicate that PD can restore the

Th1/Th2 immune balance by inhibiting the activity of both Th1 and Th2 cells, whereas LE restored Th1/Th2 immune balance by regulating the activation of Th1 and Th2 cells.

IL-1β, IL-6, and TNF-α are expressed and secreted by many inflammatory cells, including keratinocytes, Th2 cells, and mast cells; they are involved in the progression of acute and chronic AD and activate Th17 and Th22 cells [74,75]. LE treatment decreased the levels of IL-1β, IL-6, and TNF-α in the serum and splenocytes (Figures 6A–C and 7D), and the gene expression levels of IL-1β, IL-6, and TNF-α in the dorsal skin in a dose-dependent manner (Figure 9B). These results show that LE prevents the progression of acute and chronic AD through the regulation of both gene expression and protein abundance of cytokines involved in Th cell activation. These findings also suggest that LE may be involved in the restoration of immune balance by regulating the activity of Th17 and Th22 cells.

Th17 and Th22 cells show a positive correlation with Th2-mediated cytokines [47,76]. Th17 cells produce IL-17 and IL-22, while Th22 cells increase the expression and secretion of IL-22 [4,45]. IL-17 and IL-22 increase the differentiation of Th2 cells and impair skin barrier function [45]. LE decreased the expression and secretion levels of IL-17 and IL-22 in the dorsal skin and splenocytes (Figures 7E and 9C), which reduced the activity of Th17 and Th22 cells. This suggests that LE is involved in the restoration of immune balance and skin barrier function through the deactivation of Th17 and Th22 cells.

Cytokines and chemokines associated with AD, such as IL-4, IL-6, IL-13, IL-17, IL-31, IL-22, IL-33, TSLP, and MCP-1, downregulate the abundance of pro-FLG and as well as that of the skin barrier-related proteins FLG, IVL, and LOR, and the TJ proteins OCC and ZO-1 [4,31,35,47,77,78], resulting in impaired skin barrier function. It has been reported that reduction of FLG, IVL, and LOR aggravates the Th2 immune response by inducing an increase in microbiome species, such as *Staphylococcus aureus* in the skin [79]. In this study, the gene expression of pro-FLG and the abundance of skin barrier-related proteins FLG, IVL, and LOR decreased in the DNCB group compared to the normal group. However, pro-FLG gene expression and FLG, IVL, and LOR protein abundance were recovered in a dose-dependent manner by LE treatment, and slightly recovered in the PD group (Figure 10B–E). Some studies have reported that TCS increases skin barrier proteins such as FLG and LOR [80]. The restoration of pro-FLG, FLG, IVL, and LOR by LE treatment suggests that the clinical symptoms and immune balance associated with AD were recovered.

The TJ region is located below the stratum corneum, sealing the space between cells and binding cells to each other [62]. It has been reported that AD-induced TJ damage induces expansion of Langerhans cell dendrites into the TJ, triggering a Th2-mediated immune response [81]. As TJs exhibit an inverse correlation with the progression of AD, we hypothesized that TJ damage is an important feature of AD [4]. The expression of the TJ proteins OCC and ZO-1 were lower in the DNCB group compared to the normal group, however it was restored in a dose-dependent manner with LE treatment. Meanwhile, the abundance of TJ proteins was not increased in the PD group compared to the DNCB group (Figure 10F,G). A previous study reported that TCS negatively affects cell permeability by decreasing the abundance of TJ proteins, including OCC [82]. Unlike TCS, LE treatment might alleviate AD by restoring skin barrier function. Taken together, these results suggest that LE is a potential anti-AD drug candidate that functions by restoring the Th1/Th2 immune balance and enhancing skin barrier function.

5. Conclusions

Oral administration of LE ameliorated AD-related clinical symptoms, including dermatitis score, ear thickness, scratching behavior, epidermal thickness, and infiltration of inflammatory cells, in the NC/Nga AD mouse model. In addition, LE reduces scratching behavior by inhibiting the release of IgE and histamine in NC/Nga mice. LE also reduces the levels of AD-related cytokines and chemokines in the serum and restores the Th1/Th2 immune balance via regulation of IgG1 and IgG2a. The immune balance was restored through regulation of gene expression and secretion of AD-related cytokines and chemokines derived from Th1/Th2/Th17/Th22 in splenocytes and dorsal skin. Furthermore, LE increased the expression of proteins responsible for skin barrier function, restoring TJ proteins to a greater extent than PD treatment. Taken together, the findings of this study suggest that LE is a potential therapeutic agent that can alleviate AD by regulating immune balance and restoring skin barrier function.

Author Contributions: Conceptualization, S.-Y.C.; validation, J.-S.O. and S.-J.L.; formal analysis, S.-Y.C., J.-S.O. and S.-J.L.; investigation, S.-Y.C., J.-S.O. and S.-J.L.; resources, S.-Y.C., J.-S.O. and S.-J.L.; data curation, J.-S.O. and S.-J.L.; writing—original draft preparation, S.-Y.C. and J.-S.O.; writing—review and editing, S.-Y.C. and J.-S.O.; visualization, J.-S.O. and S.-J.L.; supervision, S.-Y.C.; project administration, S.-Y.C. All authors have read and agreed to the published version of the manuscript.

Funding: This research received no external funding.

Institutional Review Board Statement: The study was conducted according to the guidelines of the Declaration of Helsinki and approved by the Animal Care and Use Committee of the Kyung Hee University (approval number KHSASP-20-252, date of approval: 24 July 2020).

Informed Consent Statement: Not applicable.

Data Availability Statement: The data presented in this study are available on request from the corresponding author.

Conflicts of Interest: The authors declare no conflict of interest.

References

1. Flohr, C.; Mann, J. New insights into the epidemiology of childhood atopic dermatitis. *Allergy* **2014**, *69*, 3–16. [CrossRef]
2. Yamamoto, M.; Haruna, T.; Yasui, K.; Takahashi, H.; Iduhara, M.; Takaki, S.; Deguchi, M.; Arimura, A. A Novel Atopic Dermatitis Model Induced by Topical Application with Dermatophagoides Farinae Extract in NC/Nga Mice. *Allergol. Int.* **2007**, *56*, 139–148. [CrossRef]
3. Kang, M.C.; Cho, K.; Lee, J.H.; Subedi, L.; Yumnam, S.; Kim, S.Y. Effect of Resveratrol-Enriched Rice on Skin In-flammation and Pruritus in the NC/Nga Mouse Model of Atopic Dermatitis. *Int. J. Mol. Sci.* **2019**, *20*, 1428. [CrossRef] [PubMed]
4. Kim, S.H.; Seong, G.S.; Choung, S.Y. Fermented Morinda citrifolia (Noni) Alleviates DNCB-Induced Atopic Dermatitis in NC/Nga Mice through Modulating Immune Balance and Skin Barrier Function. *Nutrients* **2020**, *12*, 249. [CrossRef]
5. Theoharides, T.C.; Alysandratos, K.D.; Angelidou, A.; Delivanis, D.A.; Sismanopoulos, N.; Zhang, B.; Asadi, S.; Va-siadi, M.; Weng, Z.; Miniati, A.; et al. Mast cells and inflammation. *Biochim. Biophys. Acta* **2012**, *1822*, 21–33. [CrossRef] [PubMed]
6. Brandt, E.B.; Sivaprasad, U. Th2 Cytokines and Atopic Dermatitis. *J. Clin. Cell. Immunol.* **2011**, *2*, 3. [CrossRef]
7. Spergel, J.M.; Paller, A.S. Atopic dermatitis and the atopic march. *J. Allergy Clin. Immunol.* **2003**, *112*, 118–127. [CrossRef] [PubMed]
8. Nakajima, S.; Nomura, T.; Common, J.; Kabashima, K. Insights into atopic dermatitis gained from genetically defined mouse models. *J. Allergy Clin. Immunol.* **2019**, *143*, 13–25. [CrossRef] [PubMed]
9. Kim, J.E.; Kim, J.S.; Cho, D.H.; Park, H.J. Molecular Mechanisms of Cutaneous Inflammatory Disorder: Atopic Der-matitis. *Int. J. Mol. Sci.* **2016**, *17*, 1234. [CrossRef] [PubMed]
10. Kim, Y.J.; Choi, M.J.; Bak, D.H.; Lee, B.C.; Ko, E.J.; Ahn, G.R.; Ahn, S.W.; Kim, M.J.; Na, J.; Kim, B.J. Topical admin-istration of EGF suppresses immune response and protects skin barrier in DNCB-induced atopic dermatitis in NC/Nga mice. *Sci. Rep.* **2018**, *8*, 11895. [CrossRef]
11. Danso, M.O.; van Drongelen, V.; Mulder, A.; van Esch, J.; Scott, H.; van Smeden, J.; El Ghalbzouri, A.; Bouwstra, J.A. TNF-α and Th2 Cytokines Induce Atopic Dermatitis–Like Features on Epidermal Differentiation Proteins and Stratum Corneum Lipids in Human Skin Equivalents. *J. Investig. Dermatol.* **2014**, *134*, 1941–1950. [CrossRef]
12. Howell, M.D.; Fairchild, H.R.; Kim, B.E.; Bin, L.; Boguniewicz, M.; Redzic, J.S.; Hansen, K.C.; Leung, D.Y. Th2 Cytokines Act on S100/A11 to Downregulate Keratinocyte Differentiation. *J. Investig. Dermatol.* **2008**, *128*, 2248–2258. [CrossRef]
13. Noben-Trauth, N.; Hu-Li, J.; Paul, W.E. Conventional, naive CD4+ T cells provide an initial source of IL-4 during Th2 differentia-tion. *J. Immunol.* **2000**, *165*, 3620–3625. [CrossRef]
14. Paludan, S.R. Interleukin-4 and interferon-gamma: The quintessence of a mutual antagonistic relationship. *Scand. J. Immunol.* **1998**, *48*, 459–468. [CrossRef] [PubMed]
15. Poulsen, L.K.; Hummelshoj, L. Triggers of IgE class switching and allergy development. *Ann. Med.* **2007**, *39*, 440–456. [CrossRef] [PubMed]
16. Park, J.-H.; Hwang, M.H.; Cho, Y.-R.; Hong, S.S.; Kang, J.-S.; Kim, W.H.; Yang, S.H.; Seo, D.-W.; Oh, J.S.; Ahn, E.-K. Combretum quadrangulare Extract Attenuates Atopic Dermatitis-Like Skin Lesions through Modulation of MAPK Signaling in BALB/c Mice. *Molecules* **2020**, *25*, 2003. [CrossRef]

17. Badloe, F.M.S.; De Vriese, S.; Coolens, K.; Schmidt-Weber, C.B.; Ring, J.; Gutermuth, J.; Kortekaas Krohn, I. IgE auto-antibodies and autoreactive T cells and their role in children and adults with atopic dermatitis. *Clin. Transl. Allergy* **2020**, *10*, 34. [CrossRef] [PubMed]

18. Galli, S.J.; Tsai, M. IgE and mast cells in allergic disease. *Nat. Med.* **2012**, *18*, 693–704. [CrossRef] [PubMed]

19. Sonkoly, E.; Muller, A.; Lauerma, A.I.; Pivarcsi, A.; Soto, H.; Kemény, L.; Alenius, H.; Dieu-Nosjean, M.-C.; Meller, S.; Rieker, J.; et al. IL-31: A new link between T cells and pruritus in atopic skin inflammation. *J. Allergy Clin. Immunol.* **2006**, *117*, 411–417. [CrossRef]

20. Furue, M. T helper type 2 signatures in atopic dermatitis. *J. Cutan. Immunol. Allergy* **2018**, *1*, 93–99. [CrossRef]

21. Nygaard, U.; Hvid, M.; Johansen, C.; Buchner, M.; Fölster-Holst, R.; Deleuran, M.; Vestergaard, C. TSLP, IL-31, IL-33 and sST2 are new biomarkers in endophenotypic profiling of adult and childhood atopic dermatitis. *J. Eur. Acad. Dermatol. Venereol.* **2016**, *30*, 1930–1938. [CrossRef]

22. Han, H.; Roan, F.; Ziegler, S.F. The atopic march: Current insights into skin barrier dysfunction and epithelial cell-derived cytokines. *Immunol. Rev.* **2017**, *278*, 116–130. [CrossRef] [PubMed]

23. Čepelak, I.; Dodig, S.; Pavić, I. Filaggrin and atopic march. *Biochem. Med.* **2019**, *29*, 214–227. [CrossRef]

24. Honzke, S.; Wallmeyer, L.; Ostrowski, A.; Radbruch, M.; Mundhenk, L.; Schafer-Korting, M.; Hedtrich, S. Influence of Th2 Cytokines on the Cornified Envelope, Tight Junction Proteins, and ss-Defensins in Filaggrin-Deficient Skin Equivalents. *J. Investig. Dermatol.* **2016**, *136*, 631–639. [CrossRef] [PubMed]

25. Minai-Fleminger, Y.; Levi-Schaffer, F. Mast cells and eosinophils: The two key effector cells in allergic inflammation. *Inflamm. Res.* **2009**, *58*, 631–638. [CrossRef] [PubMed]

26. Kinoshita, M.; Okada, M.; Hara, M.; Furukawa, Y.; Matsumori, A. Mast Cell Tryptase in Mast Cell Granules Enhances MCP-1 and Interleukin-8 Production in Human Endothelial Cells. *Arterioscler. Thromb. Vasc. Biol.* **2005**, *25*, 1858–1863. [CrossRef] [PubMed]

27. Kawakami, T.; Ando, T.; Kimura, M.; Wilson, B.S.; Kawakami, Y. Mast cells in atopic dermatitis. *Curr. Opin. Immunol.* **2009**, *21*, 666–678. [CrossRef] [PubMed]

28. Gu, L.; Tseng, S.; Horner, R.M.; Tam-Amersdorfer, C.; Loda, M.; Rollins, B.J. Control of TH2 polarization by the chemokine monocyte chemoattractant protein-1. *Nature* **2000**, *404*, 407–411. [CrossRef]

29. Imai, T.; Chantry, D.; Raport, C.J.; Wood, C.L.; Nishimura, M.; Godiska, R.; Yoshie, O.; Gray, P.W. Macrophage-derived Chemokine Is a Functional Ligand for the CC Chemokine Receptor 4. *J. Biol. Chem.* **1998**, *273*, 1764–1768. [CrossRef]

30. Aust, G.; Simchen, C.; Heider, U.; Hmeidan, F.A.; Blumenauer, V.; Spanel-Borowski, K. Eosinophils in the human corpus luteum: The role of RANTES and eotaxin in eosinophil attraction into periovulatory structures. *Mol. Hum. Reprod.* **2000**, *6*, 1085–1091. [CrossRef] [PubMed]

31. Oh, J.; Seong, G.; Kim, Y.; Choung, S. Effects of Deacetylasperulosidic Acid on Atopic Dermatitis through Modulating Immune Balance and Skin Barrier Function in HaCaT, HMC-1, and EOL-1 Cells. *Molecules* **2021**, *26*, 3298. [CrossRef]

32. Zaniboni, M.C.; Samorano, L.P.; Orfali, R.L.; Aoki, V. Skin barrier in atopic dermatitis: Beyond filaggrin. *An. Bras. Dermatol.* **2016**, *91*, 472–478. [CrossRef] [PubMed]

33. Cabanillas, B.; Novak, N. Atopic dermatitis and filaggrin. *Curr. Opin. Immunol.* **2016**, *42*, 1–8. [CrossRef]

34. Kim, B.E.; Leung, D.Y. Epidermal Barrier in Atopic Dermatitis. *Allergy Asthma Immunol. Res.* **2012**, *4*, 12–16. [CrossRef]

35. Lee, J.W.; Wu, Q.; Jang, Y.P.; Choung, S.Y. Pinus densiflora bark extract ameliorates 2,4-dinitrochlorobenzene-induced atopic dermatitis in NC/Nga mice by regulating Th1/Th2 balance and skin barrier function. *Phytother. Res.* **2018**, *32*, 1135–1143. [CrossRef] [PubMed]

36. Sandilands, A.; Sutherland, C.; Irvine, A.D.; McLean, W.I. Filaggrin in the frontline: Role in skin barrier function and disease. *J. Cell Sci.* **2009**, *122*, 1285–1294. [CrossRef] [PubMed]

37. Cork, M.J.; Danby, S.; Vasilopoulos, Y.; Hadgraft, J.; Lane, M.E.; Moustafa, M.; Guy, R.; MacGowan, A.L.; Tazi-Ahnini, R.; Ward, S.J. Epidermal Barrier Dysfunction in Atopic Dermatitis. *J. Investig. Dermatol.* **2009**, *129*, 1892–1908. [CrossRef]

38. Kim, B.E.; Leung, D.Y.; Boguniewicz, M.; Howell, M.D. Loricrin and involucrin expression is down-regulated by Th2 cytokines through STAT-6. *Clin. Immunol.* **2008**, *126*, 332–337. [CrossRef]

39. Steinert, P.M.; Kartasova, T.; Marekov, L.N. Biochemical Evidence That Small Proline-rich Proteins and Trichohyalin Function in Epithelia by Modulation of the Biomechanical Properties of Their Cornified Cell Envelopes. *J. Biol. Chem.* **1998**, *273*, 11758–11769. [CrossRef] [PubMed]

40. Segre, J.A. Epidermal barrier formation and recovery in skin disorders. *J. Clin. Investig.* **2006**, *116*, 1150–1158. [CrossRef]

41. Candi, E.; Melino, G.; Mei, G.; Tarcsa, E.; Chung, S.-I.; Marekov, L.N.; Steinert, P.M. Biochemical, Structural, and Transglutaminase Substrate Properties of Human Loricrin, the Major Epidermal Cornified Cell Envelope Protein. *J. Biol. Chem.* **1995**, *270*, 26382–26390. [CrossRef]

42. Matsui, T.; Amagai, M. Dissecting the formation, structure and barrier function of the stratum corneum. *Int. Immunol.* **2015**, *27*, 269–280. [CrossRef] [PubMed]

43. Niessen, C.M. Tight junctions/adherens junctions: Basic structure and function. *J. Investig. Dermatol.* **2007**, *127*, 2525–2532. [CrossRef] [PubMed]

44. Fanning, A.S.; Jameson, B.J.; Jesaitis, L.A.; Anderson, J. The Tight Junction Protein ZO-1 Establishes a Link between the Transmembrane Protein Occludin and the Actin Cytoskeleton. *J. Biol. Chem.* **1998**, *273*, 29745–29753. [CrossRef]

45. Czarnowicki, T.; Krueger, J.G.; Guttman-Yassky, E. Skin Barrier and Immune Dysregulation in Atopic Dermatitis: An Evolving Story with Important Clinical Implications. *J. Allergy Clin. Immunol. Pract.* **2014**, *2*, 371–379. [CrossRef]
46. Brunner, P.M.; Guttman-Yassky, E.; Leung, D.Y. The immunology of atopic dermatitis and its reversibility with broad-spectrum and targeted therapies. *J. Allergy Clin. Immunol.* **2017**, *139*, S65–S76. [CrossRef]
47. Gittler, J.K.; Shemer, A.; Suarez-Farinas, M.; Fuentes-Duculan, J.; Gulewicz, K.J.; Wang, C.Q.; Mitsui, H.; Cardinale, I.; de Guzman Strong, C.; Krueger, J.G.; et al. Progressive activation of T(H)2/T(H)22 cytokines and selective epidermal proteins characterizes acute and chronic atopic dermatitis. *J. Allergy Clin. Immunol.* **2012**, *130*, 1344–1354. [CrossRef] [PubMed]
48. Hengge, U.R.; Ruzicka, T.; Schwartz, R.A.; Cork, M. Adverse effects of topical glucocorticosteroids. *J. Am. Acad. Dermatol.* **2006**, *54*, 1–15. [CrossRef] [PubMed]
49. Jensen, J.M.; Scherer, A.; Wanke, C.; Bräutigam, M.; Bongiovanni, S.; Letzkus, M.; Staedtler, F.; Kehren, J.; Zuehlsdorf, M.; Schwarz, T.; et al. Gene expression is differently affected by pimecrolimus and betamethasone in lesional skin of atopic dermatitis. *Allergy* **2011**, *67*, 413–423. [CrossRef] [PubMed]
50. Carr, W.W. Topical Calcineurin Inhibitors for Atopic Dermatitis: Review and Treatment Recommendations. *Pediatr. Drugs* **2013**, *15*, 303–310. [CrossRef] [PubMed]
51. Jin, H.; He, R.; Oyoshi, M.; Geha, R.S. Animal models of atopic dermatitis. *J. Investig. Dermatol.* **2009**, *129*, 31–40. [CrossRef]
52. Fujii, Y.; Takeuchi, H.; Sakuma, S.; Sengoku, T.; Takakura, S. Characterization of a 2,4-Dinitrochlorobenzene-Induced Chronic Dermatitis Model in Rats. *Skin Pharmacol. Physiol.* **2009**, *22*, 240–247. [CrossRef] [PubMed]
53. Tatsumi, K.; Ichino, T.; Onishi, N.; Shimomura, K.; Yazaki, K. Highly efficient method of Lithospermum erythrorhizon transformation using domestic Rhizobium rhizogenes strain A13. *Plant Biotechnol.* **2020**, *37*, 39–46. [CrossRef]
54. Yazaki, K. Lithospermum erythrorhizon cell cultures: Present and future aspects. *Plant Biotechnol.* **2017**, *34*, 131–142. [CrossRef]
55. Suto, H.; Matsuda, H.; Mitsuishi, K.; Hira, K.; Uchida, T.; Unno, T.; Ogawa, H.; Ra, C. NC/Nga Mice: A Mouse Model for Atopic Dermatitis. *Int. Arch. Allergy Immunol.* **1999**, *120*, 70–75. [CrossRef]
56. Takano, N.; Arai, I.; Kurachi, M. Analysis of the spontaneous scratching behavior by NC/Nga mice: A possible approach to evaluate antipruritics for subjects with atopic dermatitis. *Eur. J. Pharmacol.* **2003**, *471*, 223–228. [CrossRef]
57. Chiricozzi, A.; Maurelli, M.; Peris, K.; Girolomoni, G. Targeting IL-4 for the Treatment of Atopic Dermatitis. *ImmunoTargets Ther.* **2020**, *9*, 151–156. [CrossRef] [PubMed]
58. Voisin, T.; Chiu, I.M. Molecular link between itch and atopic dermatitis. *Proc. Natl. Acad. Sci. USA* **2018**, *115*, 12851–12853. [CrossRef]
59. Siegfried, E.C.; Hebert, A.A. Diagnosis of Atopic Dermatitis: Mimics, Overlaps, and Complications. *J. Clin. Med.* **2015**, *4*, 884–917. [CrossRef]
60. Nam, Y.; Kim, M.; Ha, I.; Yang, W. Derma-Hc, a New Developed Herbal Formula, Ameliorates Cutaneous Lichenification in Atopic Dermatitis. *Int. J. Mol. Sci.* **2021**, *22*, 2359. [CrossRef]
61. Yarbrough, K.B.; Neuhaus, K.J.; Simpson, E.L. The effects of treatment on itch in atopic dermatitis. *Dermatol. Ther.* **2013**, *26*, 110–119. [CrossRef] [PubMed]
62. Nguyen, H.; Trujillo-Paez, J.; Umehara, Y.; Yue, H.; Peng, G.; Kiatsurayanon, C.; Chieosilapatham, P.; Song, P.; Okumura, K.; Ogawa, H.; et al. Role of Antimicrobial Peptides in Skin Barrier Repair in Individuals with Atopic Dermatitis. *Int. J. Mol. Sci.* **2020**, *21*, 7607. [CrossRef] [PubMed]
63. Moreno, A.S.; McPhee, R.; Arruda, L.K.; Howell, M.D. Targeting the T Helper 2 Inflammatory Axis in Atopic Dermatitis. *Int. Arch. Allergy Immunol.* **2016**, *171*, 71–80. [CrossRef] [PubMed]
64. Chieosilapatham, P.; Kiatsurayanon, C.; Umehara, Y.; Paez, J.V.T.; Peng, G.; Yue, H.; Nguyen, L.T.H.; Niyonsaba, F. Keratinocytes: Innate immune cells in atopic dermatitis. *Clin. Exp. Immunol.* **2021**, *204*, 296–309. [CrossRef] [PubMed]
65. Wenzel, J.; Henze, S.; Wörenkämper, E.; Basner-Tschakarjan, E.; Sokolowska-Wojdylo, M.; Steitz, J.; Bieber, T.; Tüting, T. Role of the Chemokine Receptor CCR4 and its Ligand Thymus- and Activation-Regulated Chemokine/CCL17 for Lymphocyte Recruitment in Cutaneous Lupus Erythematosus. *J. Investig. Dermatol.* **2005**, *124*, 1241–1248. [CrossRef]
66. Divekar, R.; Kita, H. Recent advances in epithelium-derived cytokines (IL-33, IL-25, and thymic stromal lymphopoietin) and allergic inflammation. *Curr. Opin. Allergy Clin. Immunol.* **2015**, *15*, 98–103. [CrossRef] [PubMed]
67. Gibbs, B.F.; Patsinakidis, N.; Raap, U. Role of the Pruritic Cytokine IL-31 in Autoimmune Skin Diseases. *Front. Immunol.* **2019**, *10*, 1383. [CrossRef]
68. Guttman-Yassky, E.; Nograles, K.E.; Krueger, J.G. Contrasting pathogenesis of atopic dermatitis and psoriasis—Part II: Immune cell subsets and therapeutic concepts. *J. Allergy Clin. Immunol.* **2011**, *127*, 1420–1432. [CrossRef] [PubMed]
69. Junttila, I.S. Tuning the Cytokine Responses: An Update on Interleukin (IL)-4 and IL-13 Receptor Complexes. *Front. Immunol.* **2018**, *9*, 888. [CrossRef]
70. Bao, L.; Zhang, H.; Chan, L.S. The involvement of the JAK-STAT signaling pathway in chronic inflammatory skin disease atopic dermatitis. *Jak-Stat* **2013**, *2*, e24137. [CrossRef]
71. Nedoszytko, B.; Sokołowska-Wojdyło, M.; Ruckemann-Dziurdzińska, K.; Roszkiewicz, J.; Nowicki, R.J. Chemokines and cytokines network in the pathogenesis of the inflammatory skin diseases: Atopic dermatitis, psoriasis and skin mastocytosis. *Postępy Dermatol. Alergol.* **2014**, *2*, 84–91. [CrossRef] [PubMed]
72. Sun, L.; He, C.; Nair, L.; Yeung, J.; Egwuagu, C.E. Interleukin 12 (IL-12) family cytokines: Role in immune pathogenesis and treatment of CNS autoimmune disease. *Cytokine* **2015**, *75*, 249–255. [CrossRef]

73.   Leung, D.Y.; Boguniewicz, M.; Howell, M.D.; Nomura, I.; Hamid, Q.A. New insights into atopic dermatitis. *J. Clin. Investig.* **2004**, *113*, 651–657. [CrossRef]
74.   Miyazaki, Y.; Nakayamada, S.; Kubo, S.; Nakano, K.; Iwata, S.; Miyagawa, I.; Ma, X.; Trimova, G.; Sakata, K.; Tanaka, Y. Th22 Cells Promote Osteoclast Differentiation via Production of IL-22 in Rheumatoid Arthritis. *Front. Immunol.* **2018**, *9*, 2901. [CrossRef]
75.   Zheng, Y.; Sun, L.; Jiang, T.; Zhang, D.; He, D.; Nie, H. TNFalpha promotes Th17 cell differentiation through IL-6 and IL-1beta produced by monocytes in rheumatoid arthritis. *J. Immunol. Res.* **2014**, *2014*, 385352. [CrossRef] [PubMed]
76.   Biedermann, T.; Skabytska, Y.; Kaesler, S.; Volz, T. Regulation of T Cell Immunity in Atopic Dermatitis by Microbes: The Yin and Yang of Cutaneous Inflammation. *Front. Immunol.* **2015**, *6*, 353. [CrossRef] [PubMed]
77.   Kim, I.S.; Son, K.-H.; Park, H.-Y.; Kim, M.J.; Shin, D.-H.; Lee, J.-S. Arazyme inhibits cytokine expression and upregulates skin barrier protein expression. *Mol. Med. Rep.* **2013**, *8*, 551–556. [CrossRef] [PubMed]
78.   Furue, M. Regulation of Filaggrin, Loricrin, and Involucrin by IL-4, IL-13, IL-17A, IL-22, AHR, and NRF2: Pathogenic Implications in Atopic Dermatitis. *Int. J. Mol. Sci.* **2020**, *21*, 5382. [CrossRef]
79.   Iwamoto, K.; Moriwaki, M.; Miyake, R.; Hide, M. Staphylococcus aureus in atopic dermatitis: Strain-specific cell wall proteins and skin immunity. *Allergol. Int.* **2019**, *68*, 309–315. [CrossRef] [PubMed]
80.   Guttman-Yassky, E.; Ungar, B.; Malik, K.; Dickstein, D.; Suprun, M.; Estrada, Y.D.; Xu, H.; Peng, X.; Oliva, M.; Todd, D.; et al. Molecular signatures order the potency of topically applied anti-inflammatory drugs in patients with atopic dermatitis. *J. Allergy Clin. Immunol.* **2017**, *140*, 1032–1042. [CrossRef]
81.   Yoshida, K.; Kubo, A.; Fujita, H.; Yokouchi, M.; Ishii, K.; Kawasaki, H.; Nomura, T.; Shimizu, H.; Kouyama, K.; Ebihara, T.; et al. Distinct behavior of human Langerhans cells and inflammatory dendritic epidermal cells at tight junctions in patients with atopic dermatitis. *J. Allergy Clin. Immunol.* **2014**, *134*, 856–864. [CrossRef] [PubMed]
82.   Lee, S.E.; Choi, Y.; Kim, S.-E.; Noh, E.B.; Kim, S.-C. Differential effects of topical corticosteroid and calcineurin inhibitor on the epidermal tight junction. *Exp. Dermatol.* **2012**, *22*, 59–61. [CrossRef] [PubMed]

Article

Effects of Eggshell Membrane on Keratinocyte Differentiation and Skin Aging In Vitro and In Vivo

Kyohei Furukawa [1], Masaya Kono [1], Tetsuro Kataoka [1], Yukio Hasebe [2], Huijuan Jia [1,*] and Hisanori Kato [1,*]

[1] Health Nutrition, Department of Applied Biological Chemistry, Graduate School of Agricultural and Life Sciences, The University of Tokyo, 1-1-1 Yayoi, Bunkyo-ku, Tokyo 113-8657, Japan; k-furukawa@g.ecc.u-tokyo.ac.jp (K.F.); msy.k.1203@gmail.com (M.K.); tetu19981119@gmail.com (T.K.)

[2] ALMADO Inc., Tokyo 104-0031, Japan; yukio@almado.co.jp

* Correspondence: akakeiken@g.ecc.u-tokyo.ac.jp (H.J.); akatoq@mail.ecc.u-tokyo.ac.jp (H.K.); Tel.: +81-3-5841-5116 (H.J.); +81-3-5841-1607 (H.K.)

Abstract: Skin aging is one of the hallmarks of the aging process that causes physiological and morphological changes. Recently, several nutritional studies were conducted to delay or suppress the aging process. This study investigated whether nutritional supplementation of the eggshell membrane (ESM) has a beneficial effect on maintaining skin health and improving the skin aging process in vitro using neonatal normal human epidermal keratinocytes (NHEK-Neo) and in vivo using interleukin-10 knockout (IL-10 KO) mice. In NHEK-Neo cells, 1 mg/mL of enzymatically hydrolyzed ESM (eESM) upregulated the expression of keratinocyte differentiation markers, including keratin 1, filaggrin and involucrin, and changed the keratinocyte morphology. In IL-10 KO mice, oral supplementation of 8% powdered-ESM (pESM) upregulated the expression of growth factors, including transforming growth factor β1, platelet-derived growth factor-β and connective tissue growth factor, and suppressed skin thinning. Furthermore, voltage-gated calcium channel, transient receptor potential cation channel subfamily V members were upregulated by eESM treatment in NHEK-Neo cells and pESM supplementation in IL-10 KO mice. Collectively, these data suggest that ESM has an important role in improving skin health and aging, possibly via upregulating calcium signaling.

Keywords: eggshell membrane; keratinocyte differentiation; TRPV; skin aging; skin thickness

Citation: Furukawa, K.; Kono, M.; Kataoka, T.; Hasebe, Y.; Jia, H.; Kato, H. Effects of Eggshell Membrane on Keratinocyte Differentiation and Skin Aging In Vitro and In Vivo. *Nutrients* **2021**, *13*, 2144. https://doi.org/10.3390/nu13072144

Academic Editors: Jean Christopher Chamcheu, Anthony L. Walker and Felicite Noubissi-Kamdem

Received: 14 May 2021
Accepted: 16 June 2021
Published: 22 June 2021

Publisher's Note: MDPI stays neutral with regard to jurisdictional claims in published maps and institutional affiliations.

1. Introduction

The aging population is increasing rapidly, especially in developed countries, and the global population aged over 60 is expected to increase by 22% in 2050 [1]. Aging affects cellular metabolism, physiology and function [2], increases disease morbidity and mortality [3] and decreases the health-related quality of life (HRQoL). Skin aging, one of the hallmarks of the aging process, leads to skin wrinkling, hair graying, scalp hair thinning/baldness, thin skin, etc. These physiological changes are caused by the accumulation of macromolecular damage, impairment of tissue renewal, loss of physiological function integrity [4]. Skin aging is mainly divided into two categories; chronological aging and photo-aging [5]. Chronological aging is occurred by internal factors (such as hormone levels, genotypes, etc.) and is hard to regulate, as these factors naturally change with age. On the contrary, photo-aging is altered by external factors, such as ultraviolet radiation, nutritional levels, chemical pollution; therefore, it can be delayed by interventions of balanced nutrition, functional bioactive compounds and skincare products [4,6].

Eggshell membrane (ESM) is a natural byproduct of egg processing and is usually discarded as industrial waste, resulting in environmental load. However, as ESM is rich in fibrous protein, collagen, hyaluronic acid, chondroitin sulfate and glucosamine [7], it is possible to utilize ESM as a functional bioactive compound in health science. Previous studies indicated that oral supplementation of ESM is safe and has an anti-inflammatory and anti-obese effect in rodents [8–10]. Given that inflammation is closely linked with skin health along with the fact that collagen peptides have a positive effect against skin aging [4], we hypothesized that ESM supplementation could improve skin aging and health.

Therefore, to test this hypothesis, this study investigated the effects of powdered as well as hydrolyzed ESM on skin-derived cells using neonatal normal human epidermal keratinocytes (NHEK-Neo). Then we tested the effect of powdered ESM in interleukin-10 knockout (IL-10 KO) mice. While IL-10 KO mice are used as model for inflammatory bowel disease, it is considered as one of the frail and fragility model since it developed osteopenia and hair loss, reduces bone formation and energy metabolism, and increase levels of several inflammatory cytokines [11–14]. We explored the role and possible mechanism of ESM in keratinocyte differentiation and improved skin health and aging, which could provide an effective and functional source of bioactive compounds to improve the HRQoL of older people.

2. Materials and Methods

2.1. Cell Culture

Neonatal Normal Human Epidermal Keratinocytes (NHEK-Neo) were purchased from LONZA Japan (Tokyo, Japan). The cells were cultured at 37 °C under 95% air/5% CO_2 with KGM™ Gold Keratinocyte Growth Medium BulletKit™ (#00192060, LONZA), containing KBM™ Gold™ Basal Medium and KGM™ Gold™ SingleQuots™ supplements. The medium was replaced every 2 days during the experimental periods.

2.2. Determination of Cell Cytotoxicity

Cell cytotoxicity was determined using the lactate dehydrogenase (LDH) assay (#299-50601, Wako Pure Chemical Industries, Osaka, Japan). The cells were seeded onto Biocoat™ Collagen I 24-well plate (Corning Inc., Corning, NY, USA) at 2.5×10^4 cells/well and cultured for 2 days with KGM™ Gold Medium. The medium was replaced with KGM™ Gold Medium containing either powdered ESM (pESM; 0.1, 0.5 or 1 mg/mL), alkaline (sodium hydroxide)-hydrolyzed ESM (hESM; 1, 2 or 3 mg/mL), enzymatically-hydrolyzed ESM (eESM; 1, 2 or 3 mg/mL) or 0.2% Tween-20 (positive control). In the preparation of eESM, pESM was treated with 5.5 mg/g sodium metabisulfite for 8–10 h and heated at 90 °C for 6 h, then hydrolyzed with 3.6 mg/g papain at 60 °C for 3 h. The three types of ESMs were obtained from ALMADO Inc. (Tokyo, Japan). The amino acid compositions of each ESM are shown in Table S1. The plates were then incubated at 37 °C for 24 h, the medium was collected into 1.5 mL tubes and cellular LDH was also collected with the medium containing 1% Tween-20. The five times diluted samples were used for the LDH assay. The assay was performed according to the manufacturer's instructions, and the absorbance (Ab) was measured at 560 nm using a spectrophotometer (Thermo Scientific, Waltham, MA, USA). The cytotoxicity was calculated using the following formula.

$$\text{Cytotoxicity (\%)} = (\text{Abs in medium})/(\text{Abs in medium} + \text{Abs in cells}) \times 100$$

2.3. Morphological Changes

The cells were seeded onto Biocoat™ Collagen I 100-mm dish (Corning Inc., Corning, NY, USA) at 2.0×10^5 cells/dish and cultured for 1 day. The medium was replaced with pESM, hESM or eESM containing medium at 2 days' interval and cultured for 4 days. The morphological changes were recorded by monitoring the cells by a microscope (Leica Microsystems GmbH, Wetzlar, Germany).

2.4. Animal Experiment

Male B6.129P2-IL-10<tm1Cgn>/J mice and female C57BL/6J mice were purchased from The Jackson Laboratory and Charles River Japan, respectively. Female IL-10-Hetero KO mice (F1) obtained by mating the two aforementioned mice were further mated with male IL-10-Homo KO mice (F1). The male IL-10-Homo KO mice (F2, KO) were used in this study. The mice were individually housed in cages in a temperature-controlled (23 ± 2 °C), 12 h light-dark cycle environment. Four-week-old KO and C57BL/6J (WT) mice were fed an AIN-93-based diet or an 8% pESM-supplemented diet (KOE) for 28 weeks. The diet composition is shown in Table 1, which is the same as that reported in a previous study [8]. After the experimental period, the back skin was collected under deep anesthesia with isoflurane, removed the dermal white adipose tissue from collected skin sample, and stored at −80 °C until further analysis.

Table 1. Diet composition.

Component (%)	Control Diet	8% pESM Supplemented Diet
Casein	20.0	16.3
β-corn starch	39.7	35.4
α-corn starch	13.2	13.2
Soybean oil	7.0	7.0
Sucrose	10.3	10.3
Cellulose	5.0	5.0
Vitamin mixture *	1.0	1.0
Mineral mixture *	3.5	3.5
L-cystine	0.3	0.3
ESM powder	0.0	8.0

* AIN-93 prescription (Oriental Yeast Co., Ltd., Tokyo, Japan). ESM: Eggshell membrane.

2.5. Morphological Analysis

Hematoxylin and eosin (H&E) staining and Fontana-Masson staining were carried out to evaluate skin thickness and melanin content in the back skin samples of mice, respectively. For H&E staining, the optimal cutting temperature (OCT) compound (Sakura Finetek, Torrance, CA, USA) embedded 5-μm-thick sliced sections were used. The stained sections were then decolorized with ethanol and enclosed using Mount-Quick (Daido Sangyo, Tokyo, Japan). Tissue images were obtained using light microscopy (Olympus BX51 microscope, Olympus Optical, Tokyo, Japan), and the thickness (total of epidermis and dermis) was measured randomly at six different parts of each skin sample.

For Fontana-Masson Staining, the frozen sections were reacted with Fontana ammonia silver aqueous under shading conditions overnight at room temperature. After washing using distilled water, the sections were reacted with 0.2% gold chloride aqueous and fixing solution for 5 s and 1 min, respectively. Subsequently, the sections were washed with distilled water and reacted with Kernechtrot solution to stain nuclei. After washing, the sections were enclosed, and tissue images were obtained as described for the H&E staining.

2.6. RNA Extraction from Cell and Tissue Samples

RNA extraction from cell sample was performed using Nucleospin® RNA plus kit (Takara Bio Inc., Shiga, Japan). The cells were seeded onto a Biocoat™ Collagen I 6-well plate (Corning Inc., Corning, NY, USA) and cultured for 1 day. Then, the medium was replaced with different types of ESM-containing media and cultured until 80–90% confluence.

RNA extraction from animal samples was carried out with Nucleospin® RNA (Takara Bio Inc., Shiga, Japan). In the animal experiment, the frozen skin sample was crushed with a cool mill (AS ONE, Osaka, Japan), which was cooled in liquid nitrogen before use. Crushed samples were homogenized with 1 mL of TRIzol reagent, and then 200 μL of chloroform was added. The mixture was centrifuged at 12,000× *g*, 4 °C for 15 min, and the supernatant was mixed with Nucleospin® RNA RA1 buffer and ethanol. Further experimental procedures were per the manufacturer's instructions.

Total RNA content was measured with NanoDrop ND-1000 (Thermo Scientific, Waltham, MA, USA). For DNA microarray analysis, RNA quality was checked by Agilent 2100 Bioanalyzer (Agilent Technologies, Santa Clara, CA, USA), and all samples were confirmed to have an RNA integrity number of more than 6.0.

2.7. DNA Microarray Analysis

In total, 100 ng pooled RNA isolated from NHEK-Neo and 250 ng of pooled RNA isolated from mouse skin samples were used for microarray analysis with Clariom™ S Array, human and Affymetrix Genechip™ Mouse Genome 430 2.0 Array(Applied Biosystems™, Waltham, MA, USA), respectively. RNA samples derived from cells and mice were reverse-transcribed to single-strand cDNA using Genechip™ WT Amplification Kit (Affymetrix) and Genechip™ 3′ IVT Express Kit (Thermo Scientific, Waltham, MA, USA), respectively. The synthesis of double-strand cDNA and cRNA was performed according to the manufacturer's instructions. The cRNA samples were hybridized onto the array chip at 45 °C for 16 h and stained with streptavidin-phycoerythrin using GeneChip™ Fluidics Station 450 (Applied Biosystems™). The image data was detected with GeneChip™ Scanner 3000 (Affymetrix).

The data were converted to CEL file format using Affymetrix GeneChip® Command Console. In the cell experiment, the data were normalized with Transcriptome Analysis Console. The numerical conversion and the normalization of the animal experiment data were performed using R ver. 3.5.1 and Affymetrix Microarray Analysis Suite 5 (Affymetrix), respectively. The cut-off condition was set up as |Fold change| $\geq$ 2 in the cell experiment, and |Fold change| $\geq$ 1.5 and signal intensity $\geq$ 40 in the animal experiment. The gene sets satisfying these cut-off conditions were further subject to an ingenuity pathway analysis software (IPA, http://www.ingenuity.com/, date accessed: 29 September 2020) for the bioinformatics analysis including canonical pathway analysis, disease and function analysis and upstream analysis.

2.8. Real-Time Reverse Transcription-Polymerase Chain Reaction (Real Time RT-PCR)

To validate the microarray data, the expression of some selected genes was measured using real-time RT-PCR. RNA samples derived from the cells and mice were reverse-transcribed to cDNA using Primescript™ RT Master Mix (Takara Bio Inc.), according to the manufacturer's instructions. The cDNA was then mixed with appropriate primers and SYBR® Premix Ex Taq™ (Takara Bio Inc.), and real time RT-PCR was performed using the Thermal Cycler Dice Real Time System TP800 (Takara Bio Inc.). The primer sequences are shown in Table S2. The obtained data were normalized with B2M (Human Housekeeping Gene Primer Set, #3790, Takara Bio Inc.) and *Gapdh* in the cell and animal experiments, respectively. The results are expressed as fold-change values compared with the control or WT group.

2.9. Statistical Analysis

Results are expressed as the mean $\pm$ standard error. All data were analyzed with one-way analysis of variance (ANOVA), followed by a Tukey's test, and p-value < 0.05 was considered to indicate statistical significance.

3. Results

3.1. Effects of Different Types of ESM on Cytotoxicity and Cell Morphology

As pESM contains 91~94% indigestible protein and approximately 46% of them are digested and absorbed in the digestive tract [8], we tested the effects of three types of ESM, pESM, hESM and eESM, on cytotoxicity and cell morphology in cell culture experiments. As shown in Figure 1a, the cytotoxicity based on LDH assay was not increased by three types of ESM in the tested concentrations. However, the value was significantly decreased by 1 mg/mL of pESM, 2 and 3 mg/mL of hESM, and 1–3 mg/mL of eESM in NHEK-neo cells. Therefore, we selected 1 mg/mL supplementation for subsequent experiments in the hESM and eESM groups. In the pESM group, we chose 0.1 mg/mL supplementation, as the insoluble materials disturbed the cell observation when the medium was supplemented with 1 mg/mL pESM.

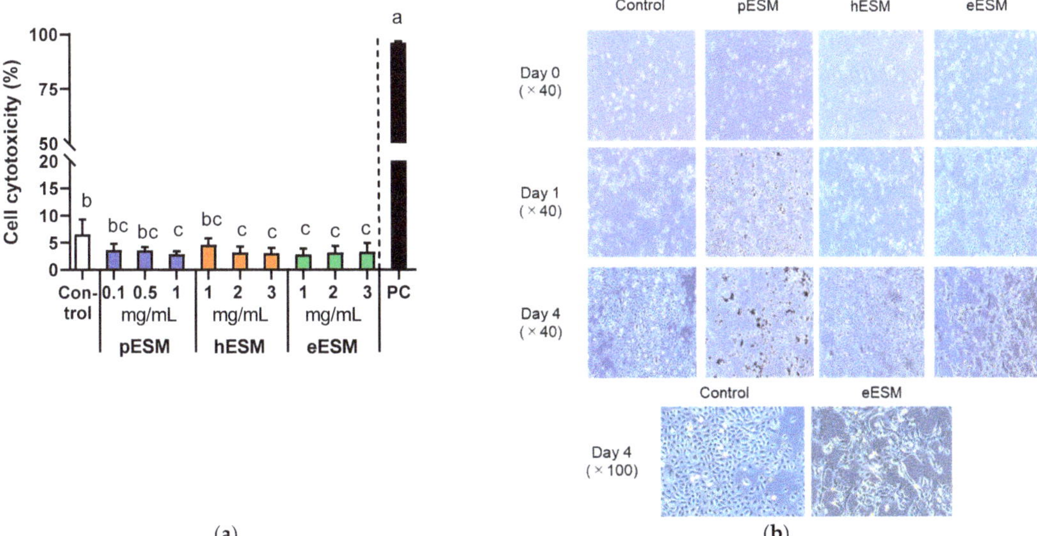

Figure 1. Effects of three types of ESMs on cell cytotoxicity and morphological characterization of NHEK-Neo cells. (**a**) Cell cytotoxicity based on LDH assay in response to different concentrations of pESM, hESM and eESM. 0.2% Tween-20 (*v/v*) was used as the positive control (PC). Data are means ± SD and expressed as % (*n* = 4). The calculation is shown in the Materials and Methods; (**b**) Morphological changes of NHEK-Neo cells in response to 0.1 mg/mL of pESM, and 1 mg/mL of hESM and eESM each. The different lowercase alphabets indicate significant differences between the tested-group (*p* < 0.05).

Figure 1b shows the time course of morphological changes in ESMs supplemented NHEK-Neo cells. pESM and hESM altered the cell morphology neither on Day 1 nor 4 compared with the control group. Notably, eESM-treated cells showed the flattening-like changes on Day 4 of treatment. These results suggested eESM might promote cell differentiation in NHEK-Neo cells.

3.2. Microarray Analysis in Cell Culture Experiments

We carried out DNA microarray analysis to further investigate the mechanism underlying eESM-induced cell differentiation in NHEK-Neo cells. As shown in Figure 2a, the expression of many genes was altered by eESM supplementation (1289 increased and 1330 decreased) compared with the control group. To clarify the positive effect of eESM, we uploaded the altered genes to IPA. Previous studies indicated that keratinocyte differentiation is accompanied by the decreases in the synthesis of nucleotide and protein [15,16]. A similar trend was also observed in our microarray analysis. As illustrated in Figure 2b, the canonical pathway analysis indicated that the pathway related to cell cycle and nucleotide biosynthesis such as "Cell Cycle Control of Chromosomal Replication", "Purine Nucleotide De Novo Biosynthesis II" and "Cyclins and Cell Cycle Regulation" were downregulated by eESM.

Furthermore, disease and function analysis identified that "Differentiation of Skin" and "Formation of Skin" were upregulated (Figure 2c). Since both of these pathways included similar gene sets, we focused on "Formation of Skin," which contained many gene sets. The major gene list is shown in Table 2. Notably, eESM upregulated the differentiation markers of keratinocytes such as involucrin (*IVL*), filaggrin (*FLG*) and keratin 1 (*KRT1*). Then, we carried out real time RT-PCR analysis to further evaluate the differentiation markers in ESM-treated keratinocytes. As shown in Figure 2d, gene markers of the basal layer, *KRT5* and *KRT14*, were not altered by any ESM supplementation. However, levels of markers of the spinous layer, *KRT1*, *KRT10* and *IVL*, were significantly increased by eESM supplementation compared to the control group, consistent with the results of microarray analysis. Additionally, hESM significantly increased the expression of these genes, while the increments were significantly lower than those in the eESM group. On the contrary, pESM treatment did not show significant changes in the expression of these genes ($p > 0.10$). The expression level of the marker of granular keratinocytes *FLG* was significantly increased by eESM supplementation, consistent with the microarray data. The expression of *FLG* was also significantly increased by pESM but not hESM ($p > 0.10$).

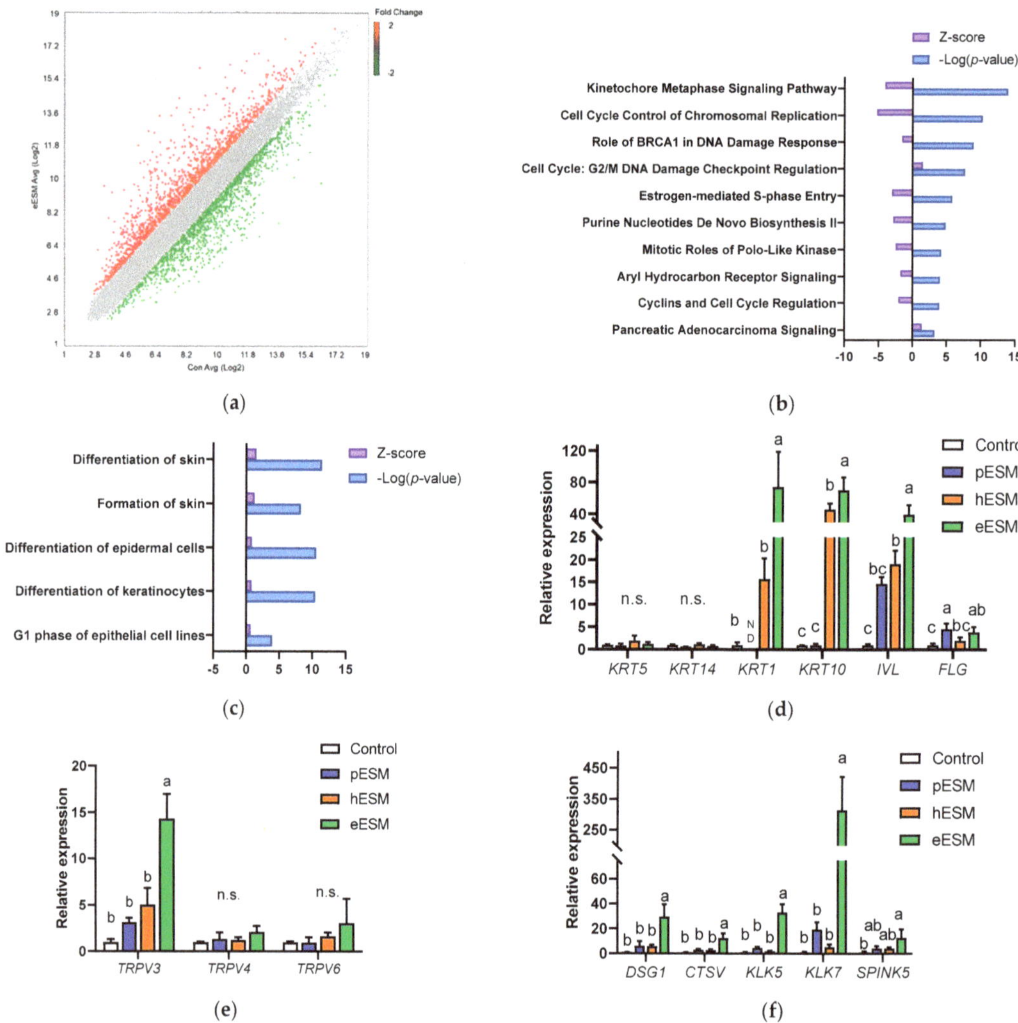

Figure 2. DNA microarray analysis of eESMtreated NHEKNeo cells. (**a**) Scatter plot of microarray data. Each dot represents the fold changes in gene expression. The values on the X and Y axes are the signal values (Log2 scaled) of the control and eESM groups, respectively; (**b,c**) Canonical analysis (**b**) and disease and function analyses (**c**) using IPA software. In both analyses, the cut-off condition was ∣Activation Z-score∣ > 1, and data was sorted by −Log (p-value); (**d–f**) results of real-time RT-PCR on keratinocyte differentiation markers (**d**), genes related to keratinocyte differentiation and calcium signal (**e**) and calcium channel (**f**). Data are means ± SD ($n = 3$) and expressed as fold change values compared with the control group. ND means the expression of the genes was not detectable in our method. The different lowercase alphabets indicate significant differences between the tested-group ($p < 0.05$).; n.s.: not significant.

Table 2. List of major genes related to "Formation of skin".

Gene Symbol	Gene Name	Fold Change (eESM vs. Control)
ALOX15B	Arachidonate 15-lipoxygenase type b	2.01
CDH1	Cadherin 1	2.17
CDSN	Corneodesmosin	48.29
CST6	Cystatin e/m	5.15
CTSV	Cathepsin V	6.69
CYLD	Cyld lysine 63 deubiquitinase	2.65
DLX3	Distal-less homeobox 3	2.08
DSG1	Desmoglein 1	15.85
FLG	Filaggrin	2.77
HBP1	HMG-box transcription factor 1	8.34
IVL	Involucrin	24.62
KLF4	Kruppel-like factor 4	3.90
KLK5	Kallikrein-related peptidase 5	22.74
KLK7	Kallikrein-related peptidase 7	39.73
KRT1	Keratin 1	2.37
KRT13	Keratin 13	78.42
NCOA3	Nuclear receptor coactivator 3	3.28
PPL	Periplakin	18.86
SPINK5	Serine protease inhibitor kazal type-5	6.77
TGM1	Transglutaminase 1	5.14
TMEM79	Transmembrane protein 79	3.11

In this category, 88 genes are listed as the altered genes. Based on the literature information, we selected 21 genes that were related to skin aging, keratinocyte differentiation and calcium signaling.

Moreover, in the microarray data, eESM upregulated Kallikrein-related peptidase (*KLK*)-5 and -7 and serine protease inhibitor kazal type (*SPINK*)-5 expression (Table 2), which are inducible in epidermal keratinocytes during cell differentiation. Similar results were observed in real-time RT-PCR analysis, while these increments were not observed in pESM or hESM-treated cells ($p > 0.10$, Figure 2e). These lines of data indicated that eESM supplementation induced cell differentiation.

3.3. Possible Involvement of Calcium Signaling in eESM-Induced Keratinocyte Differentiation

Microarray analysis indicated that *DSG1* and *CTSV* were upregulated by eESM supplementation (Table 2), and similar results were obtained by real-time RT-PCR (Figure 2e). These genes are related to keratinocyte differentiation, skin disease and calcium binding. Since the keratinocyte differentiation is triggered by extracellular calcium ion or calcium signaling [15,16], we investigated the gene expression of the transient receptor potential (TRP) channel families TRPV3, TRPV4 and TRPV6. As shown in Figure 2f, *TRPV3* expression was significantly increased by eESM supplementation compared with the control group, while no significant differences were observed in pESM- or hESM-treated cells. *TRPV4* and *TRPV6* levels were not altered by any ESM supplementation.

3.4. Effects of pESM Supplementation on Skin Aging in IL-10 KO Mice

We next investigated whether pESM supplementation could ameliorate skin aging in IL-10 KO mice. As shown in Figure 3a,b, skin thickness evaluated by H&E staining was significantly decreased in IL-10 KO mice compared with WT mice. Notably, pESM supplementation significantly improved the value near to that in WT mice. Even though the decline of melanin formation is considered to associate with the pathogenesis of silver hair, Fontana-Masson staining showed no obvious changes in response to IL-10 KO and ESM supplementation (Figure 3a). These results indicated that oral supplementation of ESM ameliorated skin thinning in IL-10 KO mice.

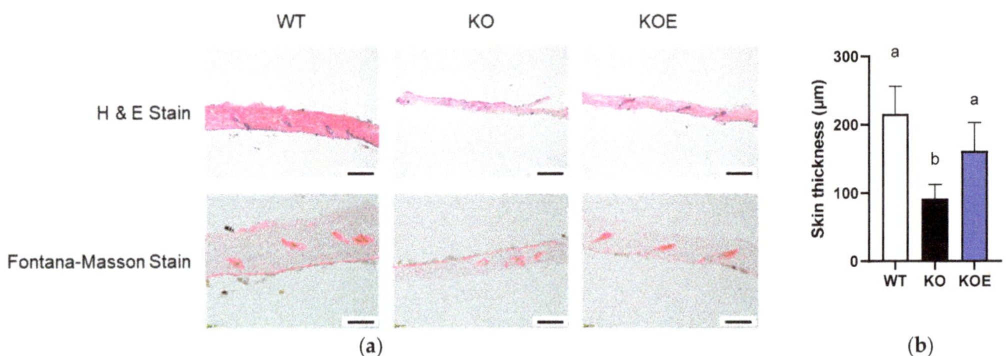

Figure 3. Effects of 8% pESM supplementation on skin morphology in IL-10 KO mice. (**a**) Results of H&E stain (upper side) and Fontana-Masson Stain (lower side); (**b**) Skin thickness based on H&E stain. Data are means ± SD, *n* = 5, 7 and 6 for WT, KO and KOE groups, respectively. The different lowercase alphabets indicate significant differences between the tested-group ($p < 0.05$). WT: Wild type, KO: IL-10 knockout mice, KOE: KO mice supplemented with 8% pESM.

3.5. Microarray Analysis in pESM-Supplemented IL-10 KO Mice

We performed the microarray analysis to investigate the molecular mechanism of pESM's effects in IL-10 KO mice. The data indicated that 2232 and 1567 probe sets were upregulated and downregulated in IL-10 KO mice compared to the WT group, respectively. Here, pESM supplementation upregulated 2067 probe sets and downregulated 195 probe sets. Then, we performed further analysis of the altered genes using IPA software. As shown in Figure 4a, in the comparison between the WT and KO groups, the inflammation-related pathways were upregulated in the KO group, such as "p38 MAPK Signaling", "Toll-like Receptor Signaling", "Acute Phase Response Signaling" and "NF-κβ Signaling". However, in comparison between the KO and KOE groups, these pathways were not listed up. As shown in Figure 4b, pESM supplementation upregulated "Calcium Signaling", which is consistent with the results of in vitro study. The list of the major genes is shown in Table 3. In this case, pESM supplementation upregulated calcium ion channel-related genes, and *Trpv6* is also listed in the gene list. Real time RT-PCR analysis showed a similar tendency, while the expression levels were not significant (Figure 4d). Other TRPVs, *Trpv3* and *Trpv4*, were also upregulated two-fold compared to the KO group; however, the increments did not reach statistical significance. The expression levels of these three genes were significantly lower in the KO and KOE groups compared with the WT group.

Furthermore, pESM supplementation upregulated "GP6 Signaling" (Figure 4b), which has a crucial role in collagen synthesis. Similar results were observed in Disease and Function analysis. As shown in Figure 4c, pESM supplementation upregulated "Proliferation of connective tissue cells", "Growth of epithelial tissue" and "Differentiation of connective tissue cells". The list of major genes related to the "Growth of epithelial tissue" is shown in Table S3. It contains collagen synthesis-related genes, such as Collagen type I alpha (*Col1a*)-1 and -2, and Collagen type XVIII alpha 1 (*Col18a1*). Fibroblast growth factor 1 (*Fgf1*) was also upregulated by pESM supplementation, which relates to collagen synthesis and skin elasticity [17]. Of these, we measured *Col1a1* expression by real time RT-PCR analysis. As shown in Figure 4e, Col1a1 was increased by pESM supplementation, consistent with the results of microarray analysis, while it did not reach the statistical significance. pESM upregulated Wnt family members (*Wnt*)-5a and -7a (Table S3), which have been identified as the skin aging biomarker and suppressive inducer against cell senescence [18,19].

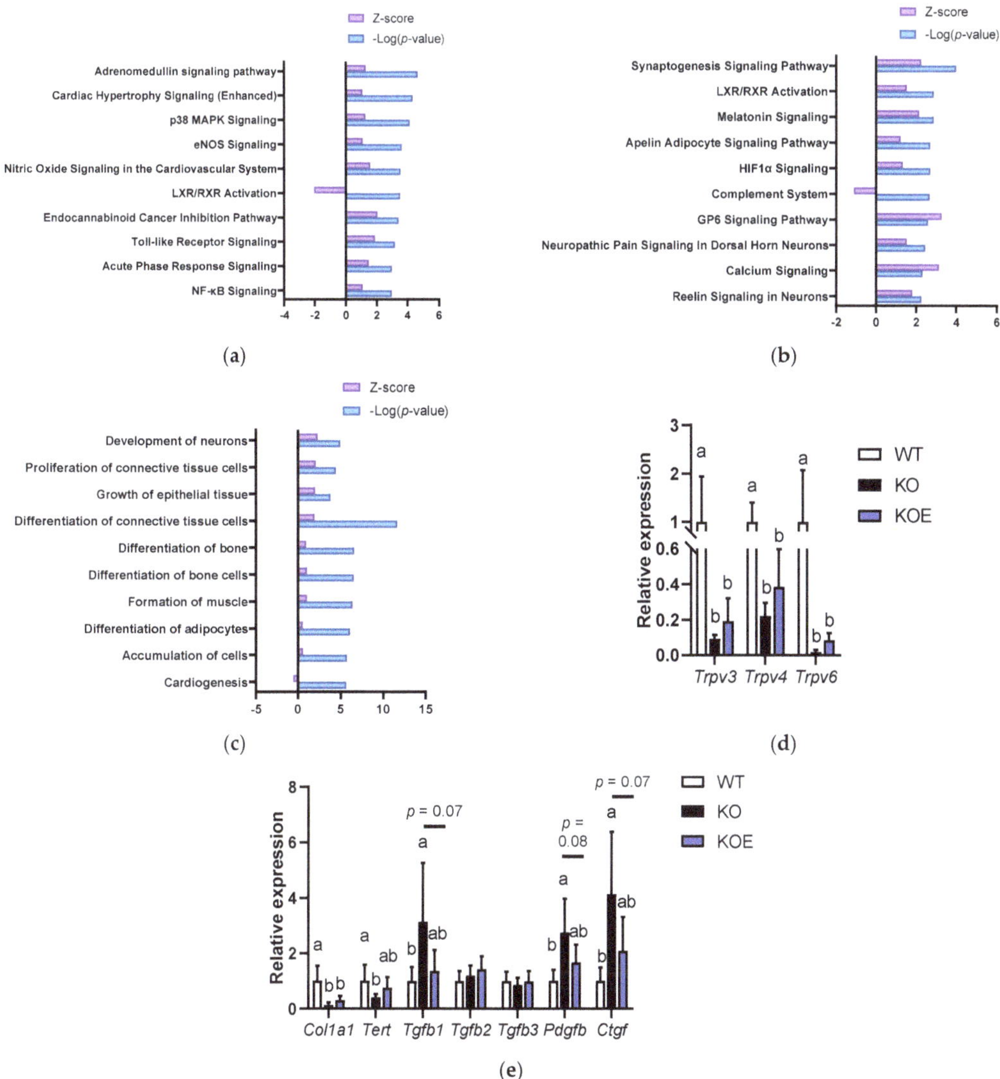

Figure 4. DNA microarray analysis on pESM supplemented IL10 KO mice. (**a**,**b**) Canonical analysis comparison between the KO and WT groups (**a**) and between the KOE and KO groups (**b**) using IPA software. In both analyses, the cut-off condition was |Activation Zscore| > 1, and data was sorted by −Log (*p*-value); (**c**) Disease and function analysis comparison between the KOE and KO groups using IPA software. The cut-off condition was |Activation Zscore| > 1, and data was sorted by −Log (*p*-value); (**d**,**e**) Results of real-time RT-PCR on calcium channel (**d**) and skin aging-related genes (**e**). Data are means ± SD, *n* = 5, 7 and 6 for WT, KO and KOE groups, respectively, and expressed as fold change values compared with the WT group. WT: Wild type, KO: IL10 knockout mice, KOE: KO mice supplemented with pESM. The different lowercase alphabets indicate significant differences between the tested-group (*p* < 0.05).

Table 3. Major gene lists of "Calcium signal".

Gene Symbol	Gene Name	Fold Change (eESM vs. Control)
Cacna1a	Calcium voltage-gated channel subunit alpha1 A	1.87
Cacna1d	Calcium voltage-gated channel subunit alpha1 D	1.59
Cacna1g	Calcium voltage-gated channel subunit alpha1 G	1.66
Camk1d	Calcium calmodulin-dependent protein kinase ID	1.75
Camk2a	Calcium calmodulin-dependent protein kinase II alpha	1.61
Casq1	calsequestrin 1	5.29
Creb3l4	cAMP responsive element binding protein 3-like 4	2.10
Micu1	Mitochondrial calcium uptake 1	1.57
Prkacb	protein kinase cAMP-activated catalytic subunit beta	1.53
Prkar1b	Protein kinase cAMP-dependent type I regulatory subunit beta	1.55
Prkar2b	Protein kinase cAMP-dependent type II regulatory subunit beta	1.59
Trpc3	Transient receptor potential cation channel subfamily C member 3	1.84
Trpv6	Transient receptor potential cation channel subfamily V member 6	1.70

In this category, 35 genes were listed up as the altered genes. Based on the literature information, we selected 13 genes that were related to the calcium signal.

We further examined the expression of skin homeostasis- and aging-related genes. The expression of telomerase reverse transcriptase (*Tert*), which induces proliferation of hair follicle bulge stem cells and hair growth [20], was significantly decreased in the KO group compared with the WT group. In the KOE group, the expression showed the intermediate value, whereas there was no significant difference compared to the KOE and KO groups ($p > 0.10$, Figure 4e). Transforming growth factor (TGF)-β signaling is considered to relate to healthy skin homeostasis, hair growth and collagen synthesis [20,21]. In our data, *Tgfb2* and *Tgfb3* were not altered in all the tested groups, while *Tgfb1* was significantly increased in the KO group, and its increment was tended to decline in the KOE group ($p = 0.07$). In addition to TGF-β1, platelet-derived growth factor (PDGF)-β is crucial in cell function in connective tissues and relates to fibrosis [22]. The expression of *Pdgfb* was significantly increased in the KO group, while the increase was tended to be suppressed in the KOE group ($p = 0.08$). TGF-β also induces connective tissue growth factor (CTGF) expression, which induces organ fibrosis, including skin [23]. Similar results were observed in *Ctgf* expression, which was significantly increased in the KO group, and its increment was declined in the KOE group ($p = 0.07$). Furthermore, based on cellular experiments, we investigated the expression of *Klk5* and *Klk7* in the microarray dataset, while the data showed that the expressions were not increased by pESM supplementation (*Klk5*, fold change (KOE vs. KO): 0.98; *Klk7*, fold change (KOE vs. KO): 0.89).

4. Discussion

The present study indicated that eESM induces keratinocyte differentiation in NHEK-Neo cells and that pESM supplementation improves the decline in skin thickness in IL-10

KO. These findings suggest that ESMs may improve skin health and aging. Previous studies indicated that polyphenols, vitamins and collagen peptides have a beneficial role in skin health, while these effects are mainly observed via anti-oxidative stress and anti-inflammation effects [4]. In contrast, our study shows the direct beneficial effect of ESM on skin health.

Our microarray analysis indicated that calcium signaling might associate with the positive effects of eESM and pESM in cell and animal experiments, respectively. Especially, TRPVs were upregulated in both experiments. Previous studies indicated that calcium or calcium ions are crucial to induce keratinocyte differentiation [15,16]. Furthermore, it has been reported that calcium concentration gradient is established in the epidermis, and it is broken down with age in mice and humans [24,25]. Therefore, the upregulation of calcium signaling could be an important factor to explain the effect of eESM and pESM in our experiment. The calcium concentrations in three types of ESM-supplemented medium using AmpliteTM Fluorimetric Calcium Quantitation Kit Red Fluorescence (AAT Bioquest®), did not vary compared with the control group (Control: 0.10 mM, pESM: 0.08 mM, hESM: 0.13 mM, and eESM: 0.23 mM) and much lower than 1.2 mM that induces keratinocyte differentiation in murine and human cell lines [26,27]. Based on these lines of information, it is suggested that at least in NHEK-Neo cells, eESM induces keratinocyte differentiation, possibly via the upregulation of calcium signaling, which might be caused by the transcription of TRPVs rather than extracellular calcium concentrations.

Notably, eESM upregulated calcium signaling and induced keratinocyte differentiation, while this effect was not observed with the same concentration of hESM. The differences could be explained by amino acids or peptides of ESM. As shown in Table S1, total amino acid concentrations of hESM and eESM were similar, but the concentrations of asparagine, histidine, threonine, citrulline, arginine, tyrosine, phenylalanine and branched-chain amino acids were higher in eESM than in hESM. Even though a previous study has shown that several nutrients and compounds have beneficial roles in skin health [28], there is little information on the relationship among amino acids, calcium signaling and keratinocyte differentiation. Moreover, functional peptides derived from natural resources have beneficial roles in regulating metabolic disorders incurred by aging [29]. Since enzymatic hydrolysis is relatively mild rather than acid hydrolysis, eESM could contain the functional peptides. This study, for the first time, suggests that the amino acids or functional peptides in eESM might associate with the upregulation of calcium signaling and keratinocyte differentiation. Nevertheless, the related mechanisms need to be investigated further.

In the animal experiment, we could not conclude the central mechanism of pESM's effect based on the DNA microarray data. However, our results imply the following possibilities. First, based on microarray data, genes related to calcium signaling and collagen synthesis might be partly involved in pESM's effect in the back skin of IL-10 KO mice. Second, based on the *Tert* result, pESM might improve telomere shorting, a hallmark of aging [30]. Third, based on the results of growth factors such as *Tgfb1*, *Pdgfb* and *Ctgf*, pESM might affect fibrosis or inflammation-related genes. However, further investigations are required to clarify the mechanism of pESM's action using other aging or frailty animal models.

Our previous study indicated that oral supplementation of pESM has several beneficial effects against intestinal inflammation and obesity [8,9]. Taken together, it can be inferred that ESM, a rich source of beneficial bioactive compounds, could improve several diseases and the aging process of the skin. Furthermore, since we indicated that ESMs did not show cell cytotoxicity in at least keratinocytes, it could be proposed that the direct application of eESM to the skin, such as skin cream, could show much impactive effect for skin health and skin aging. Since ESM is regarded as industrial waste and discarded after egg processing, it would be of significance if further in-depth studies could be carried out to explore the effective and functional usage of ESM in our life and health. Furthermore, our study could contribute for a novel cosmeceutical concept "drug repositioning" [31].

Accumulating nutritional study against skin health and disease are still important to prevent skin disorders.

5. Conclusions

This study indicates that ESM plays an important role in improving skin health and aging in NHEK-Neo cells and IL-10 KO mice. These findings could contribute to better understand the role of the beneficial compounds for improving skin health, increase HRQoL for elderly people and implement a sustainable food system.

Supplementary Materials: The following are available online at https://www.mdpi.com/article/10.3390/nu13072144/s1, Table S1: Amino acids composition of three ESMs, Table S2: Primer sequences, Table S3: Major gene list of "Growth of epithelial tissue".

Author Contributions: Conceptualization, Y.H., H.J. and H.K.; formal analysis, K.F. and M.K.; investigation, K.F., T.K. and M.K.; validation, T.K.; resources, Y.H.; data curation, K.F.; writing—original draft preparation, K.F.; writing—review and editing, H.J.; visualization, K.F.; supervision, H.J. and H.K.; funding acquisition, H.J. All authors have read and agreed to the published version of the manuscript.

Funding: This research was funded in part by a Grant-in-Aid (18K11095) from the Japan Society for the Promotion of Science (JSPS).

Institutional Review Board Statement: The animal experiment in this study was approved by the Animal Care and Use Committee of the University of Tokyo (Approval No. P17-143). All procedures were conducted according to the relevant rules and regulations.

Informed Consent Statement: Not applicable.

Data Availability Statement: The microarray data have been deposited in the GEO database (https://www.ncbi.nlm.nih.gov/geo/ (accessed on 22 June 2021)) under accession number GSE178597.

Conflicts of Interest: The authors declare no conflict of interest.

References

1. Foroushani, A.R.; Estebsari, F.; Mostafaei, D.; Ardebili, H.E.; Shojaeizadeh, D.; Dastoorpoor, M.; Jamshidi, E.; Taghdisi, M.H. The Effect of Health Promoting Intervention on Healthy Lifestyle and Social Support in Elders: A Clinical Trial Study. *Iran. Red Crescent Med. J.* **2014**, *16*, e18399. [CrossRef]
2. Catic, A. Cellular Metabolism and Aging. *Progress Mol. Biol. Transl. Sci.* **2018**, *155*, 85–107. [CrossRef]
3. Rae, M.J.; Butler, R.N.; Campisi, J.; De Grey, A.D.N.J.; Finch, C.E.; Gough, M.; Martin, G.M.; Vijg, J.; Perrott, K.M.; Logan, B.J. The Demographic and Biomedical Case for Late-Life Interventions in Aging. *Sci. Transl. Med.* **2010**, *2*, 40cm21. [CrossRef]
4. Cao, C.; Xiao, Z.; Wu, Y.; Ge, C. Diet and Skin Aging—From the Perspective of Food Nutrition. *Nutrients* **2020**, *12*, 870. [CrossRef]
5. Tobin, D.J. Introduction to skin aging. *J. Tissue Viability* **2017**, *26*, 37–46. [CrossRef]
6. Mukherjee, P.K.; Maity, N.; Nema, N.K.; Sarkar, B.K. Bioactive compounds from natural resources against skin aging. *Phytomedicine* **2011**, *19*, 64–73. [CrossRef] [PubMed]
7. Ruff, K.J.; DeVore, D.P.; Leu, M.D.; Robinson, M.A. Eggshell membrane: A possible new natural therapeutic for joint and connective tissue disorders. Results from two open-label human clinical studies. *Clin. Interv. Aging* **2009**, *4*, 235–240. [CrossRef]
8. Jia, H.; Hanate, M.; Aw, W.; Itoh, H.; Saito, K.; Kobayashi, S.; Hachimura, S.; Fukuda, S.; Tomita, M.; Hasebe, Y.; et al. Eggshell membrane powder ameliorates intestinal inflammation by facilitating the restitution of epithelial injury and alleviating microbial dysbiosis. *Sci. Rep.* **2017**, *7*, 43993. [CrossRef] [PubMed]
9. Ramli, N.S.; Jia, H.; Sekine, A.; Lyu, W.; Furukawa, K.; Saito, K.; Hasebe, Y.; Kato, H. Eggshell membrane powder lowers plasma triglyceride and liver total cholesterol by modulating gut microbiota and accelerating lipid metabolism in high-fat diet-fed mice. *Food Sci. Nutr.* **2020**, *8*, 2512–2523. [CrossRef]
10. Ruff, K.J.; Endres, J.R.; Clewell, A.E.; Szabo, J.R.; Schauss, A. Safety evaluation of a natural eggshell membrane-derived product. *Food Chem. Toxicol.* **2012**, *50*, 604–611. [CrossRef] [PubMed]
11. Dresner-Pollak, R.; Gelb, N.; Rachmilewitz, D.; Karmeli, F.; Weinreb, M. Interleukin 10-deficient mice develop osteopenia, decreased bone formation, and mechanical fragility of long bones. *Gastroenterology* **2004**, *127*, 792–801. [CrossRef] [PubMed]
12. Ko, F.; Yu, Q.; Xue, Q.-L.; Yao, W.; Brayton, C.; Yang, H.; Fedarko, N.; Walston, J. Inflammation and mortality in a frail mouse model. *AGE* **2011**, *34*, 705–715. [CrossRef] [PubMed]
13. Walston, J.; Fedarko, N.; Yang, H.; Leng, S.; Beamer, B.; Espinoza, S.; Lipton, A.; Zheng, H.; Becker, K. The Physical and Biological Characterization of a Frail Mouse Model. *J. Gerontol. Ser. A Biol. Sci. Med. Sci.* **2008**, *63*, 391–398. [CrossRef]

14. Westbrook, R.M.; Le Yang, H.; Langdon, J.M.; Roy, C.N.; Kim, J.A.; Choudhury, P.P.; Xue, Q.-L.; Di Francesco, A.; De Cabo, R.; Walston, J. Aged interleukin-10tm1Cgn chronically inflamed mice have substantially reduced fat mass, metabolic rate, and adipokines. *PLoS ONE* **2017**, *12*, e0186811. [CrossRef] [PubMed]
15. Elsholz, F.; Harteneck, C.; Muller, W.; Friedland, K. Calcium—A central regulator of keratinocyte differentiation in health and disease. *Eur. J. Dermatol. EJD* **2014**, *24*, 650–661. [CrossRef] [PubMed]
16. Bikle, D.D.; Xie, Z.; Tu, C.-L. Calcium regulation of keratinocyte differentiation. *Expert Rev. Endocrinol. Metab.* **2012**, *7*, 461–472. [CrossRef]
17. De Araújo, R.; Lôbo, M.; Trindade, K.; Silva, D.F.; Pereira, N.D.P. Fibroblast Growth Factors: A Controlling Mechanism of Skin Aging. *Ski. Pharmacol. Physiol.* **2019**, *32*, 275–282. [CrossRef]
18. Bikkavilli, R.K.; Avasarala, S.; Van Scoyk, M.; Arcaroli, J.; Brzezinski, C.; Zhang, W.; Edwards, M.G.; Rathinam, M.K.K.; Zhou, T.; Tauler, J.; et al. Wnt7a is a novel inducer of β-catenin-independent tumor-suppressive cellular senescence in lung cancer. *Oncogene* **2015**, *34*, 5317–5328. [CrossRef]
19. Makrantonaki, E.; Brink, T.C.; Zampeli, V.; Elewa, R.M.; Mlody, B.; Hossini, A.M.; Hermes, B.; Krause, U.; Knolle, J.; Abdallah, M.; et al. Identification of Biomarkers of Human Skin Ageing in Both Genders. Wnt Signalling—A Label of Skin Ageing? *PLoS ONE* **2012**, *7*, e50393. [CrossRef]
20. Kubo, C.; Ogawa, M.; Uehara, N.; Katakura, Y. Fisetin Promotes Hair Growth by Augmenting TERT Expression. *Front. Cell Dev. Biol.* **2020**, *8*, 566617. [CrossRef]
21. Liarte, S.; Bernabé-García, Á.; Nicolás, F.J. Role of TGF-β in Skin Chronic Wounds: A Keratinocyte Perspective. *Cells* **2020**, *9*, 306. [CrossRef] [PubMed]
22. Rodriguez, A.; Friman, T.; Kowanetz, M.; van Wieringen, T.; Gustafsson, R.; Sundberg, C. Phenotypical Differences in Connective Tissue Cells Emerging from Microvascular Pericytes in Response to Overexpression of PDGF-B and TGF-β1 in Normal Skin in Vivo. *Am. J. Pathol.* **2013**, *182*, 2132–2146. [CrossRef] [PubMed]
23. Kubota, S.; Takigawa, M. Cellular and molecular actions of CCN2/CTGF and its role under physiological and pathological conditions. *Clin. Sci.* **2015**, *128*, 181–196. [CrossRef]
24. Rinnerthaler, M.; Streubel, M.K.; Bischof, J.; Richter, K. Skin aging, gene expression and calcium. *Exp. Gerontol.* **2015**, *68*, 59–65. [CrossRef]
25. Streubel, M.K.; Neuhofer, C.; Bischof, J.; Steinbacher, P.; Russe, E.; Wechselberger, G.; Richter, K.; Rinnerthaler, M. From Mice to Men: An Evolutionary Conserved Breakdown of the Epidermal Calcium Gradient and Its Impact on the Cornified Envelope. *Cosmetics* **2018**, *5*, 35. [CrossRef]
26. Hennings, H.; Michael, D.; Cheng, C.; Steinert, P.; Holbrook, K.; Yuspa, S.H. Calcium regulation of growth and differentiation of mouse epidermal cells in culture. *Cell* **1980**, *19*, 245–254. [CrossRef]
27. Pillai, S.; Bikle, D.D.; Hincenbergs, M.; Elias, P.M. Biochemical and morphological characterization of growth and differentiation of normal human neonatal keratinocytes in a serum-free medium. *J. Cell. Physiol.* **1988**, *134*, 229–237. [CrossRef]
28. Michalak, M.; Pierzak, M.; Kręcisz, B.; Suliga, E. Bioactive Compounds for Skin Health: A Review. *Nutrients* **2021**, *13*, 203. [CrossRef]
29. Bhullar, K.S.; Wu, J. Dietary peptides in aging: Evidence and prospects. *Food Sci. Hum. Wellness* **2020**, *9*, 1–7. [CrossRef]
30. Ferrucci, L.; Gonzalez-Freire, M.; Fabbri, E.; Simonsick, E.; Tanaka, T.; Moore, Z.; Salimi, S.; Sierra, F.; De Cabo, R. Measuring biological aging in humans: A quest. *Aging Cell* **2020**, *19*, e13080. [CrossRef]
31. Sotiropoulou, G.; Zingkou, E.; Pampalakis, G. Redirecting drug repositioning to discover innovative cosmeceuticals. *Exp. Dermatol.* **2021**, *30*, 628–644. [CrossRef] [PubMed]

Article

Effects of *Lactobacillus plantarum* CJLP55 on Clinical Improvement, Skin Condition and Urine Bacterial Extracellular Vesicles in Patients with Acne Vulgaris: A Randomized, Double-Blind, Placebo-Controlled Study

Mi-Ju Kim [1], Kun-Pyo Kim [1], Eunhye Choi [1], June-Hyuck Yim [2], Chunpil Choi [3], Hyun-Sun Yun [4], Hee-Yoon Ahn [4], Ji-Young Oh [4] and Yunhi Cho [1,*]

[1] Department of Medical Nutrition, Graduate School of East-West Medical Science, Kyung Hee University, Yongin-si 17104, Gyeongggi-do, Korea; miju0522@gmail.com (M.-J.K.); kunpyokim@gmail.com (K.-P.K.); eunhye960718@naver.com (E.C.)
[2] Department of Dermatology, Kyung Hee University Medical Center, Seoul 02447, Korea; laurel02@nate.com
[3] Skyfeel Dermatologic Clinic, Seoul 06020, Korea; aag0001@naver.com
[4] CJ Foods R & D Center, CJ CheilJedang Corporation, Suwon-si 16495, Gyeongggi-do, Korea; hs.yun@cj.net (H.-S.Y.); hy.ahn@cj.net (H.-Y.A.); jy.oh@cj.net (J.-Y.O.)
* Correspondence: choyunhi@khu.ac.kr; Tel.: +82-31-201-3817

Citation: Kim, M.-J.; Kim, K.-P.; Choi, E.; Yim, J.-H.; Choi, C.; Yun, H.-S.; Ahn, H.-Y.; Oh, J.-Y.; Cho, Y. Effects of *Lactobacillus plantarum* CJLP55 on Clinical Improvement, Skin Condition and Urine Bacterial Extracellular Vesicles in Patients with Acne Vulgaris: A Randomized, Double-Blind, Placebo-Controlled Study. *Nutrients* **2021**, *13*, 1368. https://doi.org/10.3390/nu13041368

Academic Editors: Jean Christopher Chamcheu, Anthony L. Walker and Felicite Noubissi-Kamdem

Received: 30 March 2021
Accepted: 11 April 2021
Published: 19 April 2021

Publisher's Note: MDPI stays neutral with regard to jurisdictional claims in published maps and institutional affiliations.

Abstract: *Lactobacillus plantarum* CJLP55 has anti-pathogenic bacterial and anti-inflammatory activities in vitro. We investigated the dietary effect of CJLP55 supplement in patients with acne vulgaris, a prevalent inflammatory skin condition. Subjects ingested CJLP55 or placebo ($n = 14$ per group) supplements for 12 weeks in this double-blind, placebo-controlled randomized study. Acne lesion count and grade, skin sebum, hydration, pH and surface lipids were assessed. Metagenomic DNA analysis was performed on urine extracellular vesicles (EV), which indirectly reflect systemic bacterial flora. Compared to the placebo supplement, CJLP55 supplement improved acne lesion count and grade, decreased sebum triglycerides (TG), and increased hydration and ceramide 2, the major ceramide species that maintains the epidermal lipid barrier for hydration. In addition, CJLP55 supplement decreased the prevalence of *Proteobacteria* and increased *Firmicutes*, which were correlated with decreased TG, the major skin surface lipid of sebum origin. CJLP55 supplement further decreased the *Bacteroidetes:Firmicutes* ratio, a relevant marker of bacterial dysbiosis. No differences in skin pH, other skin surface lipids or urine bacterial EV phylum were noted between CJLP55 and placebo supplements. Dietary *Lactobacillus plantarum* CJLP55 was beneficial to clinical state, skin sebum, and hydration and urine bacterial EV phylum flora in patients with acne vulgaris.

Keywords: *Lactobacillus plantarum* CJLP55; acne vulgaris; sebum; hydration; urine bacterial extracellular vesicles

1. Introduction

Acne vulgaris is characterized by excess sebum production and follicular hyperkeratinization [1,2]. Additionally, bacterial colonization with *Cutibacterium acnes* (*C. acnes*: formerly *Propionibacterium acnes* (*P. acnes*)), which uses sebum as the main nutrient to proliferate and releases proinflammatory cytokines, contributes to the development of acne vulgaris [1,2]; therefore, topical and systemic antibiotics, often in combination each other, have been a mainstay for the treatment of acne vulgaris [3,4]. However, the nonspecific chemical eradication of *C. acnes* coupled with the emergence of bacterial resistance has led to a new insight into the utilization of probiotics for skin health improvement [3,4].

Kimchi is a Korean traditional fermented food that is made of various vegetables such as Chinese cabbage, radish, garlic, ginger, and red pepper (chili). As a well-known healthy food in the world [5], kimchi contains various nutritional components such as vitamins, minerals, fiber, and phytochemicals. Moreover, nonpathogenic lactic acid producing bacteria (LAB), involved in the process of kimchi fermentation, are considered to be responsible for its health benefits [6]. *Lactobacillus*, the major LAB species in kimchi, modulates intestinal or systemic bacterial flora and systemically exerts anti-inflammatory effects that extend beyond the gut and may even affect the skin [7,8].

Of the various *Lactobacillus* strains isolated from kimchi, *Lactobacillus plantarum* (*L. plantarum*) CJLP55 tolerates low pH and a high bile salt concentration under gastrointestinal conditions [6], which is reflective of the stability and continuity of *L. plantarum* in the intestine [9]. *L. plantarum* CJLP55 has been reported to have immunomodulation effects with altered cytokine productions [10,11]. *L. plantarum* CJLP55 increases interleukin (IL)-12 levels and decreases IL-4 levels in ovalbumin-sensitized mouse splenocytes in vitro [10]. Orally administered *L. plantarum* CJLP55 decreases the production of IL-4 and IL-5 in lymph nodes and relieves atopic dermatitis (AD)-like skin lesions in NC/Nga mice [11]. In addition, ingested or topically applied *L. plantarum* modulates bacterial flora in the intestine as well as in the skin [12–14]. However, bacteria might not be disseminated in the body and barely reach directly to the skin [15]. Instead, extracellular vesicles (EV), the nanometer-sized vesicles released from bacteria, can move freely throughout body and indirectly affect the process of skin inflammation [15,16]. In fact, orally administered *L. plantarum* CJLP55-derived EV decreases IL-4 production and inflammation in the skin of an AD-like animal model [16], but little information is available on the systemic effect of *L. plantarum* CJLP55 on acne vulgaris, a prevalent inflammatory skin disease. In this randomized, double-blind, placebo-controlled pilot study, we investigated the dietary effect of *L. plantarum* CJLP55 supplement on clinical symptoms of acne vulgaris. In addition, we further determined its effects on skin condition-related factors, skin surface lipids, and urine bacterial extracellular vesicles (EV), a potential biomarker of the altered systemic bacterial flora [16], in patients with acne vulgaris.

2. Materials and Methods

2.1. Preparation of Probiotics and Placebo Products

L. plantarum CJLP55 (KCTC 11401BP, GenBank accession number GQ336971) was provided by CJ Foods R & D Center, CJ CheilJedang Corporation (Suwon, Republic of Korea). For administration, lyophilized *L. plantarum* CJLP55 at a dosage of 1.0×10^{10} colony-forming units (CFU) was mixed with maltodextrin and glucose anhydrocrystalline per one airtight alu-bag, and stored at 4 °C until administration. Placebo was prepared with a mixture of maltodextrin and glucose anhydrocrystalline only, of which the identical appearance and taste were confirmed by CJ Foods R & D Center. One alu bag of CJLP55 or placebo products provided 10 kcal and 2 g of carbohydrates.

2.2. Study Subjects

This study was approved by the Institutional Review Board of Kyung Hee University (Yongin, Republic of Korea) (KHSIRB 2015-013) and was conducted at Kyung Hee University Medical Center, Seoul, Korea in accordance with the Declaration of Helsinki. In addition, this study was registered at the Clinical Research Information Service (No. KCT0005401), a non-profit online registration system for clinical trials established by the Korea Centers for Disease Control and Prevention (KCDC). The KCDC joined the World Health Organization's International Clinical Trials Registry Platform to become the 11th member of its Primary Registry, and the full trial protocol for this study is available there. During July 2016, male and female patients with acne vulgaris were recruited. The inclusion criteria for this study were ages of 19–39 years, mild to moderate acne vulgaris and excess sebum on the face with normal body mass index (BMI) (18.5 to 24.9 kg/m^2). Specifically, diagnosis for mild-to-moderate acne vulgaris on the face (defined as acne grade of ≥ 2.0 to <4.0 with at least 15 inflammatory and/or non-inflammatory lesions but no more than 3 nodules on the face) was based on the criteria scale of Investigator's Global Assessment (IGA). Diagnosis for excess sebum on the face (defined as > 150 µg/cm^2

sebum on the forehead) was based on the criteria scale of sebum content for oily skin by SM815 sebumeter [17]. Exclusion criteria were acne treatment of any kind including artificial ultraviolet (UV) therapy in the previous 2 months, concomitant other skin disease or systemic illnesses, use of medication that could interfere with acne, or failure to maintain normal BMI with a steady weight ($\pm$ 4 kg) throughout the study. Written informed consent forms were obtained from each subject prior to enrollment of the study.

2.3. Study Design

This was a double-blind, placebo-controlled, parallel-group clinical study. Based on our previous study [18], a sample size of 9 per group was determined to provide > 80% power at a 2-tailed α = 0.05 for a difference of 62% with a SD of 33% in levels of triglycerides (TG), a major skin surface lipid [19]. In consideration of a 10% dropout rate and 70% compliance, we aimed to recruit 15 subjects per group. For the current study 30 subjects were randomly assigned to either the CJLP55 (n = 15, 7 males and 8 females) or the placebo (n = 15, 5 males and 10 females) groups (Figure S1). Upon confirmation of the inclusion and exclusion criteria, randomization was performed by a researcher who was not involved in this study. Eligible subjects were randomly allocated into the two arms of the study in a 1:1 ratio, according to randomly permuted blocks within the strata of two assignments, the CJLP55 and placebo groups. The allocation sequence was not available to the subjects, the dermatologist, or any of the researchers involved in this study.

Subjects were instructed to ingest one bag of placebo product (placebo group) or 1.0×10^{10} CFU of *L. plantarum* CJLP55 (CJLP55 group) daily for 12 weeks from July 2015 to October 2015. Subjects were also instructed to refrain from consuming any other probiotic products 2 weeks before the beginning of the study and until the study had finished. In addition, to avoid the influence of other factors on skin, subjects were instructed to maintain their usual cleaning habits and moisturizing products throughout the study. However, to ensure the accuracy of the corneometer, sebumeter, and skin-pH meter, subjects were instructed to avoid use of any moisturizing products on the days of dermatological assessment. During the 12-week study period, compliance was monitored weekly via regular telephone interviews and all adverse events, including gastrointestinal discomfort, were recorded. Anthropometry measures, blood samples for hematology and biochemistry analyses, and a routine urine analysis with a pregnancy test (females only) were performed at 0 week (baseline) and 12 weeks. The primary outcome of this study was clinical assessment of facial acne with the acne-related lesion count and acne severity. All other measures were secondary outcomes.

2.4. Clinical Assessment of Facial Acne

Photographs of the facial area were taken under standardized conditions at baseline and 12 weeks. Clinical assessment of facial acne was performed at baseline and 12 weeks by the same dermatologist who was blind to the group assignment of subjects. The dermatologist counted acne-related lesions (noninflammatory lesions: open and closed comedones; inflammatory lesions: papule, pustule and nodule) and evaluated acne severity according to IGA scale at baseline and 12 weeks.

2.5. Measurement of Skin Condition-Related Factors

Under the standardized room conditions of 22–24 °C and 55–60% humidity, subjects were given at least 30 min to relax so that the skin condition could equilibrate. Skin sebum, hydration, and pH of forehead were measured using sebumeter (SM815), corneometer (CM825), and skin pH-meter (PH905) (all from Courage-Khazaka, Cologne, Germany) at baseline and 12 weeks, as described previously [18]. Skin sebum, hydration and pH values were the average of 5 determinations after equilibrium was attained. The temperature and humidity of the probe were also recorded.

2.6. Analysis of Skin Surface Lipids

The skin surface of the forehead was stripped using 10 tape strips (14 mm D-SQUAME® Tape; Cu-Derm Corporation, Dallas, TX, USA) at baseline and 12 weeks. All tape strips from each subject were combined and stored at −20 °C until further processing. After corneocytes were removed from the tape by sonication in methanol, the lipids were extracted and fractionated by high performance thin layer chromatography (HPTLC), as descried previously [18]. The fractions containing TG, free fatty acids (FFA), cholesterol (Chol), total ceramides (Cer), and ceramide 2 (Cer2) that comigrated with respective standards were scanned at 420 nm with a TLC III scanner (CAMAG, Muttenz, Switzerland) and were quantified by calibration curves using various concentrations of external standards of each lipid. Levels of total skin surface lipids including TG, FFA, Chol, total Cer, and Cer2 were expressed as ng/µg protein.

2.7. Metagenomic DNA Analysis of Bacterial EV Phylum in Urine

As an indirect method to analyze the alteration of systemic bacterial flora [16], metagenomic analysis of bacterial DNA was performed in urine EV [20]. Urine samples from subjects who gave written informed consent for urine sample collection ($n = 9$ in the CJLP55 group, 4 males and 5 females; $n = 12$ in placebo group, 4 males and 8 females) were collected at baseline and 12 weeks. EV were purified from urine by multiple ultracentrifugations, and sequentially, bacterial DNA was extracted according to the manufacturer's instructions (PowerSoil DNA Isolation Kit; MO Bio, Carlsbad, CA, USA), as described previously [20,21]. The extracted bacterial DNA was quantified by the QIAxpert system (QIAGEN, Hilden, Germany). The polymerase chain reaction (PCR) for variable 3 (V3) and variable 4 (V4) regions of 16S ribosomal RNA (rRNA) genes, which can be sequenced to discriminate the bacterial phylogenetic identification [15], was performed using the primer set of 16S_V3 forward primer (5′-TCGTCGGCAGCGTCAGA

TGTGTATAAGAGACAGCCTACGGGNGGCWGCAG-3′) (50mer) and 16S_V4 reverse primer (5′-GTCTCGTGGGCTCGGAGATGTGTATAAGAGACAGGACTACHVGGGT ATCTAATCC-3′) (55 mer) [20,21]. The PCR products were used for the construction of 16S ribosomal DNA gene libraries following the MiSeq System guidelines (Illumina Inc., San Diego, CA, USA). The 16S rRNA gene libraries for each sample were quantified using QIAxpert (QIAGEN, Hilden, Germany), pooled at the equimolar ratio, and used for pyrosequencing with the MiSeq System (Illumina Inc., San Diego, CA, USA) according to the manufacturer's recommendations. The operational taxonomy unit (OTU) was analyzed using UCLUST and USEARCH, with phylogenetic classification performed using QIIME based on the 16sRNA sequence database of GreenGenes 8.15.13. Based on >75% similarity, all sequences were classified for phylum profile. The proportion of each bacterial EV phylum was evaluated as a percentage of the total phylum from the sequence reads, and the *Bacteroidetes*:*Firmicutes* (B/F) ratio was then determined.

2.8. Statistics

All statistical analyses were carried out on a per-protocol basis with SPSS 21.0 for Windows (SPSS Inc., Chicago, IL, USA). Values are presented as mean ± SEM. Differences from baseline within each group were analyzed by paired Student's *t* test for normally distributed variables and Wilcoxon's signed rank test for non-normally distributed variables. Differences between groups were analyzed by unpaired Student's *t* test for normally distributed variables and Mann–Whitney *U* test for non-normally distributed variables. Raw data are available as Supplementary Materials (File S1: Raw research data).

The difference in observed numbers of male and female subjects from theoretical frequencies between groups was analyzed by a chi-square test (Table 1). Univariate analysis was performed to determine the influence of sex on percent changes of acne lesion count and acne grade over 12 weeks using a generalized linear model for inflammatory lesion count (ILC) or generalized linear mixed model for total lesion count (TLC) and acne grade (Table 2). In this analysis, acne lesion count or acne grade was used as a dependent variable and sex (females coded as 0; males coded as 1) was used as an independent variable. Principle component analysis (PCA) was performed to determine a separation of percent changes of bacterial EV phylum profiles over 12 weeks between groups. Pearson's correlation analysis was performed to determine correlations of percent changes of TG levels with those of the phyla *Proteobacteria* or *Firmicutes* proportions over 12 weeks in subjects of the CJLP55 group. Two-sided p values <0.05 were considered significant.

Table 1. Baseline characteristics by groups.

Characteristics	CJLP55 Group (n = 14)	Placebo Group (n = 14)	p Value
Sex (male/female)	7/7	5/9	0.445 [2]
Age (years)	24.29 ± 0.73 [1]	23.86 ± 0.80	0.695 [3]
Body mass index (kg/m^2)	20.74 ± 0.64	21.39 ± 0.55	0.401 [3]
Inflammatory lesion count (ILC)	17.79 ± 3.19	19.64 ± 3.81	0.874 [3]
Total lesion count (TLC)	72.93 ± 10.11	98.14 ± 15.75	0.189 [3]
Acne grade	3.21 ± 0.19	3.14 ± 0.14	0.734 [3]

[1] Mean ± SEM (all such values). [2] p value for difference in observed numbers of men and women from theoretical frequencies between the CJLP55 and placebo groups by chi-square test. [3] p value for difference between the CJLP55 and placebo groups by Student's t test (age, TLC) or Mann–Whitney U test (body mass index, ILC, acne grade).

Table 2. Altered acne lesion count and acne grade by groups.

	CJLP55 Group (n = 14, 7 Male/7 female)			Placebo Group (n = 14, 5 Male/9 Female)			% Change between Groups (95%CI)	p Value [4]
	Baseline	Week12	% Change	Baseline	Week12	% Change		
Inflammatory lesion count (ILC)								
Male subjects	24.86 ± 4.74	6.71 ± 1.36 **[1,2]	−72.99 ± 14.86	20.40 ± 6.78	15.40 ± 7.36	−24.51 ± 15.88	−48.48 (−97.84 to 0.89) [3]	0.054
Female subjects	10.71 ± 2.24	4.57 ± 1.54 *[2]	−57.33 ± 19.79	19.22 ± 4.89	14.00 ± 2.13	−27.17 ± 16.31	−30.17 (−84.67 to 24.34)	0.255
All subjects	17.79 ± 3.19	5.64 ± 1.03 ***[2]	−68.27 ± 14.83	19.64 ± 3.81	14.50 ± 2.79 *[2]	−26.18 ± 11.43	−42.09 (−80.57 to −3.61)	0.033
Total lesion count (TLC)								
Male subjects	98.00 ± 13.73	39.57 ± 7.00 **[2]	−59.62 ± 9.48	95.40 ± 28.62	78.00 ± 20.52	−18.24 ± 10.83	−41.38 (−73.67 to −9.10)	0.017
Female subjects	47.86 ± 6.67	27.57 ± 6.15 **[2]	−42.39 ± 10.41	99.67 ± 19.97	102.78 ± 20.02	3.12 ± 18.41	−45.51 (−94.67 to 3.65)	0.031
All subjects	72.93 ± 10.11	33.57 ± 4.77 ***[2]	−53.97 ± 10.04	98.14 ± 15.75	93.93 ± 14.69	−4.29 ± 12.58	−49.67 (−82.76 to −16.59)	0.002
Acne grade								
Male subjects	3.71 ± 0.18	2.86 ± 0.26 *[2]	−23.08 ± 7.02	3.20 ± 0.20	3.40 ± 0.24	6.25 ± 6.25	−29.33 (−51.36 to −7.29)	0.030
Female subjects	2.71 ± 0.18	2.00 ± 0.31 *[2]	−26.31 ± 6.79	3.11 ± 0.20	3.22 ± 0.15	3.57 ± 8.37	−29.89 (−54.04 to −5.74)	0.012
All subjects	3.21 ± 0.19	2.43 ± 0.23 **[2]	−24.44 ± 4.81	3.14 ± 0.14	3.29 ± 0.13	4.55 ± 5.64	−28.99 (−44.23 to −13.75)	0.009

CI, confidence interval. [1] Mean ± SEM for all such values. [2] p value for difference from baseline within CJLP55 or placebo groups by paired t test (total lesion count) or Wilcoxon signed rank test (inflammatory lesion count, acne grade) (* $p < 0.05$, ** $p < 0.01$, *** $p < 0.001$). [3] Mean (95% CI) for all such values. [4] p value for difference of percent change between the CJLP55 and placebo groups by Student's t test (inflammatory lesion count) or Mann–Whitney U test (total lesion count, acne grade).

3. Results

3.1. Study Subjects

Initially, 30 subjects were randomly assigned to either the CJLP55 (n = 15, 7 males and 8 females) or the placebo (n = 15, 5 males and 10 females) groups (Figure S1). During the 12-week study period, one female in the CJLP55 group dropped out for personal reasons, and one female in the placebo group who failed to maintain a steady weight and became underweight (BMI < 18.5) was excluded from analyses. Therefore, data were analyzed for 14 subjects (7 males and 7 females) in the CJLP55 group and 14 subjects (5 males and 9 females) in the placebo group (File S2: Consort 2010 checklist).

Enrolled numbers of male and female subjects (5 males and 9 females) in the placebo group were not significantly different from theoretical frequencies (7 males and 7 females) (p = 0.445) (Table 1). There were no statistical differences in ages and BMI between groups. The acne grade at baseline matched well to dermatologic assessment of ILC and TLC, all of which were not significantly different between groups. Tolerability of CJLP55 and placebo products was excellent in both groups. Changes of usual cleaning and moisturizing products and adverse events including gastrointestinal discomfort were not reported during the study period. Hematology, serum, and urine analyses of all subjects were normal at baseline and 12 weeks.

3.2. Clinical Assessment of Facial Acne

Compared to baseline, ILC, TLC, and acne grade for male and female subjects in the CJLP55 group were significantly decreased at 12 weeks; therefore, acne lesion counts and acne grade for all subjects (male and female combined) were significantly decreased at 12 weeks in the CJLP55 group (Table 2). In placebo group, ILC, TLC, and acne grade for male and female subjects remained unchanged, but ILC for all subjects was significantly decreased during the 12-week study period.

When the differences in percent changes over 12 weeks were compared between groups, the percent change of ILC for male and female subjects between CJLP55 and placebo groups was not significantly different, but it was significantly decreased by 42.09% (95% confidence interval (CI), −80.57 to −3.61; $p = 0.033$) for all subjects in the CJLP55 group compared with the placebo group. The TLC was significantly decreased by 41.38% (95% CI, −73.67 to −9.10; $p = 0.017$) for male, 45.51% (95% CI, −94.67 to 3.65; $p = 0.031$) for female, and 49.67% (95% CI, −82.76 to −16.59; $p = 0.002$) for all subjects in the CJLP55 group. In parallel, the acne grade in the CJLP55 group was also decreased significantly by 29.33% (95% CI, −51.36 to −7.29 to; $p = 0.030$) for male, 29.89% (95% CI, −54.04 to −5.74; $p = 0.012$) for female, and 28.99% (95% CI, −44.23 to −13.75; $p = 0.009$) for all subjects, compared with the placebo group.

Despite sex imbalance in the placebo group ($n = 14$, 5 males and 9 females), sex was not a significant discriminating factor for the percent changes of acne lesion count and acne grade over 12 weeks (CJLP55 group: $p = 0.539$ in ILC, $p = 0.244$ in TLC, $p = 0.383$ in acne grade; placebo group: $p = 0.917$ in ILC, $p = 0.699$ in TLC, $p = 1.000$ in acne grade) in each group. Representative photographs of acne improvement in the CJLP55 group are shown in Figure 1.

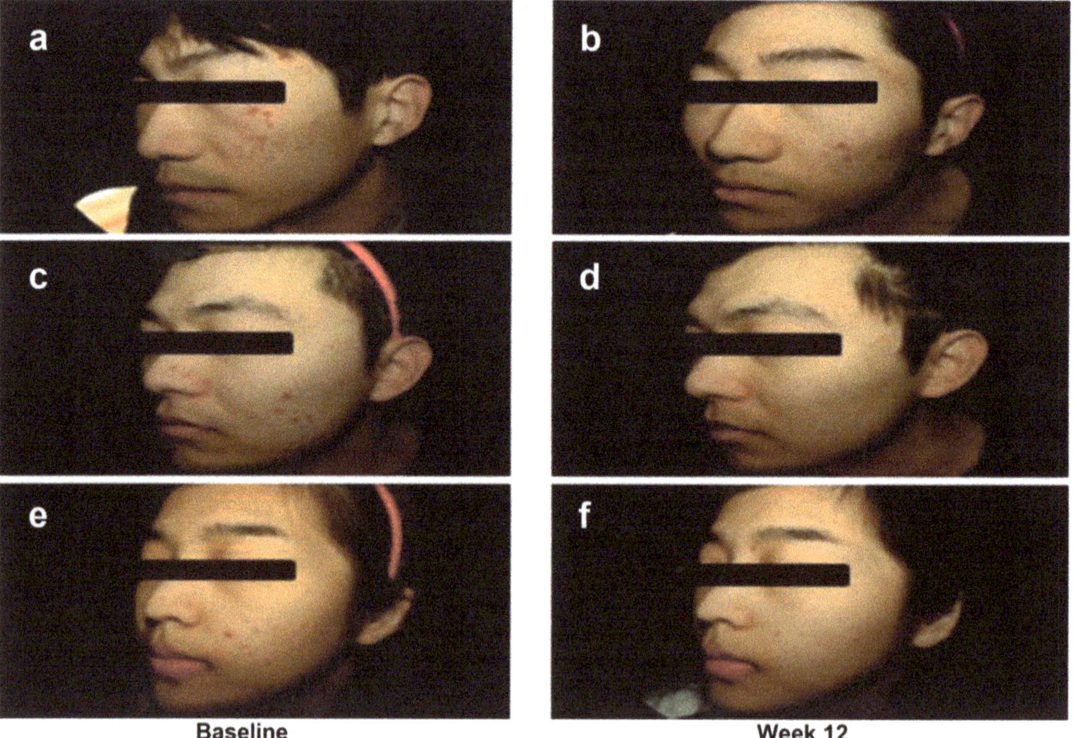

Figure 1. Photographs of acne improvement in subjects A (**a**,**b**), B (**c**,**d**), and C (**e**,**f**) in the CJLP55 group at baseline and week 12, respectively.

3.3. Measurement of Skin Condition-Related Factors

There was no sex effect on acne lesion counts and acne grade. Therefore, the data of male and female subjects for skin condition were combined as all subjects (Table 3). Compared to baseline, the skin pH of the CJLP55 and placebo groups remained unchanged at 12 weeks, with no difference in percent change between these two groups. Moreover, skin hydration of these two groups remained unchanged at 12 weeks compared to baseline; however, the positive percent change of the CJLP55 group and the negative percent change of the placebo group over 12 weeks resulted in a significant difference between these two groups. Therefore, the skin hydration in the CJLP55 group was increased by 14.52% (95% CI, 6.47 to 22.57; $p = 0.003$) compared to the placebo group. Notably, sebum content of the CJLP55 group was significantly decreased whereas that of the placebo group remained unchanged over 12 weeks, resulting in a 16.27% decrease (95% CI, -32.28 to -0.26; $p = 0.005$) of sebum content in the CJLP55 group, compared to the placebo group.

Table 3. Altered skin sebum, hydration, and pH by groups.

	CJLP55 Group (*n* = 14, 7 Male/7 Female)			Placebo Group (*n* = 14, 5 Male/9 Female)			% Change between Groups (95%CI)	*p* Value [4]
	Baseline	Week12	% Change	Baseline	Week12	% Change		
Skin sebum (μg/cm^2)								
All subjects	178.82 ± 12.98	147.11 ± 16.40 [*,1,2]	-17.74 ± 6.04	202.36 ± 15.28	199.39 ± 16.44	-1.47 ± 4.92	-16.27 (-32.28 to -0.26) [3]	0.005
Skin hydration (capacitance in au.)								
All subjects	73.47 ± 1.53	77.64 ± 1.99	5.67 ± 2.95	72.67 ± 1.73	66.24 ± 1.78	-8.85 ± 2.58	14.52 (6.47 to 22.57)	0.003
Skin pH								
All subjects	6.67 ± 0.15	5.40 ± 0.10	-18.93 ± 2.70	6.69 ± 0.11	5.61 ± 0.11	-16.17 ± 2.42	-2.76 (-10.21 to 4.69)	0.453

CI, confidence interval; au, arbitrary unit. [1] Mean ± SEM for all such values. [2] *p* value for difference from baseline within CJLP55 or placebo groups by paired *t* test (skin sebum, skin hydration) or Wilcoxon signed rank test (skin pH) (* $p < 0.05$). [3] Mean (95% CI) for all such values. [4] *p* value for difference of percent change between the CJLP55 and placebo groups by Student's *t* test (skin pH) or Mann–Whitney U test (skin sebum, skin hydration).

3.4. Analysis of Skin Surface Lipids

In HPTLC analysis, the level of total skin surface lipids in CJLP55 and placebo groups remained unchanged over 12 weeks (CJLP55 group: 209.46 ± 17.30 ng/μg protein at baseline, 175.97 ± 11.66 ng/μg protein at 12 weeks; placebo group: 232.52 ± 13.24 ng/μg protein at baseline, 231.64 ± 18.13 ng/μg protein at 12 weeks) (Figure 2), which was not consistent with the decreased sebum content in the CJLP55 group over 12 weeks (Table 3). Further fractionation of total skin surface lipids revealed that TG, FFA, Chol, total Cer and Cer2 were separated. TG were the most abundant and comprised > 57% of the total skin surface lipids of the forehead in subjects with acne in the CJLP55 and placebo groups, as reported previously [19]. FFA, Chol, and total Cer including Cer2 and other Cer species accounted for < 17% of total skin surface lipids. Levels of TG and FFA were significantly decreased in the CJLP55 group whereas the level of FFA only was significantly decreased in the placebo group over 12 weeks: The level of TG remained unchanged in the placebo group over 12 weeks, thereby resulting in a 28.44% decrease of TG specifically (95% CI, -54.08 to -2.79; $p = 0.031$), but not FFA, in the CJLP55 group compared to the placebo group. Although the level of total Cer remained unchanged in CJLP55 and placebo groups, the level of Cer2, the major Cer species of human epidermis [22], was increased over 12 weeks in the CJLP55 group. Other Cer species were barely detectable with low band intensities, of which alterations between CJLP55 and placebo groups were not apparent.

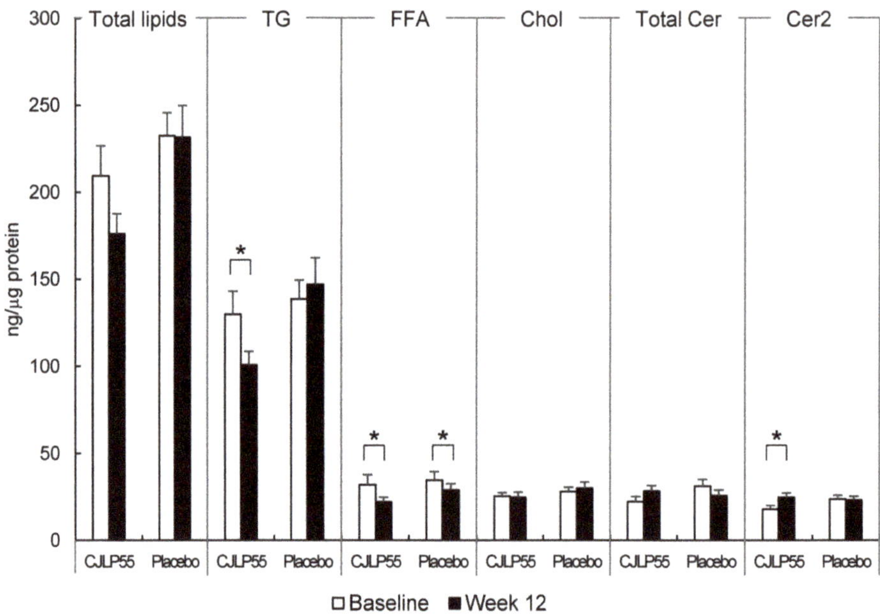

Figure 2. Altered levels of skin surface lipids in CJLP55 and placebo groups. Values are mean ± SEM (CJLP55 group: $n = 14$; placebo group: $n = 14$). Differences between baseline (white bars) and week 12 (black bars) in lipid fractions of the CJLP55 or placebo groups were evaluated using paired *t*-test test (Total lipids, Chol, Cer2) or Wilcoxon signed rank test (TG, FFA, Total Cer) (* $p < 0.05$). TG, triglycerides; FFA, free fatty acids; Chol, cholesterol; Cer, ceramides; Cer2, ceramide 2.

3.5. Metagenomic DNA Analysis of Bacterial EV Phylum in Urine

EV, the nanometer-sized vesicles released from archaea, bacterial, and eukaryotic cells, have been observed in body fluids such as serum and urine [15,16,20,21]. DNA phylum profiles of bacterial origin are similar in EV from either serum or urine [16]. Therefore, metagenomic analysis of urine bacterial EV can indirectly reflect the alteration of systemic bacterial flora. When DNA phylum profiles of urine bacterial EV were analyzed at levels of >0.5% in individual subjects of CJLP55 and placebo groups, the phyla *Proteobacteria, Bacteroidetes, Firmicutes, Actinobacteria, Verrucomicrobia*, and *Cyanobacteria* (in order from high to low proportions) were present. When the mean values of each phylum were assessed as percent of total proportion, *Proteobacteria* was the predominant phylum, comprising >35% of all bacteria, and *Cyanobacteria* was the least present at baseline of these two groups (Figure 3a,b).

Compared to baseline, the proportion of *Proteobacteria* remained unchanged at 12 weeks in the CJLP55 and placebo groups, but the negative percent change of the CJLP55 group and the positive percent change of the placebo group over 12 weeks resulted in a significant difference between these two groups. The proportion of *Firmicutes* remained unchanged in the CJLP55 group, but it was decreased in the placebo group at 12 weeks; these alterations resulted in a 40.92% decrease (95% CI, −80.78 to −1.05; $p = 0.045$) of *Proteobacteria* and a 59.66% increase (95% CI, 4.18 to 115.14; $p = 0.036$) of *Firmicutes* over 12 weeks in the CJLP55 group compared to the placebo group (Figure 3a,b). Compared to baseline, the proportion of *Verrucomicrobia* was significantly increased in the placebo group; however, its similar trend noted at 12 weeks in the CJLP55 group resulted in no difference of percent change over 12 weeks between these two groups. The proportion of other phyla, including *Bacteroidetes, Actinobacteria, Cyanobacteria* and others (phyla present at less than 0.5%), remained unchanged over 12 weeks in the CJLP55 and placebo groups, thereby resulting in no difference in percent change of these phyla over 12 weeks between these two groups.

However, despite no altered proportion of *Bacteroidetes* over 12 weeks, decreased proportion of *Firmicutes* induced a shift toward a significant increased *Bacteroidetes:Firmicutes* (B/F) ratio, a relevant biomarker of bacterial dysbiosis [16], in the placebo group, whereas its ratio remained unchanged in the CJLP55 group with no altered proportion of these two phyla over 12 weeks, therefore resulting in a 141.16% decrease (95% CI, −267.41 to −14.91; p = 0.030) of B/F ratio over 12 weeks in the CJLP55 group compared to the placebo group. The dysbiosis of urine bacterial EV phylum in the placebo group was further evident by PCA analysis. PCA scores for the CJLP55 group (filled circle in Figure 3c) were relatively separated from those for the placebo group (open circle in Figure 3c), which implies that there is a distinction of percent changes of urine bacterial EV phylum profiles between CJLP55 and placebo groups.

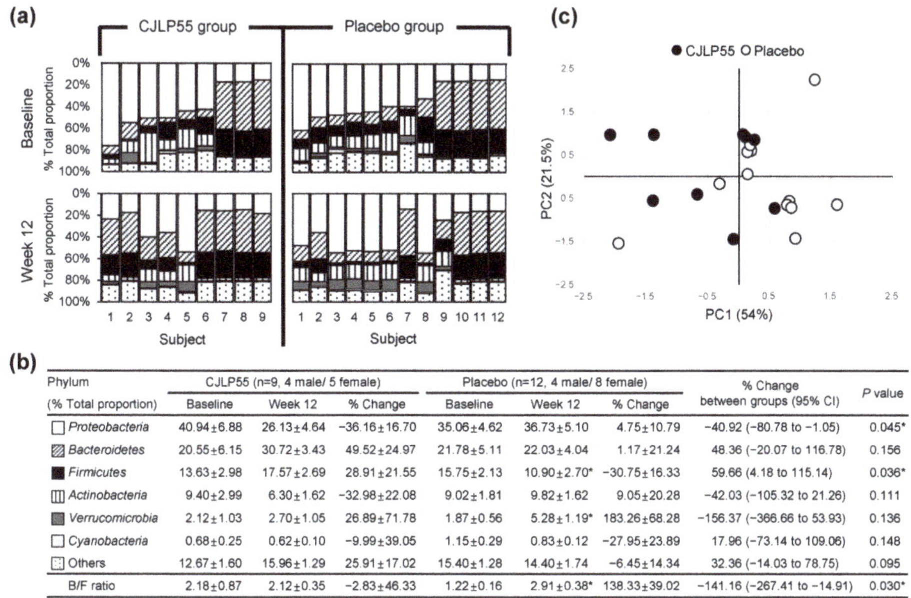

Phylum (% Total proportion)	CJLP55 (n=9, 4 male/ 5 female)			Placebo (n=12, 4 male/ 8 female)			% Change between groups (95% CI)	P value
	Baseline	Week 12	% Change	Baseline	Week 12	% Change		
Proteobacteria	40.94±6.88	26.13±4.64	−36.16±16.70	35.06±4.62	36.73±5.10	4.75±10.79	−40.92 (−80.78 to −1.05)	0.045*
Bacteroidetes	20.55±6.15	30.72±3.43	49.52±24.97	21.78±5.11	22.03±4.04	1.17±21.24	48.36 (−20.07 to 116.78)	0.156
Firmicutes	13.63±2.98	17.57±2.69	28.91±21.55	15.75±2.13	10.90±2.70*	−30.75±16.33	59.66 (4.18 to 115.14)	0.036*
Actinobacteria	9.40±2.99	6.30±1.62	−32.98±22.08	9.02±1.81	9.82±1.62	9.05±20.28	−42.03 (−105.32 to 21.26)	0.111
Verrucomicrobia	2.12±1.03	2.70±1.05	26.89±71.78	1.87±0.56	5.28±1.19*	183.26±68.28	−156.37 (−366.66 to 53.93)	0.136
Cyanobacteria	0.68±0.25	0.62±0.10	−9.99±39.05	1.15±0.29	0.83±0.12	−27.95±23.89	17.96 (−73.14 to 109.06)	0.148
Others	12.67±1.60	15.96±1.29	25.91±17.02	15.40±1.28	14.40±1.74	−6.45±14.34	32.36 (−14.03 to 78.75)	0.095
B/F ratio	2.18±0.87	2.12±0.35	−2.83±46.33	1.22±0.16	2.91±0.38*	138.33±39.02	−141.16 (−267.41 to −14.91)	0.030*

Figure 3. Altered abundance of bacterial extracellular vesicle (EV) phyla in urine of CJLP55 and placebo groups. (**a**) Altered % total proportion of urine bacterial EV phyla from baseline to 12 weeks in individual subjects of groups. The horizontal axis displays the subject number. (**b**) Altered % total proportion of urine bacterial EV phyla over 12 weeks in groups. Phyla present at less than 0.5% are noted as "others." B/F ratio: *Bacteroidetes* to *Firmicutes* ratio. All values in (**b**) are mean ± SEM (CJLP55 group: n = 9; placebo group: n = 12). Differences from baseline within CJLP55 or placebo groups in (**b**) were determined by Wilcoxon signed rank test Differences of percent change between CJLP55 and placebo groups in (**b**) were determined by unpaired Student's *t* test (*Proteobacteria, Bacteroidetes, Firmicutes, Verrucomicrobia, B/F ratio*) or Mann–Whitney *U* test (*Actinobacteria, Cyanobacteria, Others*) (* p < 0.05). (**c**) Principle component analysis (PCA) score on percent changes of bacterial EV phyla over 12 weeks in groups.

3.6. Correlations between Percent Changes in TG Levels and in Proteobacteria and Firmicutes Proportions

The level of TG, the major skin surface lipid, was selectively decreased in the CJLP55 group over 12 weeks. In parallel, compared to the placebo group, the percent changes of *Proteobacteria* and *Firmicutes* were significantly altered in the CJLP55 group over 12 weeks. Therefore, the correlations between percent changes in TG level and those in the phyla *Proteobacteria* and *Firmicutes* proportions over 12 weeks were further determined in individual subjects of the CJLP55 group. The decreased percent change of TG level was positively correlated with that of the *Proteobacteria* proportion (r = 0.758, p = 0.018), whereas it was negatively correlated with increased percent change of the *Firmicutes* proportion

($r = -0.671$, $p = 0.048$) (Figure 4). These results indicate that the selectively decreased TG level and decreased *Proteobacteria* with a concomitant increase in *Firmicutes* in urine were metabolic features that led to decreases in sebum content and acne lesion count, ultimately improving acne grade in the CJLP55 group.

(a)

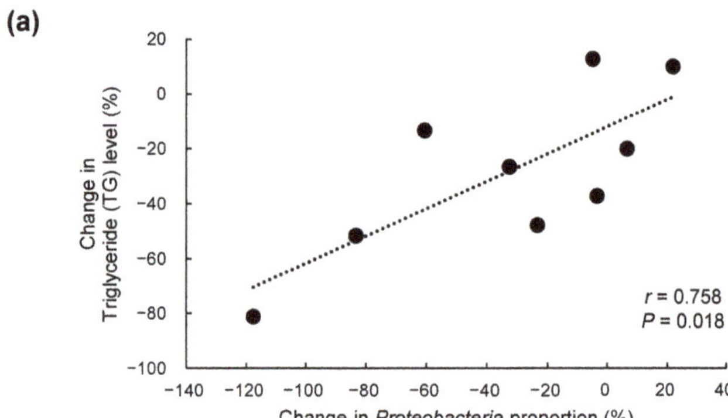

(b)

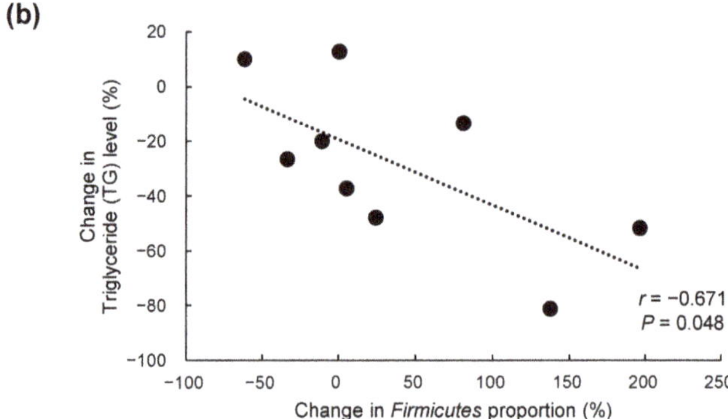

Figure 4. Correlations between percent changes in triglyceride levels and in (**a**) *Proteobacteria* and (**b**) *Firmicutes* proportions over 12 weeks in individual subjects of the CJLP55 group ($n = 9$).

4. Discussion

Lactobacillus, a well-characterized genus in LAB, is beneficial for skin health improvement [7,8]. Specifically, various strains of *L. plantarum* have been reported to improve acne vulgaris with broad anti-inflammatory and anti-pathogenic bacterial activities as well as with modulation of bacterial flora in the skin [12,14,23]. Topically applied *L. plantarum* on the face of patients with mild acne vulgaris reduces acne lesion size and erythema [12]. Treatment with *L. plantarum* THG-10 in vitro decreases the growth of *C. acnes* and nitric oxide production, which is induced by the inflammation process [23]. Furthermore, results from topically applied studies in female subjects indicate that *L. plantarum*-GMNL6 reduces erythema and the proportion of *Propionibacterium* genus in the face [14]. In this study, we demonstrated that *L. plantarum* CJLP55 supplement decreased the clinical severity of acne. Compared to baseline, ILC, TLC, and acne grade were significantly decreased in the CJLP55 group over 12 weeks. For the placebo group, ILC was decreased, but TLC and acne grade remained unchanged over the 12 weeks.

In addition to the decreased clinical severity of acne, sebum content also decreased and skin hydration improved in the CJLP55 group, but the levels of FFA significantly decreased during the 12 weeks for both the CJLP55 and placebo groups. During this period, a decrease in ILC was also detected for both groups, suggesting that the decreases in FFA and ILC might be caused by ingestion of maltodextrin and glucose anhydrocrystalline. These are the common vehicle in the CJLP55 and placebo products. High intake of hyperglycemic carbohydrates is known to aggravate acne based on the hyperinsulinemia and androgen stimulated sebum production [24]. Maltodextrin and glucose anhydrocrystalline consist of pure glucose molecules. Since the supplemented amounts of these in the CJLP55 and placebo products are low, only 2 g carbohydrate (10 kcal), the glycemic load will be very low. They are therefore not likely to affect the insulin levels. Further studies are required to determine the effects of small but long-term carbohydrate intake on ILC and FFA of skin surface lipids in acne patients. TG is the major sebum lipid [19], whereas Cer2 is the major Cer species that maintains lamellar integrity of the epidermal barrier against water loss [22]. Notably, the levels of TG decreased but the level of Cer2 increased during the 12 weeks for the CJLP55 group. These results, coupled with the reports of increased sebum production and impaired epidermal barrier with reduced Cer levels in acne patients [25], indicate that *L. plantarum* CJLP55 supplement improved acne vulgaris with selectively decreased TG of sebum, and increased Cer2 and skin hydration.

The beneficial effects of *L. plantarum* CJLP55 supplement in the skin were further accompanied by altered bacterial EV phylum in urine of the CJLP55 group. In addition to providing diverse flora in the intestine, bacteria produce EV, which have been observed in the blood and urine [16,26,27]. Over 12 weeks, there was a percent decrease of *Proteobacteria* proportion in the CJLP55 group compared to the placebo group. Since not all bacteria produce EV [28], bacterial EV phylum flora in urine could be different from intestinal or skin bacterial flora, but the majority of acne patients have an unusual presence of several genera including *Klebsiella* and *Enterobacter*, which are affiliated with phylum *Proteobacteria* in intestinal bacterial flora [29]. Furthermore, a decreased proportion of *Bifidobacterium* and *Lactobacillus*, two major beneficial anaerobic bacterial genera [30], as well as an increased proportion of *Staphylococcus aureus* (*S. aureus*), are frequently reported in intestinal or skin bacterial flora of acne patients [25,29]. There was a percent increase of *Firmicutes*, with which *Lactobacillus* genus is affiliated [30], over 12 weeks in the CJLP55 group compared to the placebo group, but the pathogenic bacteria species of *S. aureus* is also affiliated with phylum *Firmicutes* [29]. On the other hand, *Actinobacteria*, with which *Bifidobacterium* genus is affiliated [30], remained unchanged over 12 weeks in the CJLP55 group. In fact, no significant alterations of bacterial EV flora at the identified genus or species levels were observed over 12 weeks between the CJLP55 and placebo groups. Urine samples were obtained from a limited number of subjects, which might explain no alteration of bacterial EV genus or species profiles in this study. Modulation of bacterial EV profiles should be analyzed in depth with a larger number of subjects in future studies. However, compared to the place group, there was a significant decrease in the B/F ratio over 12 weeks in the CJLP55 group, and PCA analysis further confirmed a distinction in the percent changes of urine bacterial EV phylum profiles over 12 weeks between CJLP55 and placebo groups. Moreover, the decreased percent changes of TG levels were correlated with those of *Proteobacteria* proportions and a concomitantly increased percent change of *Firmicutes* proportion in urine EV of the CJLP55 group. Recent studies reported that *Proteobacteria* is predominant in the urine EV phylum of pediatric patients with AD, whereas *Firmicutes* is predominant in that of normal subjects [16], and the B/F ratio is higher in acne patients than in healthy controls [31]. Taken together, these results suggest that ingested *L. plantarum* CJLP55 is likely to restore the bacterial EV phylum profiles, specifically decreased *Proteobacteria* and increased *Firmicutes* proportions systemically, which are altered in inflammatory skin diseases such as AD or acne [16,29].

Regarding dietary or supplement intervention for acne improvement, insulin or insulin-like growth factor-1 (IGF-1)-induced activation of the phosphatidylinositol 3-kinase (PI3K)/Akt/mechanistic target of rapamycin complex 1 (mTORC1) (previously referred as mammalian target of rapamycin complex 1) signaling pathway has been recognized to play a fundamental role in the regulation of sebocyte lipogenesis and proliferation [32,33]. High intake of hyperglycemic carbohydrates as well as milk or enriched branched-chain amino acids (BCAA) induces postprandial rises of insulin and/or IGF-1 levels in serum [32,33]. As a well-established factor in acne pathogenesis, IFG-1 increases the mRNA and protein expressions of sterol response element binding protein-1 (SREBP-1), the key lipogenic factor which is the down-stream target of PI3K/Akt/mTORC1 signaling in sebocytes [32–34]. Therefore, low glycemic load in the diet has been recommended for clinical improvement and reduced sebum and inflammation in acne patients [24,35]. Notably, probiotic supplement intervention with *L. plantarum* Ln4 also decreases insulin and IGF-1 expression in the adipose tissue of mice [36] and pooled probiotic supplementation with *L. rhamnosus*, *L. acidophilus*, and *Bifidobacterium bifidumi* improves insulin sensitivity in the adipose tissue of mice, in parallel with the modulation of intestinal bacterial flora [37]. Together with these prior studies [36,37], the beneficial effect of *L. plantarum* CJLP55 supplement intervention on the skin of acne patients in this study could be explained by the inhibited effect on IFG-1-induced activation of PI3K/Akt/mTORC1 signaling. In fact, *L. rhamnosus* SP1 supplement decreases clinical severity and IGF-1 mRNA expression in the skin of acne patients [38]. However, the beneficial effects of various strains of *L. plantarum* on decreased acne severity and sebum production have been reported mostly from topically applied and in vitro studies [12,14,23], and little information is available on the relationship between ingested *L. plantarum* and altered levels of insulin, IFG-1 and BCAA in either the serum or skin of acne patients. Alternatively, *L. plantarum* CJLP55 has been reported to have immunomodulation effects with altered cytokine production [10,11]. When coupled with prior studies reported that PI3K/Akt/mTORC1 signaling pathway is induced by specific receptors of various stimuli including insulin and IGF-1 [32,33] as well as cytokines [39–41], these results suggest that after *L. plantarum* CJLP55 is ingested, systemic immunomodulation with altered cytokine production is likely to occur, ultimately inducing beneficial effects on skin such as the selectively decreased TG of sebum and increased Cer2 and skin hydration.

To explain the beneficial effect of ingested probiotics for skin health improvement, modulation of intestinal bacterial flora and their released metabolites with increased intestinal permeability were adopted for systemic immunomodulation based on gut-brain-skin axis [42]. Alternatively, from recent studies reported that bacteria produce EV, which can move freely throughout body, bacterial EV in serum or urine is implicated as a systemic bacterial factor for immunomodulation, such as altering cytokine levels [16,26,27]. Similar to orally administrated *L. plantarum* CJLP55, which decreases the production of T-helper (Th)-2 cytokines of adaptive immunity, such as IL-4 and IL-5 in lymph nodes and relieves AD-like skin lesions in NC/Nga mice [11], oral administration of *L. plantarum* CJLP55-derived EV for 25 days decreases IL-4 production and inflammation in the skin of an AD-like animal model [16]. Oral administration of *L. plantarum* CJLP133, another *L. plantarum* strain of kimchi, decreases IL-4 level and inversely increases interferon (IFN)-γ levels in the serum of NC/Nga mice [43]. Moreover, decreased lipogenesis of ingested dietary or plant alternatives with anti-acne efficacy is frequently accompanied by systemic alteration of adaptive immunity cytokines, such as IL-4, IL-10, IL-12, and IFN-γ individually or in combination [44–46]. In our previous studies [47], individual treatment of lower levels of Th-2 cytokines, such as $0.5 \times$ IL-4 (0.5 fold ($\times$) IL-4 concentration based on its normal serum concentration ($1.0 \times$) and $0.5 \times$ IL-10, and higher level of Th-1 cytokine such as $5.0 \times$ IFN-γ (5.0 fold ($\times$) IFN-γ concentration based on its normal serum concentration ($1.0 \times$)) decreases lipid content in human sebocytes, and this effect is greater with their combined treatment. Furthermore, in the context of inhibited PI3K/Akt/mTORC1 signaling with lower levels of IL-4 and IL-10 and/or chronic treatment of IFN-γ [39–41], the individual treatment of 0.5xIL-10 and the combined treatment of 0.5xIL-4, 0.5xIL-10 and 5.0xIFN-γ in human sebocytes decrease the protein expression of mature SREBP-1 and/or fatty acid synthase (FAS) [47], of which expression is upregulated by SREBP-1 and then esterified into TG [32,48,49]. Together, these studies, coupled with decreased Cer levels and increased epidermal water loss in IL-4 treated epidermis equivalent [50], suggest that ingested CJLP55 modulates adaptive immunity cytokine levels (more likely to suppress

Th-2 cytokine levels) systemically via altered bacterial EV phylum flora, which may lower TG of sebum and improve skin hydration with increased Cer levels, thereby decreasing follicular plugging and inflammation from *C. acnes* bacteria in acne lesions. Future studies with a larger number of acne patients would facilitate evaluation of systemic alterations in adaptive immunity cytokines as well as in insulin, IFG-1 and BCAA levels and intestinal and skin bacterial flora after *L. plantarum* CJLP55 supplement intervention.

5. Conclusions

This randomized, double-bind, placebo-controlled study demonstrated that a 12-week supplementation of CJLP55 decreased the clinical severity of acne vulgaris, *Proteobacteria* proportion in urine EV phylum, and TG of sebum. The last feature was inversely correlated to increases in the *Firmicutes* proportion of urine EV phylum, Cer, and hydration of skin. CJLP55 supplement was beneficial to clinical state, skin sebum, and hydration and urine bacterial EV phylum flora in patients with acne vulgaris. Dietary *Lactobacillus plantarum* CJLP55 may be a potential alternative therapy or may serve as an adjunct to conventional therapies for the treatment of acne vulgaris.

Supplementary Materials: The following are available online at https://www.mdpi.com/article/10.3390/nu13041368/s1. Figure S1: Consort flow diagram. File S1: Raw research data. File S2: Consort 2010 checklist.

Author Contributions: Conceptualization: H.-S.Y., H.-Y.A., J.-Y.O., and Y.C.; validation: C.C.; formal analysis and investigation: M.-J.K., K.-P.K., E.C., J.-H.Y., and H.-S.Y.; resources: H.-Y.A. and J.-Y.O.; writing—original draft: M.-J.K. and H.-S.Y.; writing—review and editing: Y.C.; visualization: M.-J.K. and K.-P.K.; supervision, project administration and funding acquisition: Y.C. All authors have read and agreed to the published version of the manuscript.

Funding: This study was supported by CJ Foods R & D Center, CJ CheilJedang Corporation in Suwon-si, Gyeongggi-do, Republic of Korea (KHU Grant No. 20151196).

Institutional Review Board Statement: The study was conducted according to the guidelines of the Declaration of Helsinki, and approved by the Institutional Review Board of Kyung Hee University (Yongin, Republic of Korea) (KHSIRB 2015-013).

Informed Consent Statement: Written informed consent has been obtained from all subjects to publish this paper.

Data Availability Statement: The data presented in this study are available in Supplementary Materials (File S2: Raw research data).

Conflicts of Interest: Hyun Sun Yun, Hee Yoon Ahn and Ji Young Oh are employed by CJ Foods R & D Center, CJ CheilJedang Corporation. *L. plantarum* CJLP55 (KCTC 11401BP, GenBank accession number GQ336971) was provided by CJ Foods R & D Center, CJ CheilJedang Corporation (Suwon, Republic of Korea). Other authors declare that there are no potential conflicts of interests with respect to the authorship and/or publication of this article.

References

1. Kurokawa, I.; Danby, F.W.; Ju, Q.; Wang, X.; Xiang, L.F.; Xia, L.; Chen, W.C.; Nagy, I.; Picardo, M.; Suh, D.H.; et al. New developments in our understanding of acne pathogenesis and treatment. *Exp. Dermatol.* **2009**, *18*, 821–832. [CrossRef] [PubMed]
2. Toyoda, M.; Morohashi, M. Pathogenesis of acne. *Med. Electron. Microsc.* **2001**, *34*, 29–40. [CrossRef] [PubMed]
3. Bowe, W.P.; Logan, A.C. Clinical implications of lipid peroxidation in acne vulgaris: Old wine in new bottles. *Lipids Health Dis.* **2010**, *9*, 141. [CrossRef] [PubMed]
4. Patel, M.; Bowe, W.P.; Heughebaert, C.; Shalita, A.R. The development of antimicrobial resistance due to the antibiotic treatment of acne vulgaris: A review. *J. Drugs Dermatol.* **2010**, *9*, 655–664.
5. Lee, H.; Yoon, H.; Ji, Y.; Kim, H.; Park, H.; Lee, J.; Shin, H.; Holzapfel, W. Functional properties of Lactobacillus strains isolated from kimchi. *Int. J. Food Microbiol.* **2011**, *145*, 155–161. [CrossRef] [PubMed]
6. Ji, Y.; Kim, H.; Park, H.; Lee, J.; Lee, H.; Shin, H.; Kim, B.; Franz, C.M.A.P.; Holzapfel, W.H. Functionality and safety of lactic bacterial strains from Korean kimchi. *Food Control.* **2013**, *31*, 467–473. [CrossRef]
7. Kober, M.M.; Bowe, W.P. The effect of probiotics on immune regulation, acne, and photoaging. *Int. J. Women's Dermatol.* **2015**, *1*, 85–89. [CrossRef] [PubMed]

8.   Ouwehand, A.C.; Salminen, S.; Isolauri, E. Probiotics: An overview of beneficial effects. *Antonie Van Leeuwenhoek* **2002**, *82*, 279–289. [CrossRef]

9.   Dunne, C.; O'Mahony, L.; Murphy, L.; Thornton, G.; Morrissey, D.; O'Halloran, S.; Feeney, M.; Flynn, S.; Fitzgerald, G.; Daly, C.; et al. In vitro selection criteria for probiotic bacteria of human origin: Correlation with in vivo findings. *Am. J. Clin. Nutr.* **2001**, *73*, 386s–392s. [CrossRef]

10.   Won, T.J.; Kim, B.; Song, D.S.; Lim, Y.T.; Oh, E.S.; Lee, D.I.; Park, E.S.; Min, H.; Park, S.Y.; Hwang, K.W. Modulation of Th1/Th2 balance by Lactobacillus strains isolated from Kimchi via stimulation of macrophage cell line J774A.1 in vitro. *J. Food Sci.* **2011**, *76*, H55–H61. [CrossRef]

11.   Won, T.J.; Kim, B.; Lim, Y.T.; Song, D.S.; Park, S.Y.; Park, E.S.; Lee, D.I.; Hwang, K.W. Oral administration of Lactobacillus strains from Kimchi inhibits atopic dermatitis in NC/Nga mice. *J. Appl. Microbiol.* **2011**, *110*, 1195–1202. [CrossRef]

12.   Muizzuddin, N.; Maher, W.; Sullivan, M.; Schnittger, S.; Mammone, T. Physiological effect of a probiotic on skin. *J. Cosmet. Sci.* **2012**, *63*, 385–395.

13.   Han, Y.; Kim, B.; Ban, J.; Lee, J.; Kim, B.J.; Choi, B.S.; Hwang, S.; Ahn, K.; Kim, J. A randomized trial of *Lactobacillus plantarum* CJLP133 for the treatment of atopic dermatitis. *Pediatr. Allergy Immunol.* **2012**, *23*, 667–673. [CrossRef]

14.   Tsai, W.H.; Chou, C.H.; Chiang, Y.J.; Lin, C.G.; Lee, C.H. Regulatory effects of lactobacillus plantarum-GMNL6 on human skin health by improving skin microbiome. *Int. J. Med. Sci.* **2021**, *18*, 1114–1120. [CrossRef] [PubMed]

15.   Pyun, B.Y. Extracellular vesicle: An unknown environmental factor for causing airway disease. *Allergy Asthma Immunol. Res.* **2016**, *8*, 179–180. [CrossRef] [PubMed]

16.   Kim, M.H.; Choi, S.J.; Choi, H.I.; Choi, J.P.; Park, H.K.; Kim, E.K.; Kim, M.J.; Moon, B.S.; Min, T.K.; Rho, M.; et al. Lactobacillus plantarum-derived extracellular vesicles protect atopic dermatitis induced by Staphylococcus aureus-derived Extracellular Vesicles. *Allergy Asthma Immunol. Res.* **2018**, *10*, 516–532. [CrossRef] [PubMed]

17.   Courage + Khazaka Electronic GmbH, Cologne, G. The Sebumeter®SM815. Available online: https://www.courage-khazaka.de/en/scientific-products/all-products/16-wissenschaftliche-produkte/alle-produkte/151-sebumeter-e (accessed on 31 May 2016).

18.   Kim, J.; Ko, Y.; Park, Y.K.; Kim, N.I.; Ha, W.K.; Cho, Y. Dietary effect of lactoferrin-enriched fermented milk on skin surface lipid and clinical improvement of acne vulgaris. *Nutrition* **2010**, *26*, 902–909. [CrossRef]

19.   Clarys, P.; Barel, A. Quantitative evaluation of skin surface lipids. *Clin. Dermatol.* **1995**, *13*, 307–321. [CrossRef]

20.   Yoo, J.Y.; Rho, M.; You, Y.A.; Kwon, E.J.; Kim, M.H.; Kym, S.; Jee, Y.K.; Kim, Y.K.; Kim, Y.J. 16S rRNA gene-based metagenomic analysis reveals differences in bacteria-derived extracellular vesicles in the urine of pregnant and non-pregnant women. *Exp. Mol. Med.* **2016**, *48*, e208. [CrossRef]

21.   Lee, Y.; Park, J.Y.; Lee, E.H.; Yang, J.; Jeong, B.R.; Kim, Y.K.; Seoh, J.Y.; Lee, S.H.; Han, P.L.; Kim, E.J. Rapid assessment of microbiota changes in individuals with autism spectrum disorder using bacteria-derived membrane vesicles in urine. *Exp. Neurobiol.* **2017**, *26*, 307–317. [CrossRef]

22.   Chen, H.C.; Mendelsohn, R.; Rerek, M.E.; Moore, D.J. Fourier transform infrared spectroscopy and differential scanning calorimetry studies of fatty acid homogeneous ceramide 2. *Biochim. Biophys. Acta Biomembr.* **2000**, *1468*, 293–303. [CrossRef]

23.   Cha, H.M.; Kim, S.K.; Kook, M.C.; Yi, T.H. Lactobacillus paraplantarum THG-G10 as a potential anti-acne agent with anti-bacterial and anti-inflammatory activities. *Anaerobe* **2020**, *64*, 102243. [CrossRef] [PubMed]

24.   Kwon, H.H.; Yoon, J.Y.; Hong, J.S.; Jung, J.Y.; Park, M.S.; Suh, D.H. Clinical and histological effect of a low glycaemic load diet in treatment of acne vulgaris in Korean patients: A randomized, controlled trial. *Acta Derm. Venereol.* **2012**, *92*, 241–246. [CrossRef] [PubMed]

25.   Rocha, M.A.; Bagatin, E. Skin barrier and microbiome in acne. *Arch. Dermatol. Res.* **2018**, *310*, 181–185. [CrossRef]

26.   Admyre, C.; Telemo, E.; Almqvist, N.; Lötvall, J.; Lahesmaa, R.; Scheynius, A.; Gabrielsson, S. Exosomes—Nanovesicles with possible roles in allergic inflammation. *Allergy* **2008**, *63*, 404–408. [CrossRef]

27.   Shin, T.S.; Kim, J.H.; Kim, Y.S.; Jeon, S.G.; Zhu, Z.; Gho, Y.S.; Kim, Y.K. Extracellular vesicles are key intercellular mediators in the development of immune dysfunction to allergens in the airways. *Allergy* **2010**, *65*, 1256–1265. [CrossRef]

28.   Kang, C.S.; Ban, M.; Choi, E.J.; Moon, H.G.; Jeon, J.S.; Kim, D.K.; Park, S.K.; Jeon, S.G.; Roh, T.Y.; Myung, S.J.; et al. Extracellular Vesicles Derived from Gut Microbiota, Especially Akkermansia muciniphila, Protect the Progression of Dextran Sulfate Sodium-Induced Colitis. *PLoS ONE* **2013**, *8*, e76520. [CrossRef]

29.   Volkova, L.; Khalif, I.; Kabanova, I. Impact of the impaired intestinal microflora on the course of acne vulgaris. *Klin. Med.* **2001**, *79*, 39–41.

30.   Wallace, T.C.; Guarner, F.; Madsen, K.; Cabana, M.D.; Gibson, G.; Hentges, E.; Sanders, M.E. Human gut microbiota and its relationship to health and disease. *Nutr. Rev.* **2011**, *69*, 392–403. [CrossRef] [PubMed]

31.   Deng, Y.; Wang, H.; Zhou, J.; Mou, Y.; Wang, G.; Xiong, X. Patients with acne vulgaris have a distinct gut microbiota in comparison with healthy controls. *Acta Derm. Venereol.* **2018**, *98*, 783–790. [CrossRef]

32.   Melnik, B.C. Acne vulgaris: The metabolic syndrome of the pilosebaceous follicle. *Clin. Dermatol.* **2018**, *36*, 29–40. [CrossRef] [PubMed]

33.   Cong, T.X.; Hao, D.; Wen, X.; Li, X.H.; He, G.; Jiang, X. From pathogenesis of acne vulgaris to anti-acne agents. *Arch. Dermatol. Res.* **2019**, *311*, 337–349. [CrossRef] [PubMed]

34.   Smith, T.M.; Gilliland, K.; Clawson, G.A.; Thiboutot, D. IGF-1 induces SREBP-1 expression and lipogenesis in SEB-1 sebocytes via activation of the phosphoinositide 3-kinase/Akt pathway. *J. Investig. Dermatol.* **2008**, *128*, 1286–1293. [CrossRef] [PubMed]

35. Çerman, A.A.; Aktaş, E.; Altunay, I.K.; Arici, J.E.; Tulunay, A.; Ozturk, F.Y. Dietary glycemic factors, insulin resistance, and adiponectin levels in acne vulgaris. *J. Am. Acad. Dermatol.* **2016**, *75*, 155–162. [CrossRef]
36. Lee, E.; Jung, S.R.; Lee, S.Y.; Lee, N.K.; Paik, H.D.; Lim, S. Il Lactobacillus plantarum strain ln4 attenuates diet-induced obesity, insulin resistance, and changes in hepatic mRNA levels associated with glucose and lipid metabolism. *Nutrients* **2018**, *10*, 643. [CrossRef]
37. Bagarolli, R.A.; Tobar, N.; Oliveira, A.G.; Araújo, T.G.; Carvalho, B.M.; Rocha, G.Z.; Vecina, J.F.; Calisto, K.; Guadagnini, D.; Prada, P.O.; et al. Probiotics modulate gut microbiota and improve insulin sensitivity in DIO mice. *J. Nutr. Biochem.* **2017**, *50*, 16–25. [CrossRef]
38. Fabbrocini, G.; Bertona, M.; Picazo, O.; Pareja-Galeano, H.; Monfrecola, G.; Emanuele, E. Supplementation with Lactobacillus rhamnosus SP1 normalises skin expression of genes implicated in insulin signalling and improves adult acne. *Benef. Microbes* **2016**, *7*, 625–630. [CrossRef]
39. Weichhart, T.; Säemann, M.D. The PI3K/Akt/mTOR pathway in innate immune cells: Emerging therapeutic applications. *Ann. Rheum. Dis.* **2008**, *67*, iii70–iii74. [CrossRef]
40. Park, H.J.; Lee, S.J.; Kim, S.H.; Han, J.; Bae, J.; Kim, S.J.; Park, C.G.; Chun, T. IL-10 inhibits the starvation induced autophagy in macrophages via class I phosphatidylinositol 3-kinase (PI3K) pathway. *Mol. Immunol.* **2011**, *48*, 720–727. [CrossRef]
41. Su, X.; Yu, Y.; Zhong, Y.; Giannopoulou, E.G.; Hu, X.; Liu, H.; Cross, J.R.; Rätsch, G.; Rice, C.M.; Ivashkiv, L.B. Interferon-γ regulates cellular metabolism and mRNA translation to potentiate macrophage activation. *Nat. Immunol.* **2015**, *16*, 838–849. [CrossRef]
42. Bowe, W.P.; Patel, N.B.; Logan, A.C. Acne vulgaris, probiotics and the gut-brain-skin axis: From anecdote to translational medicine. *Benef. Microbes* **2014**, *5*, 185–199. [CrossRef]
43. Won, T.J.; Kim, B.; Lee, Y.; Bang, J.S.; Oh, E.S.; Yoo, J.S.; Hyung, K.E.; Yoon, J.; Hwang, S.; Park, E.S.; et al. Therapeutic potential of *Lactobacillus plantarum* CJLP133 for house-dust mite-induced dermatitis in NC/Nga mice. *Cell. Immunol.* **2012**, *277*, 49–57. [CrossRef] [PubMed]
44. Jung, J.Y.; Kwon, H.H.; Hong, J.S.; Yoon, J.Y.; Park, M.S.; Jang, M.Y.; Suh, D.H. Effect of dietary supplementation with omega-3 fatty acid and gamma-linolenic acid on acne vulgaris: A randomised, double-blind, controlled trial. *Acta Derm. Venereol.* **2014**, *94*, 521–526. [CrossRef] [PubMed]
45. Nielsen, A.A.; Jørgensen, L.G.M.; Nielsen, J.N.; Eivindson, M.; Grønbæk, H.; Vind, I.; Hougaardt, D.M.; Skogstrand, K.; Jensen, S.; Munkholm, P.; et al. Omega-3 fatty acids inhibit an increase of proinflammatory cytokines in patients with active Crohn's disease compared with omega-6 fatty acids. *Aliment. Pharmacol. Ther.* **2005**, *22*, 1121–1128. [CrossRef] [PubMed]
46. Oláh, A.; Tóth, B.I.; Borbíró, I.; Sugawara, K.; Szöllõsi, A.G.; Czifra, G.; Pál, B.; Ambrus, L.; Kloepper, J.; Camera, E.; et al. Cannabidiol exerts sebostatic and antiinflammatory effects on human sebocytes. *J. Clin. Investig.* **2014**, *124*, 3713–3724. [CrossRef] [PubMed]
47. Shin, J.; Kim, K.-P.; Ahn, H.Y.; Kim, B.; Cho, Y. Alterations in IL-4, IL-10 and IFN-γ levels synergistically decrease lipid content and protein expression of FAS and mature SREBP-1 in human sebocytes. *Arch. Dermatol. Res.* **2019**, *311*, 563–571. [CrossRef] [PubMed]
48. Peterson, T.R.; Sengupta, S.S.; Harris, T.E.; Carmack, A.E.; Kang, S.A.; Balderas, E.; Guertin, D.A.; Madden, K.L.; Carpenter, A.E.; Finck, B.N.; et al. mTOR complex 1 regulates lipin 1 localization to control the srebp pathway. *Cell* **2011**, *146*, 408–420. [CrossRef]
49. Rosignoli, C.; Nicolas, J.C.; Jomard, A.; Michel, S. Involvement of the SREBP pathway in the mode of action of androgens in sebaceous glands in vivo. *Exp. Dermatol.* **2003**, *12*, 480–489. [CrossRef] [PubMed]
50. Hatano, Y.; Katagiri, K.; Arakawa, S.; Fujiwara, S. Interleukin-4 depresses levels of transcripts for acid-sphingomyelinase and glucocerebrosidase and the amount of ceramide in acetone-wounded epidermis, as demonstrated in a living skin equivalent. *J. Dermatol. Sci.* **2007**, *47*, 45–47. [CrossRef]

Article

Perilla frutescens Extracts Enhance DNA Repair Response in UVB Damaged HaCaT Cells

Hyuna Lee and Eunmi Park *

Department of Food and Nutrition, Hannam University, Daejeon 34054, Korea; hyuna916@naver.com
* Correspondence: eunmi_park@hnu.kr; Tel.: +82-42-629-8793

Abstract: Physiological processes in skin are associated with exposure to UV light and are essential for skin maintenance and regeneration. Here, we investigated whether the leaf and callus extracts of *Perilla frutescens (Perilla)*, a well-known Asian herb, affect DNA damage response and repair in skin and keratinocytes exposed to Untraviolet B (UVB) light. First, we examined the protective effects of *Perilla* leaf extracts in UVB damaged mouse skin in vivo. Second, we cultured calluses using plant tissue culture technology, from *Perilla* leaf explant and then examined the effects of the leaf and callus extracts of *Perilla* on UVB exposed keratinocytes. HaCaT cells treated with leaf and callus *Perilla* extracts exhibited antioxidant activities, smaller DNA fragment tails, and enhanced colony formation after UVB exposure. Interestingly, keratinocytes treated with the leaf and callus extracts of *Perilla* showed G1/S cell cycle arrest, reduced protein levels of cyclin D1, Cyclin Dependent Kinase 6 (CDK6), and γH_2AX, and enhanced levels of phosphorylated checkpoint kinase 1 (pCHK1) following UVB exposure. These observations suggest that the leaf and callus extracts of *Perilla* are candidate nutraceuticals for the prevention of keratinocyte aging.

Keywords: *Perilla frutescens*; keratinocytes; cell proliferation; ultraviolet radiation; DNA repair

Citation: Lee, H.; Park, E. *Perilla frutescens* Extracts Enhance DNA Repair Response in UVB Damaged HaCaT Cells. *Nutrients* **2021**, *13*, 1263. https://doi.org/10.3390/nu13041263

Academic Editor: Jean Christopher Chamcheu, Anthony L. Walker and Felicite Noubissi-Kamdem

Received: 19 March 2021
Accepted: 9 April 2021
Published: 12 April 2021

Publisher's Note: MDPI stays neutral with regard to jurisdictional claims in published maps and institutional affiliations.

1. Introduction

Skin is primarily composed of epidermis and dermis and exposed to a variety of environmental factors including UV radiation. UV exposure is a major factor of age-related changes such as pigmentary changes, thinning, wrinkling, and skin cancer development [1]. UV radiation is composed of Ultraviolet A (UVA), Ultraviolet B (UVB), and Ultraviolet C (UVC), which are defined by wavelength [2]. Exposure to UVB causes inflammation and immune changes that induce cellular senescence, direct DNA damage, and increases levels of oxidative stress, which can indirectly induce DNA mutagenesis. In response to DNA damage, the DNA response and repair pathway is activated in keratinocytes [3,4].

If not repaired properly, DNA damage in keratinocytes may result in the accumulation of mutations that prompt cell cycle arrest and promote skin aging. In particular, when exposed to UVB, phosphorylated checkpoint kinase 1 (pCHK1) is activated, this causes G1/S cycle arrest and reduces the proportion of cells in the S phase and skin repair response [5].

Perilla frutescens (Perilla) is an annual herb widely cultured in Asia. *Perilla* leaves are known to have antioxidant, anti-inflammatory, and anti-allergic effects and to be rich in polyphenols [6,7]. Calluses (as known as plant stem cells) are disorganized, non-differentiated plant cell masses and are induced using hormones, such as auxin or cytokinin from plant explants [8–10]. Calluses contain phenolic acid, which acts as an antioxidant, and other derivative sources. Recently, callus induction has been further developed to produce bioactive nutrient compounds [11,12].

Here, we investigated whether *Perilla* leaf extracts can induce DNA damage response and repair in UVB exposed mouse skin. In addition, we describe an efficient, rapid means of *Perilla* callus induction from *Perilla* leaves in different media in the presence of the growth regulators, 2,4-dichlorophenoxy acetic acid (2,4-D) kinetin, and organic additives. We then examined the effects of the leaf and callus extracts of *Perilla* on DNA damage response and repair in UVB exposed keratinocytes. This study shows that *Perilla* extracts are potential nutraceutical reagents for maintaining keratinocyte homeostasis after sun exposure.

2. Materials and Methods

2.1. UVB Irradiated Mouse Experiment

In the present study, Friend leukemia virus B (FVB) female mice (10 per group) were used as previously described [4]. After removing dorsal skin hair, *Perilla* leaf extracts of 5% w/v were applied in acetone to mouse skin. Controls were treated with acetone alone. Mice were then placed in separate compartments of a modified cage and 30 min after extract application, their backs were UVB irradiated. This process was performed 4 times/week for 2 weeks. The mount of UVB administered was progressively increased by increasing exposure times (i.e., 30, 60, 90, 120 s during the first week and 240, 270, 300, and 330 s during the second week (up to 100 mJ/cm^2). Epidermal thickness was measured using photomicrographs (100 μM) of 5 sections per mouse from 10 mice. The software program used was Leica application suite version 4.8. All animal experiments were approved beforehand by the Institutional Animal Use and Care Committee (HNU 2016-6).

2.2. Preparation of Perilla Callus Induction

Perilla frutescens were collected from Geumsan, Chungcheongnamdo, South Korea. To induce callus growth, *Perilla* leaves were first disinfected in 5 mM salicylic acid solution for 2 min, 70% ethanol for 10 min, and 1% bleach (sodium hypochlorite) for 10 min, and then washed 3 times in sterile distilled water for 15 min. Sterilized leaves were cut into 1×1 cm^2 pieces and inoculated into one of three media. For media preparation, agar 8 g/L, sucrose 10.09 g/L, and 3-(N-morpholino) propane sulfonic acid (MOPS) 0.5 g/L were added to each of; (1) Murashige & Skoog medium (M&S medium), (2) Murashige & Skoog modified medium (M&S modified medium), or (3) 2,4-dichlorophenoxy acetic acid medium (2,4-D medium). All media were adjusted to pH 5.7 and autoclaved at 121 °C for 15 min before use. Sterilized leaves were cultivated in media at room temperature for about 4 weeks. Calluses were then harvested and extracted with 70% ethanol. Extracts were then filtered, vacuum concentrated, and freeze-dried. The powdered extracts of calluses obtained using the three different media were used in the experiments detailed below.

2.3. ORAC (Oxygen Radical Absorbance Capacity) Assay

The peroxyl radical scavenging abilities of *Perilla* leaf and callus extracts were assessed using an ORAC assay [13]. Briefly, fluorescein (100 μL/80 nM) in 75 mM potassium phosphate buffer (pH 7.4) was added to triplicate wells in a black, flat-bottom, 96-well microplate. 2,2′-Azobis (2-amidino-propane) dihydrochloride (AAPH; 50 μL/80 mM), a peroxyl radical generator, was then added to the microplate, which was then immediately inserted into a SpectraMax i3x Platform (Molecular Devices, Lagerhausstrasse, San Jose, CA, USA). ORAC were measured every 2 min at 37 °C and expressed as Trolox equivalents (TE; μM). One ORAC unit was the equivalent to the net area provided by 1 μM Trolox.

2.4. Cell Culture and UV Irradiation

HaCaT cells (a human keratinocyte cell line) were a generous gift of Dr. Dae Joon Kim (University of Texas Health Rio Grande Valley) and cultured in Dulbecco's modified Eagle's medium containing 10% fetal bovine serum and 1% penicillin/streptomycin in a 5% CO$_2$ atmosphere at 37 °C [14]. Samples were pretreated 12 h before UVB irradiation. Cells were washed twice with phosphate buffered saline (PBS) and irradiated with UVB using a UV source (Spectrolinker Xl-1000B UV crosslinker; New York, NY, USA) once at 30 mJ/cm^2. After UVB irradiation, the medium was replaced with fresh serum-free medium. Cells were left for 3 h and then harvested for experiments.

2.5. Cell Viability Assay (MTT Assay)

The effects of *Perilla* leaf and callus extracts on HaCaT viability were determined using an 3-(4,5-Dimethylthiazol-2-yl)-2,5-Diphenyltetrazolium Bromide (MTT) assay [15]. Cells were plated at 5×10^5 cells/well into 12-well plates, cultured for 24 h, and then treated with leaf and callus of *Perilla* for 12 h. MTT solution (5 mg/mL) was then added, and cells were incubated for 1 h. After removing media, formazan crystals were dissolved with DMSO and absorbances were measured at 570 nm.

2.6. Colony Formation Assay

Colony formation assays were used to measure cell viability by counting colonies formed by single cells [16]. HaCaT cells were treated for 12 h with *Perilla* leaf and callus extracts, irradiated with UVB, and then incubated in fresh Dulbecco's Modified Eagle's Medium (DMEM) medium at 37 °C in a 5% humidified CO_2 atmosphere for 3 h. Cells were harvested with trypsin-Ethylenediaminetetraacetic acid (trypsin-EDTA), counted using a hemocytometer, and seeded in 6-well plates or 100 mm cell culture dishes. After incubation for 14 days, colonies were fixed with methanol, stained with crystal violet, counted manually, and photographed.

2.7. Comet Assay

Comet assays were used to assess DNA damage levels in cells [17]. HaCaT cells were treated with *Perilla* leaf or callus extracts for 12 h, exposed to UVB, and cultured for 3 or 12 h. Cells were then trypsinized, mixed with low melting agarose gel (LMA), and dispersed onto precoated slides with normal melting agarose (NMA). Cells were then covered with cover glasses and stored at 4 °C, and cover glasses were removed after the gel had hardened. Slides were soaked in pre-chilled alkali lysis buffer (2.5 M NaCl, 100 mM Na_2EDTA, 10 mM Tris, 1% N-lauryl-sarcosinate, pH 10; 1% Triton-X-100 and 10% DMSO) at 4 °C for 60 min in the dark, and then in pre-chilled electrophoresis buffer (300 mM NaOH, 100 mM Na_2 EDTA, pH > 13) at 4 °C for 40 min in the dark. Slides were then placed in a horizontal electrophoresis chamber filled with cold electrophoresis buffer, electrophoresed at 25 V and 300 mA for 20 min, washed with tris buffer (pH 7.4), dried, stained with ethidium bromide (20 µL/mL), and observed under a fluorescence microscope (Leica, Wetzlar, Germany). Images were taken using a CCD camera (Nikon, Tokyo, Japan) and analyzed using a Comet image analysis program (Kinetic image 4.0, Caliper Life Sciences, Alameda, CA, USA). Degrees of DNA damage in HaCaT cells were quantified using DNA fragment percentages as DNA tails.

2.8. Fluorescence-Activated Cell Sorting (FACS) Analysis

Cell cycle distributions were determined by propidium iodide staining [18]. HaCaT cells were cultured in 6-well plates, treated with *Perilla* leaf or callus extracts (0.1 µg/mL) for 12 h, UV irradiated, and incubated in fresh DMEM medium at 37 °C for 3 h. Harvested cells were immediately fixed in 70% ethanol at 4 °C overnight, washed once with PBS, stained with propidium iodide/RNase staining solution (Invitrogen, Waltham, MA, USA), and analyzed by flow cytometry (Beckman Coulter, Brea, CA, USA). The percentages of cells in phases of the cell cycle were calculated using CytExpert 1.0 software (Beckman Coulter, Brea, CA, USA).

2.9. Western Blot Analysis

HaCaT cells were cultured in 6-well plates, treated with *Perilla* leaf or callus extracts for 12 h, UVB-irradiated, and then incubated for 3 h at 37 °C [14]. Following incubation, cells were harvested with trypsin-EDTA, resuspended in lysis buffer on ice for 40 min, centrifuged at $13{,}714 \times g$ for 20 min at 4 °C, and protein concentrations in supernatants were determined using a NanoDrop (NanoDrop Lite spectrophotometer, Waltham, MA, USA). Equal amounts of proteins were subjected to SDS-PAGE and then transferred to nitrocellulose membranes, which were reacted with primary antibodies; pCHK1 (Serine

345), γH2AX (both from Cell Signaling, Danvers, MA, USA), CyclinD1, Cell division protein kinase 6 (CDK6), and β-actin (all antibodies from Santa Cruz Bicycles, Santa Cruz, CA, USA) at 4 °C overnight. The following day, membranes were incubated with anti-mouse or anti-rabbit IgG horseradish peroxidase-conjugated secondary antibodies for 40 min at room temperature. Blots were detected using the West Pico chemiluminescent kit (Thermo, Rockford, IL, USA) and visualized using LAS 4000 chemiluminescent image analyzer (Fuji, Tokyo, Japan).

2.10. Statistics

Results were analyzed by ANOVA with Tukey's test. The analysis was conducted using SPSS-PC ver. 23.0 (SPSS Inc., Chicago, IL, USA). For the mouse experiments, results are presented as the means $\pm$ standard errors (SEs) of 10 mice per group. Other results are presented as means $\pm$ SEs or SDs of three independent experiments, as indicated. Data shown are representative of three independent experiments. Statistical significance was accepted for p values < 0.05.

3. Results

3.1. The Effect of Perilla Leaf Extracts on Mouse Skin Exposed to UVB

To determine the effects of the extracts on UV-induced skin damage, the dorsal skins of FVB mice were shaved and treated with *Perilla* leaf, callus extracts (5% w/v in acetone), or acetone (controls) before UVB exposure once daily for two weeks. Two weeks after UVB treatment, mice were sacrificed and skin tissue sections were fixed for hematoxylin and eosin (H&E) staining. Epidermal thickness was measured by microscopy. Acetone treatment (the control group) increased epidermal thickness after 2 weeks of treatment (Figure 1A,B). Interestingly, *Perilla* leaf extract reduced UVB-induced epidermal thickness increases ($p < 0.001$). Neither acetone nor *Perilla* leaf extract influenced epidermis thickness when the skin was not UVB exposed.

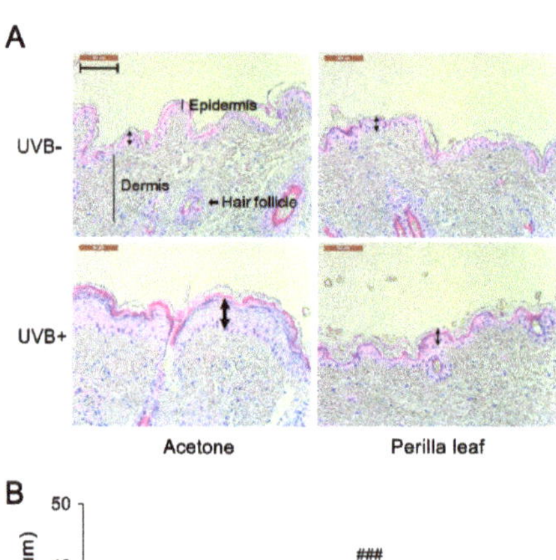

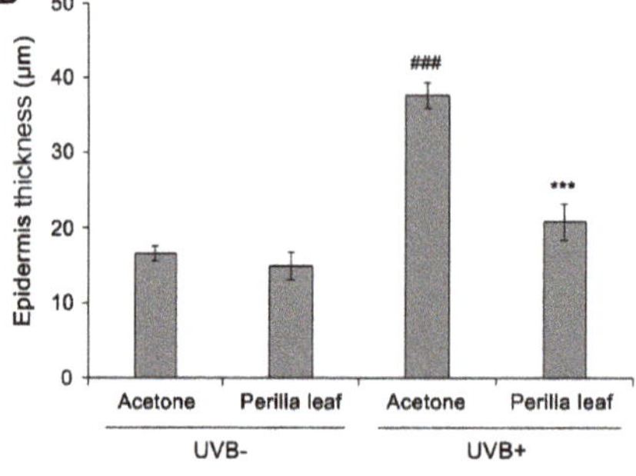

Figure 1. *Perilla* leaf extract suppressed UVB-induced epidermal thickening in FVB mice. (**A**) H&E-

stained section of mouse dorsal skin after Ultraviolet B (UVB) exposure. *Perilla* leaf extract (5% w/v in acetone) was applied to mouse skin, which was then irradiated with UVB for 30 min. UVB-induced epithelial cell proliferation was evaluated in each indicated group (ten mice per group), after 2 weeks of UVB irradiation was completed. The photograph shows skin from a representative mouse. Bold arrows indicate epidermis. The scale bar represents 100 μm. (**B**) Quantification of epidermis thickness in UVB-induced epithelial layers. Data are expressed as means ± SDs. ### $p < 0.001$ as compared with non-radiated mice by the t-test. *** $p < 0.001$ as compared with the UVB-irradiated control group by the *t*-test.

3.2. Induction of Perilla Callus Formation Using Different Media

Next, we established a means of culturing *Perilla* calluses, a well-known nutraceutical [12]. Calluses were induced from *Perilla* leaves over four weeks using three media types: M&S medium, M&S modified medium, and 2,4-D medium. Interestingly, *Perilla* callus induction was observed after five days of culture on 2,4-D medium and proliferated after nine days (Figure 2A,B). In contrast, induction of *Perilla* calluses was observed after nine days of culture on M&S modified medium, but *Perilla* calluses barely proliferated. Callus formation was more rapid in 2,4-D medium than in the other two media ($p < 0.001$, Figure 2A). *Perilla* calluses produced on 2,4-D medium, M&S modified medium, or M&S medium in five weeks were weighed (Figure 2C), and the amount of callus produced was larger on 2,4-D medium than on M&S modified medium (20.38 mg versus 5.78 mg, respectively). Callus formation was not induced in M&S medium. These observations showed 2,4-D medium containing 2,4-dichlorophenoxy acetic acid (2,4-D) provided an effective medium for *Perilla* callus induction and suggested that 2,4-D may be crucial for *Perilla* callus induction.

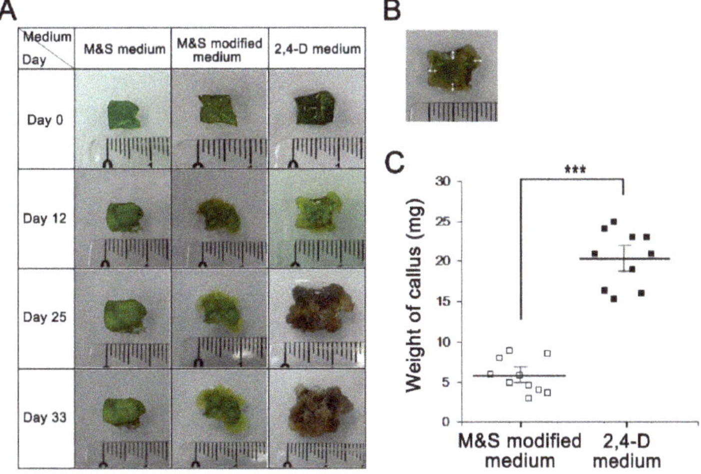

Figure 2. The effect of *Perilla* callus formation in 2,4-D and two M&S media. (**A**) *Perilla* callus induction dependence on the three medium types. Representative photographs showing *Perilla* calluses on culture days 0 to 33. (**B**) Calluses that formed from the four corners of *Perilla* leaves were measured using a ruler. (**C**) Weights of calluses by medium type. We measured the biomasses of *Perilla* calluses after culture for 5 weeks. Dots and bars represent the means and SDs of three independent samples. *** $p < 0.001$ as determined by the t-test.

3.3. Perilla Leaf and Callus Extracts Had Antioxidant Effects and Promoted DNA Repair in UVB Exposed Keratinocytes

The antioxidative activities of *Perilla* leaf and callus extracts were assessed using ORAC assays using ascorbic acid as an antioxidant. We found the antioxidant activities of

Perilla leaf extract was greater than those of callus extracts produced using 2,4-D medium or M&S modified medium (see Figure 3A). Interestingly, the antioxidant activities of *Perilla* callus extract induced using 2,4-D medium were greater than those induced using M&S modified medium ($p < 0.005$, Figure 3A). Therefore, we examined the effects of *Perilla* callus extract from calluses induced using 2,4-D medium and *Perilla* leaf extract in subsequent studies.

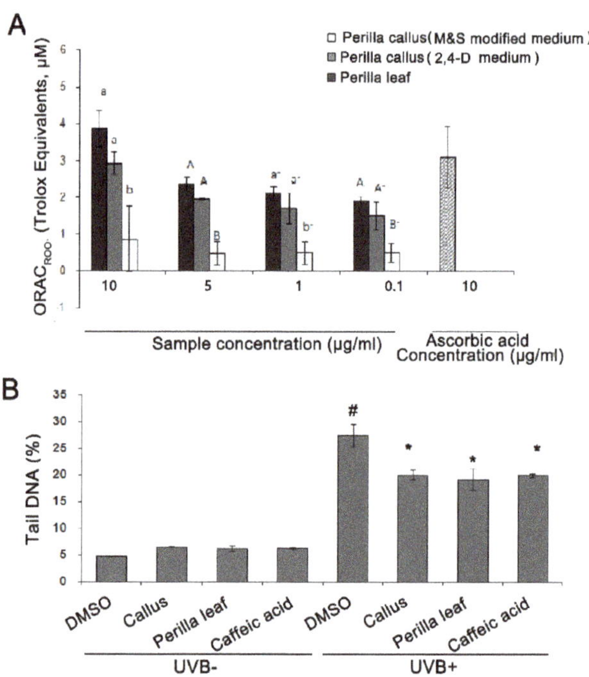

Figure 3. *Perilla* leaf and callus extracts exhibited antioxidant effects and enhanced DNA repair as determined by the Comet assay in UVB exposed HaCaT cells. (**A**) Antioxidant activities of *Perilla* leaf and callus extracts as determined by the oxygen radical absorbance capacity (ORAC)$_{ROO.}$ assay in HaCaT cells. The results presented are the means ± SDs of triplicate determinations. Data shown are representative of three independent experiments. Values were compared with those obtained for *Perilla* callus extracts induced by MS modified medium and 2,4-dichlorophenoxy acetic acid (2,4-D) medium, and leaf extracts at the same concentrations, as 1 uM trolox equivalent. $p < 0.01$ as determined by Duncan's test. (**B**) DNA damage was assessed using the Comet assay. Tail DNA percentages were determined. Comet images revealed different degrees of DNA damage. HaCaT cells were cultured in 6-well plates and pretreated with *Perilla* leaf or callus extracts (at 0.1 µg/mL), caffeic acid (150 µM; the positive control), or DMSO (the negative control) for 12 h, exposed to UVB (50 mJ/cm^2), and then cultured for 12 h later. The bar graph was calculated from three independent experiments. Images of fluorescence intensities were obtained using the Comet assay. # $p < 0.05$; significant versus UVB non-treated and DMSO controls. * $p < 0.05$; significant versus UVB treated and DMSO controls. Different letters indicate significant intergroup differences ($p < 0.01$).

The effects of the *Perilla* leaf and callus extracts on UVB-induced DNA damage were investigated using human keratinocytes (HaCaT cells). In Comet assays, the tail region containing DNA fragments provides an index of DNA damage. We observed that UVB (50 mJ/cm^2) treatment induced longer DNA tails than UVB untreatment in keratinocytes (27.5% versus 6.6 % for UVB+ versus UVB- in the DMSO treatment, as a negative control group; $p < 0.001$, Figure 3B). Moreover, we found that HaCaT cells treated with *Perilla* leaf or callus extract (0.1 µg/mL) and then UVB, had smaller Comet tails than cells treated with

DMSO and then UVB ($p < 0.001$, Figure 3B). We also treated HaCaT cells with caffeic acid, a derivative of cinnamic acid and a bioactive constituent of *Perilla* leaves, as a positive control. We found that *Perilla* leaf, callus extract (0.1 µg/mL), or caffeic acid (150 µM) reduced DNA tails after UVB exposure (Figure 3B, $p < 0.001$; Supplementary Figure S1). Our findings suggest that both *Perilla* leaf and callus extracts have potential use for repairing UV-induced DNA damage.

3.4. Perilla Leaf and Callus Extracts Enhanced UVB Irradiated Keratinocyte Survival

First, we measured HaCaT cell viabilities after exposure to different doses of *Perilla* leaf and callus extracts using an MTT assay. HaCaT cells were treated with extracts at concentrations of 0.01, 0.1, 1, 10, or 100 µg/mL (see Figure 4A). We found 85.8% of keratinocytes remained viable after treatment with *Perilla* leaf extract at 10 µg/mL, and that 90% and 81.4% of keratinocytes remained viable after treatment with *Perilla* callus extract at 10 or 100 µg/mL, respectively (Figure 4A). The observations suggest that *Perilla* leaf and callus extracts had no toxic effects on keratinocytes.

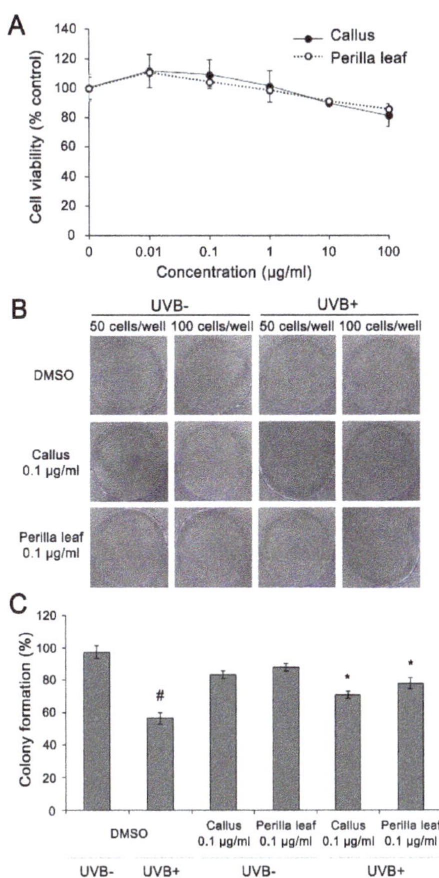

Figure 4. ***Perilla* leaf and callus extracts prevented the suppression of HaCaT proliferation by UVB.** (**A**) 3-(4,5-Dimethylthiazol-2-yl)-2,5-Diphenyltetrazolium Bromide (MTT) assay of HaCaT cells treated

with *Perilla* leaf and callus extracts. Results are expressed as means $\pm$ SDs. Different letters indicate significantly different ($p < 0.05$) by Duncan's test. (**B,C**) HaCaT cells were treated with *Perilla* leaf or callus extracts (0.1 µg/mL) overnight, irradiated with UVB (30 mJ/cm^2), harvested, and reseeded at 50 or 100 cells/well in 6 well plates. Colony formation was observed 14 days after seeding. (**B**) Representative colony-forming assay images after methanol fixation and crystal blue staining. (**C**) Colony numbers were counted and recorded. Colony formation rates of HaCaT cells (determined by analyzing 50 cells/well). Results are presented as the means $\pm$ SDs of three plates. # $p < 0.05$ versus UVB non-treated and DMSO controls. * $p < 0.05$ versus the UVB treated and DMSO controls.

Then, colony formation assays were performed to determine whether *Perilla* leaf and callus extracts influence cellular survival after exposure to UVB. DMSO treatment post UVB irradiation (controls) significantly reduced HaCaT cell colony formation (99% and 60% for UVB untreated and UVB treated cells, $p < 0.05$, Figure 4B,C). However, *Perilla* leaf and callus extracts inhibited UVB-induced reductions in colony formation ($p < 0.05$, Figure 4C).

3.5. Perilla Leaf and Callus Extracts Enhanced DNA Repair Signaling and Protected HaCaT cells from UVB-Induced Cell Cycle Changes

Perilla leaf and callus extracts enhanced the levels of pCHK1(S345) protein (an inducer of DNA repair signaling) more than DMSO treatment in UVB exposed HaCaT cells (Figure 5A,B; Supplementary Figure S2). Moreover, *Perilla* callus extracts reduced levels of γH2AX protein (a hallmark of DNA damage) more than *Perilla* leaf extract in UVB exposed keratinocytes (Figure 5B). These data suggest that *Perilla* callus extracts reduce DNA damage after UVB irradiation in keratinocytes.

Increased pCHK1(S345) protein expression affects the cell cycle. Therefore, we examined cell cycle profiles to investigate the protective effects of *Perilla* leaf and callus extracts on UVB irradiated keratinocytes.

Synchronous culture with chemical agents are in the same growth stage and provide less variation data over non-synchronous cells for studying cellular levels [19]. However, the synchronized drugs used resulted in side effects. We chose an asynchronized method, and its analysis of changes (Δ) in the cell cycle involved less perturbation of biological systems and drug interaction between chemical reagents and *Perilla* extracts. HaCaT cells were treated with *Perilla* leaf or callus extract overnight, irradiated with UVB (30 mJ/cm^2), and 3hrs later, PI staining and flow cytometry analysis were used to determine cell cycle profiles (Figure 5C,D).

UVB increased the percentage of HaCaT cells in the G0/G1 phase (from 50.5% to 60.9%), indicating that it inhibited cell growth by causing G1/S arrest. We observed that the percentage of cells in the G0/G1 phase after UVB exposure was increased by *Perilla* leaf and callus extracts (Figure 5D, $p < 0.05$). Furthermore, 0.1 µg/mL of *Perilla* callus extract increased the change of proportion of HaCaT cells in the G0/G1 phase, as compared with DMSO treatment after UV treatment (Figure 5D, 10.4 $\pm$ 0.8 versus 19.1 $\pm$ 1.3 of change (Δ) of DMSO treatment versus change (Δ) of 0.1 µg/mL callus extract, $p < 0.05$).

We also found that 0.1 µg/mL of *Perilla* callus extract decreased the proportion of cells in the S phase as compared with DMSO treatment after UV exposure (Figure 5D, -7.4 ± 0.9 versus -12.9 ± 1.1 of change (Δ) of DMSO treatment versus change (Δ) of 0.1 µg/mL callus extract, $p < 0.05$). Our observations suggest that *Perilla* extract, especially callus, arrested UVB-induced cell cycle phase at G1/S in HaCaT cells.

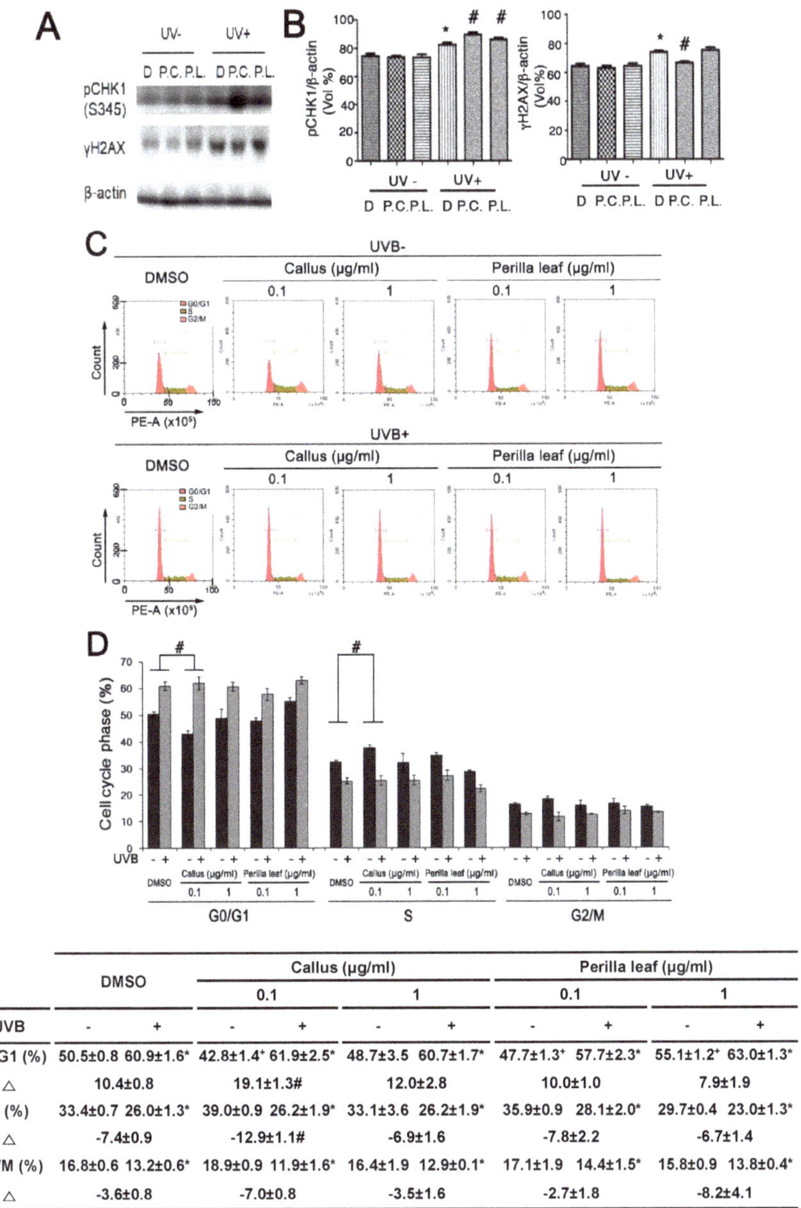

	DMSO		Callus (µg/ml)				Perilla leaf (µg/ml)			
			0.1		1		0.1		1	
UVB	-	+	-	+	-	+	-	+	-	+
G0/G1 (%)	50.5±0.8	60.9±1.6*	42.8±1.4⁺	61.9±2.5*	48.7±3.5	60.7±1.7*	47.7±1.3⁺	57.7±2.3*	55.1±1.2⁺	63.0±1.3*
△	10.4±0.8		19.1±1.3#		12.0±2.8		10.0±1.0		7.9±1.9	
S (%)	33.4±0.7	26.0±1.3*	39.0±0.9	26.2±1.9*	33.1±3.6	26.2±1.9*	35.9±0.9	28.1±2.0*	29.7±0.4	23.0±1.3*
△	-7.4±0.9		-12.9±1.1#		-6.9±1.6		-7.8±2.2		-6.7±1.4	
G2/M (%)	16.8±0.6	13.2±0.6*	18.9±0.9	11.9±1.6*	16.4±1.9	12.9±0.1*	17.1±1.9	14.4±1.5*	15.8±0.9	13.8±0.4*
△	-3.6±0.8		-7.0±0.8		-3.5±1.6		-2.7±1.8		-8.2±4.1	

Figure 5. **Effect of *Perilla* leaf and callus extracts on the cell cycle distribution of UVB exposed HaCaT cells.** (**A**) Representative Western blot for phosphorylated checkpoint kinase 1 (pCHK1) (S345) and γH2AX proteins in HaCaT cells treated with *Perilla* leaf or callus extracts at 0.1 µg/mL (D: DMSO treatment; P.C; *Perilla* callus; P.L: *Perilla* leaf). β-Actin was used as the loading control. Proteins were analyzed using the Fusion image system (Fuji, Japan). Cells were treated with *Perilla* extracts overnight, washed with Phosphate-buffered saline (PBS), and irradiated with UVB (30 mJ/cm^2), and 3 h later, pellets were harvested for Western blotting. (**B**) Results are presented as the means ± SDs of three biological replicates in two independent measurements (please see all images in supplementary Figure S2). * $p < 0.05$ versus UVB non-treated

controls. # $p < 0.05$; versus UVB treated comparisons. (**C,D**) Cell cycle analysis was performed by propidium iodide staining in asynchronized HaCaT cells. The proportions of cells in different phases of the cell cycle were determined by flow cytometry. (**C**) Representative histograms show cell cycle distributions. HaCaT cells were subjected to fluorescence-activated cell sorting (FACS) analysis. Histograms show side scatter (SS) and forward scatter (FS) for all events. The parameters of FACS analysis are referred to as FSC-A: Forward Scatter-Area; FSC-H: Forward Scatter-Height and SSC-A: Side Scatter-A. (**D**) Bar chart presentations of cell cycle distributions after treatment with *Perilla* leaf or callus extract and UVB (30 mJ/cm^2). The cell cycle distribution table shows percentages of cells in the three phases (G0/G1, S, and G2/M). Results are presented as the means $\pm$ SDs of triplicate plates per sample. * $p < 0.05$; significant versus UVB non-treated controls. + $p < 0.05$; significant versus UVB non-treated controls on G0/G1 phase. Δ; change of UVB treated from UVB untreated in each cell cycle phase. # $p < 0.05$; significant versus change (Δ) of DMSO treatment.

3.6. Perilla Leaf and Callus Extracts Regulated G1/S Cell Cycle Genes in UVB Exposed HaCaT Cells

At the G1 phase of the cell cycle, the cyclin D1 and CDK6 genes are directly involved in cell cycle regulation. Moreover, G1/S cell cycle arrest is associated with reductions in cyclin D1 and CDK6 protein levels after UVB exposure [19]. We found that cyclin D1 and CDK6 protein expressions were reduced in HaCaT cells irradiated with 30 mJ/cm^2 of UVB (Figure 6), and that *Perilla* leaf and callus extracts reduced significantly (Figure 6, $p < 0.05$). These results suggest that *Perilla* extracts may cause G1/S cell cycle arrest after UVB exposure by reducing cyclin D1 and CDK6 protein levels.

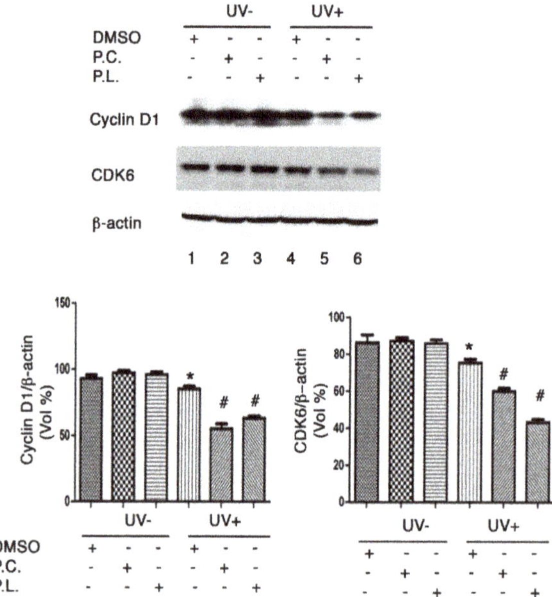

Figure 6. Western blot analysis of G1/S cell cycle regulatory protein expressions in HaCaT cells treated with *Perilla* leaf or callus extracts. Effects of *Perilla* leaf and callus extracts on the UVB-induced protein levels of Cyclin D1 and CDK6 in HaCaT cells. Cells were treated with *Perilla* leaf, callus extracts (0.1 μg/mL), or DMSO (negative control) overnight, washed with PBS, irradiated with UVB (30 mJ/cm^2), and harvested 3 h later. Pellets were subjected to Western blotting using β-actin as a loading control. (Top panel) Representative Western blot for cyclin D1 and CDK6 proteins in HaCaT cells treated with Perilla leaf or callus extracts at 0.1 μg/mL (D: DMSO treatment; P.C; Perilla callus; P.L: Perilla leaf). (Bottom panel) Proteins were analyzed using the Fusion image system (Fuji, Japan). Data are representative of three independent experiments and expressed as means $\pm$ SDs. * $p < 0.05$; significant versus UVB non-treated controls. # $p < 0.05$; significant versus UVB treated controls.

4. Discussion

It has been reported that UV-induced damage increases the levels of pCHK1protein expression during DNA damage and repair response [5,20]. In this study, we examined the effects of *Perilla* leaf and callus extracts in UVB exposed keratinocytes.

Previous studies have investigated the roles played by ubiquitin-specific protease 1 (USP1) in DNA damage response and on cellular functions and genetic instability in keratinocytes [14]. Activation of the DNA response and repair pathway induces cell proliferation and protects against senescence, and UVB radiation-mediated DNA damage reduces colony-formation potential and cell viability and destabilizes DNA in keratinocytes. Our results show that UVB significantly induced DNA breaks and reduced colony formation by HaCaT cells, whereas *Perilla* leaf and callus extracts both suppressed these UVB induced effects. Persistent exposure to UVB also increases DNA lesions, fragments, and mutations, and leads to premature skin aging, and interestingly, the present study shows *Perilla* leaf extract protects against DNA damage and enhanced cell proliferation in UVB exposed keratinocytes.

Links between UVB exposure, skin cell proliferation, and aging can be observed when the DNA repair and response pathway becomes dysfunctional. Defects in the DNA repair and response pathway enhance genomic instability in keratinocytes during aging [1]. Furthermore, the CHK1 gene activates the DNA repair systems that repair UVB-induced DNA lesions [20]. We found *Perilla* leaf and callus extracts enhanced pCHK1 protein expression and reduced γH2AX protein expression in UVB exposed keratinocytes. Accordingly, our findings suggest that both extracts increase keratinocyte proliferation by upregulating pCHK1 protein levels.

UV-induced reactive oxygen species (ROS) are responsible for DNA damage in skin. ROS are produced endogenously by metabolic processes and exogenously by the effect of UV on skin. Moreover, persistently elevated ROS and oxidative DNA damage levels are associated with poorer colony-forming ability [21]. In the present study, *Perilla* leaf and callus extracts effectively reduced ROS levels, increased DNA damage repair, and keratinocyte colony-forming ability, and were also found to have antioxidant effects. Thus, *Perilla* extracts may be a useful modality for the prevention of UV-induced keratinocyte aging.

DNA damage simultaneously triggers the cell cycle checkpoint and activates the DNA repair pathway [20]. Checkpoint proteins such as cyclin A, D, E, and K are activated by UV exposure and elicit cell cycle arrest by G1/S or G2/M signaling and facilitate DNA repair [19]. We found *Perilla* callus extracts induced G1/S cell cycle arrest by reducing cyclin D1 and CDK6 levels, and that both extracts, especially callus extract, effectively regulated cell cycle checkpoint proteins and triggered DNA repair response in UVB exposed keratinocytes. Thus, our findings indicate that *Perilla* extracts may effectively regulate the cell cycle and the G1/S cell cycle checkpoint.

Interestingly *Perilla* extracts also increased keratinocyte proliferation, which suggests that they might protect against keratinocyte aging. The antioxidant effects of *Perilla* leaf extracts have been previously studied in pharmacological properties [7]. *Perilla* leaf has been used as a traditional medicine to treat depression, anxiety, tumors, coughs, allergies, and other conditions, and our findings suggest that *Perilla* leaf and callus extracts could be used to treat keratinocyte damage and UV exposure-induced keratinocyte conditions [6,22]. *Perilla* extracts have been reported to contain the following active compounds: phenolic acids, flavonoids, anthocyanins, volatile compounds, triterpenes, phytosterols, fatty acid tocopherols, policosanols, rosmarinic acid, luteolin, and tormentic acid [22]. Although our preclinical data have yielded promising results, further studies are needed to determine the effects of active compounds in *Perilla* extracts.

2,4-D is a well-known plant growth regulator and an analog of auxin (a plant growth hormone). It has been reported that callus induction and lycopene production by the leaves of *Barringtonia racemose* were dependent on 2,4-D concentration in media [23]. Our study was conducted to establish a means of culturing calluses and regenerating *Perilla* plants using different plant hormones. *Perilla* leaf extract exhibited a better callus response when 2,4-D medium was supplemented with 2,4-dichlorophenoxy acetic acid (2,4-D), than when it was supplemented with kinetin, and a maximum callus score of 4.28 was recorded when leaves were cultured on MS modified media containing naphthaleneacetic acid. Furthermore, given increasing demands for natural bioactive products, plant cell cultures using plant growth regulators offer an attractive means of overcoming the limitations of extraction-based methods for obtaining biomaterials for nutraceutical studies.

It should be noted the amount of callus produced in the present study was relatively small and not enough to examine its effects on skin in vivo. Nevertheless, both *Perilla* leaf and callus extracts were found to have potent antioxidant effects and to increase DNA repair response in keratinocytes, and thus, to offer potential means of preventing and treating keratinocyte photoaging. Finally, *Perilla* extract, leaf or callus, presents better potential as source of keratocyte antiaging products.

Supplementary Materials: The following are available online at https://www.mdpi.com/article/10.3390/nu13041263/s1, Figure S1: DNA damage was assessed using the Comet assay. Tail DNA percentages were determined, Figure S2: All images of western blot for pCHK1(S345) and γH2AX proteins in HaCaT cells treated with Perilla leaf or callus extracts at 0.1 μg/mL (D: DMSO treatment; P.C; Perilla callus; P.L: Perilla leaf).

Author Contributions: Conceptualization, E.P.; methodology, E.P. and H.L.; resources, E.P.; data curation, H.L.; Original draft writing preparation, E.P. and H.L.; funding acquisition, E.P. All authors have read and agreed to the published version of the manuscript.

Funding: This study was supported by Basic Science Research Program through the National Research of Korea (NRF) funded by the Ministry of Science, ICT & Future Planning (Grant No. NRF-2019R1H1A2039746).

Institutional Review Board Statement: All animal experiments were approved beforehand by the Hannam University Institutional Animal Use and Care Committee (protocol code HNU 2016-6 and date of approval; March 2016).

Acknowledgments: We gratefully acknowledge the experimental assistance of Eunji Yeo.

Conflicts of Interest: The authors declare no conflict of interest.

References

1. Rittie, L.; Fisher, G.J. UV-light-induced signal cascades and skin aging. *Ageing Res. Rev.* **2002**, *1*, 705–720. [CrossRef]
2. Krutmann, J.; Morita, A.; Chung, J.H. Sun exposure: What molecular photodermatology tells us about its good and bad sides. *J. Investig. Dermatol.* **2012**, *132*, 976–984. [CrossRef]
3. Ribezzo, F.; Shiloh, Y.; Schumacher, B. Systemic DNA damage responses in aging and diseases. *Semin. Cancer Biol.* **2016**, *37–38*, 26–35. [CrossRef]
4. Xia, X.; Park, E.; Liu, B.; Willette-Brown, J.; Gong, W.; Wang, J.; Mitchell, D.; Fischer, S.M.; Hu, Y. Reduction of IKKalpha expression promotes chronic ultraviolet B exposure-induced skin inflammation and carcinogenesis. *Am. J. Pathol.* **2010**, *176*, 2500–2508. [CrossRef]
5. Capasso, H.; Palermo, C.; Wan, S.; Rao, H.; John, U.P.; O'Connell, M.J.; Walworth, N.C. Phosphorylation activates Chk1 and is required for checkpoint-mediated cell cycle arrest. *J. Cell Sci.* **2002**, *115*, 4555–4564. [CrossRef]
6. Ahmed, H.M. Ethnomedicinal, Phytochemical and Pharmacological Investigations of *Perilla* frutescens (L.) Britt. *Molecules* **2018**, *24*, 102. [CrossRef]
7. Kim, H.R.; Kim, S.Y. *Perilla* frutescens Sprout Extract Protect Renal Mesangial Cell Dysfunction against High Glucose by Modulating AMPK and NADPH Oxidase Signaling. *Nutrients* **2019**, *11*, 356. [CrossRef] [PubMed]
8. Coenen, C.; Lomax, T.L. Auxin-cytokinin interactions in higher plants: Old problems and new tools. *Trends Plant Sci.* **1997**, *2*, 351–356. [CrossRef]
9. Hwang, I.; Sheen, J.; Muller, B. Cytokinin signaling networks. *Annu. Rev. Plant Biol.* **2012**, *63*, 353–380. [CrossRef]
10. Ikeuchi, M.; Sugimoto, K.; Iwase, A. Plant callus: Mechanisms of induction and repression. *Plant Cell* **2013**, *25*, 3159–3173. [CrossRef]
11. Maatta, K.R.; Kamal-Eldin, A.; Torronen, A.R. High-performance liquid chromatography (HPLC) analysis of phenolic compounds in berries with diode array and electrospray ionization mass spectrometric (MS) detection: Ribes species. *J. Agric. Food Chem.* **2003**, *51*, 6736–6744. [CrossRef] [PubMed]

12. Maneechai, S.; De-Eknamkul, W.; Umehara, K.; Noguchi, H.; Likhitwitayawuid, K. Flavonoid and stilbenoid production in callus cultures of Artocarpus lakoocha. *Phytochemistry* **2012**, *81*, 42–49. [CrossRef]

13. Yang, S.Y.; Lee, S.H.; Tai, B.H.; Jang, H.D.; Kim, Y.H. Antioxidant and Anti-Osteoporosis Activities of Chemical Constituents of the Stems of Zanthoxylum piperitum. *Molecules* **2018**, *23*, 457. [CrossRef]

14. Park, E.; Kim, H.; Kim, J.M.; Primack, B.; Vidal-Cardenas, S.; Xu, Y.; Price, B.D.; Mills, A.A.; D'Andrea, A.D. FANCD2 activates transcription of TAp63 and suppresses tumorigenesis. *Mol. Cell* **2013**, *50*, 908–918. [CrossRef]

15. Choi, Y.E.; Park, E. Curcumin enhances poly(ADP-ribose) polymerase inhibitor sensitivity to chemotherapy in breast cancer cells. *J. Nutr. Biochem.* **2015**, *26*, 1442–1447. [CrossRef] [PubMed]

16. Park, E.; Kim, J.M.; Primack, B.; Weinstock, D.M.; Moreau, L.A.; Parmar, K.; D'Andrea, A.D. Inactivation of Uaf1 causes defective homologous recombination and early embryonic lethality in mice. *Mol. Cell Biol.* **2013**, *3*, 360–4370.

17. Kang, S.; Lim, Y.; Kim, Y.J.; Jung, E.S.; Suh, D.H.; Lee, C.H.; Park, E.; Hong, J.; Velliquette, R.A.; Kwon, O.; et al. Multivitamin and Mineral Supplementation Containing Phytonutrients Scavenges Reactive Oxygen Species in Healthy Subjects: A Randomized, Double-Blinded, Placebo-Controlled Trial. *Nutrients* **2019**, *11*, 101. [CrossRef]

18. Cho, J.; Park, E. Ferulic acid maintains the self-renewal capacity of embryo stem cells and adipose-derived mesenchymal stem cells in high fat diet-induced obese mice. *J. Nutr. Biochem.* **2020**, *77*, 108327. [CrossRef]

19. Meyerson, M.; Harlow, E. Identification of G1 kinase activity for cdk6, a novel cyclin D partner. *Mol. Cell. Biol.* **1994**, *14*, 2077–2086. [CrossRef]

20. Liu, Q.; Guntuku, S.; Cui, X.S.; Matsuoka, S.; Cortez, D.; Tamai, K.; Luo, G.; Carattini-Rivera, S.; DeMayo, F.; Bradley, A.; et al. Chk1 is an essential kinase that is regulated by Atr and required for the G(2)/M DNA damage checkpoint. *Genes Dev.* **2000**, *14*, 1448–1459.

21. Cadet, J.; Wagner, J.R. DNA base damage by reactive oxygen species, oxidizing agents, and UV radiation. *Cold Spring Harb. Perspect. Biol.* **2013**, *5*, a012559. [CrossRef] [PubMed]

22. Yu, H.; Qiu, J.F.; Ma, L.J.; Hu, Y.J.; Li, P.; Wan, J.B. Phytochemical and phytopharmacological review of *Perilla* frutescens L. (Labiatae), a traditional edible-medicinal herb in China. *Food Chem. Toxicol.* **2017**, *108*, 375–391. [CrossRef] [PubMed]

23. Zhao, T.; Wang, Z.; Su, L.; Sun, X.; Cheng, J.; Zhang, L.; Karungo, S.K.; Han, Y.; Li, S.; Xin, H. An efficient method for transgenic callus induction from Vitis amurensis petiole. *PLoS ONE* **2017**, *12*, e0179730. [CrossRef] [PubMed]

Article

Propolis Suppresses UV-Induced Photoaging in Human Skin through Directly Targeting Phosphoinositide 3-Kinase

Da Hyun Kim [1,†], Joong-Hyuck Auh [2,†], Jeongyeon Oh [1], Seungpyo Hong [3], Sungbin Choi [4], Eun Ju Shin [3], Soon Ok Woo [5], Tae-Gyu Lim [6] and Sanguine Byun [1,*]

[1] Department of Biotechnology, Yonsei University, Seoul 03722, Korea; dahyun.kim@yonsei.ac.kr (D.H.K.); ojo1661@naver.com (J.O.)
[2] Department of Food Science and Technology, Chung-Ang University, Anseong 456-756, Korea; jhauh@cau.ac.kr
[3] Korea Food Research Institute, Wanju 55365, Korea; hsp@kfri.re.kr (S.H.); sej296@naver.com (E.J.S.)
[4] Division of Bioengineering, Incheon National University, Incheon 22012, Korea; 201921148@inu.ac.kr
[5] Department of Agricultural Biology, National Institute of Agricultural Science, Rural Development Administration, Wanju 55365, Korea; wooso1@korea.kr
[6] Department of Food Science & Biotechnology, Sejong University, Seoul 05006, Korea; tglim@sejong.ac.kr
*   Correspondence: sanguine@yonsei.ac.kr
†   These authors contributed equally to this work.

Received: 21 October 2020; Accepted: 8 December 2020; Published: 10 December 2020

Abstract: Propolis is a resinous substance generated by bees using materials from various plant sources. It has been known to exhibit diverse bioactivities including anti-oxidative, anti-microbial, anti-inflammatory, and anti-cancer effects. However, the direct molecular target of propolis and its therapeutic potential against skin aging in humans is not fully understood. Herein, we investigated the effect of propolis on ultraviolet (UV)-mediated skin aging and its underlying molecular mechanism. Propolis suppressed UV-induced matrix metalloproteinase (MMP)-1 production in human dermal fibroblasts. More importantly, propolis treatment reduced UV-induced MMP-1 expression and blocked collagen degradation in human skin tissues, suggesting that the anti-skin-aging activity of propolis can be recapitulated in clinically relevant conditions. While propolis treatment did not display any noticeable effects against extracellular signal-regulated kinase (ERK), p38, and c-jun *N*-terminal kinase (JNK) pathways, propolis exerted significant inhibitory activity specifically against phosphorylations of phosphoinositide-dependent protein kinase-1 (PDK1) and protein kinase B (Akt). Kinase assay results demonstrated that propolis can directly suppress phosphoinositide 3-kinase (PI3K) activity, with preferential selectivity towards PI3K with p110α and p110δ catalytic subunits over other kinases. The content of active compounds was quantified, and among the compounds identified from the propolis extract, caffeic acid phenethyl ester, quercetin, and apigenin were shown to attenuate PI3K activity. These results demonstrate that propolis shows anti-skin-aging effects through direct inhibition of PI3K activity.

Keywords: propolis; skin; matrix metalloproteinase-1; UV; phosphoinositide 3-kinase

---

1. Introduction

Ultraviolet (UV) radiation causes skin photoaging, which is characterized by wrinkle formation, dyspigmentation, and increased fragility [1]. Reduction in collagen and elastic fibers is the major contributing factor for wrinkling of the skin [2]. UV radiation induces signal transduction pathways that lead to the transactivation of matrix metalloproteinases (MMPs), which are enzymes responsible for the degradation of extracellular matrix proteins [3]. MMP-1 in particular is known to majorly contribute in the UV-mediated collagen degradation process causing skin wrinkle formation [4]. As degradation of collagen is known to be the primary reason for UV-mediated skin aging, blocking MMP-1 activity or expression has been recognized as a promising strategy for preventing skin aging [2].

Phosphoinositide 3-kinases (PI3Ks) are lipid kinases that produce phosphatidylinositol 3,4,5-trisphosphate (PIP3) by phosphorylating phosphatidylinositol 4, 5-bisphosphate (PI 4,5-P2). PIP3 can initiate signaling events leading to the activation of phosphoinositide-dependent protein kinase-1 (PDK1) and protein kinase B (Akt) [5]. PI3K and its downstream PDK1 and Akt pathways have been known to play a crucial role in maintaining homeostasis, and abnormal regulation of these signaling pathways are implicated in various pathological conditions, including cancer, immunological disorders, metabolic syndrome, and aging [6]. Importantly, activation of PI3K is also reported to participate in the process of skin aging [7]. As PI3K can promote the expression of MMP-1 leading to collagen breakdown [8], targeting PI3K can provide a therapeutic window to effectively regulate MMP-1 levels and prevent skin aging.

Propolis is produced by bees using substances collected from various parts of plants. Multiple studies have identified the constituents of propolis, which belong to diverse groups of compounds: flavonoids, terpenoids, phenylpropanoids, stilbenes, lignans, sugars, and amino acids [9,10]. Propolis extracts have been known to have applications for the treatment of various diseases due to their anti-oxidant, anti-inflammatory, anti-bacterial, anti-diabetic, and anti-cancer effects [9,11,12], and thus have been widely used as nutraceutical ingredients. However, the effect of propolis against photoaging in human skin is not fully understood, and its molecular target is largely unknown. Therefore, in the current study, we sought to investigate the protective effects of propolis against UV-induced skin aging using human skin tissues and elucidate the molecular mechanism and active components responsible.

2. Materials and Methods

2.1. Reagents

Crude propolis was purchased from a local market in Chungju, Korea. Antibodies to detect phosphorylated extracellular signal-regulated kinase (ERK) 1/2 (Thr202/Tyr204), phosphorylated c-jun *N*-terminal kinase (JNK) (Thr183/Tyr185), phosphorylated p38 (Thr180/Tyr182), phosphorylated PDK1 (Ser241), phosphorylated Akt (Thr308), and total Akt were purchased from Cell Signaling Biotechnology (Beverly, MA, USA). Antibodies against total ERK, total JNK, total p38, and vinculin were obtained from Santa Cruz Biotechnology (Dallas, TX, USA). Antibody for MMP-1 was obtained from R&D systems (Minneapolis, MN, USA). The reference standards of apigenin, caffeic acid, caffeic acid phenethyl ester (CAPE), ferulic acid, gallic acid, naringenin, and quercetin for high-performance liquid chromatography (HPLC) analysis were purchased from Sigma-Aldrich (St. Louis, MO, USA).

2.2. Extraction of Propolis

One gram of propolis was immersed in 10 mL of 80% EtOH and extracted by stirring for 48 h. The extracted propolis was filtered through a non-woven fabric and a filter paper (Watmann # 2). The extracts were then subjected to a 0.22 μm filter and concentrated.

2.3. Cell Culture and UV Irradiation

Hs68 human dermal fibroblast cells were purchased from the American Type Culture Collection (ATCC, Manassas, VA, USA). Cells were cultured in Dulbecco's modified Eagle's medium (DMEM) (Corning Inc., Corning, NY, USA) containing 10% fetal bovine serum (FBS) (Sigma, St. Louis, MO, USA) and 1% penicillin/streptomycin (Corning Inc., New York, NY, USA) at 37 °C in a 5% CO_2 incubator. Cells were irradiated using a UVB cross-linker (6 × 8 W, 312 nm, Model Bio-link BLX, Vilber lourmat, Paris, France). The dosage which produced the highest MMP-1 level without cytotoxicity was chosen as the optimal UVB irradiating condition.

2.4. Excised Human Skin and UV Irradiation

Human skin tissues were purchased from Biopredic International (Rennes, France). The human skin tissues had been obtained following the French Law L.1245-2 CSP. Biopredic International has been approved by the French Ministry of Higher Education and Research for the acquisition, transformation, sale, and export of human biological material to be used in research (AC-2013-1754). This study complied with all principles set forth in the Helsinki Declaration. Human skin tissues were incubated with DMEM containing 10% fetal bovine serum with penicillin/streptomycin in a 5% CO_2 incubator at 37 °C. Ex vivo human skin tissues were treated with propolis extracts (PPE) and exposed to UV once a day for 10 days.

2.5. Cell Cytotoxicity Assay

Cells were seeded in 96-well plates and incubated for 24 h before being cultured in serum-free DMEM. The indicated concentration of PPE was added, and 48 h after the treatment, cell cytotoxicity was measured using the CellTox™ Green Cytotoxicity Assay (Promega, Madison, WI, USA). The fluorescence was measured using the Varioskan multimode microplate reader (Thermo Fisher Scientific, Waltham, MA, USA).

2.6. Enzyme-Linked Immunosorbent Assay (ELISA)

Cells were seeded in 6-well plates at a density of 1.8×10^5 cells/well. When cells reached 80–90% confluence, the medium was replaced with serum-free DMEM. Cells were treated with PPE (2.5, 5, 10 and 20 μg/mL) and incubated for 1 h. After UVB irradiation ($0.03\ J/cm^2$), cells were incubated for 48 h. Media were centrifuged and supernatants were collected and stored at −80 °C. The concentration of MMP-1 protein in the medium was determined using human total MMP-1 DuoSet ELISA kits (R&D systems Inc., Minneapolis, MN, USA), according to the manufacturer's protocol. The absorbance was measured using the Varioskan multimode microplate reader (Thermo Fisher Scientific, Waltham, MA, USA).

2.7. Quantitative Real-Time PCR

Total RNA was extracted from cells using a NucleoSpin RNA isolation kit (Macherey-Nagel, Düren, Germany). RNA purity and concentration were determined by measuring the absorbance at both 260 and 280 nm. cDNA was synthesized from 0.5 μg of total RNA using a ReverTra Ace® qPCR RT Master Mix with gDNA Remover (TOYOBO, Osaka, Japan). MMP-1 mRNA expression levels were analyzed using real-time PCR with Step One Plus Real-Time PCR system (Thermo Fisher Scientific, Waltham, MA, USA) and THUNDERBIRD® SYBR® qPCR Mix (TOYOBO, Osaka, Japan). Primers used in the reaction were MMP-1 (Forward, GCATATCGATGCTGCTCTTTC; Reverse, ACTTTGTGGCCAATTCCAGG) and Glyceraldehyde-3-Phosphate Dehydrogenase (GAPDH) (FW, CCATCACCATCTTCCAGGAG; RV, ACAGTCTTCTGGGTGGCAGT).

2.8. Histological Analysis

Formalin-fixed tissue samples were embedded in paraffin, and 3~4 μm sections were cut. Tissue sections were cut on glass slides, de-paraffinized with xylene, and rehydrated through a series of graded ethanol baths. The sections were then stained with a Masson's trichrome satin kit (Sigma, St. Louis, MO, USA). After the washing step, slides were dehydrated and cleared in xylene before mounting. Stained slides were then photographed using a light microscope.

2.9. Immunoblot Assay

Cells were lysed with RIPA lysis buffer containing a protease and a phosphatase inhibitor cocktail (Sigma-Aldrich, St. Louis, MO, USA). The concentrations of protein in the lysates were determined using a Pierce BCA Protein Assay Kit (Thermo Fisher Scientific, Waltham, MA, USA). Samples were separated using 10% sodium dodecyl sulfate-polyacrylamide gel electrophoresis (SDS-PAGE) gels with 5% stacking gels and transferred to a nitrocellulose (NC) membrane (PALL Corporation,

Port Washington, NY, USA). After transfer, the NC membranes were incubated with the specific primary antibodies at 4 °C overnight. Horseradish peroxidase (HRP)-conjugated immunoglobulin G (IgG) was used as a secondary antibody. The protein expression levels were visualized using the ChemiDocTM XRS + System (Biorad, Hercules, CA, USA) or film.

2.10. High-Performance Liquid Chromatography (HPLC) Analysis

Dionex Ultimate 3000 HPLC system (Thermo Fisher Scientific, Waltham, MA, USA) and INNO C-18 Column (250 mm × 4.6 mm, 5 μm, Youngjinbiochrom, Seongnam, Korea) were used to analyze the chemical composition of propolis. The mobile phase consisted of 0.3% trifluoroacetic acid (Buffer A) and acetonitrile (Buffer B), which were applied in gradient elution as follows: 0–1 min, 5% B; 1–35 min, 5–35% B; 35–50 min, 35–100% B; 50–55 min, 100% B; 55–56 min, 100–5% B; 56–60 min, 5% B at a flow rate of 0.8 mL/min. The injection volume was kept at 10 μL, and the oven temperature was 45 °C. The compounds detected were identified by comparing with reference standards, and the detection wavelengths were 280 and 340 nm (190–400 nm diode array detector (DAD) scanning.

2.11. Kinase Assay

The SelectScreen Kinase Profiling Service (Thermo Fisher Scientific, Waltham, MA, USA) was used to examine kinase activities.

2.12. Modeling the Structures of Chemical Compounds on PI3K

Five highest-resolution crystal structures of human PI3K, p110α catalytic subunit, were retrieved from the Protein Data Bank (PDB ID: 6PYS, 4JPS, 5DXT, 4WAF, 5UBR) [13]. In addition, the structures of porcine PI3K, p110γ catalytic subunit, co-crystallized with ATP (PDB ID: 1E8X) and quercetin (PDB ID: 1E8W) were retrieved from PDB [14]. All the structures were superimposed to that of the human PI3K (PDB ID: 6PYS) before further analysis. The structures of CAPE, quercetin, apigenin, gallic acid, caffeic acid, ferulic acid, and naringenin were retrieved from the PubChem database with PubChem compound identifiers of 5281787, 5280343, 5280443, 370, 689043, 445858, and 932, respectively [15]. The structure of PI3Ks and chemical compounds were converted into the PDBQT format using the prepare_receptor4.py and prepare_ligand4.py scripts of the AutoDock Tools, respectively [16].

The chemical compounds were docked onto the PI3K structures using AutoDock Vina (Version 1.1.2) [17]. The docking simulations were executed within a cubic box with size 20 Å. The box was centered at the coordinate of the N9 atom of the ATP molecule, which is located near the center of the ATP. During the docking simulation, up to 20 docking poses were retrieved for each PI3K and chemical compound pair. The docking structures were inspected on the PyMOL Molecular Graphics System, Version 2.5.0 (Schrödinger, LLC, New York, NY, USA).

3. Results

3.1. Propolis Extract Suppresses UV-Mediated MMP-1 Expression and Collagen Degradation in Human Skin

In order to examine the preventive potential of propolis against skin aging, we examined the effect of propolis on UV-induced MMP-1 levels in Hs68 human dermal fibroblasts (HDFs). Propolis treatment displayed significant inhibitory activity against UV-induced MMP-1 production in HDFs (Figure 1A). Notably, propolis extract did not show cytotoxicity in HDFs at the concentrations tested (Figure 1B). Next, the effect of propolis on mRNA expression of MMP-1 was examined. Propolis completely blocked UV-induced MMP-1 mRNA levels, indicating that the propolis-mediated reduction in MMP-1 levels is due to transcriptional control of MMP-1 expression (Figure 1C).

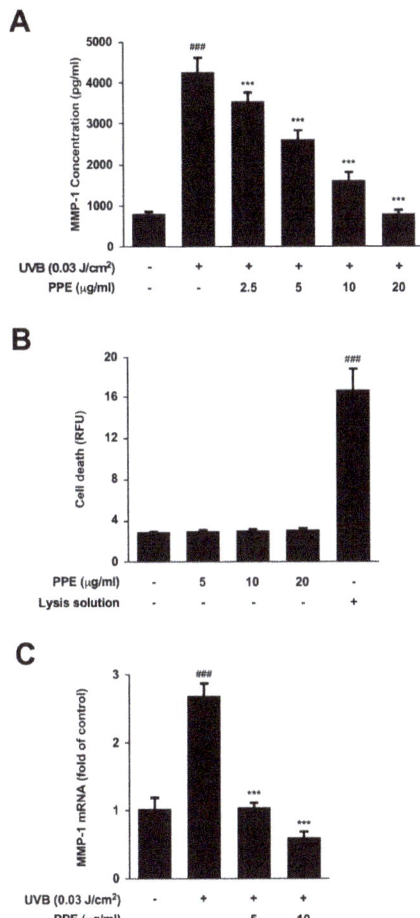

Figure 1. Effect of propolis extract on UV-induced matrix metalloproteinase (MMP)-1 expression in human dermal fibroblasts (HDFs). (**A**) Cytotoxicity of propolis extracts (PPE) was measured using the CellTox Green Cytotoxicity Assay kit. Lysis solution was used as a positive control for cell death; (**B**) Hs68 cells were pre-treated with various concentrations of PPE for 1 h before UVB irradiation (0.03 J/cm^2). Forty-eight hours after UVB exposure, media was collected and MMP-1 levels in the media were determined by ELISA; (**C**) MMP-1 mRNA levels were examined by RT-PCR. Data are shown as mean ± S.D. ### $p < 0.001$ compared with untreated control; *** $p < 0.001$ compared with the group of UVB exposure alone.

To further confirm the efficacy of propolis against skin aging, we applied propolis extract on skin tissues from two different donors. We irradiated human skin tissues with UV for 10 days and investigated the protective effect of propolis (Figure 2A). Propolis treatment effectively suppressed UV-induced MMP-1 expression in human skin measured by immunoblotting of MMP-1 (Figure 2B). In addition, propolis prevented UV-induced collagen degradation in human skin tissues (Figure 2C), suggesting that the inhibitory activity of propolis against skin aging can be observed not only in human skin cells, but also in actual human skin tissues. These results demonstrate that propolis can display anti-skin-aging effects at physiological conditions.

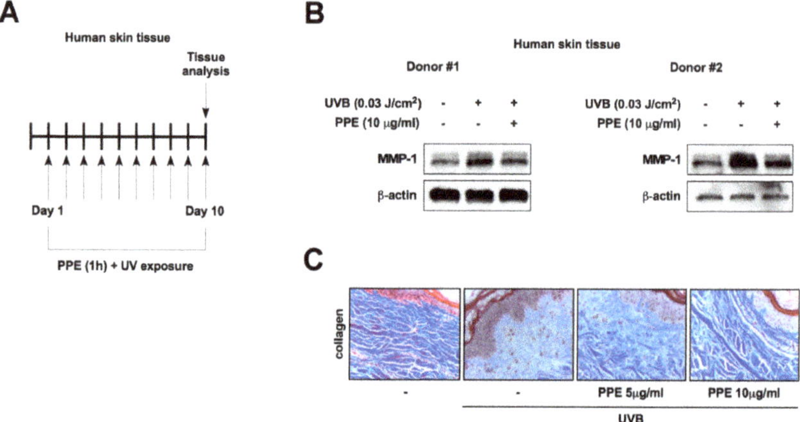

Figure 2. Propolis extract prevents UV-induced matrix metalloproteinase (MMP)-1 expression and collagen degradation in human skin tissues. (**A**) Human skin tissues from two donors were treated with propolis extracts (PPE) and exposed to UVB for 10 consecutive days. At the end of the study, the tissues were fixed in formalin or lysed with lysis buffer for protein extraction; (**B**) MMP-1 expression in human skin tissue lysates was assessed by immunoblotting; (**C**) Collagen fibers were detected with Masson's trichrome staining.

3.2. Propolis Specifically Downregulates Akt and PDK1 Signaling Pathways

As propolis reduced UV-mediated MMP-1 mRNA levels (Figure 1C), we next examined the effect of propolis on upstream regulators of MMP-1 expression. The MAPKs, including ERK, JNK, and p38, are well-known major players in the UV-mediated signal transduction pathway contributing to skin aging [18]. However, treatment of propolis extract did not show any noticeable effect against UV-induced phosphorylations of ERK, JNK, and p38 (Figure 3A). Phosphoinositide 3-kinase (PI3K) and Akt have also been recognized to mediate UV-driven photoaging [8]. Interestingly, propolis extract exerted a significant inhibitory effect against UV-induced phosphorylation of Akt (Figure 3B). PI3K activates PDK1, which in turn phosphorylates Akt (Thr 308) [19]. Therefore, we questioned if propolis would also affect PDK1 activation. Propolis extract dose-dependently inhibited UV-induced phosphorylation of PDK1 (Figure 3B), suggesting that propolis extract could be targeting the PI3K-PDK1-Akt signaling axis.

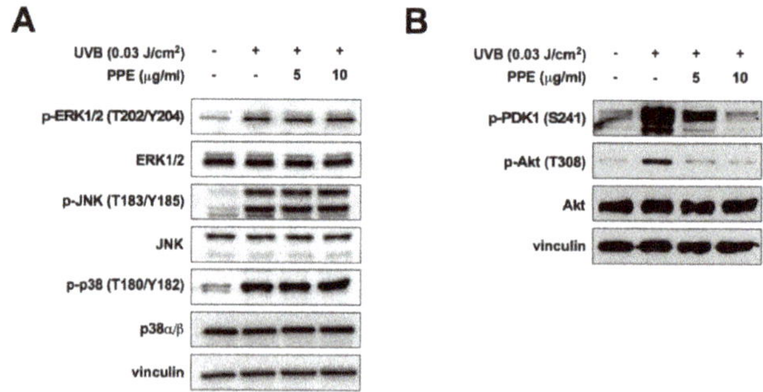

Figure 3. Effect of propolis extract on UV-induced signaling pathway in human dermal fibroblasts. Hs68 cells were pre-treated with propolis extracts (PPE) for 1 h before UVB stimulation (0.03 J/cm^2). Immunoblot analysis of phosphorylated and total (**A**) extracellular signal-regulated kinase (ERK), c-jun *N*-terminal kinase (JNK), and p38; (**B**) phosphoinositide-dependent protein kinase-1 (PDK1) and protein kinase B (Akt) expression in cell lysates. Vinculin was used as a loading control.

3.3. Propolis Directly Targets Phosphoinositide 3-Kinase (PI3K)

Since propolis downregulated phospho-PDK1 and phospho-Akt, we examined the effect of propolis extract against PI3K activity. Kinase assay results against a panel of five PI3K isoforms demonstrated that propolis extract can directly inhibit PI3K activity (Figure 4A). PI3Ks containing the catalytic subunit p110α or p110δ were the most sensitive PI3K isoforms towards propolis. To further confirm the specificity of propolis against PI3K, we assessed the effect of propolis extract against several other kinases. Propolis displayed little or no effect against c-Raf, MEK1, ERK1, PDK1, and p38 kinase activity (Figure 4B–F). Collectively, these results indicate that propolis is a selective and potent inhibitory agent against PI3K.

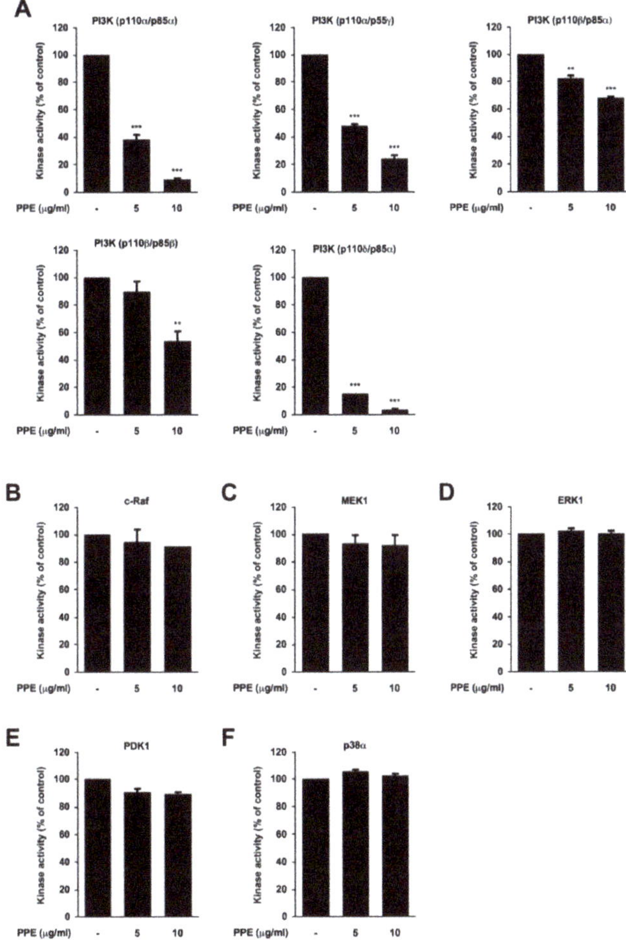

Figure 4. Propolis extract selectively attenuates phosphoinositide 3-kinase (PI3K) activity. Kinase activity of (**A**) PI3K isoforms; (**B–F**) c-Raf, mitogen-activated protein kinase (MAPK)/ERK kinase 1 (MEK1), extracellular signal-regulated kinase 1 (ERK1), phosphoinositide-dependent protein kinase-1 (PDK1), and p38α are shown after propolis extracts (PPE) treatment at the indicated concentrations. Data are shown as mean ± S.D. ** $p < 0.01$, *** $p < 0.001$ compared with untreated control.

3.4. Caffeic Acid Phenethyl Ester (CAPE), Quercetin, and Apigenin Are the Bioactive Components of Propolis in Targeting PI3K

In order to identify the active components responsible for the marked suppression of PI3K activity, we analyzed the chemical composition of propolis. The amounts of apigenin, caffeic acid, CAPE, ferulic acid, gallic acid, naringenin, and quercetin were quantified by HPLC analysis (Table 1). We tested each compound individually against PI3K. CAPE, quercetin, and apigenin showed strong inhibitory effects against PI3K activity, whereas the other compounds displayed relatively minor effects on PI3K activity (Figure 5). As propolis was able to selectively suppress PI3K activity (Figure 4A), CAPE, quercetin, and apigenin appear to be major contributing components to the bioactivity of propolis.

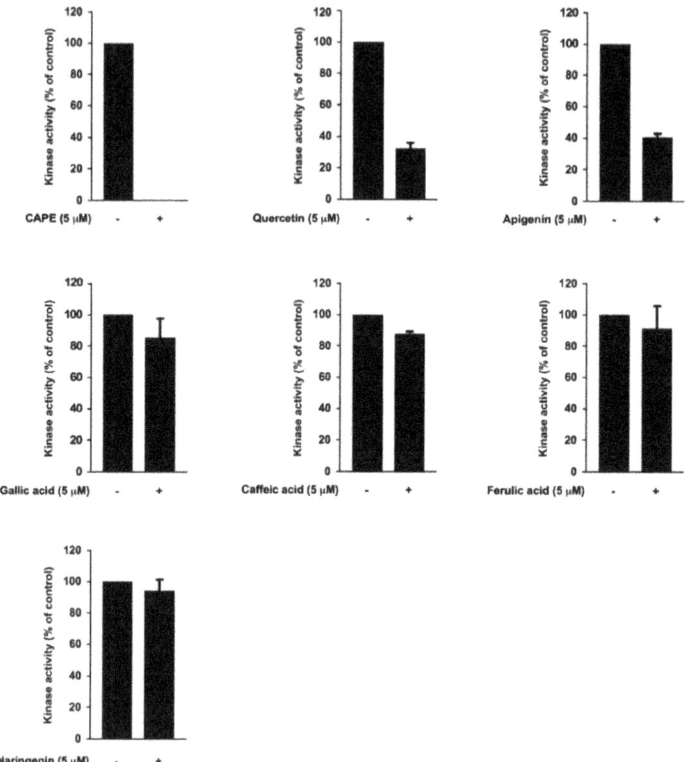

Figure 5. Differential inhibitory activity of propolis compounds against phosphoinositide 3-kinase (PI3K). Effects of caffeic acid phenethyl ester (CAPE), quercetin, apigenin, gallic acid, caffeic acid, ferulic acid, and naringenin on PI3K activity are shown. All compounds were treated at 5 μM. Data are shown as mean ± S.D.

Table 1. Chemical composition of propolis extract identified by HPLC.

No.	Compound	mg/kg of Extract
1	Gallic acid	290.88
2	Protocatechuic acid	69.61
3	Catechin	248.37
4	Caffeic acid	868.34
5	4-Coumaric acid	2170.78
6	Ferulic acid	691.84
7	Quercetin	1070.33
8	Naringenin	504.65
9	Apigenin	2597.77
10	Caffeic acid phenethyl ester (CAPE)	9153

4. Discussion

In the current study, we have discovered that propolis exhibits protective effects against UV-induced skin aging. Notably, the anti-skin-aging efficacy of propolis was also confirmed in ex vivo human skin tissues, suggesting that propolis has the potential to exert its effects when applied to human subjects. These results provide a rationale to further develop propolis as a main ingredient for skin products designed to prevent photoaging.

Analysis of the molecular mechanism of propolis led to the identification of PI3K as a direct target of propolis. PI3K has been reported to regulate MMP-1 expression and skin aging [7,8]. UVB activates the PI3K/Akt signaling pathway, which induces MMP-1 in human dermal fibroblasts [8]. Propolis was able to downregulate UVB-induced PDK1 and Akt activation, in addition to directly attenuating PI3K activity. To validate the specificity, we also tested the inhibitory effect of propolis against the activity of several other kinases involved in UV signaling. Propolis appears to selectively inhibit PI3K activity compared to other kinases. In addition, based on our results, we believe that propolis has a preference for PI3Ks containing the catalytic subunit p110α or p110δ. Previous reports have shown that PI3K can display different functions in an isoform-specific manner [20,21]. Moreover, structural analysis on the catalytic subunits of PI3K demonstrates that each isoform has distinct structural features [5]. Thus, the differential inhibitory effects observed from propolis against PI3K isoforms may be attributed to the unique structure of each isoform.

Flavonoids have been recognized as major active ingredients in various plants with health-promoting effects [22,23]. Due to their structural features, multiple studies have reported that flavonoids can directly target kinases [24]. In our study, we investigated the PI3K inhibitory potency of flavonoids that were detected in the propolis extract. Interestingly, among the tested compounds, CAPE, quercetin, and apigenin attenuated PI3K activity. To further understand how CAPE, quercetin, and apigenin modulated the activity of PI3K, we have built structural models of these compounds on PI3K. Quercetin and apigenin have the benzopyran moiety, and CAPE has the phenyl moiety conjugated with the adjacent alkene moiety (Figure 6A). These moieties form planar structures that resemble the adenine moiety of ATP. Molecular docking simulations located these moieties at the hydrophobic cleft of PI3K, and their polar atoms can form hydrogen bonds with PI3K polar atoms deep in the cleft, similar to the interaction observed between ATP and PI3K (Figure 6B). In addition, the structural models showed that the phenyl group of CAPE can interact with the nearby hydrophobic surface of PI3K, and the phenyl group of quercetin and apigenin can be placed at the hydrophobic cleft. Quercetin had been successfully co-crystallized with porcine PI3K, and its derivatives were developed as PI3K inhibitors [14]. Collectively, it appears that CAPE, quercetin, and apigenin directly bind to the p110 submit of PI3K to modulate its kinase activity.

Although other compounds from propolis displayed structural similarities with CAPE, quercetin, or apigenin, they exhibited minimal inhibitory effects against PI3K activity. Computational modeling suggested that gallic acid, caffeic acid, and ferulic acid may bind to the PI3K ATP binding site, but with reduced binding affinity (Figure S1). They have relatively small sizes and may not be able to bind strongly enough to compete with ATP. Naringenin is structurally similar to quercetin and apigenin, and it can be modeled to bind to the ATP binding site (Figure S1). However, it does not have a benzopyran moiety, and the pyran ring is reduced. This change is likely to abolish the conjugated planar structure and reduce its binding affinity to PI3K, which could explain its lack of effect against PI3K.

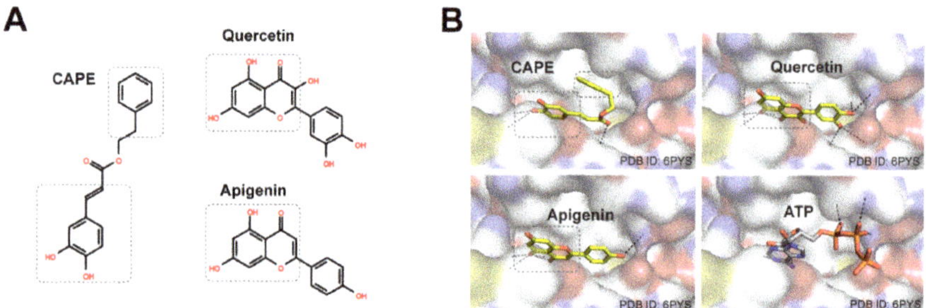

Figure 6. Structures of bioactive compounds and their structural models on phosphoinositide 3-kinase (PI3K) (p110α catalytic subunit). (**A**) The 2D structures of caffeic acid phenethyl ester (CAPE), quercetin, and apigenin. The moieties interacting with PI3K were highlighted with dashed boxes; (**B**) The structural models of CAPE, quercetin, and apigenin on PI3K (p110α catalytic subunit). ATP was located on human PI3K by superimposing the structure of porcine PI3K co-crystallized with ATP to the human PI3K and displaying its ATP part.

Previous reports have suggested the involvement of some of the active compounds in regulating PI3K or MMP-1. Apigenin was reported to suppress UVA-mediated MMP-1 expression in the HaCaT cell line and promote collagen synthesis in dermal fibroblasts [25,26]. Previous studies suggested that CAPE and quercetin can reduce UV-induced MMP-1 expression in cells and skin tissues [27,28]. Apigenin and quercetin have been shown to suppress PI3K activity [29,30]. Based on these findings, we propose that CAPE, quercetin, and apigenin may function as major contributing factors for the anti-skin-aging and PI3K inhibitory effects exerted by propolis.

In the current study, we have found that propolis can show anti-skin-aging effects in human skin tissues, suggesting that the anti-skin-aging activity of propolis can be exhibited in physiological conditions. More importantly, we have identified that propolis directly suppresses the kinase activity of phosphoinositide 3-kinase (PI3K) and its downstream signaling pathway (Figure 7). Discovering PI3K as the molecular target and identifying active components of propolis will aid in expanding the therapeutic applications of propolis.

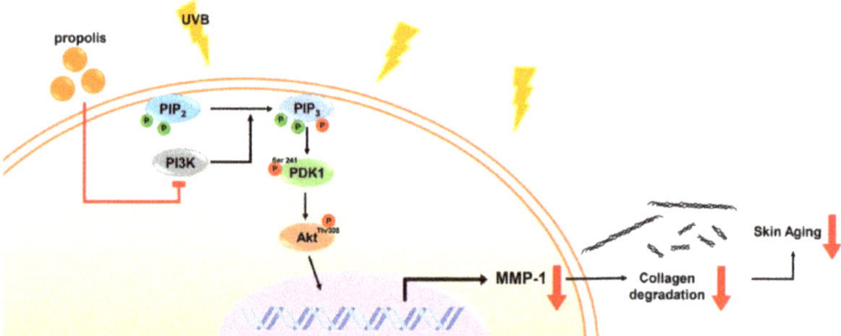

Figure 7. Schematic diagram summarizing the molecular mechanism of propolis. PIP₂, Phosphatidylinositol 4,5-bisphosphate; PIP₃, Phosphatidylinositol (3,4,5)-trisphosphate; PI3K, phosphoinositide 3-kinase; PDK1, phosphoinositide-dependent protein kinase-1; Akt, protein kinase B; MMP-1, matrix metalloproteinase-1.

Supplementary Materials: The following are available online at http://www.mdpi.com/2072-6643/12/12/3790/s1, Figure S1: Distribution of binding scores.

Author Contributions: Conceptualization, S.B.; formal analysis, D.H.K., J.-H.A., J.O., S.H., S.C., E.J.S., S.O.W., T.-G.L.; investigation, D.H.K., J.-H.A., J.O., S.H., S.C., E.J.S., S.O.W., T.-G.L.; resources, S.O.W., T.-G.L., S.B.; writing—original draft preparation, D.H.K., S.B.; writing—review and editing, S.B.; visualization, D.H.K., J.-H.A., J.O., S.C.; supervision, S.B.; project administration, S.B.; funding acquisition, S.B. All authors have read and agreed to the published version of the manuscript.

Funding: This work was supported by the National Research Foundation of Korea (NRF) grant funded by the Korean government (MSIP) (NRF-2020R1A2C1010703), and by the Yonsei University Research Fund of 2020-22-0073. This research was supported in part by the Brain Korea 21 (BK21) program; D.H.K. is a fellowship awardee by BK21 program.

Conflicts of Interest: The authors declare no conflict of interest.

References

1. Rittié, L.; Fisher, G.J. UV-light-induced signal cascades and skin aging. *Ageing Res. Rev.* **2002**, *1*, 705–720. [CrossRef]

2. Philips, N.; Auler, S.; Hugo, R.; Gonzalez, S. Beneficial Regulation of Matrix Metalloproteinases for Skin Health. *Enzym. Res.* **2011**, *2011*, 427285. [CrossRef] [PubMed]

3. Fisher, G.J.; Kang, S.; Varani, J.; Bata-Csorgo, Z.; Wan, Y.; Datta, S.; Voorhees, J.J. Mechanisms of photoaging and chronological skin aging. *Arch. Dermatol.* **2002**, *138*, 1462–1470. [CrossRef]

4. Talwar, H.S.; Griffiths, C.E.; Fisher, G.J.; Hamilton, T.A.; Voorhees, J.J. Reduced type I and type III procollagens in photodamaged adult human skin. *J. Investig. Dermatol.* **1995**, *105*, 285–290. [CrossRef]

5. Amzel, L.M.; Huang, C.H.; Mandelker, D.; Lengauer, C.; Gabelli, S.B.; Vogelstein, B. Structural comparisons of class I phosphoinositide 3-kinases. *Nat. Rev. Cancer* **2008**, *8*, 665–669. [CrossRef]

6. Fruman, D.A.; Chiu, H.; Hopkins, B.D.; Bagrodia, S.; Cantley, L.C.; Abraham, R.T. The PI3K Pathway in Human Disease. *Cell* **2017**, *170*, 605–635. [CrossRef]

7. Noh, E.M.; Park, J.; Song, H.R.; Kim, J.M.; Lee, M.; Song, H.K.; Hong, O.Y.; Whang, P.H.; Han, M.K.; Kwon, K.B.; et al. Skin Aging-Dependent Activation of the PI3K Signaling Pathway via Downregulation of PTEN Increases Intracellular ROS in Human Dermal Fibroblasts. *Oxid. Med. Cell. Longev.* **2016**, *2016*, 6354261. [CrossRef]

8. Oh, J.H.; Kim, A.; Park, J.M.; Kim, S.H.; Chung, A.S. Ultraviolet B-induced matrix metalloproteinase-1 and -3 secretions are mediated via PTEN/Akt pathway in human dermal fibroblasts. *J. Cell. Physiol.* **2006**, *209*, 775–785. [CrossRef]

9. Wagh, V.D. Propolis: A wonder bees product and its pharmacological potentials. *Adv. Pharmacol. Sci.* **2013**, *2013*, 308249. [CrossRef]

10. Huang, S.; Zhang, C.P.; Wang, K.; Li, G.Q.; Hu, F.L. Recent advances in the chemical composition of propolis. *Molecules* **2014**, *19*, 19610–19632. [CrossRef]

11. Ambi, A.; Bryan, J.; Borbon, K.; Centeno, D.; Liu, T.; Chen, T.P.; Cattabiani, T.; Traba, C. Are Russian propolis ethanol extracts the future for the prevention of medical and biomedical implant contaminations? *Phytomedicine* **2017**, *30*, 50–58. [CrossRef] [PubMed]

12. Chen, L.H.; Chien, Y.W.; Chang, M.L.; Hou, C.C.; Chan, C.H.; Tang, H.W.; Huang, H.Y. Taiwanese Green Propolis Ethanol Extract Delays the Progression of Type 2 Diabetes Mellitus in Rats Treated with Streptozotocin/High-Fat Diet. *Nutrients* **2018**, *10*, 503. [CrossRef] [PubMed]

13. Berman, H.M.; Westbrook, J.; Feng, Z.; Gilliland, G.; Bhat, T.N.; Weissig, H.; Shindyalov, I.N.; Bourne, P.E. The Protein Data Bank. *Nucleic Acids Res.* **2000**, *28*, 235–242. [CrossRef] [PubMed]

14. Walker, E.H.; Pacold, M.E.; Perisic, O.; Stephens, L.; Hawkins, P.T.; Wymann, M.P.; Williams, R.L. Structural determinants of phosphoinositide 3-kinase inhibition by wortmannin, LY294002, quercetin, myricetin, and staurosporine. *Mol. Cell* **2000**, *6*, 909–919. [CrossRef]

15. Kim, S.; Chen, J.; Cheng, T.; Gindulyte, A.; He, J.; He, S.; Li, Q.; Shoemaker, B.A.; Thiessen, P.A.; Yu, B.; et al. PubChem 2019 update: Improved access to chemical data. *Nucleic Acids Res.* **2019**, *47*, D1102–D1109. [CrossRef]

16. Morris, G.M.; Huey, R.; Lindstrom, W.; Sanner, M.F.; Belew, R.K.; Goodsell, D.S.; Olson, A.J. AutoDock4 and AutoDockTools4: Automated docking with selective receptor flexibility. *J. Comput. Chem.* **2009**, *30*, 2785–2791. [CrossRef]

17. Trott, O.; Olson, A.J. AutoDock Vina: Improving the speed and accuracy of docking with a new scoring function, efficient optimization, and multithreading. *J. Comput. Chem.* **2010**, *31*, 455–461. [CrossRef]
18. Bode, A.M.; Dong, Z. Mitogen-activated protein kinase activation in UV-induced signal transduction. *Sci. STKE* **2003**, *2003*, re2. [CrossRef]
19. Manning, B.D.; Toker, A. AKT/PKB Signaling: Navigating the Network. *Cell* **2017**, *169*, 381–405. [CrossRef]
20. Matheny, R.W., Jr.; Adamo, M.L. PI3K p110 alpha and p110 beta have differential effects on Akt activation and protection against oxidative stress-induced apoptosis in myoblasts. *Cell Death Differ.* **2010**, *17*, 677–688. [CrossRef]
21. Utermark, T.; Rao, T.; Cheng, H.; Wang, Q.; Lee, S.H.; Wang, Z.C.; Iglehart, J.D.; Roberts, T.M.; Muller, W.J.; Zhao, J.J. The p110α and p110β isoforms of PI3K play divergent roles in mammary gland development and tumorigenesis. *Genes Dev.* **2012**, *26*, 1573–1586. [CrossRef] [PubMed]
22. Bondonno, N.P.; Lewis, J.R.; Blekkenhorst, L.C.; Bondonno, C.P.; Shin, J.H.; Croft, K.D.; Woodman, R.J.; Wong, G.; Lim, W.H.; Gopinath, B.; et al. Association of flavonoids and flavonoid-rich foods with all-cause mortality: The Blue Mountains Eye Study. *Clin. Nutr.* **2020**, *39*, 141–150. [CrossRef] [PubMed]
23. Panche, A.N.; Diwan, A.D.; Chandra, S.R. Flavonoids: An overview. *J. Nutr. Sci.* **2016**, *5*, e47. [CrossRef] [PubMed]
24. Farrand, L.; Byun, S. Induction of Synthetic Lethality by Natural Compounds Targeting Cancer Signaling. *Curr. Pharm. Des.* **2017**, *23*, 4311–4320. [CrossRef] [PubMed]
25. Hwang, Y.P.; Oh, K.N.; Yun, H.J.; Jeong, H.G. The flavonoids apigenin and luteolin suppress ultraviolet A-induced matrix metalloproteinase-1 expression via MAPKs and AP-1-dependent signaling in HaCaT cells. *J. Dermatol. Sci.* **2011**, *61*, 23–31. [CrossRef] [PubMed]
26. Zhang, Y.; Wang, J.; Cheng, X.; Yi, B.; Zhang, X.; Li, Q. Apigenin induces dermal collagen synthesis via smad2/3 signaling pathway. *Eur. J. Histochem.* **2015**, *59*, 2467. [CrossRef] [PubMed]
27. Shin, E.J.; Jo, S.; Choi, H.K.; Choi, S.; Byun, S.; Lim, T.G. Caffeic Acid Phenethyl Ester Inhibits UV-Induced MMP-1 Expression by Targeting Histone Acetyltransferases in Human Skin. *Int. J. Mol. Sci.* **2019**, *20*, 3055. [CrossRef] [PubMed]
28. Shin, E.J.; Lee, J.S.; Hong, S.; Lim, T.G.; Byun, S. Quercetin Directly Targets JAK2 and PKCdelta and Prevents UV-Induced Photoaging in Human Skin. *Int. J. Mol. Sci.* **2019**, *20*, 5262. [CrossRef]
29. Lee, W.J.; Chen, W.K.; Wang, C.J.; Lin, W.L.; Tseng, T.H. Apigenin inhibits HGF-promoted invasive growth and metastasis involving blocking PI3K/Akt pathway and beta 4 integrin function in MDA-MB-231 breast cancer cells. *Toxicol. Appl. Pharmacol.* **2008**, *226*, 178–191. [CrossRef]
30. Maurya, A.K.; Vinayak, M. PI-103 and Quercetin Attenuate PI3K-AKT Signaling Pathway in T-Cell Lymphoma Exposed to Hydrogen Peroxide. *PLoS ONE* **2016**, *11*, e0160686. [CrossRef]

Publisher's Note: MDPI stays neutral with regard to jurisdictional claims in published maps and institutional affiliations.

Article

Anti-Melanogenesis Effects of Lotus Seedpod In Vitro and In Vivo

Jen-Ying Hsu [1,†], Hui-Hsuan Lin [2,†], Ting-Shuan Li [1], Chaio-Yun Tseng [1], Yueching Wong [1] and Jing-Hsien Chen [1,3,*]

1   Department of Nutrition, Chung Shan Medical University, Taichung City 40201, Taiwan;
    jyhsu0530@gmail.com (J.-Y.H.); summer6279@yahoo.com.tw (T.-S.L.); winnielovejb3131@gmail.com (C.-Y.T.);
    wyc@csmu.edu.tw (Y.W.)
2   Department of Medical Laboratory and Biotechnology, Chung Shan Medical University,
    Taichung City 40201, Taiwan; linhh@csmu.edu.tw
3   Department of Medical Research, Chung Shan Medical University Hospital, Taichung City 40201, Taiwan
*   Correspondence: cjh0828@csmu.edu.tw; Tel.: +886-4-24730022 (ext. 12195); Fax: +886-4-23248175
†   These authors contributed equally to this work.

Received: 14 October 2020; Accepted: 15 November 2020; Published: 18 November 2020

Abstract: Melanogenesis has many important physiological functions. However, abnormal melanin production causes various pigmentation disorders. Melanin synthesis is stimulated by α-melanocyte stimulating hormone (α-MSH) and ultraviolet (UV) irradiation. Lotus seedpod extract (LSE) has been reported as possessing antioxidative, anti-aging, and anticancer activities. The present study examined the effect of LSE on melanogenesis and the involved signaling pathways in vitro and in vivo. Results showed that non-cytotoxic doses of LSE and its main component epigallocatechin (EGC) reduced both tyrosinase activity and melanin production in the α-MSH-induced melanoma cells. Western blotting data revealed that LSE and EGC inhibited expressions of tyrosinase and tyrosinase-related protein 1 (TRP-1). Phosphorylation of p38 and protein kinase A (PKA) stimulated by α-MSH was efficiently blocked by LSE treatment. Furthermore, LSE suppressed the nuclear level of cAMP-response element binding protein (CREB) and disturbed the activation of melanocyte inducing transcription factor (MITF) in the α-MSH-stimulated B16F0 cells. The in vivo study revealed that LSE inhibited melanin production in the ear skin of C57BL/6 mice after exposure to UVB. These findings suggested that the anti-melanogenesis of LSE involved both PKA and p38 signaling pathways. LSE is a potent novo natural depigmenting agent for cosmetics or pharmaceutical applications.

Keywords: melanogenesis; α-MSH; UVB irradiation; lotus seedpod extract; epigallocatechin

1. Introduction

Melanin produced by melanocytes through melanogenesis directly affects the color of skin, hair, and eyes. Besides, melanin protects the skin from ultraviolet (UV) stimulation. However, abnormal melanogenesis causes hypo- or hyper-pigmentation disease and impacts the patient's quality of life. Melanogenesis is influenced by intrinsic and extrinsic factors. Intrinsic factors of melanogenesis are resulted from overproduction of second messenger cyclic adenosine monophosphate (cAMP) which might occur during pregnancy and inflammation [1]. The major extrinsic factor of melanogenesis is UV radiation acting through the melanocortin-1 receptor (MC1R) that is activated by α-melanocyte-stimulating hormone (α-MSH) [2].

Melanogenesis is triggered by UV radiation or hormone stimulation such as α-MSH and adrenocorticotropic hormone (ACTH) or inflammation. Tyrosinase is the rate-limiting enzyme of melanin synthesis which is responsible for catalyzing L-tyrosine hydroxylation to levodopa (L-DOPA) [3]. Aside from tyrosinase, tyrosinase-related protein-1 (TRP-1) and tyrosinase-related protein-2 (TRP-2), resided in melanosome, also regulate melanin production [3]. The expressions of these enzymes were regulated by melanocyte inducing transcription factor (MITF), which has its specific DNA binding sequence to modulate melanocyte proliferation, differentiation, and melanogenesis [4]. The expression and activity of MITF were modulated by cAMP-responsive binding protein (CREB), mitogen-activated protein kinase (MAPK) family proteins, ribosomal S6 kinase (RSK), glycogen synthase kinase-3β (GSK3β), and p38 [5]. The binding of α-MSH to its receptor, MC1R, triggered intracellular cAMP levels, and induced cAMP-dependent signaling pathways [6]. It has been well established that cAMP elevation induced protein kinase A (PKA) activity and MAPK family proteins, including extracellular signal-regulated kinase (ERK), c-Jun N terminal kinase (JNK)1/2, and p38, were involved in melanogenesis signaling [6,7].

Nelumbo nucifera Gaertn., more commonly known as lotus, is a perennial hydrophyte which is mainly cultivated in high temperature and humidity climate patterns such as Taiwan, Japan, and some African countries. Most parts of lotus could be used as food or traditional Chinese medicine. However, lotus seedpod is a non-edible part of lotus and is usually discarded after harvest. Several studies have revealed that extract of lotus seedpods contained bioactive compounds. Proanthocyanin has been identified in lotus seedpods extract and was reported to exert antioxidative activity [8], antitumor activity [9], ameliorating cognitive impairment [10], inhibiting glycation [11], anti-inflammation [12], and anti-lipotoxicity [13]. According to the analysis of previous studies, EGC is abundant in LSE [12,14] that has been reported to have a depigmenting effect [15]. The present study used EGC as a reference standard to compare its efficacy with LSE.

The study is aimed to investigate the melanin synthesis effect and mechanism of LSE through in vivo and in vitro studies. Murine melanoma cells were used for the in vitro study under the induction of α-MSH. LSE or its main component EGC in various concentrations were applied to α-MSH-induced melanocyte to investigate the melanin synthesis effect and their possible mechanisms. Furthermore, LSE was applied topically to mice ears to confirm the melanin synthesis effect under repetitive UVB exposure.

2. Materials and Methods

2.1. Chemical and Reagents

α-MSH, EGC, H89, SB203580, and TRIzol reagent used in the present study were purchased from Sigma-Aldrich (St. Louis, MO, USA). Dulbecco's modified Eagle's Medium (DMEM), fetal bovine serum, penicillin/streptomycin, L-glutamine, and trypsin-EDTA (ethylenediaminetetraacetic acid) were purchased from Hyclone Laboratories (UT, USA). Antibodies including CREB, ERK (extracellular signal-regulated kinases), JNK1 (c-Jun N-terminal kinases 1), JNK2 (c-Jun N-terminal kinases 2), laminin B, MC1R, MITF, p-CREB, TRP-1, TRP-2, and tyrosinase were purchased from Santa Cruz Biotechnology (Dallas city, TX, USA). p38 MAPK, phospho-p38 (p-p38) MAPK, phospho-JNK, PKA, and p-PKA were purchased from Cell Signaling Technology (Beverly, MA, USA). Phospho-ERK and β-actin were purchased from Novus Biologicals (Littleton, CO, USA). The second antibody conjugated with horseradish peroxidase was purchased from Santa Cruz and Sigma-Aldrich.

2.2. Lotus Seedpod Extract (LSE) Preparation

The extraction of lotus seedpod was described in the previous study [12]. Briefly, dried lotus seedpods without lotus seeds were obtained from Baihe District, Tainan City, Taiwan. Dried lotus seedpods (100 g) were simmered in 4 L water at 95 °C for 2 h. After cooling, the aqueous extract of lotus seedpods was evaporated under vacuum at −80 °C. The decoction was concentrated and lyophilized as lotus seedpods extract powder (LSE) with a yield of about 18.4% from dried material. LSE powder was stored at −20 °C before use. The compounds of LSE have been identified and reported in the previous study [12].

2.3. Cell Culture

Murine melanoma cells B16F0 were purchased from Bioresource Collection and Research Center (Food Industry Research and Development Institute, Hsinchu City, Taiwan). B16F0 cells were cultured in DMEM supplemented with 10% FBS, 1% penicillin/streptomycin, and 1% glutamine at 37 °C in a 5% CO_2 humidified incubator.

2.4. Cell Viability Assay

B16F0 cells were seeded in a 6-well plate (2×10^5 cells/well) for 24 h. LSE was dissolved in deionized water with an equal volume of ethanol before use. EGC was prepared in deionized water. The cells were treated with/without LSE (5, 10, 15, 20, 25 µg/mL) or EGC (5, 10, 15, 20, 25 µM) for 48 h. Cell viability assay was analyzed by trypan blue, and live cells were measured and expressed as % of cell growth.

2.5. Tyrosinase Activity Assay

Tyrosinase activity assay was modified from the method described previously [16]. B16F0 cells were seeded in a 6-well plate (2×10^5 cells/well) and treated with 1 µM α-MSH with/without LSE (10, 15, 20 µg/mL) and 15 µM EGC. After 48 h of incubation, cells were harvested by trypsin-EDTA and collected into a centrifuge tube. After being centrifuged, the cell pellet was washed with PBS twice and homogenized with PBS buffer containing 1% Triton X-100. Cell lysate (100 µL) and 2 µM L-DOPA in equal volume were mixed in a 96-well plate at 37 °C for 60 min. Tyrosinase activity in each treatment was determined at absorbance 490 nm with a SpectraMax M5/M5 Microplate Reader (Molecular Devices, Sunnyvale, CA, USA). Chemiluminescence (ECL) reagent was obtained from Millipore (Burlington, MA, USA). The tyrosinase activity of α-MSH-induced B16F0 cells with LSE, H89, and SB203580 treatment was assessed by the same method. The concentration of H89 or SB203580 was used according to the previous studies [17,18].

2.6. Melanin Content Assay

Melanin content assay was measured according to the method of Park et al. [19]. α-MSH-induced B16F0 cells were treated with/without LSE, EGC, H89, or SB 203580 for 48 h. The cell pellet was washed with PBS twice, and then 500 µL 1N NaOH was added and incubated at 80 °C for 1 h. Melanin content was measured at absorbance 409 nm with a microplate reader (SpectraMax M5/M5 Microplate Reader, Molecular Devices, Sunnyvale, CA, USA) and calculated from the melanin standard curve using synthetic melanin.

2.7. Cyclic AMP Assay

The intracellular cAMP level was measured by Cyclic AMP ELISA Kit (cat#581001, Cayman Chemical Company, Ann Arbor, MI, USA). According to the manufacturer's instruction, cell lysate and cAMP-acetylcholinesterase (AChE) conjugate (cAMP tracer) were added and reacted in the pre-coated microplate. After washout with wash buffer, Ellman's reagent, which contained the substrate of cAMP tracer, was added to wells and incubated for 90 min. The absorbance was measured at 412 nm with a microplate reader (SpectraMax M5/M5 Microplate Reader, Molecular Devices, Sunnyvale, CA, USA).

2.8. Real-Time PCR for Analyzing mRNA Levels

Primer sequences used in the present study were shown in Table 1. Total RNA was isolated from B16F0 cells by TRIzol reagent according to the manufacturer's instructions. cDNA was synthesized from total RNA using the GoScript™ Reverse Transcriptase System (Thermo Scientific, Rockford, IL, USA). The amplification of cDNA was performed using Power SYBR Green Master Mix (Applied Biosystems, Foster City, CA, USA) and detected by the StepOne™ Real-Time System using the comparative Ct (cycle threshold) method. Ct value in each sample was determined and normalized by the Ct value of *Gapdh* (ΔCt). The gene expression levels were calculated by the $2^{-\Delta\Delta Ct}$ method. Every experiment was conducted in triplicate.

Table 1. Primer sequence of the target genes for real-time PCR.

Gene	Sequence
Tyrosinase	Forward: 5′-CACCTGAGGGACCACTATTACG-3′ Reverse: 5′-GGCAGTTCTATCCATTGATCCAG-3′
Trp-1	Forward: 5′-ATGAAATCTTACAACGTCCTCCC-3′ Reverse: 5′-TGGCACACTCTCGTGGAAACTGA-3′
Trp-2	Forward: 5′-CTTCAACCGGACATGCAAATGC-3′ Reverse: 5′-GCTTCTTCCGATTACAGTCGGG-3′
Mc1r	Forward: 5′-CTATGCGCTGCGTTATCACAG-3′ Reverse: 5′-AAAGAAAGTGACGAGGCAGAG-3′
Gapdh	Forward: 5′-AGGTCGGTGTGAACGGATTTG-3′ Reverse: 5′-GGGGTCGTTGATGGCAACAAT-3′

2.9. Electrophoretic Mobility Shift Assay (EMSA)

Cell nucleus was isolated by the Mitochondria isolation kit for mammalian cells (cat#89874, Thermo Scientific, Rockford, IL, USA). The DNA-binding activity of CREB and MITF was assessed by EMSA. The sequences of DNA probe were listed in Table 2. Cell nuclear protein (10 μL) was mixed with 1 μL poly (dI-dC), 2 μL biotin-DNA probe, loading buffer, and binding buffer for the binding reaction. Protein and DNA complexes were separated by 8% negative-TBE (tris borate EDTA) PAGE (polyacrylamide gel electrophoresis) and transferred to a positive-charged nylon membrane. Crosslinking of the transferred membrane was performed under 1200 mJ/cm^2 of UV light. The band shift was visualized by chemiluminescence and quantified via ImageQuant™ LAS 4000 mini (GE Healthcare Bio-Sciences AB, Uppsala, Sweden).

Table 2. Sequence of DNA probe.

Gene	Sequence
CREB	Forward: AGAGA TTGCC TGACG TCAGA GAGGT AG-5′ Biotin Reverse: CTAGC TCTCT GACGT CAGGC AATCT CT-5′ Biotin
MITF	Forward: ACTAC AACAC GTGTA GGCCA-5′ Biotin Reverse: TGGCC TACAC GTGTT GTAGT 5′-Biotin

2.10. Animal and Experimental Design

Male C57BL/6 mice were purchased from the National Laboratory Animal Center (National Science Council, Taipei City, Taiwan). The animal procedure was approved and complied with the Chung Shan Medical University Animal Care Committee (IAUCC: 1255). Mice were housed in a constant humidity (55% ± 2%) and temperature (22 ± 2 °C) control room with a 12-h light/dark cycle. Food and drink (water) were given ad libitum during the experiment. Mice were randomly divided into five groups (n = 5) after adaptation for a week. The experimental groups were as follows: control group, UVB group, UVB + LSE 1.25 mg/cm^2 group, UVB + LSE 2.5 mg/cm^2 group, and UVB + EGC 2 mM/cm^2 group. LSE was dissolved in an equal volume of water and alcohol and EGC in DMSO (dimethyl sulfoxide) before use. Mice ears were topically applied with LSE (1.25 mg/cm^2 and 2.5 mg/cm^2) or 2 mM/cm^2 EGC before exposure to UVB. After about 30 min for the moisture to evaporate, mice were exposed to UVB irradiation (100 mJ/cm^2). The UVB irradiation was conducted three times a week for 8 weeks and LSE or EGC was applied before UVB exposure. After 8 weeks of treatment, mice were sacrificed by carbon-dioxide asphyxiation followed by exsanguinations, and ear tissues were collected for melanin content and Western blot analysis.

2.11. Western Blotting

Mice ear skin tissue (1.5 g) was homogenized by RIPA buffer (radioimmunoprecipitation assay buffer) and centrifuged at 12,000× g for 10 min at 4 °C. The supernatant was transferred to a new microtube and stored at −80 °C before use. Protein concentrations in cell lysates or animal tissues were quantified by a Dual-Range™ BCA Protein Assay Kit (Energenesis Biomedical Co., LTD, Taipei city, Taiwan). Proteins (20–50 µg) were separated by SDS-PAGE and transferred to nitrocellulose paper. The membranes were blocked with 5% skim milk at 4 °C for 1–2 h before being incubated with primary antibody overnight at 4 °C. The appropriate horseradish peroxidase-conjugated secondary antibodies were incubated for 2 h at 4 °C. β-actin and laminin B were used as internal controls. Protein bands were visualized by enhanced chemiluminescence reagent. Protein levels were quantified by ImageQuant™ LAS 4000 mini (GE Healthcare Bio-Sciences AB, Uppsala, Sweden).

2.12. Statistical Analysis

Three or more separate experiments were performed. Data are reported as means ± standard deviation (SD) of three independent experiments. Student's *t*-test was used for the analysis between two groups with only one factor involved. For the experiments of dose response of LSE or EGC, one-way analysis of variance (ANOVA) with post-hoc Dunnett's test was used to calculate the *p*-value for each dose treatment compared to the α-MSH-treated group (α-MSH alone, without LSE or EGC), and regression was used to test the *p*-value of the dependency of a parameter to dosage. Significant differences were established at $p < 0.05$.

3. Results

3.1. Effect of LSE and EGC on Cell Viability in B16F0 Cells

The cell viability of B16F0 cells treated with different LSE and ECG concentrations is shown in Figure 1a,b. After being treated for 48 h, the survival rate was significantly decreased by 27% in 25 µg/mL LSE (Figure 1a). The survival rate of EGC treatment was significantly reduced by 36% in the 25 µM EGC group (Figure 1b). The concentrations of LSE below 20 µg/mL and EGC below 20 µM showed no cytotoxicity and were selected for investigating melanin formation.

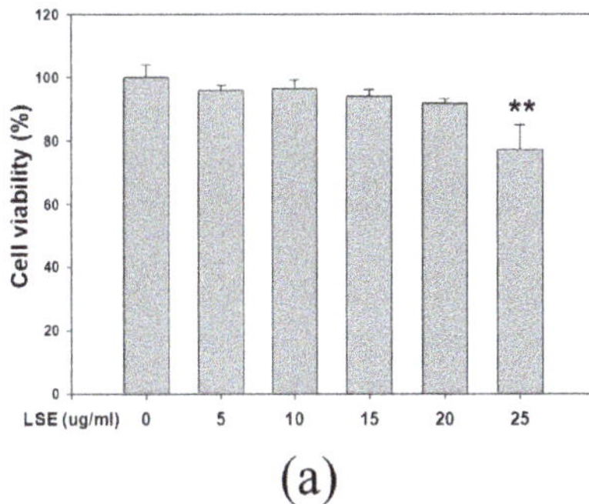
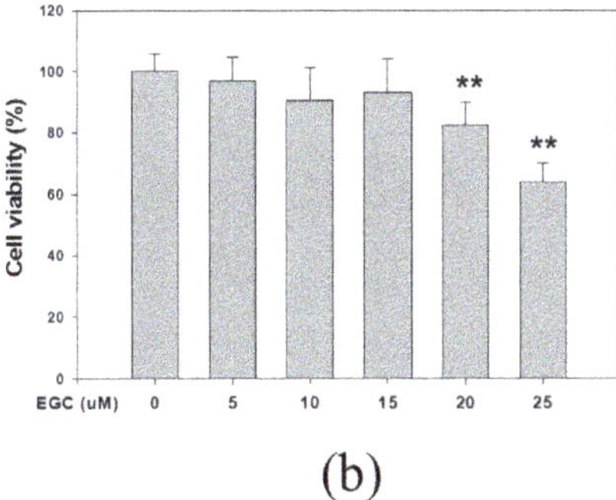

(a)   (b)

Figure 1. Effect of LSE or EGC on B16F0 cell viability. B16F0 cells were treated with various concentrations of (**a**) LSE (0, 5, 10, 15, 20, 25 µg/mL) and (**b**) EGC (0, 5, 10, 15, 20, 25 µM) for 48 h. The number of cells was counted by trypan blue dye exclusion assay. LSE: lotus seedpod extract. EGC: epigallocatechin. Values represent the mean ± SD of three independent experiments. ** $p < 0.01$, compared with control via one-way ANOVA with post-hoc Dunnett's test.

3.2. Effect of LSE and EGC on Melanin Formation and Tyrosinase Activity in α-MSH-Induced B16F0 Cells

Melanin formation in the α-MSH-induced B16F0 cells treated with various LSE and EGC concentrations is revealed in Figure 2a,b. LSE of 10, 15, and 20 µg/mL doses significantly suppressed the melanin production in α-MSH-induced B16F0 cells, especially in 15 and 20 µg/mL LSE treatments whose cell pellets showed lighter colors (Figure 2a). Treatments of 10 µM and 15 µM EGC had significantly decreased melanin content when compared to the α-MSH group. Tyrosinase activities in LSE and EGC treatments were assessed and the results are shown in Figure 2b. According to the results, 10, 15, and 20 µg/mL LSE significantly inhibited 65%, 76%, and 112% tyrosinase activity in the α-MSH-induced B16F0 cells (Figure 2c). EGC treatment (15 µM) suppressed 47% of tyrosinase activity (Figure 2c). The results suggested that LSE and EGC reduced melanin production through repressing tyrosinase activity. Tyrosinase, TRP-1, and TRP-2 are important enzymes for catalyzing melanin formation in melanosome. The protein and mRNA levels in each group were analyzed and revealed in Figure 2d,e. Tyrosinase and TRP-1 protein and mRNA levels were significantly increased by α-MSH stimulation when compared with the control group, while TRP-2 protein and mRNA levels were similar to the control group. Tyrosinase protein and mRNA levels in 15 and 20 µg/mL LSE and 15 µM EGC treatments were significantly decreased in α-MSH-induced B16F0 cells. TRP-1 protein levels were significantly reduced by 66% and 69% by 15 and 20 µg/mL LSE (Figure 2d). LSE in 10, 15, and 20 µg/mL as well as 15 µM EGC treatments significantly inhibited TRP-1 mRNA levels by 196%, 207%, 256%, and 153% respectively, in α-MSH-induced B16F0 cells (Figure 2e). TRP-2 protein and mRNA expression remained unchanged after LSE and EGC treatments. Based on these results, LSE and EGC inhibited melanin formation through inhibiting tyrosinase and TRP-1.

3.3. Effect of LSE and EGC on cAMP/PKA Signaling Pathway and MAPK Pathway in α-MSH-Induced B16F0 Cells

The MC1R protein and mRNA levels were investigated in LSE and EGC treatments. As shown in Figure 3a, α-MSH induced MC1R protein expressions but were markedly reduced by 126%, 142%, and 106% in 15 and 20 µg/mL LSE and 15 µM EGC treatments, respectively. MC1R gene expressions were analyzed by RT-PCR to evaluate the inhibitory effects of LSE and EGC. The results revealed that LSE and EGC treatments significantly repressed MC1R gene expressions by 143%, 153%, 168%, and 120% (Figure 3b). The binding of α-MSH to its receptor stimulated intracellular cAMP and PKA activation, and the downstream signal transduction of MC1R [20]. In agreement with the previous study, intracellular cAMP was significantly elevated by 70% in the α-MSH group when compared with the control group (Figure 3c). Treatments of 10, 15, and 20 µg/mL LSE markedly suppressed 66%, 95%, and 94% of cAMP level compared to that of the α-MSH group (Figure 3c). The cAMP level was decreased by 69% in the 15 µM EGC group. The ratio of phosphorylated PKA (p-PKA)/PKA was significantly elevated in the α-MSH group but reduced in 15 µg/mL and 20 µg/mL LSE and 15 µM EGC treatments (Figure 3d). In addition to the cAMP/PKA signaling pathway, the involvement of the MAPK pathway in the regulation was investigated by assessing ERK, p38, and JNK in α-MSH-induced B16F0 cells in the presence or absence of LSE and EGC. As shown in Figure 3e, p-ERK/ERK, p-JNK1/JNK1, and p-JNK2/JNK2 in the α-MSH-induced B16F0 cells revealed no change by LSE or EGC treatments. The p-p38/p38 level was increased by 31% in the α-MSH group but significantly reduced by 91% and 73% in 15 and 20 µg/mL LSE treatments. These data indicated that LSE treatment, especially in 15 and 20 µg/mL, suppressed melanogenesis in the α-MSH-induced B16F0 cells through regulating the cAMP/PKA signaling pathway and p-p38.

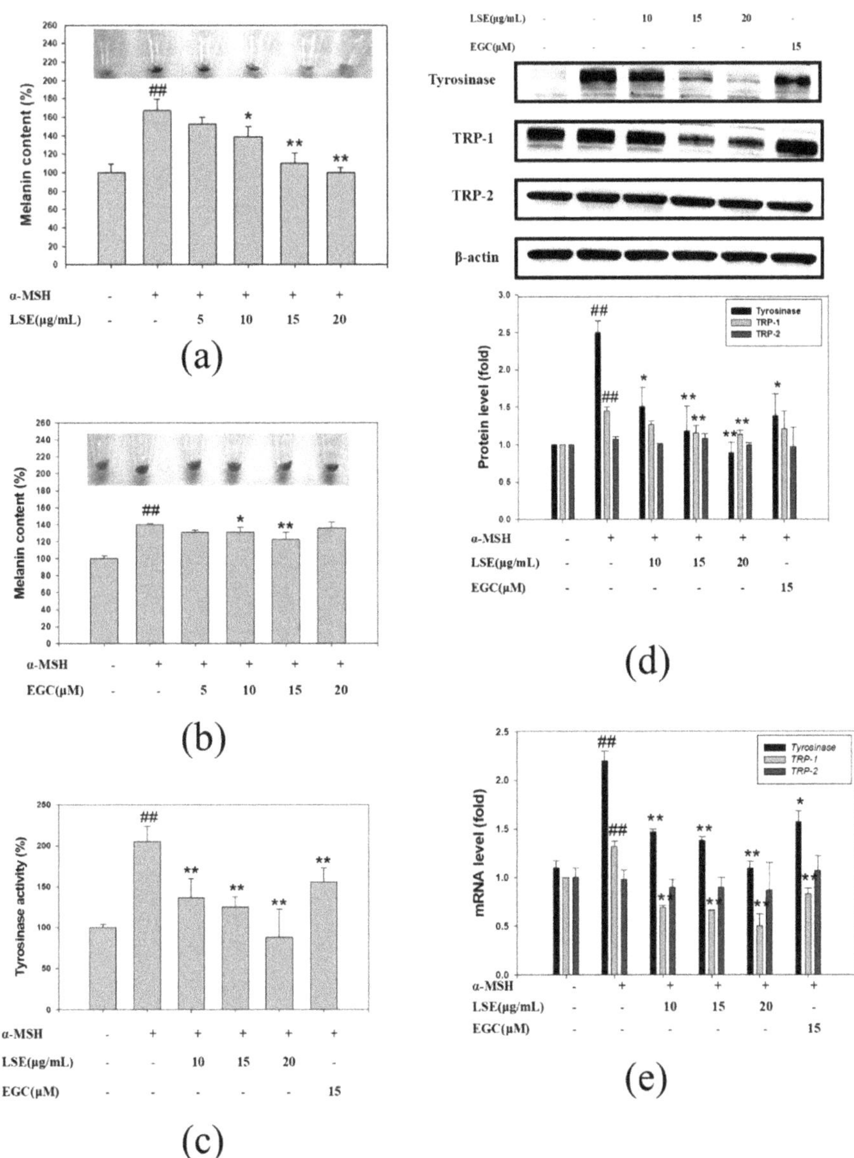

Figure 2. Effect of LSE or EGC on α-MSH-induced melanogenesis and tyrosinase activity in B16F0 melanoma cells. Cell pellets and melanin content of α-MSH-induced B16F0 cells treated with different concentrations of (**a**) LSE and (**b**) EGC. (**c**) Tyrosinase activity of B16F0 cells treated with LSE and EGC in various concentrations. (**d**) Tyrosinase, TRP-1, and TRP-2 proteins were analyzed by Western blotting. β-actin served as an internal control. (**e**) Tyrosinase, Trp-1, Trp-2 mRNA levels were analyzed by real-time PCR. α-MSH: α-melanocyte-stimulating hormone. LSE: lotus seedpod extract. EGC: epigallocatechin. TRP: tyrosinase-related protein. Values represent the mean ± SD of three independent experiments. $^{\#\#}$ $p < 0.01$ compared with control via student's *t*-test. * $p < 0.05$, ** $p < 0.01$, compared with α-MSH-treated group via one-way ANOVA with post-hoc Dunnett's test or via student's *t*-test. +: added. -: non-added.

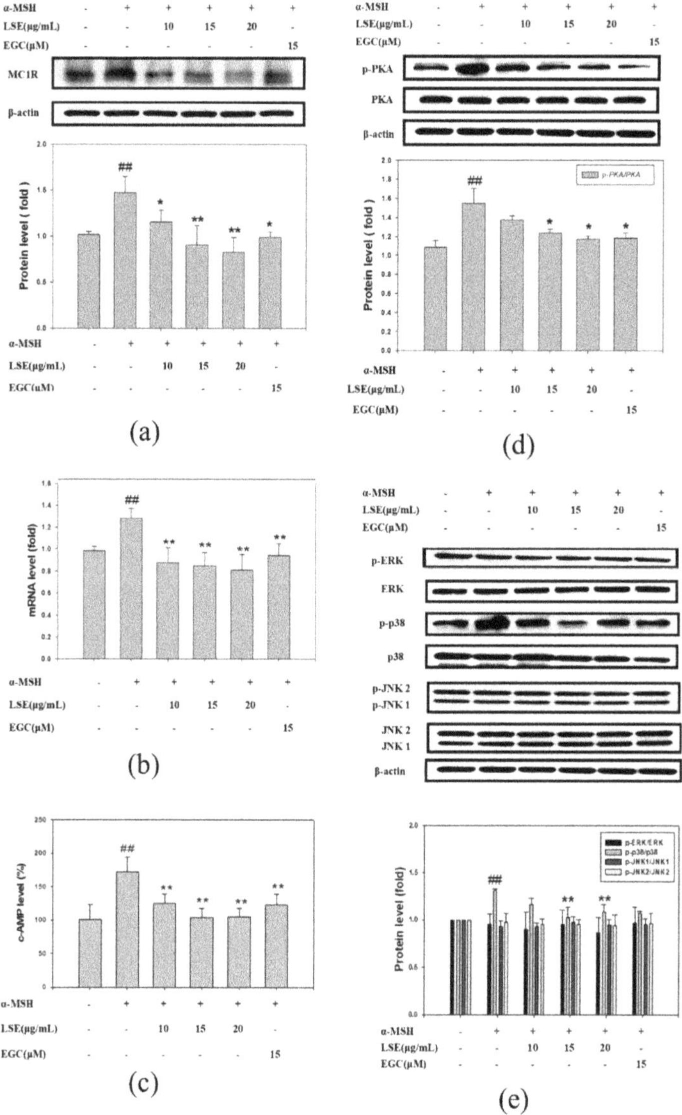

(a)

(b)

(c)

(d)

(e)

Figure 3. Effect of LSE and EGC on α-MSH-stimulated MC1R/cAMP/PKA signaling pathway in B16F0 cells. (**a**) MC1R protein and (**b**) Mc1r mRNA level in α-MSH-induced B16F0 cells treated with LSE or EGC in various concentrations. (**c**) Intracellular cAMP level was affected in LSE or EGC treatment. (**d**) p-PKA and PKA protein levels and (**e**) MAPK family proteins were analyzed by Western blotting. β-actin served as an internal control. α-MSH: α-melanocyte-stimulating hormone. LSE: lotus seedpod extract. EGC: epigallocatechin. MC1R: melanocortin-1 receptor. PKA: protein kinase A. ERK: extracellular signal-regulated kinases. JNK: c-Jun N-terminal kinases. cAMP: cyclic adenosine monophosphate. Values represent mean ± SD of three independent experiments. $^{\#\#}$ $p < 0.01$ compared with control via student's *t*-test. * $p < 0.05$, ** $p < 0.01$, compared with α-MSH-treated group via one-way ANOVA with post-hoc Dunnett's test or via student's *t*-test. +: added. -: non-added.

3.4. Effect of LSE and EGC on MITF and CREB in α-MSH-Induced B16F0 Cells

Based on the results, LSE inhibited melanin synthesis through the cAMP/PKA pathway and p-p38. A previous study revealed that activated PKA phosphorylated CREB and MITF to promote melanogenesis gene expressions [21]. The present study analyzed CREB and MITF in the nucleus extracts of B16F0 cells. As shown in Figure 4a, the ratios of phosphorylated CREB (p-CREB)/CREB and MITF were significantly higher in the α-MSH-induced B16F0 cells than in the control group, while nucleus p-CREB/CREB levels were significantly reduced by 75%, 77%, and 98% in 10, 15, and 20 μg/mL LSE groups, and nucleus MITF levels were decreased in the LSE groups, especially or 15 μg/mL and 20 μg/mL, by 98% and 124%, as compared to the α-MSH group (Figure 4a). The EGC (15 μM)-treated group suppressed nucleus p-CREB/CREB protein levels but not MITF expression. To further confirm whether LSE or EGC treatment interfered with CREB or MITF binding to their promoters, EMSA was performed for analysis, and the results are shown in Figure 4b,c. The second lane in Figure 4b,c reveals that the biotin-labeled CREB or MITF probe was bound to the respective promoter and formed a DNA complex. According to the results, α-MSH stimulated CREB and MITF binding to the promoters 23% and 118% more than the control group (Figure 4b,c, lane 3). Compared with α-MSH (lane 3), 20 μg/mL LSE and 15 μM EGC significantly reduced CREB/DNA complex to 129% and 90%, respectively (Figure 4b, lanes 5 and 6). Besides, the MITF/DNA complex was significantly inhibited by 10, 15, and 20 μg/mL LSE and 15 μM EGC (Figure 4c, lanes 3 to 6). These results indicated that LSE and EGC interfered with the CREB and MITF binding to their promoters to regulate tyrosinase and *Tyrp1* gene expressions.

3.5. LSE Repressed Melanogenesis through Inhibiting PKA and p38 Signling Pathways

To further confirm the involvement of PKA and p-p38 pathways in the melanogenesis inhibition of LSE, PKA inhibitor, H89, and p38 inhibitor, SB203580, were used for investigation. To begin with, 10 μM H89 or 5 μM SB203580 was treated along with α-MSH to confirm the inhibition of PKA and p38. As shown in Figure 5a, H89 and SB203580 significantly inhibited p-PKA/PKA and p-p38/p38 levels in the α-MSH-induced B16F0 cells. LSE alone reduced p-PKA/PKA to 38% and p-p38/p38 to 60% of those of α-MSH treatment (Figure 5a). LSE (15 μg/mL) along with H89 and SB203580 significantly repressed PKA and p38 activation by 221% and 148%, respectively (Figure 5a). Next, melanin content and tyrosinase activity in various treatments were shown, as in Figure 5b,c. The ability to reduce melanin content and tyrosinase activity by 15 μg/mL LSE was similar to H89 (Figure 5b,c). Tyrosinase and TRP-1 protein levels were significantly reduced by LSE with/without inhibitors in the α-MSH-induced B16F0 cells (Figure 5d). Nucleus extract of each treatment was analyzed to assess if LSE or inhibitors altered CREB and MITF expressions in the nucleus. The results are shown in Figure 5e: LSE alone or with inhibitors significantly decreased CREB phosphorylation and MITF levels in the α-MSH-induced B16F0 cells. According to these results, 15 μg/mL LSE inhibited melanin synthesis via repressing PKA and p38 and their downstream signaling pathways.

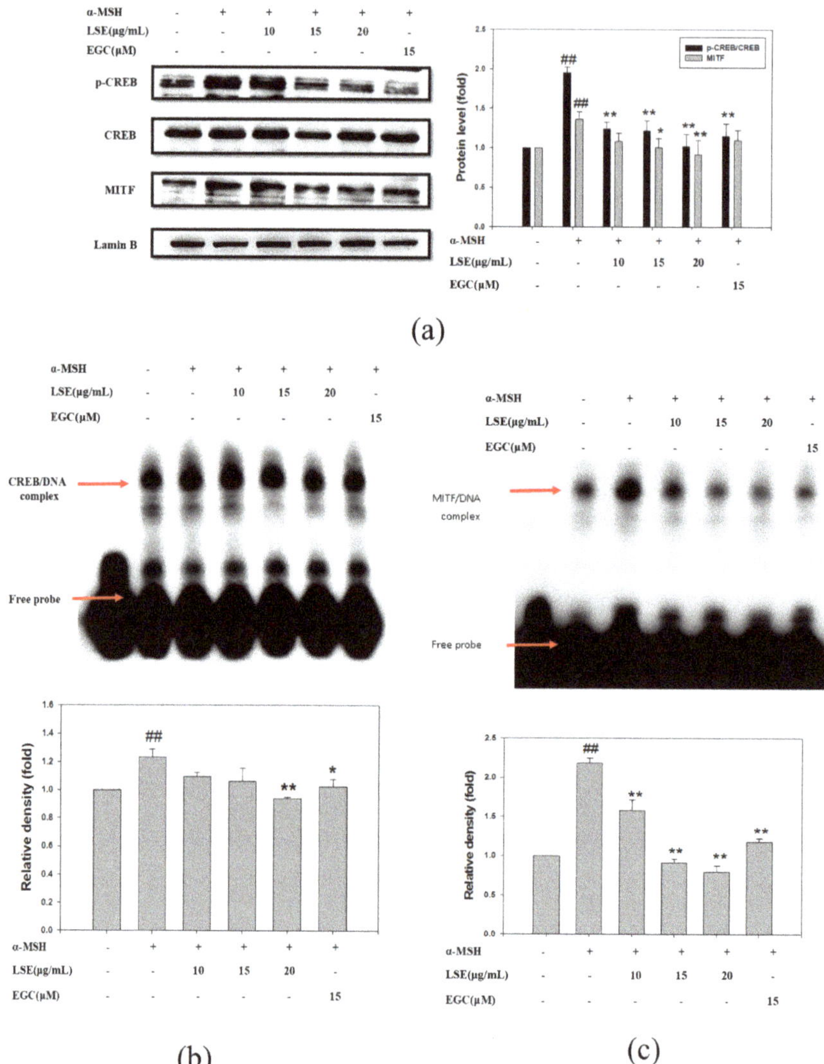

Figure 4. Effects of LSE on α-MSH-induced nuclear levels of p-CREB and MITF in B16F0 cells. (**a**) p-CREB/CREB and MITF levels were affected by LSE and EGC treatment, which was analyzed via Western blotting. Lamin B served as an internal control of nuclear fraction. (**b**) DNA-binding CREB activity of nucleus extracts which used biotin-labelled CREB-specific oligonucleotides by EMSA. (**c**) The nuclear extracts were analyzed for DNA-binding MITF activity of nucleus extracts, which used biotin-labelled MITF-specific oligonucleotides by EMSA. Lane 1 represents nuclear extracts incubated with unlabeled oligonucleotide (free probe) to confirm the specificity of binding. α-MSH: α-melanocyte-stimulating hormone. LSE: lotus seedpod extract. EGC: epigallocatechin. CREB: cAMP-response element binding protein. MITF: melanocyte inducing transcription factor. The results represent mean ± SD of three independent experiments and the significant difference was expressed as $^{\#\#}$ $p < 0.01$ compared with control via student's *t*-test. * $p < 0.05$, ** $p < 0.01$, compared with the α-MSH-treated group via one-way ANOVA with post-hoc Dunnett's test or via student's *t*-test. +: added. -: non-added.

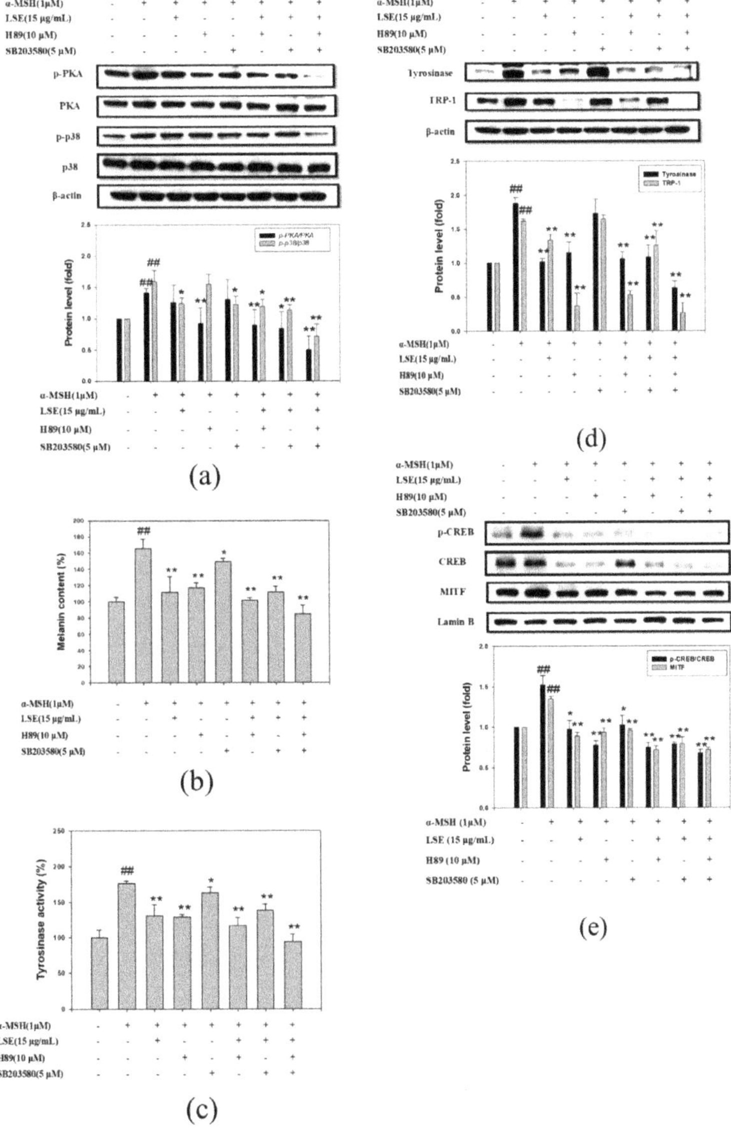

Figure 5. Effects of LSE, H89, and SB203580 on α-MSH-induced melanogenesis and tyrosinase in B16F0 cells. α-MSH-stimulated B16F0 cells were treated with 15 µg/mL of LSE or 10 µM of H89 (PKA inhibitor) or 5 µM of SB203580 (p38 inhibitor) for 48 h. (**a**) p-PKA, PKA, p-p38, and p38 protein levels were analyzed by Western blotting. β-actin served as an internal control. (**b**) Melanin content and (**c**) tyrosinase activity were determined in various treatments. (**d**) Tyrosinase and TRP-1 proteins were analyzed by Western blotting. β-actin served as an internal control. (**e**) The expression of p-CREB and MITF in the nucleus were determined by Western blotting. Lamin B served as an internal control of nuclear fraction. α-MSH: α-melanocyte-stimulating hormone. LSE: lotus seedpod extract. EGC: epigallocatechin. PKA: protein kinase A. TRP: tyrosinase-related protein. CREB: cAMP-response element binding protein. MITF: melanocyte inducing transcription factor. Values represent mean ± SD of three independent experiments. ## $p < 0.01$ compared with control via student's *t*-test. * $p < 0.05$, ** $p < 0.01$, compared with the α-MSH-treated group via student's *t*-test. +: added. -: non-added.

3.6. LSE and EGC Suppressed Melanin Synthesis Induced by UVB-Irridiation in Mice

According to the anti-melanogenesis mechanism of LSE and EGC in α-MSH-induced B16F0 cells, an animal study was designed to investigate the in vivo effect of LSE and EGC on melanin synthesis induced by UVB radiation. The dose of LSE applied on mouse ears was calculated by dividing 10 µg/mL, 20 µg/mL LSE, and 15 µM EGC treated to B16F0 cells by the culture dish area and to infer the amount of LSE from each cell exposed, then magnify it thousand times. Thus, the doses of LSE and EGC were 1.25 mg/cm^2, 2.5 mg/cm^2, and 2 mM/cm^2, respectively. Mouse ear skin was analyzed by western blot to evaluate the molecular mechanism of LSE in melanin synthesis. Mouse ear samples were treated with 1 M NaOH at 80 °C for 1 h to dissolve melanin for content determination (Figure 6a). As shown in Figure 6a, melanin content in UVB irradiation treatment was significantly increased by 79% when compared with the control group. LSE of 1.25 mg/cm^2 and 2.5 mg/cm^2 treatments before UVB irradiation markedly reduced melanin content by 70% and 93% when compared with the UVB group (Figure 6a). Melanin content in 2 mM/cm^2 EGC treatment was 74% lower than that of the UVB group. To determine the molecular pathway of melanin synthesis, tyrosinase, TRP-1 and 2, PKA, and p38 protein levels were analyzed by Western blot. Similar to the in vitro studies, when compared with the UVB group, tyrosinase and TRP-1 expressions in 1.25 mg/cm^2 and 2.5 mg/cm^2 LSE were significantly decreased, while there was no change in TRP-2 (Figure 6b). EGC (2 mM/cm^2) revealed similar results with 2.5 mg/cm^2 LSE treatment (Figure 6b). PKA and p38 phosphorylation were stimulated after UVB irradiation but were suppressed by 1.25 mg/cm^2 and 2.5 mg/cm^2 LSE as well as 2 mM/cm^2 EGC treatment (Figure 6c). Consistent with the in vitro study, the in vivo study revealed that LSE and EGC treatment suppressed PKA and p38 phosphorylation that consequently reduced melanin synthesis after UVB irradiation.

4. Discussion

The present study investigated the anti-melanogenic effect of LSE intervention through in vitro and in vivo studies. It is the first study of LSE on melanogenesis. We investigated the non-toxic dose of LSE in B16F0 cells and selected 10, 15, and 20 µg/mL LSE, which reduced the melanin content without cytotoxicity. The possible anti-melanogenesis molecular pathways were analyzed in 15 and 20 µg/mL LSE and 15 µM EGC. We further confirmed the anti-melanogenetic effect of LSE via topical application to mouse ears. The results of both studies revealed that LSE reduced melanin synthesis by regulating CREB phosphorylation and MITF to inhibit tyrosinase and TRP-1 activity and expressions.

Our previous study analyzed LSE by liquid chromatography and found that its most abundant polyphenol was EGC [12]. The concentrations of LSE (10, 15, and 20 µg/mL) used in the present study were equivalent to 5.39, 7.1, and 8.9 µM EGC, respectively. The depigmenting effect of EGC in 10 and 20 µM has been reported to be significant in suppressing tyrosinase activity [15]. However, EGC 20 µM showed cytotoxicity when compared with the control group (Figure 1b). EGC 15 µM is equivalent to 25 µg/mL LSE, which revealed cytotoxicity to B16F0 cells. Hence, 15 µM EGC was used in the present study as a reference standard and we compared its efficacy on melanogenesis with LSE. According to the results, LSE and EGC treatment reduced melanin content in the α-MSH-induced B16F0 cells by suppressing tyrosinase activity and the expressions of tyrosinase and TRP-1, both of which were critical enzymes of melanin synthesis and were modulated by MC1R-induced downstream signaling pathways [21,22]. The binding of α-MSH to its receptor MC1R induced cAMP elevation, as the second messenger, to activate PKA [23]. Besides, p38 has been reported, whose activity was cAMP-dependent and regulated melanogenesis and proliferation of the α-MSH-induced melanoma cells [24]. Consistent with previous studies, the inhibition of melanin synthesis by LSE involved the cAMP/PKA signaling pathway and p38 phosphorylation (Figure 3) in the α-MSH-induced B16F0 cells. Activated PKA increased CREB phosphorylation that bound to its specific sequence to promote MITF expression [2]. MITF is an important transcription factor of melanogenesis including *Tyrosinase* and *Tyrp1* [23], proliferation, and differentiation [25]. The activation or degradation of MITF was concerned with the MAPK pathway [25]. The present study found that LSE reduced CREB phosphorylation and MITF levels in the nucleus, and whose DNA binding complexes were significantly decreased in the α-MSH-induced melanocyte (Figure 3). These results suggested that the inhibitory effects of LSE on cAMP/PKA signaling and p38 phosphorylation contributed to the downregulation of CREB and MITF activity binding to their promotor sites, that consequently decreased *Tyrosinase*, *Tyrp1*, and *Mc1r* expression levels and resulted in reducing melanin content. Compared with EGC, LSE in 15 and 20 μg/mL showed stronger inhibition effects in melanin production. This result indicates that there are still unknown compounds in LSE which are potent in depigmentation or synergistically working with EGC on melanin synthesis.

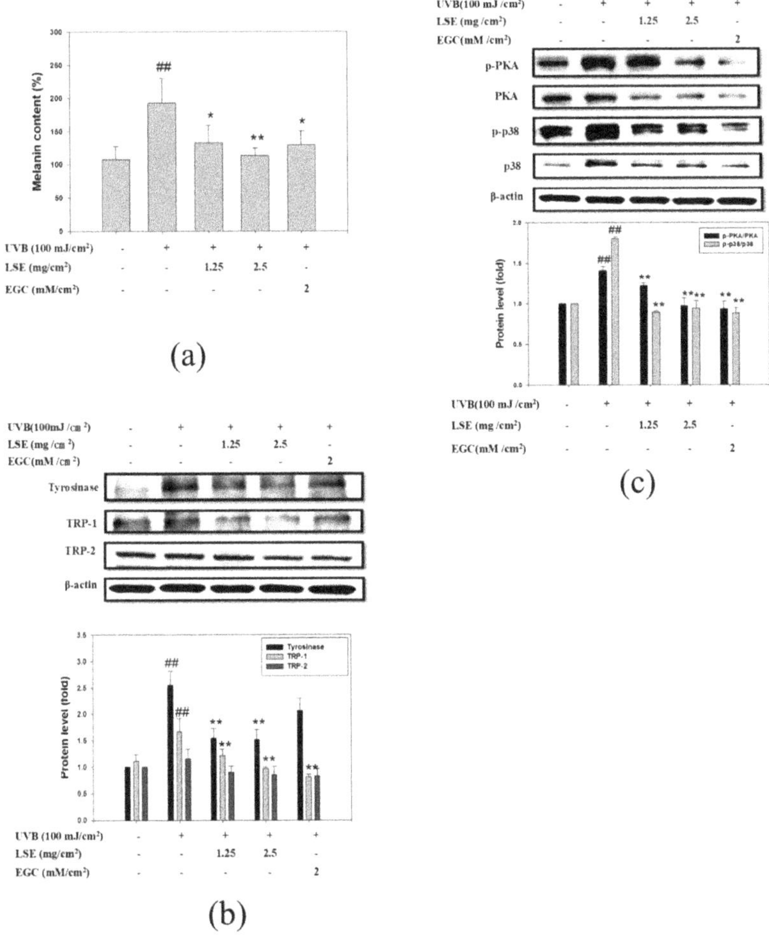

Figure 6. Effect of LSE and EGC on UVB-induced melanogenesis in mice ears. (**a**) Melanin content contained in mice ears with different treatments. (**b**) Tyrosinase, TRP-1, and TRP-2 proteins and (**c**) p-PKA, PKA, p-p38, and p38 proteins in ear tissues were analyzed by Western blotting. β-actin served as an internal control. UVB: ultraviolet radiation b. LSE: lotus seedpod extract. EGC: epigallocatechin. PKA: protein kinase A. TRP: tyrosinase-related protein. Values represent mean ± SD of three independent experiments. $^{\#\#}$ $p < 0.01$ compared with control via student's *t*-test. * $p < 0.05$, ** $p < 0.01$, compared with UVB-treated group via one-way ANOVA with post-hoc Dunnett's test or via student's *t*-test. +: added. -: non-added.

To investigate the potency of LSE under the inhibitor of PKA or p38 on the mechanism of melanin synthesis, 15 μg/mL LSE were combined with/without H89 and SB203580 for the treatments. Based on the results, the PKA inhibition effect of 15 μg/mL LSE was less than H89, while equivalent to SB203580 on p38 inhibition. The pPKA/PKA level was reduced in the combination of 15 μg/mL LSE and SB203580, which indicated that LSE could suppress both PKA and p38 activation (Figure 5a). Although 15 μg/mL LSE and H89 inhibited tyrosinase and TRP-1 expressions in the α-MSH-induced melanocyte, 15 μg/mL LSE was equivalent to H89 on suppressing tyrosinase but weaker than H89 on TRP-1 inhibition (Figure 5d). SB203580 revealed no effect on both but significantly reduced when treated with LSE (Figure 5d). The changes of tyrosinase and TRP-1 levels in different treatments were in a similar trend as the results in Figure 5b,c. The melanin synthesis inhibition of 15 μg/mL LSE was equivalent to H89 but was more potent than SB203580. The inhibition effect of 15 μg/mL LSE on CREB and MITF expressions in the nucleus revealed equivalent outcomes as H89 or SB203580 alone (Figure 5e). Besides, the combination of LSE and H89 or SB203580 showed a stronger inhibitory effect on CREB and MITF. Based on the experiment, the anti-melanogenesis effect of LSE was mainly through reducing PKA and p38 activity that consequently decreased CREB and MITF on regulating melanin synthesis-related gene expressions.

To further confirm the anti-melanogenesis potential of the LSE in vivo, the well-established animal model of UV-induced hyperpigmentation in the C57BL/6 mouse skin was utilized to analyze the effect of extract on melanin production and melanogenesis signaling. The process of melanogenesis, including melanin synthesis and distribution, not only constitutes a complex series of enzymatic and chemical reactions, but is also implicated in the cellular interaction of skin layers [7,26]. It has been reported that pigments formed in melanocyte melanosomes are then stored in the basal layer of epidermal cells, as well as in dermal macrophages, which become melanophores [27]. In agreement with these reports, the results of the in vivo experiments in this study showed that UVB induced melanin synthesis, whereas LSE or EGC treatments decreased melanin content in the mixture of epidermis and dermis of the mouse ear (Figure 6a). Furthermore, the key enzymes of melanogenesis signaling, tyrosinase and TRP-1, decreased in the 1.25 mg/cm^2 and 2.5 mg/cm^2 LSE treatments. The changes of tyrosinase and TRP-1 but not of TRP-2 in the LSE treatment were observed in both in vivo and in vitro studies. This is probably related to different functions of TRP-1 and TRP-2, although whose amino acid had high structural and sequence similarity with tyrosinase [28]. It has been reported that TRP-1 was involved in stabilizing and activating tyrosinase and melanosome during melanin synthesis [3]. The abnormal changes of TRP-1 led to pigmentation disorders such as Vitiligo and Oculocutaneous albinism [21]. TRP-2 was also required for melanin synthesis [3], but its biological function still needs to be clarified. It has been suggested that there is a requirement of TRP-1 for melanin synthesis. LSE suppressed both tyrosinase and TRP-1, which demonstrated that LSE effectively inhibited pigmentation. A previous study reported that UVB-induced melanogenesis was mediated by PKA and p38 pathways [29]. Based on the results, phosphorylated PKA and p38 were increased after repetitive UVB exposure, whereas both were reduced by 1.25 mg/cm^2 and 2.5 mg/cm^2 LSE treatment. The activity of PKA and p38 regulated CREB and MITF in the nucleus and thus affected pigmenting gene expressions. Consistent with the in vitro study, 1.25 mg/cm^2 and 2.5 mg/cm^2 LSE treatments suppressed melanogenesis by inhibiting PKA and p38 activation. Earlier studies have shown that the melanin production in melanocytes is influenced by keratinocytes and fibroblasts [30,31]. Primary melanocytes and keratinocytes have some drawbacks such as the limited proliferation capacity and the donor variability [31]. Primary murine melanocytes defective at the Mc1r locus have been historically difficult to culture in vitro [32]. Their poor growth has been attributed to faulty MSH responsiveness and decreased survival. Accordingly, current protocols for culturing primary murine melanocytes rely on pharmacologic manipulation [32]. Thus, the use of immortalized cell lines would be more suitable for investigating the effects of pigmentation regulators. B16 melanoma cells have been used widely in the pigmentation research model for normal melanocytes' behavior. In addition, B16F0 cells were less malignant compared with other B16 melanoma sublines, suggesting that this clone was similar to normal melanocytes. Consistent with this is the demonstration by Bellei et al., showing a similar response of B16F0 and normal human melanocytes to promote melanogenesis when stimulating with various GSK3β-specific inhibitors [33]. Hwang et al. used the human epidermal melanocytes to explore the possible mechanism of maclurin, a natural xanthone, on melanin synthesis inhibition, including p38- and PKA-mediated signaling [34]. As the present study was a preliminary one to investigate the effects of LSE on melanogenesis, the effect of the extract upon normal melanocytes and UV irradiation on epidermal melanocytes separated from the dermis would be interesting to validate the photoprotective mechanism and thereby, needs to be explored in the future.

5. Conclusions

In summary, the present study first identified the melanin synthesis inhibition mechanism of LSE in the α-MSH-induced B16F0 cells through inhibiting PKA and p38 signaling pathways to repress melanin synthesis-related gene expressions. Consistent with the in vitro study, the inhibition effect of LSE was further confirmed in mouse ear skin. In conclusion, LSE possesses the potent efficacy of anti-melanogenesis and could be a novo depigmenting agent for cosmetics or pharmaceutical applications.

Author Contributions: Conceptualization, investigation, project administration, supervision, H.-H.L. and J.-H.C.; methodology, data curation, formal analysis, and validation, J.-Y.H., T.-S.L., C.-Y.T., and Y.W.; resources, J.-H.C.; visualization, J.-Y.H., H.-H.L., and J.-H.C.; writing—original draft preparation, J.-Y.H. and T.-S.L.; writing—review and editing, H.-H.L. and J.-H.C. All authors have read and agreed to the published version of the manuscript.

Funding: This work was supported by the grant from the National Science Council (NSC102-2320-B-040-004), Taiwan.

Conflicts of Interest: The authors declare no conflict of interest.

References

1. Yamaguchi, Y.; Hearing, V.J. Physiological factors that regulate skin pigmentation. *Biofactors* **2009**, *35*, 193–199. [CrossRef] [PubMed]
2. Lin, J.Y.; Fisher, D.E. Melanocyte biology and skin pigmentation. *Nature* **2007**, *445*, 843–850. [CrossRef] [PubMed]
3. Park, H.Y.; Kosmadaki, M.; Yaar, M.; Gilchrest, B.A. Cellular mechanisms regulating human melanogenesis. *Cell. Mol. Life Sci.* **2009**, *66*, 1493–1506. [CrossRef]
4. Vance, K.W.; Goding, C.R. The transcription network regulating melanocyte development and melanoma. *Pigment. Cell Res.* **2004**, *17*, 318–325. [CrossRef]
5. Kawakami, A.; Fisher, D.E. The master role of microphthalmia-associated transcription factor in melanocyte and melanoma biology. *Lab. Investig.* **2017**, *97*, 649–656. [CrossRef]
6. D'Mello, S.A.N.; Finlay, G.J.; Baguley, B.C.; Askarian-Amiri, M.E. Signaling Pathways in Melanogenesis. *Int. J. Mol. Sci.* **2016**, *17*, 1144. [CrossRef]
7. Wu, L.C.; Lin, Y.Y.; Yang, S.Y.; Weng, Y.T.; Tsai, Y.T. Antimelanogenic effect of c-phycocyanin through modulation of tyrosinase expression by upregulation of ERK and downregulation of p38 MAPK signaling pathways. *J. Biomed. Sci.* **2011**, *18*, 74. [CrossRef]
8. Ling, Z.Q.; Xie, B.J.; Yang, E.L. Isolation, characterization, and determination of antioxidative activity of oligomeric procyanidins from the seedpod of Nelumbo nucifera Gaertn. *J. Agric. Food Chem.* **2005**, *53*, 2441–2445. [CrossRef]
9. Duan, Y.; Zhang, H.; Xu, F.; Xie, B.; Yang, X.; Wang, Y.; Yan, Y. Inhibition effect of procyanidins from lotus seedpod on mouse B16 melanoma in vivo and in vitro. *Food Chem.* **2010**, *122*, 84–91. [CrossRef]
10. Duan, Y.; Wang, Z.; Zhang, H.; He, Y.; Fan, R.; Cheng, Y.; Sun, G.; Sun, X. Extremely low frequency electromagnetic field exposure causes cognitive impairment associated with alteration of the glutamate level, MAPK pathway activation and decreased CREB phosphorylation in mice hippocampus: Reversal by procyanidins extracted from the lotus seedpod. *Food Funct.* **2014**, *5*, 2289–2297. [CrossRef]
11. Wu, Q.; Li, S.; Li, X.; Sui, Y.; Yang, Y.; Dong, L.; Xie, B.; Sun, Z. Inhibition of Advanced Glycation Endproduct Formation by Lotus Seedpod Oligomeric Procyanidins through RAGE-MAPK Signaling and NF-kappaB Activation in High-Fat-Diet Rats. *J. Agric. Food Chem.* **2015**, *63*, 6989–6998. [CrossRef] [PubMed]
12. Tseng, H.C.; Tsai, P.M.; Chou, Y.H.; Lee, Y.C.; Lin, H.H.; Chen, J.H. In Vitro and In Vivo Protective Effects of Flavonoid-Enriched Lotus Seedpod Extract on Lipopolysaccharide-Induced Hepatic Inflammation. *Am. J. Chin. Med.* **2019**, *47*, 153–176. [CrossRef]
13. Liu, Y.T.; Lai, Y.H.; Lin, H.H.; Chen, J.H. Lotus Seedpod Extracts Reduced Lipid Accumulation and Lipotoxicity in Hepatocytes. *Nutrients* **2019**, *11*, 2895. [CrossRef]

14. Xiao, J.S.; Xie, B.J.; Cao, Y.P.; Wu, H.; Sun, Z.D.; Xiao, D. Characterization of oligomeric procyanidins and identification of quercetin glucuronide from lotus (Nelumbo nucifera Gaertn.) seedpod. *J. Agric. Food Chem.* **2012**, *60*, 2825–2829. [CrossRef] [PubMed]

15. Sato, K.; Toriyama, M. Depigmenting effect of catechins. *Molecules* **2009**, *14*, 4425–4432. [CrossRef]

16. Shen, T.; Heo, S.I.; Wang, M.H. Involvement of the p38 MAPK and ERK signaling pathway in the anti-melanogenic effect of methyl 3,5-dicaffeoyl quinate in B16F10 mouse melanoma cells. *Chem. Biol. Interact.* **2012**, *199*, 106–111. [CrossRef] [PubMed]

17. Bizzozero, L.; Cazzato, D.; Cervia, D.; Assi, E.; Simbari, F.; Pagni, F.; De Palma, C.; Monno, A.; Verdelli, C.; Querini, P.R.; et al. Acid sphingomyelinase determines melanoma progression and metastatic behaviour via the microphtalmia-associated transcription factor signalling pathway. *Cell Death Differ.* **2014**, *21*, 507–520. [CrossRef]

18. Smalley, K.S.; Eisen, T.G. Differentiation of human melanoma cells through p38 MAP kinase is associated with decreased retinoblastoma protein phosphorylation and cell cycle arrest. *Melanoma Res.* **2002**, *12*, 187–192. [CrossRef]

19. Park, S.Y.; Jin, M.L.; Kim, Y.H.; Kim, Y.; Lee, S.J. Aromatic-turmerone inhibits alpha-MSH and IBMX-induced melanogenesis by inactivating CREB and MITF signaling pathways. *Arch. Dermatol. Res.* **2011**, *303*, 737–744. [CrossRef]

20. Mas, J.S.; Gerritsen, I.; Hahmann, C.; Jimenez-Cervantes, C.; Garcia-Borron, J.C. Rate limiting factors in melanocortin 1 receptor signalling through the cAMP pathway. *Pigment. Cell Res.* **2003**, *16*, 540–547. [CrossRef]

21. Videira, I.F.; Moura, D.F.; Magina, S. Mechanisms regulating melanogenesis. *An. Bras. Dermatol.* **2013**, *88*, 76–83. [CrossRef] [PubMed]

22. Del Marmol, V.; Beermann, F. Tyrosinase and related proteins in mammalian pigmentation. *FEBS Lett.* **1996**, *381*, 165–168. [CrossRef]

23. Busca, R.; Ballotti, R. Cyclic AMP a key messenger in the regulation of skin pigmentation. *Pigment. Cell Res.* **2000**, *13*, 60–69. [CrossRef] [PubMed]

24. Smalley, K.; Eisen, T. The involvement of p38 mitogen-activated protein kinase in the alpha-melanocyte stimulating hormone (alpha-MSH)-induced melanogenic and anti-proliferative effects in B16 murine melanoma cells. *FEBS Lett.* **2000**, *476*, 198–202. [CrossRef]

25. Levy, C.; Khaled, M.; Fisher, D.E. MITF: Master regulator of melanocyte development and melanoma oncogene. *Trends Mol. Med.* **2006**, *12*, 406–414. [CrossRef] [PubMed]

26. Abdel-Malek, Z.; Swope, V.B.; Suzuki, I.; Akcali, C.; Harriger, M.D.; Boyce, S.T.; Urabe, K.; Hearing, V.J. Mitogenic and melanogenic stimulation of normal human melanocytes by melanotropic peptides. *Proc. Natl. Acad. Sci. USA* **1995**, *92*, 1789–1793. [CrossRef]

27. Maranduca, M.A.; Branisteanu, D.; Serban, D.N.; Branisteanu, D.C.; Stoleriu, G.; Manolache, N.; Serban, I.L. Synthesis and physiological implications of melanic pigments. *Oncol Lett.* **2019**, *17*, 4183–4187. [CrossRef]

28. Sturm, R.A.; O'Sullivan, B.J.; Box, N.F.; Smith, A.G.; Smit, S.E.; Puttick, E.R.; Parsons, P.G.; Dunn, I.S. Chromosomal structure of the human TYRP1 and TYRP2 loci and comparison of the tyrosinase-related protein gene family. *Genomics* **1995**, *29*, 24–34. [CrossRef]

29. Bolognia, J.; Murray, M.; Pawelek, J. UVB-induced melanogenesis may be mediated through the MSH-receptor system. *J. Investig. Dermatol.* **1989**, *92*, 651–656. [CrossRef]

30. Lei, T.C.; Virador, V.M.; Vieira, W.D.; Hearing, V.J. A melanocyte-keratinocyte coculture model to assess regulators of pigmentation in vitro. *Anal. Biochem.* **2002**, *305*, 260–268. [CrossRef]

31. Liu, S.H.; Chu, I.M.; Pan, I.H. Effects of hydroxybenzyl alcohols on melanogenesis in melanocyte-keratinocyte co-culture and monolayer culture of melanocytes. *J. Enzym. Inhib. Med. Chem.* **2008**, *23*, 526–534. [CrossRef] [PubMed]

32. Scott, T.L.; Wakamatsu, K.; Ito, S.; D'Orazio, J.A. Purification and growth of melanocortin 1 receptor (Mc1r)-defective primary murine melanocytes is dependent on stem cell factor (SFC) from keratinocyte-conditioned media. *Vitr. Cell. Dev. Biol. Anim.* **2009**, *45*, 577–583. [CrossRef] [PubMed]

33. Bellei, B.; Flori, E.; Izzo, E.; Maresca, V.; Picardo, M. GSK3beta inhibition promotes melanogenesis in mouse B16 melanoma cells and normal human melanocytes. *Cell. Signal.* **2008**, *20*, 1750–1761. [CrossRef] [PubMed]

34. Hwang, Y.S.; Oh, S.W.; Park, S.H.; Lee, J.; Yoo, J.A.; Kwon, K.; Park, S.J.; Kim, J.; Yu, E.; Cho, J.Y.; et al. Melanogenic Effects of Maclurin Are Mediated through the Activation of cAMP/PKA/CREB and p38 MAPK/CREB Signaling Pathways. *Oxidative Med. Cell Longev.* **2019**, *2019*, 9827519. [CrossRef]

Publisher's Note: MDPI stays neutral with regard to jurisdictional claims in published maps and institutional affiliations.

Article

Cornus officinalis Ethanolic Extract with Potential Anti-Allergic, Anti-Inflammatory, and Antioxidant Activities

Yixian Quah [1,†], Seung-Jin Lee [1,2,†], Eon-Bee Lee [1], Biruk Tesfaye Birhanu [1], Md. Sekendar Ali [1,3,4], Muhammad Aleem Abbas [1], Naila Boby [1], Zi-Eum Im [5] and Seung-Chun Park [1,*]

[1] Laboratory of Veterinary Pharmacokinetics and Pharmacodynamics, College of Veterinary Medicine, Kyungpook National University, Daegu 41566, Korea; im.yixianquah@gmail.com (Y.Q.); dvmleesj@naver.com (S.-J.L.); eonbee@gmail.com (E.-B.L.); btbtes@gmail.com (B.T.B.); alipharm2000@gmail.com (M.S.A.); syedaleemabbas77@gmail.com (M.A.A.); nailaboby1584@gmail.com (N.B.)
[2] Development and Reproductive Toxicology Research Group, Korea Institute of Toxicology, Daejeon 34114, Korea
[3] Department of Biomedical Science and Department of Pharmacology, School of Medicine, Brain Science and Engineering Institute, Kyungpook National University, Daegu 41944, Korea
[4] Department of Pharmacy, International Islamic University Chittagong, Kumira, Chittagong 4318, Bangladesh
[5] Forest Resources Development Institute of Gyeongsangbuk-do, Andong-si, Gyeongsangbuk-do 36605, Korea; zium78@korea.kr
*   Correspondence: parksch@knu.ac.kr; Tel.: +82-53-950-5964
†   These authors contributed equally to this work.

Received: 8 September 2020; Accepted: 27 October 2020; Published: 29 October 2020

Abstract: Atopic dermatitis (AD) is an allergic and chronic inflammatory skin disease. The present study investigates the anti-allergic, antioxidant, and anti-inflammatory activities of the ethanolic extract of *Cornus officinalis* (COFE) for possible applications in the treatment of AD. COFE inhibits the release of β-hexosaminidase from RBL-2H3 cells sensitized with the dinitrophenyl-immunoglobulin E (IgE-DNP) antibody after stimulation with dinitrophenyl-human serum albumin (DNP-HSA) in a concentration-dependent manner (IC_{50} = 0.178 mg/mL). Antioxidant activity determined using 2,2-diphenyl-1-picrylhydrazyl (DPPH) radical scavenging activity, ferric reducing antioxidant power assay, and 2,2′-azino-bis(3-ethylbenzothiazoline-6-sulfonic acid) (ABTS) scavenging activity, result in EC_{50} values of 1.82, 10.76, and 0.6 mg/mL, respectively. Moreover, the extract significantly inhibits lipopolysaccharide (LPS)-induced nitric oxide (NO) production and the mRNA expression of iNOS and pro-inflammatory cytokines (IL-1β, IL-6, and TNF-α) through attenuation of NF-κB activation in RAW 264.7 cells. COFE significantly inhibits TNF-α-induced apoptosis in HaCaT cells without cytotoxic effects ($p < 0.05$). Furthermore, 2-furancarboxaldehyde and loganin are identified by gas chromatography/mass spectrometry (GC-MS) and liquid chromatography with tandem mass spectrometry (LC-MS/MS) analysis, respectively, as the major compounds. Molecular docking analysis shows that loganin, cornuside, and naringenin 7-O-β-D-glucoside could potentially disrupt the binding of IgE to human high-affinity IgE receptors (FceRI). Our results suggest that COFE might possess potential inhibitory effects on allergic responses, oxidative stress, and inflammatory responses.

Keywords: anti-inflammatory activity; antioxidant activity; atopic dermatitis; *Cornus officinalis*; molecular docking; human high-affinity IgE receptors

---

1. Introduction

Atopic dermatitis (AD) is an allergic and chronic skin inflammation condition characterized by pruritic eczema and mechanical skin injury caused by scratching, elevated serum IgE levels, and markedly increased immune cells levels (eosinophils, mast cells, and lymphocytes) [1]. Particularly, T helper 2 (Th2) cells are involved in AD development [2]. Th2 cells release cytokines such as interleukin (IL)-4, IL-5, IL-10, and IL-13, resulting in the activation and proliferation of eosinophils and mast cells. The resultant itching, skin scratching, mechanical skin barrier injury, and activation of immune cells, lead to further Th2 cell activation and itching [3].

Cornus officinalis is a dogwood species native to Eastern Asia (Korea, China, and Japan). Its fruit has been used in traditional medicine for backache, hypertension, and polyuria in Korea [4]. The protective effect of *C. officinalis*, used in herbal mixture with four other plant extracts, against atopic dermatitis in BALB/C mice was recently reported [5]. The major component of the reported herbal mixture was *C. officinalis* extract, constituting 50% of the formulation. Besides, a recent review on this medicinal plant discussed its biological properties such as antioxidant, anti-inflammatory, renal and hepatic protective, antimicrobial, and immunomodulatory effects [6] which all play a substantial role in modulating pathological mechanism in AD.

Previous studies almost exclusively focused on the anti-atopic, antioxidant, and anti-inflammatory effects of the herbal mixture but not on *C. officinalis* used individually in the in vitro studies for AD. This study is the first to report the potency and efficacy of the anti-atopic effects of the *C. officinalis* extract in an in vitro model of AD.

2. Materials and Methods

2.1. Sample Preparation and Extraction

To prepare the ethanolic extract of *C. officinalis*, dried fruits of *C. officinalis* (50 g) (Sansooyoo, Korea) were extracted three times with 70% ethanol (500 mL) at 65 °C for 4 h using a heating mantle. The filtered supernatant was concentrated using a rotary evaporator. The dense supernatant was solidified with a vacuum freeze dryer. The yield of the ethanolic extraction of *C. officinalis* (COFE) was approximately 14.33% of the dried material. COFE was dissolved in dimethyl sulfoxide (DMSO, Sigma-Aldrich, St. Louis, MO, USA) and stored as a stock solution at −20 °C until use.

2.2. β-Hexosaminidase Release

The inhibition of β-hexosaminidase release from a rat basophilic leukemia cell line (RBL-2H3) was evaluated using the method described by Juckmeta et al. [7] with some modifications. Briefly, RBL-2H3 cells were seeded at 5×10^5 cells/mL and incubated for 1.5 h. RBL-2H3 cells were sensitized with anti-dinitrophenyl-immunoglobulin E (anti-DNP IgE; 0.45 µg/mL) and incubated for 24 h. The cells were washed with Siraganian buffer. Siraganian buffer (160 µL) was added and the cells were incubated for 10 min at 37 °C. Different concentrations of COFE (20 µL) were added to each well to reach a final concentration of 0, 0.003, 0.01, 0.03, 0.1, and 0.3 mg/mL, then the cells were incubated for 10 min. Next, 20 µL of dinitrophenyl-human serum albumin (DNP-HSA, final concentration 10 µg/mL) was added to each well and the cells were incubated for another 20 min to stimulate cell degranulation. The supernatants were transferred to 96-well plates at 50 µL/well and incubated with 50 µL of *p*-nitrophenyl-*N*-acetyl-b-D-glucosaminide (PNAG, 1 mM) in 0.1 M citrate buffer (pH 4.5) at 37 °C for 3 h. The reaction was stopped by the addition of 200 µL of the stop solution. The absorbance was measured at 405 nm. The percentage of β-hexosaminidase release was calculated as a percentage of the total β-hexosaminidase content. A brief experimental timeline of this assay is illustrated in Supplementary Figure S1.

2.3. Antioxidant Activity

2.3.1. 2,2-Diphenyl-1-picrylhydrazyl (DPPH) Free Radical Scavenging Activity

The extracts scavenging activity against DPPH free radicals was evaluated using the method described by Chang et al. [8]; ascorbic acid was used as a positive control. The presence of DPPH free radicals was measured at 517 nm.

2.3.2. Ferric Reducing Antioxidant Power (FRAP) Assay

The ferric reducing ability of tissues was determined using a modified FRAP method [9]. The absorbance of the reaction mixture was read at 595 nm. The results were expressed as μM Trolox equivalent (TE), as E_{max} value, which is the value at which the maximal effect of the sample was obtained, and as EC_{50} value, which is the TE of the sample required to cause a 50% reduction in the FRAP value.

2.3.3. Trolox Equivalent Antioxidant Capacity (TEAC) Assay

The TEAC levels were determined with the method of Rufino et al. [10], with modifications. Trolox (10–1000 μmol TE) was used to generate the standard curve. Ten microliters of each sample were added to 90 μL of an ABTS$^\bullet$ solution and the absorbance was measured at 734 nm. The results were expressed in μmol TE, as E_{max} value, and EC_{50} value, which is the TE of the sample required for the inhibition of 50% ABTS free radical activity.

2.4. Nitric Oxide (NO) Assay

RAW 264.7 cells (Korean Cell Line Bank, Seoul, Korea) at 4×10^5 cells/well were allowed to adhere for 24 h until 80% confluency was reached. After 30 min of treatment of different extract concentrations, the cells were incubated with lipopolysaccharide (LPS) treatment (1 μg/mL) for 18 h. NO production in the cell culture medium was measured as nitrite (NO_2) at 540 nm and quantified from a standard curve generated from sodium nitrite ($NaNO_2$). A brief experimental timeline of this assay was illustrated in Supplementary Figure S1. The cytotoxicity of the extracts in RAW 264.7 cells was determined using an 3-(4,5-dimethylthiazol-2-yl)-2,5-diphenyl tetrazolium bromide (MTT) assay [11].

2.5. Total RNA Extraction and Quantitative RT-PCR

Total RNA was extracted from cells using TRIzol reagent and used for cDNA synthesis with the cDNA EcoDry Premix (Takara, Shiga, Japan). Relative mRNA levels were determined by cycler CFX96 (Bio-Rad). PCR amplification was performed in triplicate with the following specific primers: IL-1β forward (F): TGAGCACCTTCTTTTCCTTCA and reverse (R): TTGTCTAATGGGAACGTCACAC; IL-6 F: TAATTCATATCTTCAACCAAGAGG and R: TGGTCCTTAGCCACTCCTTC; TNF-α F: CTGTAG CCCACGTCGTAGC and R: GGTTGTCTTTGAGATCCATGC; iNOS F: TGTGGCTACCACATTG AAGAA and R: TCATGATAACGTTTCTGGCTCTT; and β-actin F: GTCATCACTATTGGCAACGAG and R: TTGGCATAGAGGTCTTTACGG. Real-time PCR (RT-PCR) was performed using the following cycling conditions: enzyme activation and initial denaturation for 5 min at 95 °C and 40 cycles of amplification at 95 °C for 10 s followed by 55 °C (IL-1 β and IL-6), 58 °C (iNOS), or 62 °C (TNF-α) for 20 s.

2.6. Western Blot Analysis

RAW 264.7 cells were treated for 60 min with COFE (0–0.3 mg/mL) and then treated with LPS (1 μg/mL) for 30 min (for IkBα and p-p65 protein determination), or treated with LPS (1 μg/mL) for 18 h (for iNOS protein determination) (Supplementary Figure S1). At the end of the LPS treatment, Western blot analysis was conducted according to a reported method [12], with slight modification. Briefly, after washing cells with ice-cold PBS, total proteins were extracted from the cell pellets using the PRO-PREP protein extraction solution (iNtRON Biotechnology, Seongnam, Korea) according to the manufacturer's instruction. The lysates were centrifuged at 16,000$\times$ g for 10 min at 4 °C and stored at −20 °C until use. Protein was then measured using the Pierce™ BCA assay kit (ThermoFisher Scientific, Waltham, United States), with bovine serum albumin as a standard and then separated by sodium dodecyl sulfate polyacrylamide gel electrophoresis (SDS-PAGE) in 10% gels, and transferred to polyvinylidene fluoride (PVDF) membranes. Membranes were blocked with 5% bovine serum albumin (BSA) in tris-buffered saline with tween-20 (TBST) for 1 h at room temperature followed by incubating with appropriate primary antibodies for 90 min at room temperature. Membranes were washed three times and incubated for an additional 60 min with horseradish peroxidase-conjugated secondary antibody. The expression of IkBα and p-p65 (1:1000) (both were purchased from Cell Signaling Technology, Danvers, MA, USA) and β-actin (Santa Cruz Biotechnology, Dallas, TX, USA) as a loading control were visualized using Thermo Scientific™ Pierce™ ECL Western Blotting Substrate (ThermoFisher Scientific, Waltham, MA, USA).

2.7. Flow Cytometric Analysis

The MTT assay was used to examine the cytotoxicity of COFE in HaCaT cells [11]. The absorbance was measured at 570 nm. After TNF-α (20 ng/mL) treatment for 1 h, the cells were incubated with different extract concentrations for 18 h. The activation of caspases 3 and 7 in HaCaT cells was detected using the Muse Caspase-3/7 assay kit with a Muse Cell Analyzer (EMD Millipore, Billerica, MA, USA) according to the manufacturer's protocol. The data from the Muse Cell Analyzer were analyzed using Muse 1.4 software (Luminex Corporation, Austin, TX, USA). A brief experimental timeline of this assay was illustrated in Supplementary Figure S1.

2.8. Gas Chromatography/Mass Spectrometry (GC-MS) and Liquid Chromatography with Tandem Mass Spectrometry (LC-MS/MS) Analysis

The GC-MS analysis of COFE was performed by using an HP 6890 Plus GC gas chromatograph with a mass selective detector (MSD; HP 5973, Hewlett-Packard, California, United States). The samples were diluted 1:1000 (*v:v*) with HPLC-grade dichloromethane. The samples (1 µL) were injected into an HP-5 column. The GC oven temperature was set at 50 °C for 4 min, increased to 280 °C at a rate of 4 °C/min, and held at the final temperature for 2 min. The velocity of the carrier gas, 99.99% He, was 0.7 mL/min. Quantitative analysis was performed by using the area normalization method.

LC-MS analysis was performed on a Thermo Accela UHPLC system (Thermo Fisher Scientific, San Jose, CA, USA). The samples were separated on a Waters BEH C18 column (2.1 × 150 mm, 1.7 µm) at room temperature. The mobile phase consisted of water (A) and acetonitrile (B), both with 0.1% formic acid added. The elution gradient was set as follows: 5% B (0 min), 5% B (1 min), 70% B (20 min), 100% B (24 min), and 100% B (27 min). The flow rate was 400 µL/min and the sample loading volume was 1 µL. The UHPLC was coupled to an LTQ-Orbitrap XL hybrid mass spectrometer (Thermo Electron, Bremen, Germany) via an ESI interface. The samples were analyzed in positive ion mode and the conditions of the ESI source were the same as previously used [13].

2.9. In Vivo Toxicity Evaluation

Ten female 8-week-old Sprague-Dawley rats were obtained from Orient Bio Inc. (Gyeonggi-do, Korea). The acute toxicity study was performed by using a previously reported method [14] with slight modifications. Five rats per group were randomly assigned to the control and test groups. A single dose of COFE (2000 mg/kg of body weight) was administered intragastrically according to the OECD test guideline 423 [15]. A standard pellet diet (Hyochang Science, Daegu, Korea) and distilled water were provided ad libitum. The animals were under constant observation for abnormal signs and symptoms for the first 12 h after COFE administration and subsequently observed once per day for 2 weeks. The changes in body weight and food and water intake were measured twice a week for 14 days after treatment. The animals were sacrificed after the experimental period and the major organs were collected and inspected for gross lesions.

2.10. Molecular Docking

To investigate the binding mode of the loganin, cornuside, and naringenin 7-O-β-D-glucoside to human high-affinity IgE receptors (FceRI), a molecular docking analysis was performed using the FceRI-IgE complex structure data with the protein data bank (PDB) [16] ID 2Y7Q [17]. The conformations of loganin (ZINC000003978792) and naringenin 7-O-β-D-glucoside (ZINC4097895) were generated using a conformational search against the ZINC docking database, University of California, San Francisco (UCSF). The cornuside (InChI Key: SMTKSCGLXONVGL-UHFFFAOYSA-N) conformation was searched against the PubChem database. The ligand file in Structure Data File (SDF) format was converted to MOL format using the Openbabel software. The co-crystallized structures were prepared using UCSF Chimera (Chimera, Version 1.12, RBVI, San Francisco, CA, USA) and iGEMDOCK (Version 2.1; NCTU, Hsinchu City, Taiwan). Molecular docking was performed using iGEMDOCK with the accurate docking mode. The best-docked poses were further analyzed, and 3D structure images were prepared using UCSF Chimera.

2.11. Statistical Analysis

The IC_{50} and EC_{50} values were calculated using Prism (GraphPad Software Inc., La Jolla, CA, USA). Statistical differences were assessed by one-way analysis of variance and Duncan's multiple comparisons test using Statistical Analysis Systems software (SAS Institute, Cary, NC, USA). Values of $p < 0.05$ were considered statistically significant.

3. Results

3.1. Effect of COFE on DNP/IgE-Induced Degranulation in RBL-2H3 Cells

COFE suppressed the release of β-hexosaminidase from RBL-2H3 cells in a concentration-dependent manner (Figure 1A). A dose of 0.3 mg/mL of COFE significantly inhibited β-hexosaminidase release (46%) ($p < 0.05$). The IC_{50} value of COFE for β-hexosaminidase release inhibition was 0.178 mg/mL, showing that the potential anti-atopic activity occurred through the suppression of the IgE-induced degranulation of RBL-2H3 cells.

3.2. Antioxidant Activity of COFE

Supplementary Figure S2 shows the DPPH radical scavenging activity of COFE. COFE showed concentration-dependent DPPH radical scavenging activity from 0.1 to 10 mg/mL. Its EC_{50} value was 1.82 mg/mL, which is approximately six times higher than that of ascorbic acid (0.30 mg/mL). The E_{max} values of COFE were 3.15 mM TE and 3.23 mM TE in the FRAP assay and ABTS assay, respectively. Its EC_{50} values were 10.76 mg/mL and 0.60 mg/mL in the FRAP assay and ABTS assay, respectively.

3.3. Effect of COFE on LPS-Induced NO Production and Pro-Inflammatory Cytokines in RAW 264.7 Cells

COFE significantly inhibited LPS-induced NO production in RAW 264.7 cells in a concentration-dependent manner (0.01–0.3 mg/mL) ($p < 0.05$) with an EC_{50} value of 0.74 mg/mL (Figure 1B). Similarly, COFE attenuated LPS-induced iNOS mRNA expression in RAW 264.7 cells in a concentration-dependent manner, indicating a positive correlation between the inhibition of NO production and the suppression of iNOS mRNA expression (Figure 1C). At the tested concentrations, COFE did not exert any cytotoxic effects on RAW 264.7 cells (data not shown). Furthermore, COFE inhibited LPS-induced IL-1β, IL-6, and TNF-α mRNA expression in RAW 264.7 cells in a concentration-dependent manner (0.01–0.3 mg/mL), with E_{max} values ranging from 73% to 90% (Figure 1D). Furthermore, the EC_{50} values for the inhibition of IL-1β, IL-6, and TNF-α were 0.0007, 0.039, and 0.006 mg/mL, respectively.

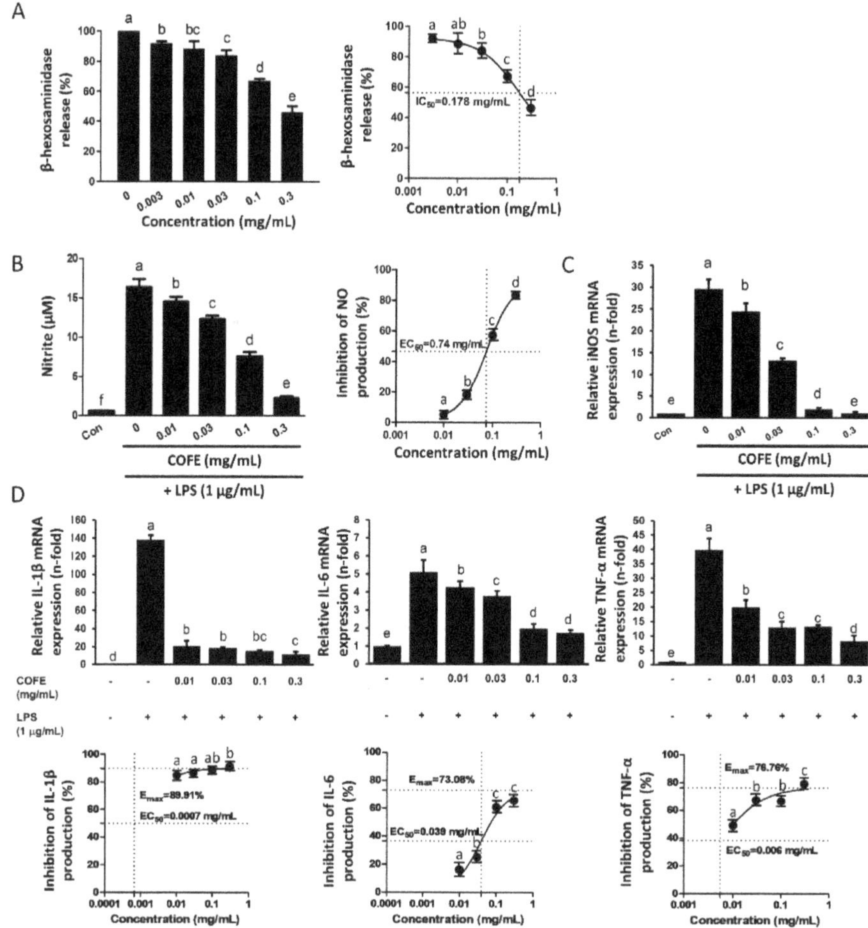

Figure 1. Anti-allergic and anti-inflammatory activity of ethanolic extract of *Cornus officinalis* (COFE). (**A**) Evaluation of the β-hexosaminidase release from RBL-2H3 cells treated with COFE (left) and the IC_{50} of COFE for β-hexosaminidase release (right). (**B**) Effect of COFE on lipopolysaccharide (LPS)-induced nitric oxide (NO) production. (**C**) iNOS mRNA expression. (**D**) Anti-inflammatory cytokine gene expression. Values are expressed as the mean ± SD of three independent experiments. Letters (a–e) indicate significantly different values ($p < 0.05$), as determined by Duncan's multiple comparison test.

3.4. Effect of COFE on NF-κB Stimulation

Western blot analysis showed that LPS reduced the cytosolic IκBα expression but COFE increased it in a concentration-dependent manner (Figure 2A). Comparably, COFE inhibited LPS-induced cytosolic p-p65 expression. The expression of iNOS protein also showed significant reduction compared to the LPS induced group (Supplementary Figure S3). Thus, the anti-inflammatory activity of COFE occurs through the inhibition of NF-κB activation.

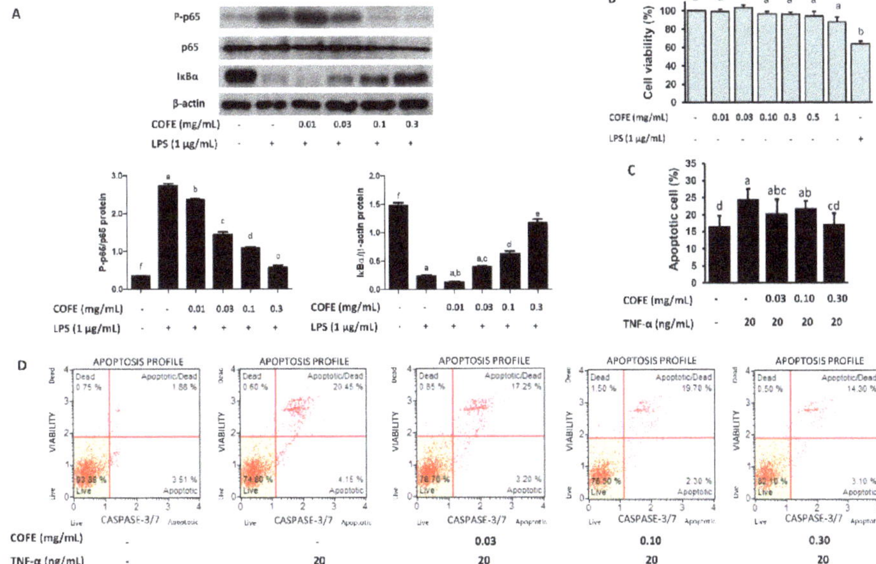

Figure 2. Effect of COFE on LPS-induced NF-κB activation in RAW 264.7 cells (**A**), cell viability (**B**), and TNF-α-induced apoptotic death in HaCaT cells (**C,D**). (**A**) Cells were treated with COFE (0–0.3 mg/mL) for 60 min and then treated with LPS (1 μg/mL) for 30 min. The levels of total IκBα and phospho-p65 (P-p65) were determined by Western blotting of the total protein of cell lysates (n = 3). β-actin was used as a loading control. (**B**) The viability of HaCaT cells treated with increasing concentrations of COFE (0–1 mg/mL) for 24 h was evaluated using an MTT assay; LPS was used as a control. Each sample was assayed in triplicate at each concentration. (**C**) After TNF-α (20 ng/mL) treatment for 1 h, cells were incubated with the indicated concentrations of extract for 18 h. TNF-α-induced apoptosis was detected by flow cytometric analysis. The data show healthy cells (annexin-V negative and caspase 3/7 and 7-ADD negative (lower left)), early apoptotic cells (positive for annexin-V and caspase 3/7 and negative for 7-ADD (lower right)), late apoptotic/dead cells (both annexin V and caspase 3/7 and 7-ADD positive (upper right)), and necrotic cells (only 7-ADD positive (upper left)) (n = 3). (**D**) Total apoptotic cells (early apoptotic and late apoptotic cells) were quantified by positive staining for annexin-V or caspase 3/7 activity. Values are expressed as the mean ± SD of three independent experiments. Letters (a–f) indicate significantly different values ($p < 0.05$), as determined by Duncan's multiple comparison test.

3.5. Effect of COFE on TNF-α Induced Apoptosis in HaCaT Cells

To assess the suitability of COFE for the treatment of human AD, cytotoxicity and apoptosis were examined in human keratinocytes. Treating HaCaT cells with COFE (0–1 mg/mL) for 24 h did not affect their viability (Figure 2B). The impact of COFE on TNF-α-induced apoptosis in HaCaT cells was evaluated through cytofluorometric analysis. TNF-α-induced apoptosis was estimated through the net change in the percentage of total apoptotic cells compared with the controls (Figure 2C). After caspase 3/7 antibody treatment, TNF-α-treated samples exhibited approximately 1.5× apoptotic/dead and apoptotic cells in comparison with untreated cells (Figure 2D). COFE significantly reduced TNF-α-induced cell apoptosis ($p < 0.05$).

3.6. Identification of the Constituent Compounds of COFE

Table 1 shows the main constituents of COFE identified by GC-MS analysis, namely 4,6-cycloheptatrien-1-one, malic acid, 2-furancarbox-aldehyde, and 2,3-dihydro-3,5-dihydroxy-6-methyl-4H-pyran-4-one. The compounds responsible for the physiological effects in COFE were analyzed by LC-MS/MS and identified as loganin, cornuside, and naringenin 7-O-β-D-glucoside (Figure 3A). Loganin, a natural flavonoid, was the major component of COFE.

Table 1. Gas chromatography/mass spectrometry (GC-MS) analysis of compounds of the ethanolic extract of *C. officinalis*.

RT (Retention Time)	Formula	%	ID	Activity
34.21	$C_6H_8O_4$	4.58	2,3-dihydro-3,5-dihydroxy-6-methyl-4H-pyran-4-one	Antimicrobial, anti-inflammatory, antiproliferative
38.23	$C_6H_6O_3$	39.9	2-furancarboxaldehyde	Antimicrobial, preservative
43.38	$C_4H_6O_5$	9.71	Malic acid	Antimicrobial
50.41	C_7H_6O	1.41	2,4,6-cycloheptatrien-1-one	Antioxidant

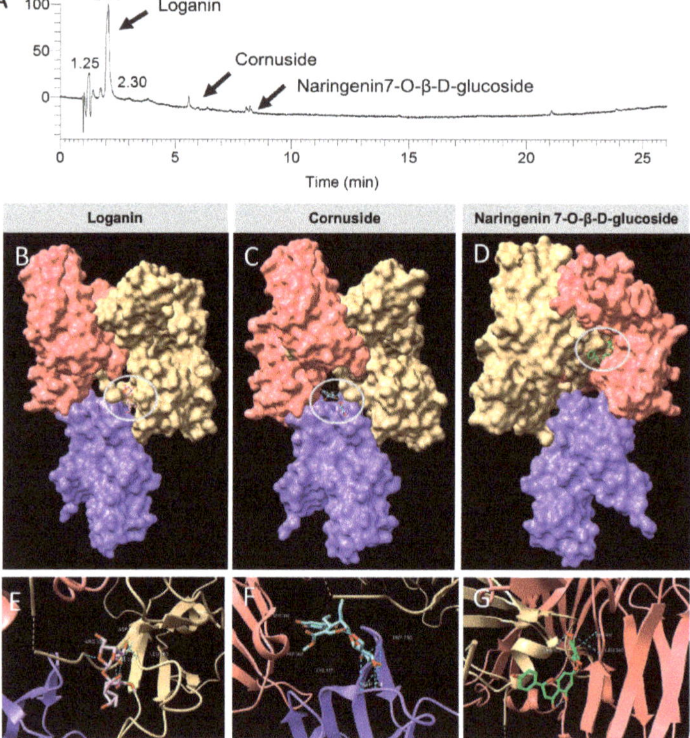

Figure 3. Liquid chromatography with tandem mass spectrometry (LC-MS/MS) COFE (**A**) and molecular docking analysis of the FceRI-IgE complex (**B–G**). The best-docked position showing the interaction sites of loganin (**B**), cornuside (**C**), and naringenin 7-O-β-D-glucoside (**D**) in the FceRI-IgE complex. The blue dotted lines represent the hydrogen bonds between the amino acid residues of the FceRI-IgE complex and loganin (ARG334, VAL336, ASP362, LEU363, ALA364), cornuside (SER337, ASP362, LYS117, ASP114, TRP156), and naringenin 7-O-β-D-glucoside (LYS302, ARG342, LEU340, ARG431, ALA338) (**E–G**).

3.7. In Vivo Toxicity Evaluation of COFE

Animals displayed no clinical signs or mortality up to 14 days after the administration. The same procedure was carried out on three female rats and a similar observation was obtained. Therefore, the acute toxic class method, following the flow chart of LD_{50} cut-off, confirmed COFE as a category 5 substance in the Globally Harmonized System of Classification and Labeling of Chemicals (GHS).

3.8. Molecular Docking

By integrating the data obtained from LC-MS analysis with the in silico molecular docking analysis, the major bioactive components (loganin, cornuside, and naringenin 7-O-β-D-glucoside) found in COFE were found to interact with the FceRI-IgE complex. Figure 3B–D shows the loganin, cornuside, and naringenin 7-O-β-D-glucoside binding to the 3D structure of the FceRI-IgE complex. Figure 3E–G shows the amino acid residues associated with the binding of the compounds through hydrogen bonding. The binding energies of the FceRI-IgE complex with the compounds according to molecular docking analysis are shown in Table 2.

Table 2. Binding energies (Kcal) between the FceRI-IgE complex and the bioactive compounds identified in COFE.

Compound	Binding Energy (Kcal)	Van der Waals	Hydrogen Bond	Electrostatic
Loganin	−116.9	−90.5332	−26.3531	0
Cornuside	−141.1	−101.645	−39.4547	0
Naringenin 7-O-β-D-glucoside	−125.7	−96.4237	−29.3185	0

4. Discussion

The anti-allergic potential of COFE was assessed with RBL-2H3 cells (Figure 1), which are commonly used to study allergy and inflammation reactions because they express high-affinity IgE receptors (FceRI) [5]. β-Hexosaminidase was used as an indicator of degranulation in RBL-2H3 cells sensitized by exposure to IgE-DNP antibody after stimulation with DNP-HSA [18]. Our results show COFE inhibits β-Hexosaminidase release with an IC_{50} of 0.178 mg/mL which is about 60- and 1.4-fold lower than that of ethanol extracts from mulberry fruit [19] and *Arctium lappa* fruit [20], respectively. Antioxidants potentially exert beneficial effects in the treatment of AD by inhibiting the oxidative stress which is involved in the pathophysiology of the acute exacerbation of AD [21]. However, the antioxidant activity of COFE was lower than that reported by Hwang and co-workers [4]. This may be due to the concentration of bioactive compounds in the plant, which could be influenced by photoperiodic, temperature, and geographical differences [22].

Macrophages play crucial roles in acute inflammation and persistence of proinflammatory cytokines, which lead to chronic inflammation conditions including AD [23]. We found that COFE significantly inhibits LPS-induced NO production, iNOS mRNA expression (Figure 1B,C), and iNOS protein expression in RAW 264.7 cells (Figure S3). Besides, COFE consistently suppresses the expression of IL-1β, IL-6, and TNF-α, suggesting that the upstream signaling molecules (such as NF-κB) could be the determining step in the anti-inflammatory response induced by COFE (Figure 4). Further studies are needed to confirm the use of COFE as a therapeutic agent for AD and various inflammatory diseases through the reduction of pro-inflammatory cytokines. Several plant extracts with potential anti-inflammatory activity may be suitable for the treatment of immune disorders like AD [24].

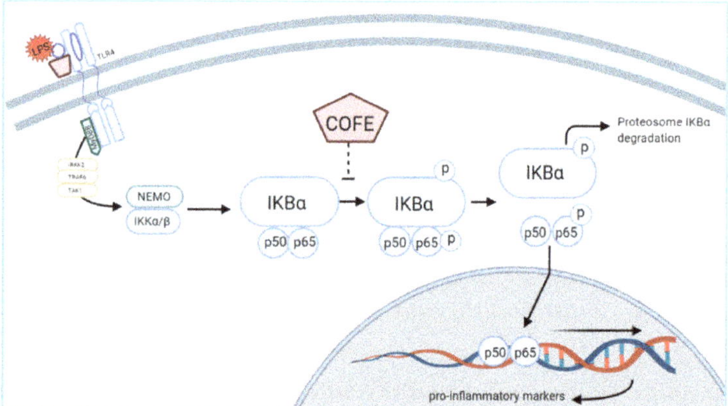

Figure 4. Illustration of the mechanism of action of COFE through NF-κB suppression in RAW 264.7 cells treated with LPS. TLR4: Toll-like receptor 4; IRAK2: Interleukin 1 Receptor Associated Kinase 2; TRAF6: Tumor necrosis factor receptor (TNFR)-associated factor 6; TAK1: Transforming growth factor β-activated kinase 1; MyD88: Myeloid differentiation primary response 88; NEMO: nuclear factor kappa-B essential modulator; IKKα/β: inhibitor of nuclear factor kappa-B kinase subunit α/β; IKBα: nuclear factor of kappa light polypeptide gene enhancer in B-cells inhibitor alpha; p65/p50: nuclear factor NF-kappa-B p65/p50 subunit.

COFE significantly inhibited TNF-α-induced apoptosis in HaCaT cells. TNF-α and pro-inflammatory cytokines are secreted by macrophages and mast cells and promote apoptosis in HaCaT cells by binding to the TNF-receptor 1 [25]. Considering the deterioration of skin lesions by inflammatory mediators that are released from various skin cells during the inflammatory process, COFE could be useful for the management of AD without acute oral toxicity.

Oxidative stress is associated with chronic skin inflammation and plays a role in the pathogenesis of AD [26]. Reactive oxygen species (ROS) can modify DNA, lipids, and proteins [26]. NO production is one of the biomarkers of lipid peroxidation resulting from oxidative stress [27]. A previous report stated that the increase in oxidative stress in AD was due to an increase in lipid peroxidation and a decrease in antioxidant levels [21]. The antioxidant properties of COFE (Figure S2) could reduce oxidative stress and thus disrupt the pro-inflammatory microenvironment through ROS-mediated signaling events.

Loganin, the main bioactive constituent in COFE, synergistically contributes to the inhibition of oxidative stress, inflammation, and metabolic disorders [28]. It may directly promote the differentiation of MC3T3-E1 cells and inhibit their apoptosis [29]. Cornuside attenuates LPS-induced inflammatory cytokines through the inhibition of NF-κB activation in RAW 264.7 cells [30]. Moreover, cornuside shows remarkable antioxidant activity and inhibits isoproterenol-induced myocardial cell necrosis [31]. Interestingly, naringenin 7-*O*-β-D-glucoside (identified and isolated from *Marrubium globosum* Montbr. et Auch. ex Benth. ssp. *libanoticum* Boiss. (Lamiaceae), a medicinal plant used to treat inflammatory diseases and asthma), shows antibacterial activity in several bacterial strains [32]. Besides, naringenin 7-*O*-β-D-glucoside has been reported for its antioxidant activity [33]. These compounds may act synergistically to confer various bioactivities to COFE. Molecular docking analysis was carried out to elucidate the interaction between the FceR1-IgE complex and the bioactive compounds identified in COFE.

The best poses of FceR1-IgE docked with loganin, cornuside, and naringenin 7-*O*-β-D-glucoside were determined. Interestingly, the docking analysis showed that loganin, being the major component found in this extract, binds to IgE through hydrogen bonds with ARG334, VAL336, ASP362, LEU363, and ALA364 (Figure 3E). It is noteworthy that, in humans, the disruption of the glycosylation site at asparagine-394 (ASN394) in the Cε3 domain nullifies the binding of IgE to FceR1 [34]. From our simulation analysis, the binding site of loganin to the complex was the nearest to ASN394, which might indirectly disrupt the formation of the FceR1-IgE complex. Conversely, cornuside possessed the lowest binding energy (−141.1 Kcal) among the compounds tested (Table 2). A lower energy value represents a higher receptor-ligand binding affinity. In other words, cornuside forms a more stable receptor-ligand complex with FceR1-IgE compared to loganin and naringenin 7-*O*-β-D-glucoside. The region where naringenin 7-*O*-β-D-glucoside interacts with IgE was further analyzed by the protein sequences in ChimeraX. The protein sequences were then searched against the conserved domain database (CDD) [35]. We found that naringenin 7-*O*-β-D-glucoside binds to the CH2 domain (second constant Ig domain of the heavy chain) of IgE, cd05847. Therefore, the binding of naringenin 7-*O*-β-D-glucoside could prevent the conformation change of IgE, subsequently disrupt the engagement of the second Cε3 domain to FceR1, and lead to a reduction of skin inflammation.

Our findings suggest diverse potential dermatologic and cosmetic applications for COFE. In silico findings suggest that the bioactive compounds found in COFE impede skin inflammation by disrupting the binding of IgE to the human high-affinity IgE receptors (FceRI). Since our study relies only on in vitro assays, further studies using animal models or artificial skin models would confirm the therapeutic effects and mechanisms of action of COFE.

5. Conclusions

Our results show that COFE might exert inhibitory effects on oxidative stress, allergic responses, and inflammatory responses. Additionally, loganin, cornuside, and naringenin 7-*O*-β-D-glucoside were identified as compounds potentially responsible for these effects. COFE showed antioxidant, anti-allergic, and anti-inflammatory effects that could potentially be leveraged to treat AD, yet did not exert cytotoxicity or acute oral toxicity. Our findings suggest that COFE could be used in the development of preventative and treatment therapies for AD.

Supplementary Materials: The following are available online at http://www.mdpi.com/2072-6643/12/11/3317/s1, Supplementary Figure S1: The experimental timeline for β-Hexosaminidase release assay using RBL-2H3 cells (A); NO assay and isolation of iNOS protein (B) and isolation of IkBα and p-p65 protein for Western blot analysis (C) in RAW264.7 cells; flow cytometric analysis for apoptosis assay using HaCaT cells (D). Supplementary Figure S2: Evaluation of the extracts' 2,2-diphenyl-1-picrylhydrazyl (DPPH) radical scavenging activity (left). After the incubation of the samples and the DPPH solution at 37 °C for 30 min, the absorbance at 517 nm was measured; ascorbic acid was used as a positive control. For the ferric reducing antioxidant power (FRAP) assay (middle), after the incubation of the samples and FRAP reagent at 37 °C for 30 min, the absorbance at 595 nm was measured. For the ABTS assay (right), after mixing the ABTS solution and the samples, the absorbance at 734 nm was measured. The results are expressed as μM Trolox equivalent (TE), Emax, and EC50 (n = 3). Letters (a–c) indicate significantly different values ($p < 0.05$), as determined by Duncan's multiple comparison test. Supplementary Figure S3: Cells were treated with COFE (0–0.3 mg/mL) for 30 min and then treated with LPS (1 μg/mL) for 18 h. The levels of iNOS were determined by Western blotting of the total protein of cell lysates. β-actin was used as a loading control. Values are expressed as the mean ± SD of three independent experiments. Letters (a–d) indicate significantly different values ($p < 0.05$), as determined by Duncan's multiple comparison test. Reserved DOI:10.5281/zenodo.4065293

Author Contributions: Conceptualization, S.-C.P., Y.Q., and S.-J.L.; methodology, S.-J.L., Y.Q., E.-B.L., B.T.B., M.S.A., M.A.A., N.B. and Z.-E.I.; software, S.-J.L. and Y.Q.; validation, B.T.B. and Z.-E.I.; investigation, S.-J.L., Y.Q., E.-B.L.; resources, Z.-E.I.; data curation, B.T.B.; writing—original draft preparation, S.-J.L. and Y.Q.; writing—review and editing, Y.Q.; supervision, S.-C.P.; funding acquisition, S.-C.P. All authors have read and agreed to the published version of the manuscript.

Funding: This research was supported in part by the Forest Resources Development Institute of Gyeongsangbuk-do and in part The National Research Foundation of Korea (NRF) funded in part by the Ministry of Education (2019R1A2C2006277), Republic of Korea. The funders had no role in study design, data collection, and interpretation, or the decision to submit the work for publication.

Acknowledgments: The authors wish to thank Yuan Yee Lee and Muhammad Irfan for providing technical support.

Conflicts of Interest: The authors declare no conflict of interest.

References

1. Brandt, E.B.; Sivaprasad, U. Th2 Cytokines and atopic dermatitis. *J. Clin. Cell. Immunol.* **2011**, *2*, 110.
2. Leung, D.Y.M.; Bieber, T. Atopic dermatitis. *Lancet* **2003**, *361*, 151–160.
3. Cooper, K.D. Atopic dermatitis: Recent trends in pathogenesis and therapy. *J. Investig. Dermatol.* **1994**, *102*, 128–137.

4. Hwang, K.A.; Hwang, Y.J.; Song, J. Antioxidant activities and oxidative stress inhibitory effects of ethanol extracts from *Cornus officinalis* on raw 264.7 cells. *BMC Complement. Altern. Med.* **2016**, *16*, 196.

5. Mechesso, A.F.; Lee, S.J.; Park, N.H.; Kim, J.Y.; Im, Z.E.; Suh, J.W.; Park, S.C. Preventive effects of a novel herbal mixture on atopic dermatitis-like skin lesions in BALB/c mice. *BMC Complement. Altern. Med.* **2019**, *19*, 25.

6. Czerwińska, M.E.; Melzig, M.F. Cornus mas and Cornus officinalis—Analogies and differences of two medicinal plants traditionally used. *Front. Pharmacol.* **2018**, *9*, 894.

7. Juckmeta, T.; Thongdeeying, P.; Itharat, A. Inhibitory effect on β -Hexosaminidase release from RBL-2H3 cells of extracts and some pure constituents of Benchalokawichian, a Thai herbal remedy, used for allergic disorders. *Evid.-Based Complement. Altern. Med. ECAM* **2014**, *2014*, 828760.

8. Chang, Z.Q.; Hwang, M.H.; Rhee, M.H.; Kim, K.S.; Kim, J.C.; Lee, S.P.; Jo, W.S.; Park, S.C. The in vitro anti-platelet, antioxidant and cellular immunity activity of *Phellinus gilvus* fractional extracts. *World J. Microbiol. Biotechnol.* **2008**, *24*, 181–187.

9. Benzie, I.F.; Strain, J.J. The ferric reducing ability of plasma (FRAP) as a measure of "antioxidant power": The FRAP assay. *Anal. Biochem.* **1996**, *239*, 70–76.

10. Rufino, M.D.S.M.; Alves, R.E.; de Brito, E.S.; Pérez-Jiménez, J.; Saura-Calixto, F.; Mancini-Filho, J. Bioactive compounds and antioxidant capacities of 18 non-traditional tropical fruits from Brazil. *Food Chem.* **2010**, *121*, 996–1002.

11. Lee, S.J.; Hossaine, M.D.A.; Park, S.C. A potential anti-inflammation activity and depigmentation effect of *Lespedeza bicolor* extract and its fractions. *Saudi J. Biol. Sci.* **2016**, *23*, 9–14. [CrossRef]

12. Endale, M.; Park, S.C.; Kim, S.; Kim, S.H.; Yang, Y.; Cho, J.Y.; Rhee, M.H. Quercetin disrupts tyrosine-phosphorylated phosphatidylinositol 3-kinase and myeloid differentiation factor-88 association, and inhibits MAPK/AP-1 and IKK/NF-κB-induced inflammatory mediators production in RAW 264.7 cells. *Immunobiology* **2013**, *218*, 1452–1467. [CrossRef] [PubMed]

13. Xu, W.; Zhang, J.; Zhu, D.; Huang, J.; Huang, Z.; Bai, J.; Qiu, X. Rapid separation and characterization of diterpenoid alkaloids in processed roots of *Aconitum carmichaeli* using ultra high performance liquid chromatography coupled with hybrid linear ion trap-Orbitrap tandem mass spectrometry. *J. Sep. Sci.* **2014**, *37*, 2864–2873. [PubMed]

14. Hossain, M.A.; Lee, S.J.; Park, J.Y.; Reza, M.A.; Kim, T.H.; Lee, K.J.; Suh, J.W.; Park, S.C. Modulation of quorum sensing-controlled virulence factors by *Nymphaea tetragona* (water lily) extract. *J. Ethnopharmacol.* **2015**, *174*, 482–491. [CrossRef]

15. OECD/OCDE. *Acute Oral Toxicity-Acute Toxic Class. Method OECD Guideline for the Testing of Chemicals*; OECD/OCDE: Paris, France, 2001; pp. 1–14.

16. Berman, H.M.; Westbrook, J.; Feng, Z.; Gilliland, G.; Bhat, T.N.; Weissig, H.; Shindyalov, I.N.; Bourne, P.E. The Protein Data Bank. *Nucleic Acids Res.* **2000**, *28*, 235–242. [PubMed]

17. Holdom, M.D.; Davies, A.M.; Nettleship, J.E.; Bagby, S.C.; Dhaliwal, B.; Girardi, E.; Hunt, J.; Gould, H.J.; Beavil, A.J.; McDonnell, J.M.; et al. Conformational changes in IgE contribute to its uniquely slow dissociation rate from receptor FcεRI. *Nat. Struct. Mol. Biol.* **2011**, *18*, 571–576. [CrossRef] [PubMed]

18. Chen, B.-H.; Hung, M.-H.; Chen, J.Y.-F.; Chang, H.-W.; Yu, M.-L.; Wan, L.; Tsai, F.J.; Wang, T.-P.; Fu, T.-F.; Chiu, C.-C. Anti-allergic activity of grapeseed extract (GSE) on RBL-2H3 mast cells. *Food Chem.* **2012**, *132*, 968–974. [CrossRef]

19. Yoo, J.M.; Kim, N.Y.; Seo, J.M.; Kim, S.J.; Lee, S.Y.; Kim, S.K.; Kim, H.D.; Lee, S.W.; Kim, M.R. Inhibitory effects of mulberry fruit extract in combination with naringinase on the allergic response in IgE-activated RBL-2H3 cells. *Int. J. Mol. Med.* **2014**, *33*, 469–477. [CrossRef]

20. Yoo, J.M.; Yang, J.H.; Yang, H.J.; Cho, W.K.; Ma, J.Y. Inhibitory effect of fermented *Arctium lappa* fruit extract on the IgE-mediated allergic response in RBL-2H3 cells. *Int. J. Mol. Med.* **2016**, *37*, 501–508. [CrossRef]

21. Sivaranjani, N.; Rao, S.V.; Rajeev, G. Role of reactive oxygen species and antioxidants in atopic dermatitis. *J. Clin. Diagn. Res.* **2013**, *7*, 2683–2685. [CrossRef]

22. Hwang, Y.J.; Lee, E.J.; Kim, H.R.; Hwang, K.A. In vitro antioxidant and anticancer effects of solvent fractions from *Prunella vulgaris* var. lilacina. *BMC Complement. Altern. Med.* **2013**, *13*, 310. [CrossRef]

23. Valledor, A.F.; Comalada, M.; Santamaría-Babi, L.F.; Lloberas, J.; Celada, A. Macrophage proinflammatory activation and deactivation: A question of balance. *Adv. Immunol.* **2010**, *108*, 1–20.

24. Shin, J.H.; Chung, M.J.; Seo, J.G. A multistrain probiotic formulation attenuates skin symptoms of atopic dermatitis in a mouse model through the generation of CD4+Foxp3+ T cells. *Food Nutr. Res.* **2016**, *60*, 32550. [CrossRef] [PubMed]

25. Gilfillan, A.M.; Beaven, M.A. Regulation of mast cell responses in health and disease. *Crit. Rev. Immunol.* **2011**, *31*, 475–529. [CrossRef] [PubMed]

26. Bertino, L.; Guarneri, F.; Cannavò, S.P.; Casciaro, M.; Pioggia, G.; Gangemi, S. Oxidative stress and atopic dermatitis. *Antioxidants* **2020**, *9*, 196. [CrossRef]

27. Yidong, T.; Taihao, Q. Oxidative stress and human skin connective tissue aging. *Cosmetics* **2016**, *3*, 28.

28. Yokozawa, T.; Kang, K.S.; Park, C.H.; Noh, J.S.; Yamabe, N.; Shibahara, N.; Tanaka, T. Bioactive constituents of Corni Fructus: The therapeutic use of morroniside, loganin, and 7-O-galloyl-D-sedoheptulose as renoprotective agents in type 2 diabetes. *Drug Discov. Ther.* **2010**, *4*, 223–234. [PubMed]

29. Li, M.; Wang, W.; Wang, P.; Yang, K.; Sun, H.; Wang, X. The pharmacological effects of morroniside and loganin isolated from Liuweidihuang Wan, on MC3T3-E1 cells. *Molecules* **2010**, *15*, 7403–7414. [CrossRef]

30. Choi, Y.H.; Jin, G.Y.; Li, G.Z.; Yan, G.H. Cornuside suppresses lipopolysaccharide-induced inflammatory mediators by inhibiting nuclear factor-kappa B activation in RAW 264.7 macrophages. *Biol. Pharm. Bull.* **2011**, *34*, 959–966. [CrossRef]

31. Huang, F.H.; Zhang, X.Y.; Zhang, L.Y.; Li, Q.; Ni, B.; Zheng, X.L.; Chen, A.J. Mast cell degranulation induced by chlorogenic acid. *Acta Pharmacol. Sin.* **2010**, *31*, 849–854. [CrossRef]

32. Rigano, D.; Formisano, C.; Basile, A.; Lavitola, A.; Senatore, F.; Rosselli, S.; Bruno, M. Antibacterial activity of flavonoids and phenylpropanoids from *Marrubium globosum* ssp. libanoticum. *Phytother. Res.* **2007**, *21*, 395–397.

33. Orhan, D.D. Novel flavanone glucoside with free radical scavenging properties from *Galium fissurense*. *Pharm. Biol.* **2003**, *41*, 475–478.

34. Balbino, B.; Conde, E.; Marichal, T.; Starkl, P.; Reber, L.L. Approaches to target IgE antibodies in allergic diseases. *Pharmacol. Ther.* **2018**, *191*, 50–64. [CrossRef] [PubMed]

35. Lu, S.; Wang, J.; Chitsaz, F.; Derbyshire, M.K.; Geer, R.C.; Gonzales, N.R.; Gwadz, M.; Hurwitz, D.I.; Marchler, G.H.; Song, J.S.; et al. CDD/SPARCLE: The conserved domain database in 2020. *Nucleic Acids Res.* **2020**, *48*, D265–D268. [CrossRef]

Publisher's Note: MDPI stays neutral with regard to jurisdictional claims in published maps and institutional affiliations.

Article

Effect of Konjac Glucomannan (KGM) on the Reconstitution of the Dermal Environment against UVB-Induced Condition

Kyung Ho Choi [1,†], Sung Tae Kim [2,3,†], Bum Ho Bin [1,*] and Phil June Park [4,*]

1   Department of Applied Biology, Ajou University, 206 World Cup-ro, Yeongtong-gu, Suwon-si, Gyeonggi-do 16499, Korea; c457854@naver.com
2   Department of Pharmaceutical Engineering, Inje University, Gimhae-si, Gyeongsangnam-do 50834, Korea; stkim@inje.ac.kr
3   Department of Nanoscience and Engineering, Inje University, Gimhae-si, Gyeongsangnam-do 50834, Korea
4   AMOREPACIFIC R&D Center, 1920 Yonggu-daero, Giheung-gu, Yongin-si, Gyeonggi-do 17074, Korea
*   Correspondence: bhb@ajou.ac.kr (B.H.B.); mosme@amorepacific.com (P.J.P.); Tel.: +82-31-219-2618 (B.H.B.); +82-31-280-5639 (P.J.P.)
†   These authors contributed equally to this work.

Received: 11 August 2020; Accepted: 10 September 2020; Published: 11 September 2020

Abstract: Skin layers serve as a barrier against unexpected critical changes in the body due to environmental factors. Excessive ultraviolet (UV) B exposure increases the levels of age-related factors, leading to senescent cells and damaged skin tissues. Widely used as a dietary supplement, konjac (*Amorphophallus konjac*) glucomannan (KGM) has shown skin regeneration potential in patch or sheet form with anti-inflammatory or immunosuppressive effects. However, the ability of KGM to reconstitute senescent/damaged skin following UV radiation has not been explored. Here, we demonstrate that KGM alleviates skin damage by increasing the proportion of young cell populations in UVB-exposed senescent human epidermal primary melanocytes. Young cell numbers increased depending on KGM dosage, but the senescent cells were not removed. Real-time quantitative polymerase chain reaction (RT-qPCR) and Western blot analysis showed that mRNA and protein levels of age- and pigmentation-related factors decreased in a manner dependent on the rate at which new cells were generated. Moreover, an analysis of mRNA and protein levels indicated that KGM facilitated youth by increasing cell proliferation in UVB-damaged human fibroblasts. Thus, KGM is a highly effective natural agent for maintaining skin homeostasis by promoting the reconstitution of the dermal environment against UVB-induced acute senescence or skin damage.

Keywords: konjac glucomannan; ultraviolet B; human epidermal primary melanocytes; human embryonic fibroblasts; senescence

---

1. Introduction

The human skin, comprising almost 16% of the body, maintains a certain level of physiological and biological function (homeostasis) to ensure that the various skin cells beyond the dermis layer are preserved to form a solid barrier [1–3]. In particular, the epidermis layer of the skin is a critical component for maintaining the physiological functions of the body in the face of various external environmental changes as well as internal factors [4–7].

Among the diverse external factors that cause changes in the skin, ultraviolet radiation (UVR) is a well-known major contributor to human skin aging acceleration by inducing DNA damage (photo-aging) or oncogene activation [8–10]. UVR is classified according to wavelength as UVA (315–400 nm), UVB (280–315 nm), and UVC (<280 nm) rays, each of which has distinctive skin penetration depth due to their respective retention energies [11,12]. A recent study demonstrated that exposure to UVR, especially UVB, led to pigment accumulation and senescence in human epidermal primary melanocytes (HEMns) [13]. In addition, UVB-induced changes in p53 gene expression were identified as a major contributor to pigmentation [13]. Skin aging caused by long-term UVB exposure cannot be easily reverted to normal homeostatic conditions, and the accumulated damage can easily activate aging-related factors even upon exposure to weak stimuli. Thus, accelerated aging-related factors can increase the chance of forming abnormal cell clusters, leading to critical pathological conditions such as melanoma [14–16].

Konjac (*Amorphophallus konjac*) is a perennial plant that grows in the subtropical regions of South East Asia and Africa [17]. Konjac glucomannan (KGM), a heteropolysaccharide produced from the tubers of *A. konjac*, is widely used in various foods and medicines owing to its unique physical and chemical properties [18–21]. KGM is composed of β-1,4-linked D-glucose and D-mannose at a 1:1.6 molar ratio and has a characteristic randomly acetylated structure at the C-6 position per approximately every 19 sugar units [22–24]. Due to its water absorptivity, safety, and stability, KGM has recently been demonstrated to be a valuable source of hydrocolloidal dietary fiber and has been proposed as a useful nutritional supplement for the treatment and prevention of obesity, diabetes, and related symptoms [25–27]. In addition to its broad use in food and medicine, the applications and popularity of KGM have been increasing with recognition of its inherent biocompatibility, biodegradability, renewability, and nontoxicity [28–31]. Moreover, studies regarding KGM in patch form have demonstrated its strong cell regeneration capabilities, suggesting its potential in the treatment of various dermatological conditions, including wound healing [32,33].

However, the ability of KGM to reconstitute the dermal environment against UVR-induced senescence/damage and the underlying mechanism of action remain unclear. Therefore, the aim of the present study was to investigate the effect of KGM on HEMns and human embryonic fibroblasts (HEFs) after UVB exposure. We focused on the changes in the extent of UVB-induced senescence/damage to the skin cells, after two weeks of culture in KGM. Moreover, we explored whether hyperpigmentation is regulated by KGM in UVB-induced acute senescence in HEMns. Furthermore, we evaluated the relationship between physiological changes and related factors by measuring mRNA and protein expression levels. Our results highlight KGM as a novel natural anti- aging agent in addition to its currently well-known health and nutritional benefits.

2. Materials and Methods

2.1. Chemicals

Purified konjac glucomannan (KGM) powder (95% content) was purchased from Konson Konjac Co. (Wuhan, Hubei, China).

2.2. Cell Culture and UVB Irradiation

Neonatal foreskin-derived HEMns and HEFs were purchased from Life Technologies (Carlsbad, CA, USA). HEMns were cultured with Medium 254 (M254; Life Technologies) supplemented with human melanocyte growth supplement (Life Technologies), and HEFs were cultured in Dulbecco's modified Eagle medium (Thermo Fisher Scientific, Waltham, MA, USA) with 5% fetal bovine serum (Thermo Fisher Scientific) in a humidified incubator under 5% CO_2 conditions. Cells at passages 2–5 were used in subsequent experiments.

To accelerate the senescence or damage of HEMns and HEFs, UVB irradiation was applied as previously described [13]. Briefly, plated cells were rinsed twice with phosphate-buffered saline (PBS) containing calcium and magnesium (Corning Life Science, Glendale, AZ, USA), and 1 mL of PBS was added to the cells to prevent them from drying out during UV exposure. Cells were irradiated with UVB (20 mJ/cm^2) twice at 24-h intervals using a Bio-Sun irradiator (Vilber Lourmat, Marne-la-Vallée, France). Then, the PBS was removed immediately and replaced every 2 days with growth medium with or without KGM. The cells were then cultured for 2 more weeks. Control cells were sustained in the same culture conditions without UVB exposure.

2.3. Cell Viability Assay

The viability of HEMns after KGM treatment was measured using the EZ-Cytox assay kit (Daeil Lab Service, Seoul, Korea) according to the manufacturer's instructions. HEMns were cultured for 1 day in 96-well plates and treated with various concentrations (0.01–100 μg/mL) of KGM for 24 h and 72 h, respectively. EZ-Cytox solution (10 μL) was then added to each well, followed by incubation at 37 °C for 2 h. Absorbance at 450 nm was measured with a spectrophotometer (Synergy H2; BioTek, VT, USA). The experiments were performed in triplicate, and the data were presented as the percentage of absorbance values.

2.4. SA-β-Galactosidase Assay and Staining

The degree of senescence was determined using the mammalian β-Gal assay kit (Thermo Fisher Scientific, Waltham, MA, USA) according to the manufacturer's instructions. Following 2 weeks of cultivation, cells were washed twice with cold PBS and collected for protein extraction using M-PER mammalian extraction reagent (Thermo Fisher Scientific, Waltham, MA, USA). An equivalent amount of β-Gal assay reagent was added to 50 μL of the supernatant obtained after centrifugation at 15,000 rpm for 15 min. Following incubation for 30 min at 37 °C, absorbance was measured at 405 nm using a Synergy H2 microplate reader (BioTek, Winoosk, VT, USA). The experiment was performed in triplicate, and the data are presented as the percentage of absorbance values.

Microscopy images of cells were stained using the senescence cell histochemical staining kit (#CS0030; Sigma-Aldrich, St. Louis, MO, USA) according to the manufacturer's instructions. Following 1 week of cultivation, cells were washed twice with cold PBS, fixed with 1 mL 4% formaldehyde solution (Wako, Kyoto, Japan) and incubated for 2 h at room temperature. To visualize the physiological changes, images and number of β SA-β-Gal stained cells in a given area were obtained using an optical microscope, IX-75 (Olympus, Tokyo, Japan).

2.5. RNA Extraction and Real-Time Quantitative Polymerase Chain Reaction (RT-qPCR)

Total RNA was extracted using the ReliaPrep™ RNA Cell Miniprep System (Promega, Madison, WI, USA) according to the manufacturer's instructions, and cDNA was synthesized from approximately 1 μg total RNA using the RevertAid First Strand cDNA Synthesis Kit (Thermo Fisher Scientific, Waltham, MA, USA). The cDNA was used as a template for RT-qPCR analysis on a 7500 Fast Real-Time PCR System (Life Technologies) using the following TaqMan probes: Tyrosinase-related protein 1 (TRP1; Hs00167051_m1), tyrosinase-related protein 2 (DCT; Hs01098278_m1), tyrosinase (TYR; Hs00165976_m1), interleukin 1 beta (IL-1β; Hs01555410_m1), cyclin dependent kinase inhibitor 2A (p16^{INK4A}; Hs00923894_m1), and glyceraldehyde 3-phosphate dehydrogenase (GAPDH; #4352339E). Data were obtained from three independent experiments and are presented as the fold change relative to GAPDH level in the sample.

2.6. Western Blot Analysis

Proteins were extracted from each sample, at 5 days and 2 weeks after UVB exposure. Total protein (15 μg) was separated by sodium dodecyl sulfate–polyacrylamide gel electrophoresis and blotted with anti-H2A histone family member X (H2AX), anti-γ-H2AX, anti-p53, anti-phospho-p53 (Ser15), anti-acetyl-p53 (Lys382), and anti-p21CIP1 antibodies (Cell Signaling Technology, Danvers, MA, USA), and with anti-TYR, anti-TYRP1, anti-TYRP2, and anti-GAPDH antibodies (Santa Cruz Biotechnology, Dallas, TX, USA), respectively.

2.7. Statistical Analysis

All data are presented as the mean ± SD. Two-tailed Student's *t*-tests were used to analyze the differences between pairs of groups, and differences among multiple groups were analyzed using one-way ANOVA. The threshold for statistical significance was set at $p < 0.05$.

3. Results

3.1. UVB-Induced Senescence in Human Epidermal Melanocytes Is Alleviated by KGM Treatment

KGM used in these experiments was extracted from the roots of *A. konjac* (Figure 1). The specifications of KGM as supplied by the manufacturer are presented in Table 1. Since these specifications indicated the presence of minimal levels of ash or heavy metals such as lead and arsenic (Table 1), we performed cell viability assays to obtain the appropriate KGM concentrations for culturing HEMns. The results indicated that KGM had no notable changes in cell survival at 1, 2, 5, 10, and 20 µg/mL (Figure S1A), and these concentrations were therefore used in subsequent experiments.

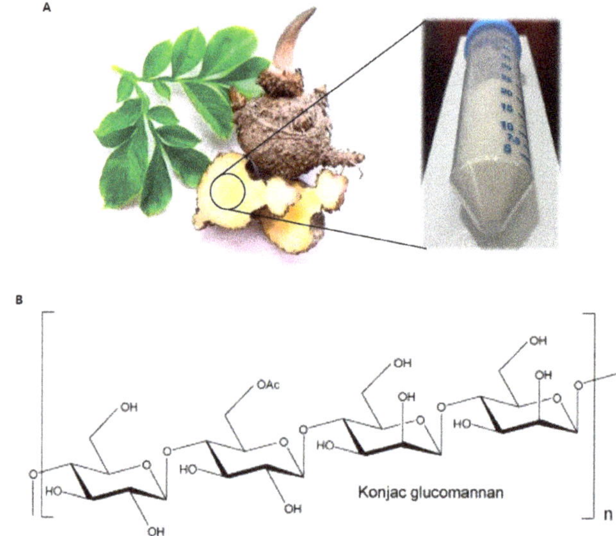

Figure 1. The shape and main composition of konjac plants. (**A**) Image of a konjac plant and the starchy root, which is known as the corm. (**B**) The structure of β-1,4-linked D-glucose and D-mannose repeating units in konjac glucomannan (KGM).

Table 1. Specifications of the konjac glucomannan (KGM) used in this study.

Category	Viscosity (mPa.s)	Glucomannan (%) ≥	pH (1% in DW)	Water (%) ≤	Ash (%) ≤	Lead (Pb) (mg/kg) ≤	Sulfate (g/kg) ≤	Arsenic (As) (mg/kg) ≤	Aflatoxin B1 (ug/kg) ≤
Purified KGM	36,000	95	5.0~7.0	8	2	0.8	0.004	2	5

DW, distilled water.

To induce senescence, cells were exposed to UVB (20 mJ/cm^2) twice and cultured with KGM containing media for two weeks (Figure 2A) as described previously [13]. Microscopy analysis of morphological changes in confluent cells revealed densely packed and healthy cells in unexposed normal control (CTL) cells (Figure 2B, upper left panel). In contrast, the UVB group exhibited spatial separation marked by cell death, and surviving cells displayed morphological abnormalities characterized by enlarged sizes and numerous dendrites (Figure 2B, upper right panel). These morphological changes were consistent with melanocyte senescence reported in a previous study [13]. Notably, culturing of HEMns with 20 µg/mL KGM for two weeks compensated for the cell death-mediated spacing (Figure 2B, lower right panel). Specifically, the population of the UVB- exposed HEMns gradually increased in the presence of KGM in a dose-dependent manner (Figure 2B, lower panel), suggesting that KGM induces cell proliferation in UVB-exposed HEMn.

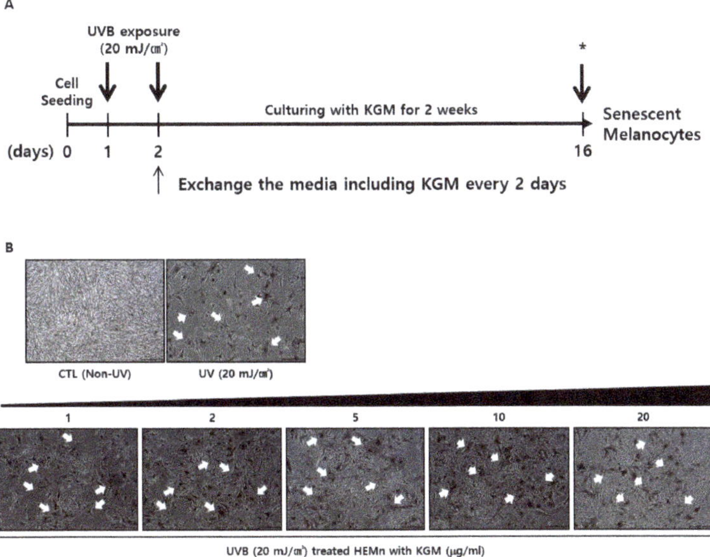

Figure 2. Effects of KGM treatment on UVB-induced senescence in human epidermal primary melanocytes (HEMns). (**A**) Schematic illustration showing the induction of senescence in HEMns by UVB irradiation. The asterisks (*) indicate cell harvesting times for the analysis of senescence after 2 weeks of culture. (**B**) Morphological changes in HEMns (n = 3) after UVB irradiation and cultivation with various concentrations of KGM were visualized by microscopy. Senescent HEMns displayed abnormal morphologies following UVB exposure, including flat shapes, darkening, numerous dendrites and enlarged bodies (arrows). Scale bar = 50 μm. CTL, control; UVB, ultraviolet (UV) B; KGM, konjac glucomannan.

3.2. KGM Promotes the Growth of UVB-Induced Senescent Human Melanocytes in a Dose-Dependent Manner, but Does Not Eliminate Dead Cells

To verify whether KGM restored proliferation of cells affected by UVB exposure or induced the growth of new and healthy cells that had not been affected, UVB-irradiated HEMns were cultured for one week with varying concentrations of KGM and stained using senescence-associated β-galactosidase (SA-β-Gal), a common aging marker [34,35], for microscopy visualization (Figure 3A). SA-β-Gal-positive cells were absent in the CTL group but were clearly observed in UVB-exposed cells (Figure 3A, upper panel). In the UVB exposed KGM-treated group, the intercellular spaces created by cell death were gradually filled with newly proliferated non-stained cells (Figure 3A, lower panel). The growth of the new cell population was dependent on KGM concentration (Figure 3A, lower panel).

To quantitatively assess these results, total cell counts were measured on the 14th day following UVB irradiation. The number of the UVB-exposed KGM untreated cells decreased to about 57% of the number of cells in the CTL group (Figure 3B). However, total cell counts gradually increased with increasing KGM concentrations in the UVB-exposed KGM-treated group. Moreover, there were no significant differences in cell numbers between HEMns treated with 20 μg/mL KGM following UVB exposure and UVB-unexposed CTL cells. In addition, UVB exposure induced the expression of the senescence marker, SA-β-Gal (Figure 3C), whose activity gradually reduced in the presence of KGM in a dose-dependent manner, as the total cell count increased (Figure 3C). The number of cells that stained positive for SA-β-Gal-reached a maximum after UVB exposure and remained constant despite the increase in the KGM concentration, suggesting that KGM had no effect on the removal of SA-β-Gal-stained senescent cells (Figure 3D). As shown in Figure 3E, the change in the proportion of SA-β-Gal-positive cells following KGM treatment after UVB exposure reflected an anti-aging effect of KGM under UVB-induced acute senescence, likely caused by promoting the proliferation of young cells without removing senescent cells.

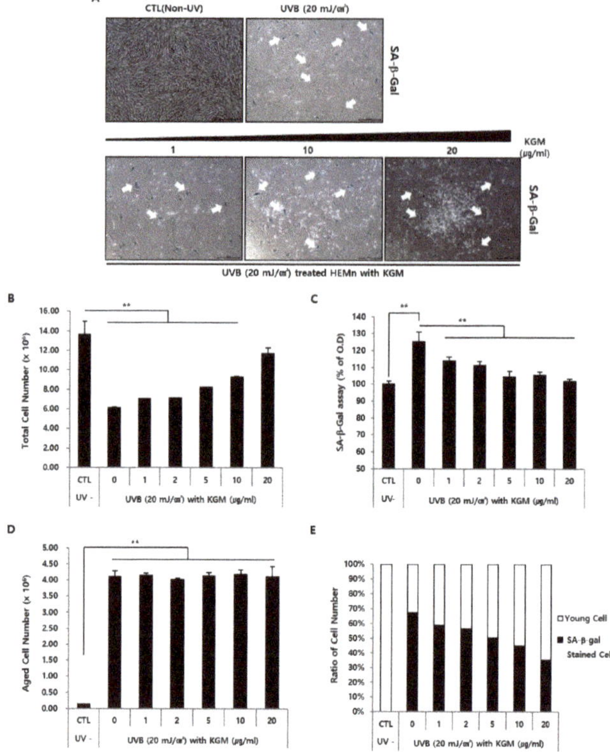

Figure 3. UVB-induced senescence in HEMns indicated by an increase the number of SA-β-Gal-positive cells, and reduced proportion of senescent cells in the presence of KGM. After UVB irradiation, HEMns (*n* = 3) were cultured with various concentrations of KGM for 1 week. (**A**) Senescent HEMns were visualized by microscopy using SA-β-Gal staining (arrow). Scale bar = 500 µm. (**B**) Total numbers of HEMns (*n* = 3) were measured by cell counting. (**C**) Senescence activity (*n* = 3) was determined using the SA-β-Gal assay. The data are presented as the percentage of absorbance values. (**D**) The number of senescent HEMns (*n* = 3) was measured by counting SA- β- Gal- positive cells. The data are presented as mean ± SD (** *p* < 0.01; unpaired Student's *t*-test). (**E**) Cellular state in SA-β-Gal-stained and unstained HEMns (*n* = 3).

3.3. KGM Treatment Suppresses Hyperpigmentation in UVB-Induced Senescent Human Melanocytes

A previous study reported that UVB irradiation induced senescence and hyperpigmentation in HEMns through the activation of p53-dependent melanogenic-related factors [13]. Therefore, we sought to verify whether hyperpigmentation is regulated by KGM in UVB-induced acute senescence in HEMns.

KGM-dependent physiological changes in senescence were evaluated by measuring melanin accumulation and content level using image analysis and absorbance measurements, respectively (Figure 4A). The pigmentation changes of dissolved melanin strongly increased following UVB exposure compared to CTL, but gradually decreased in a dose-dependent manner in the presence of KGM (Figure 4A, upper panel). Similarly, the melanin content also significantly increased after UVB exposure and decreased upon KGM treatment (Figure 4A, lower panel).

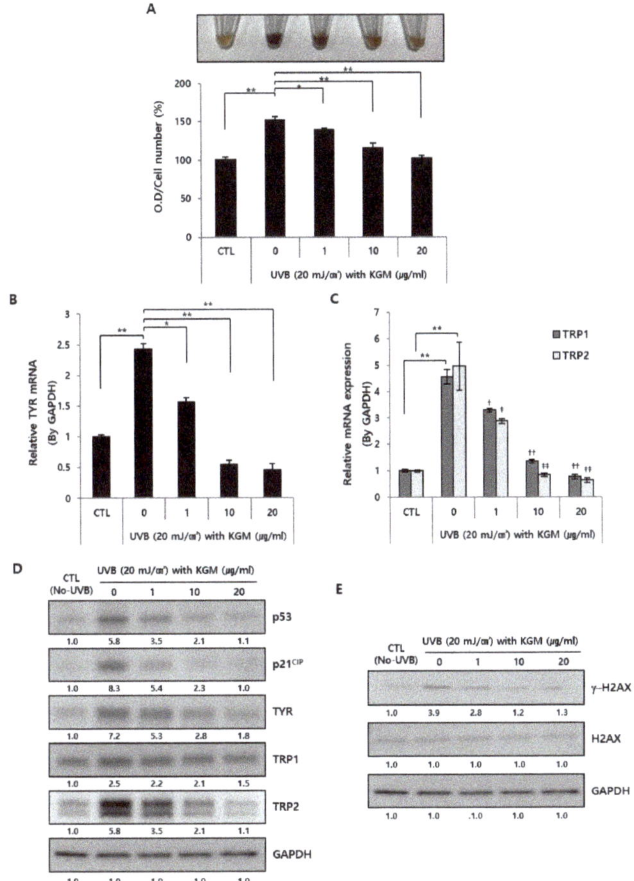

Figure 4. The acceleration of hyperpigmentation in UVB-induced senescent HEMns was regulated by KGM in a dose-dependent manner. The senescence of HEMns was induced by UVB irradiation twice followed by cell culturing with various concentrations of KGM. Proteins and melanins were extracted by lysis from each group of HEMns. (**A**) The melanin content was visualized (upper) or quantified by measuring the absorbance (O.D, optical density) at 450 nm and normalized by total cell count numbers (lower) ($n = 3$). (**B**) mRNA expression levels were determined by Real-time quantitative polymerase chain reaction (RT-qPCR) using specific Taqman probes for tyrosinase (*TYR*), (**C**) tyrosinase- related protein 1 (*TRP1*), and *TRP2*. The data are presented as mean ± SD (* $p < 0.05$, ** $p < 0.01$ compared to CTL ($n = 3$): †/‡ $p < 0.05$, ††/‡‡ $p < 0.01$ compared to 0 group ($n = 3$) [untreated after UVB exposure] for each gene; unpaired Student's *t*-test). (**D**,**E**) Protein expression levels were analyzed by Western blotting using specific antibodies for aging-related factors, p53 and p21CIP, and pigmentation-related factors, TYR, TRP1, and TRP2 and DNA damage marker, gamma-H2A histone family member X (γ-H2AX). Glyceraldehyde 3-phosphate dehydrogenase (GAPDH) was used as a control ($n = 3$).

To determine the relationship between the physiological changes and pigmentation-related factors, mRNA expression levels of melanogenesis-related factors, tyrosinase (TYR), tyrosinase- related protein 1 (TRP1), and TRP2 [36], were analyzed after 48 h of cultivation using RT-qPCR (Figure 4B,C). Results showed that the expression levels of pigmentation-related genes were significantly increased by UVB exposure in the untreated group compared to the CTL group, and were gradually suppressed following KGM treatment in a dose-dependent manner (Figure 4B,C). Subsequently, the changes in pigmentation, aging-related factors and degree of DNA damage at day 5 of cultivation with KGM after UVB exposure were analyzed by Western blotting. Specifically, we focused on the expression levels of pigmentation related factors, TYR, TYRP1, and TYRP2, and aging-related factors, p53 together with cyclin dependent kinase inhibitor 1A (CDKN1A; p21 cell cycle-inhibitory protein, p21CIP), and gamma H2AX (γ-H2AX), a DNA damage marker confirmed in a previous study [13]. Overall, the UVB-induced senescent group showed strongly increased expression of aging, pigmentation-related factors, and DNA damage markers compared to the CTL group (Figure 4D,E). Furthermore, the expression of each protein decreased gradually depending on the concentration of KGM, suggesting that UVB-induced senescence and hyperpigmentation improved as cell population increased. Additionally, KGM-treated HEMns promoted the expression of Ki-67, a cell growth marker, inhibited one of senescence associated secretory phenotype (SASP), interleukin 1 beta (IL-1β), and senescence marker, p16^{INK4A}, also the known cyclin-dependent kinase inhibitor 2A (CDKN2A), respectively (Figure S2A–C). Notably, an evaluation of changes in pigmentation and melanin content by image comparison and quantitative analysis revealed that KGM had no effect on unexposed normal HEMns (Figure S2D). Additionally, there were no changes in the mRNA expression levels of pigmentation-related genes, TYR, TRP1, and TRP2 (Figure S2E).

Taken together, these results suggest that KGM might have the positive effect of anti- pigmentation on skin homeostasis by promoting the cell proliferation in UVB-induced acute senescent HEMns rather than by suppressing pigmentation in normal HEMns.

3.4. KGM Reconstitutes UVB-Damaged Human Primary Fibroblasts

After confirming that KGM-dependent cell growth promotion induced anti-pigmentation in UVB-induced accelerated senescence in HEMns, we then investigated whether this restorative function of KGM was universally applicable to human skin cells using HEFs, a key model for UV irradiation-induced cell damage [37,38].

As previously described in the Methods, physiological and biological changes in UVB-irradiated HEFs cultured in KGM-containing media for two weeks were evaluated by analyzing morphological changes using microscopy (Figure 5A, upper panel). UVB-damaged HEFs displayed serious phenotypic changes, including intercellular spaces due to cell elimination. The remaining cells also had abnormal shapes and sizes distinct from those of unaffected cells. Cultivation with media containing KGM refilled the empty spaces with new HEFs depending on KGM concentration (Figure 5A, lower panel). Notably, the effect of KGM on UVB-damaged HEFs mirrored that of KGM on HEMns in the previous experiment.

Quantitative analysis showed that the number of HEFs was reduced by UVB exposure, but gradually increased in a concentration-dependent manner in the presence of KGM (Figure 5B). Additionally, whereas the number of SA-β-Gal-positive cells increased following UVB exposure, the number of cells staining positive for SA-β-Gal remained constant regardless of the increase in KGM concentration (Figure 5C). Moreover, unexposed HEFs cultured with KGM were relatively less damaged than after UVB exposure, likely due to KGM-mediated increase in the total number of cells, even as the number of SA-β-Gal-positive cells was maintained (Figure 5D).

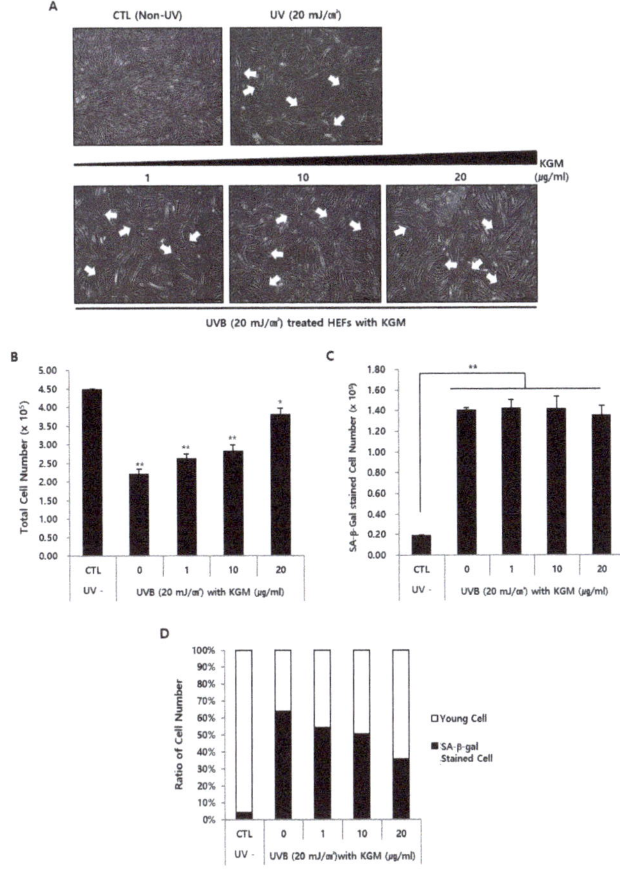

Figure 5. UVB-induced damage in human embryonic fibroblasts (HEFs) displaying abnormal shapes, and improvement of the overall cellular state by KGM. After UVB irradiation twice, HEFs were cultured with various concentrations of KGM for an additional 2 weeks. (**A**) Morphological changes of UVB-exposed HEFs ($n = 3$) were visualized using microscopy. Arrows indicate the abnormal cell shape. Scale bar = 200 μm. (**B**) The change in total number of HEFs and (**C**) SA-β-Gal-positive HEFs was measured by cell counting. The data are presented as mean ± SD (* $p < 0.05$, ** $p < 0.01$; unpaired Student's *t*-test) ($n = 3$). (**D**) The cellular state was analyzed by comparing SA-β-Gal-stained HEFs ($n = 3$) and unstained HEFs ($n = 3$).

4. Discussion

Previous studies have shown that the aging of the human skin is not only inherently caused by internal factors, but also by drastic environmental changes and external factors such as UVB exposure, which strongly impede skin homeostasis [4–7]. The main challenge of UV-induced skin damage is that the injury is thought to be irreversible. Thus, it is difficult to restore the skin to its previous state. Due to widespread social and health problems caused by UV exposure, there have been various attempts at commercial product development and pre-emptive requirements for regulation [8–11]. The present study suggests the possibility of KGM to overcome the sudden disruption of skin homeostasis caused by UV irradiation in HEMns and HEFs, which are particularly sensitive components of the skin to environmental changes.

Despite the various applications of KGM [18–20,28–31], many studies related to the effects of KGM on the human skin have only focused on a sheet- or patch-type polymer of KGM rather than on the properties of the material itself [19–21]. There are two main reasons for this. The first is that the unique characteristics of KGM with respect to water absorption or clustering may pose a challenge to elucidating its functions as an individual material. The second reason is that the various effects of KGM, including its therapeutic potential, have already been generally demonstrated in studies using a patch or cluster form. In particular, the clinical applications of KGM polymers in improving inflammation after skin damage or acne have been reported [32,33]. Given this background, in the present study, we focused on the biological effectiveness of non-clustered KGM, rather than on physical approaches using conventional patches or sheet polymers.

First, we established the appropriate KGM concentrations for evaluating its effect as a solution rather than as the commercially-available patch or sheet forms. We confirmed that a wide range of KGM concentrations (1–20 µg/mL) was harmless to cell viability, suggesting that KGM could be considered a safe natural material for HEMns. However, as KGM is not readily dissolved in water at concentrations of 3% or higher [22,23], determining the suitable concentrations for manufacturing is necessary since the concentrations used in this study are set in vitro.

The changes in cell morphology and related factors after the exposure of HEMns to UVB were consistent with those of a previous report [13]. In addition, microscopy image analysis showed that the induction of cell growth was dependent on KGM concentration. Consistent with these results, analysis of SA-β-Gal staining and degree of senescence revealed that the number of stained cells was not affected by KGM treatment, even though the cell morphology had been altered. Furthermore, the inability of KGM to clear cells damaged by UVB exposure was a typical challenge that could not be overcome. In previous studies, anti-aging researchers have argued that the elimination of individual aging cells took precedence over the inhibition of aging-related factors. However, our results suggest that culturing UV-exposed cells with KGM results in significant cell regrowth, so that the population appears to recover to the levels prior to UVB exposure, even though in the absence of direct improvement in aging-related factors. Therefore, the overall increase in the cell count could be due to the inhibition of senescence or damage during UVB-induced acute injury. In other words, KGM helps overcome UVB-induced damage by increasing the cell count (without the removal of aging cells) in a concentration-dependent manner.

Acute changes in primary melanocytes due to UVB exposure cause not only external physiological changes such as melanin accumulation but also disrupt the inherent production of melanin by melanocytes [13]. Our results confirmed that UVB exposure caused an increment of melanin synthesis and gene and protein expression of pigmentation-related factors. These changes suggested that KGM inhibits pigmentation after UVB exposure. In this study, the efficacy of KGM in HEMns was validated only under harsh UVB exposure conditions. However, we also confirmed that KGM itself has no direct anti-pigmentation effect under normal conditions. That is, KGM slightly induced the growth of normal HEMns in a concentration-dependent manner (data not shown). Moreover, the normalized cell counts showed that KGM had no effect on the pigmentation-related factors. These results suggest that KGM itself likely does not suppress pigmentation, but could lead to changes in the cellular environment by inducing cell regrowth in UVB-induced acute senescence.

Many previous studies on UVB-mediated senescence or aging-like damage with accompanying physiological changes in the skin have used keratinocytes or dermal fibroblasts rather than melanocytes [8,37,38]. Because UVR-induced physiological changes depend on wavelength, skin penetration depth, and the outermost position of the skin layer, we focused on HEMns, which are located near the basal layer in the innermost layer of the skin [11–13]. In addition, there is a lack of clear criteria for skin senescence or aging in keratinocytes or fibroblasts. Therefore, these results must be interpreted carefully since the detailed mechanism of human skin aging after UV exposure remains to be fully elucidated.

As with HEMns, we found that KGM also somewhat improved the overall physiological changes to HEFs caused by UVB exposure, and improved the SA-β-Gal staining ratio. However, SA-β-Gal activity by itself is not a sufficient marker of aging. Therefore, although it has been applied in several studies on aging and aging-related factors, the interpretations and contexts are limited and need careful considerations [13,39] even though the mRNA expression level of another senescence marker, IL-1β was investigated. The lack of change in the number of SA-β-Gal-stained cells in addition to a change in the ratio of stained cells suggested that KGM has a general positive effect on the cellular environment. Although whether UVR induces overall senescence or the aging of HEFs remains to be elucidated, the increase in the proportion of stained cells strongly suggests that UVB promoted a senescence-like environment such as aging acceleration, and that KGM might have a reconstitution effect in a concentration-dependent manner.

Taken together, the present results presumably suggest that KGM could transform the skin environment from a UVR-induced acute senescence/damaged cellular state to a normal cellular condition via the promotion of cell growth. However, further detailed experiments are required to investigate the reconstitution effects of KGM on the dermal environment and the underlying mechanism, including identification of a KGM receptor and Sry-related HMG-Box 10 (Sox10), the proliferation-associated gene in HEMns thought to mediate the observed effects [40–42]. Unlike conventional anti-aging materials, which are typically designed to inhibit or eliminate aging-related factors in individual cells, KGM appears to have a different mechanism of action by improving the overall cell environment to overcome UVR-induced changes. Combining KGM with other functional substances that downregulate non-functional proteins deposited in skin, such as denatured elastin on the upper dermis of solar elastosis [43,44] might enhance the effectiveness of this material. In conclusion, KGM is an attractive natural material with potential as an anti-aging product, which further expands the industrial value of this beneficial nutrient.

Supplementary Materials: The following are available online at http://www.mdpi.com/2072-6643/12/9/2779/s1, Figure S1: The KGM has no significant effect on cell viability, Figure S2. The mRNA expression levels of cell growth, senescence were investigated in UVB-induced senescent HEMns and melanogenesis-related factors were investigated in normal HEMns after KGM treatment.

Author Contributions: K.H.C., S.T.K., B.H.B. and P.J.P. conceived, designed, and performed the experiments. K.H.C. and P.J.P. analyzed the data and wrote the paper. All authors have read and agreed to the published version of the manuscript.

Funding: This research received no external funding.

Conflicts of Interest: The authors declare no conflict of interest.

References

1. Fuchs, E. Scratching the surface of skin development. *Nature* **2007**, *445*, 834–842. [CrossRef]
2. Slominski, A.T.; Zmijewski, M.A.; Skobowiat, C.; Zbytek, B.; Slominski, R.M.; Steketee, J.D. Sensing the environment: Regulation of local and global homeostasis by the skin neuroendocrine system. In *Advances in Anatomy, Embryology and Cell Biology*; Springer: Berlin/Heidelberg, Germany, 2012; Volume 212.
3. Sotiropoulou, P.A.; Blanpain, C. Development and homeostasis of the skin epidermis. *Cold Spring Harb. Perspect. Biol.* **2012**, *4*, a008383. [CrossRef]
4. Elwood, J.M.; Jopson, J. Melanoma and sun exposure: An overview of published studies. *Int. J. Cancer* **1997**, *73*, 198–203. [CrossRef]
5. Lowe, N.J. An overview of ultraviolet radiation, sunscreens, and photo-induced dermatoses. *Dermatol. Clin.* **2006**, *24*, 9–17. [CrossRef]
6. Slominski, A.; Wortsman, J. Neuroendocrinology of the skin. *Endocr. Rev.* **2000**, *21*, 457–487. [CrossRef]
7. Slominski, A.; Wortsman, J.; Luger, T.; Paus, R.; Solomon, S. Corticotropin releasing hormone and proopiomelanocortin involvement in the cutaneous response to stress. *Physiol. Rev.* **2000**, *80*, 979–1020. [CrossRef]
8. D'Orazio, J.; Jarrett, S.; Amaro-Ortiz, A.; Scott, T. UV radiation and the skin. *Int. J. Mol. Sci.* **2013**, *14*, 12222–12248. [CrossRef]
9. Svobodová, A.; Psotová, J.; Walterová, D. Natural phenolics in the prevention of UV-induced skin damage. *Biomed. Pap. Med. Fac. Univ. Palacky Olomouc. Czech Repub.* **2003**, *147*, 137–145. [CrossRef]
10. Nakanishi, M.; Niida, H.; Murakami, H.; Shimada, M. DNA damage responses in skin biology—Implications in tumor prevention and aging acceleration. *J. Dermatol. Sci.* **2009**, *56*, 76–81. [CrossRef]
11. Godar, D.E. UV doses worldwide. *Photochem. Photobiol.* **2005**, *81*, 736–749. [CrossRef]
12. Doré, J.F.; Chignol, M.C. Tanning salons and skin cancer. *Photochem. Photobiol. Sci.* **2012**, *11*, 30–37. [CrossRef]

13. Choi, S.Y.; Bin, B.H.; Kim, W.; Lee, E.; Lee, T.R.; Cho, E.G. Exposure of human melanocytes to UVB twice and subsequent incubation leads to cellular senescence and senescence-associated pigmentation through the prolonged p53 expression. *J. Dermatol. Sci.* **2018**, *90*, 303–312. [CrossRef]

14. Baker, L.A.; Horbury, M.D.; Greenough, S.E.; Ashfold, M.N.; Stavros, V.G. Broadband ultrafast photoprotection by oxybenzone across the UVB and UVC spectral regions. *Photochem. Photobiol. Sci.* **2015**, *14*, 1814–1820. [CrossRef]

15. Amaro-Ortiz, A.; Yan, B.; D'Orazio, J.A. Ultraviolet radiation, aging and the skin: Prevention of damage by topical cAMP manipulation. *Molecules* **2014**, *19*, 6202–6219. [CrossRef]

16. Gonzaga, E.R. Role of UV light in photodamage, skin aging, and skin cancer: Importance of photoprotection. *Am. J. Clin. Dermatol.* **2009**, *1*, 19–24. [CrossRef] [PubMed]

17. Chua, M.; Baldwin, T.C.; Hocking, T.J.; Chan, K. Traditional uses and potential health benefits of *Amorphophallus konjac* K. Koch ex N.E.Br. *J. Ethnopharmacol.* **2010**, *128*, 268–278. [CrossRef]

18. Zhou, Y.; Cao, H.; Hou, M.; Nirasawa, S.; Tatsumi, E.; Foster, T.J.; Cheng, Y. Effect of konjac glucomannan on physical and sensory properties of noodles made from low-protein wheat flour. *Food Res. Int.* **2013**, *51*, 879–885. [CrossRef]

19. Huang, C.Y.; Zhang, M.Y.; Peng, S.S.; Hong, J.R.; Wang, X.; Jiang, H.J.; Zhang, F.L.; Bai, Y.X.; Liang, J.Z.; Yu, Y.R.; et al. Effect of Konjac food on blood glucose level in patients with diabetes. *Biomed. Environ. Sci.* **1990**, *3*, 123–131.

20. Wu, J.; Peng, S.S. Comparison of hypolipidemic effect of refined konjac meal with several common dietary fibers and their mechanisms of action. *Biomed. Environ. Sci.* **1997**, *10*, 27–37.

21. Xiong, G.; Cheng, W.; Ye, L.; Du, X.; Zhou, M.; Lin, R.; Geng, S.; Chen, M.; Corke, H.; Cai, Y.Z. Effects of konjac glucomannan on physicochemical properties of myofibrillar protein and surimi gels from grass carp (*Ctenopharyngodon idella*). *Food Chem.* **2009**, *116*, 413–418. [CrossRef]

22. Wang, Y.; Wu, K.; Xiao, M.; Riffat, S.B.; Su, Y.; Jiang, F. Thermal conductivity, structure and mechanical properties of konjac glucomannan/starch based aerogel strengthened by wheat straw. *Carbohydr. Polym.* **2018**, *197*, 284–291. [CrossRef] [PubMed]

23. Zhang, C.; Chen, J.D.; Yang, F.Q. Konjac glucomannan, a promising polysaccharide for OCDDS. *Carbohydr. Polym.* **2014**, *104*, 175–181. [CrossRef] [PubMed]

24. Katsuraya, K.; Okuyama, K.; Hatanaka, K.; Oshima, R.; Sato, T.; Matsuzaki, K. Constitution of konjac glucomannan: Chemical analysis and 13C NMR spectroscopy. *Carbohydr. Polym.* **2003**, *53*, 183–189. [CrossRef]

25. Zhang, Y.Q.; Xie, B.; Gan, X. Advance in the applications of konjac glucomannan and its derivatives. *Carbohydr. Polym.* **2005**, *60*, 27–31. [CrossRef]

26. Behera, S.S.; Ray, R.C. Nutritional and potential health benefits of konjac glucomannan, a promising polysaccharide of elephant foot yam, *Amorphophallus konjac* K. Koch: Review. *Food Rev. Int.* **2017**, *33*, 22–43. [CrossRef]

27. Saha, D.; Bhattacharya, S. Hydrocolloids as thickening and gelling agents in food: A critical review. *J. Food Sci. Technol.* **2010**, *47*, 587–597. [CrossRef]

28. Pan, T.; Peng, S.; Xu, Z.; Xiong, B.; Wen, C.; Yao, M.; Pang, J. Synergetic degradation of konjac glucomannan by γ-ray irradiation and hydrogen peroxide. *Carbohydr. Polym.* **2013**, *93*, 761–767. [CrossRef]

29. Chen, H.L.; Cheng, H.C.; Wu, W.T.; Liu, Y.J.; Liu, S.Y. Supplementation of konjac glucomannan into a low-fiber Chinese diet promoted bowel movement and improved colonic ecology in constipated adults: A placebo-controlled, diet-controlled trial. *J. Am. Coll. Nutr.* **2008**, *27*, 102–108. [CrossRef]

30. Chen, M.; Wang, S.; Liang, X.; Ma, D.; He, L.; Liu, Y. Effect of dietary acidolysis-oxidized Konjac glucomannan supplementation on serum immune parameters and intestinal immune-related gene expression of *Schizothorax prenanti*. *Int. J. Mol. Sci.* **2017**, *18*, 2258. [CrossRef]

31. Hu, Y.; Tian, J.; Zou, J.; Yuan, X.; Li, J.; Liang, H.; Zhan, F.; Li, B. Partial removal of acetyl groups in konjac glucomannan significantly improved the rheological properties and texture of konjac glucomannan and κ-carrageenan blends. *Int. J. Biol. Macromol.* **2019**, *123*, 1165–1171. [CrossRef]

32. Chen, H.; Cheng, J.; Ran, L.; Yu, K.; Lu, B.; Lan, G.; Dai, F.; Lu, F. An injectable self-healing hydrogel with adhesive and antibacterial properties effectively promotes wound healing. *Carbohydr. Polym.* **2018**, *201*, 522–531. [CrossRef] [PubMed]

33. Dong, R.; Zhao, X.; Guo, B.; Ma, P.X. Self-healing conductive injectable hydrogels with antibacterial activity as cell delivery carrier for cardiac cell therapy. *ACS Appl. Mater. Interfaces* **2016**, *8*, 17138–17150. [CrossRef] [PubMed]

34. Park, P.J.; Lee, T.R.; Cho, E.G. Substance P stimulates endothelin 1 secretion via endothelin-converting enzyme 1 and promotes melanogenesis in human melanocytes. *J. Investig. Dermatol.* **2015**, *135*, 551–559. [CrossRef]

35. Bennett, D.C. Human melanocyte senescence and melanoma susceptibility genes. *Oncogene* **2003**, *22*, 3063–3069. [CrossRef] [PubMed]

36. Dimri, G.P.; Lee, X.; Basile, G.; Acosta, M.; Scott, G.; Roskelley, C.; Medrano, E.E.; Linskens, M.; Rubelj, I.; Pereira-Smith, O. A biomarker that identifies senescent human cells in culture and in aging skin in vivo. *Proc. Natl. Acad. Sci. USA* **1995**, *92*, 9363–9367. [CrossRef]

37. Yamaba, H.; Haba, M.; Kunita, M.; Sakaida, T.; Tanaka, H.; Yashiro, Y.; Nakata, S. Morphological change of skin fibroblasts induced by UV Irradiation is involved in photoaging. *Exp. Dermatol.* **2016**, *3*, 45–51. [CrossRef]

38. Kim, Y.J.; Lee, E.H.; Cho, E.B.; Kim, D.H.; Kim, B.O.; Kang, I.K.; Jung, H.Y.; Cho, Y.J. Protective effects of galangin against UVB irradiation-induced photo-aging in CCD-986sk human skin fibroblasts. *Appl. Biol. Chem.* **2019**, *62*, 40. [CrossRef]

39. Lee, B.Y.; Han, J.A.; Im, J.S.; Morrone, A.; Johung, K.; Goodwin, E.C.; Kleijer, W.J.; DiMaio, D.; Hwang, E.S. Senescence-associated β-galactosidase is lysosomal β-galactosidase. *Aging Cell* **2006**, *5*, 187–195. [CrossRef]

40. Szolnoky, G.; Bata-Csörgö, Z.; Kenderessy, A.S.; Kiss, M.; Pivarcsi, A.; Novák, Z.; Nagy, N.K.; Michel, G.; Ruzicka, T.; Maródi, L.; et al. A mannose-binding receptor is expressed on human keratinocytes and mediates killing of Candida albicans. *J. Investig. Dermatol.* **2001**, *117*, 205–213. [CrossRef]

41. Sheikh, H.; Yarwood, H.; Ashworth, A.; Isacke, C.M. Share. Endo180, an endocytic recycling glycoprotein related to the macrophage mannose receptor is expressed on fibroblasts, endothelial cells and macrophages and functions as a lectin receptor. *J. Cell Sci.* **2000**, *113*, 1021–1032.

42. Graf, S.A.; Busch, C.; Bosserhoff, A.K.; Besch, R.; Berking, C. SOX10 promotes melanoma cell invasion by regulating melanoma inhibitory activity. *J. Investig. Dermatol.* **2014**, *134*, 2212–2220. [CrossRef]

43. Sellheyer, K. Pathogenesis of solar elastosis: Synthesis or degradation? *J. Cutan. Pathol.* **2003**, *30*, 123–127. [CrossRef]

44. Heng, J.K.; Aw, D.C.; Tan, K.B. Solar elastosis in its papular form: Uncommon, mistakable. *Case Rep. Dermatol.* **2014**, *6*, 124–128. [CrossRef]

Article

Thymocid®, a Standardized Black Cumin (*Nigella sativa*) Seed Extract, Modulates Collagen Cross-Linking, Collagenase and Elastase Activities, and Melanogenesis in Murine B16F10 Melanoma Cells

Huifang Li [1,2], Nicholas A. DaSilva [2], Weixi Liu [3], Jialin Xu [2,4], George W. Dombi [3], Joel A. Dain [3], Dongli Li [1], Jean Christopher Chamcheu [5], Navindra P. Seeram [2] and Hang Ma [1,2,*]

[1] School of Biotechnology and Health Sciences, Wuyi University, International Healthcare Innovation Institute (Jiangmen), Jiangmen 529020, China; huifang_li@uri.edu (H.L.); wyuchemldl@126.com (D.L.)
[2] Bioactive Botanical Research Laboratory, Department of Biomedical and Pharmaceutical Sciences, College of Pharmacy, University of Rhode Island, Kingston, RI 02881, USA; NickDasilva91@gmail.com (N.A.D.); jialin_xu@mail.neu.edu.cn (J.X.); nseeram@uri.edu (N.P.S.)
[3] Department of Chemistry, University of Rhode Island, Kingston, RI 02881, USA; weixi_liu@my.uri.edu (W.L.); gdombi@chm.uri.edu (G.W.D.); jdain@chm.uri.edu (J.A.D.)
[4] Institute of Biochemistry and Molecular Biology, College of Life and Health Sciences, Northeastern University, Shenyang 110819, China
[5] School of Basic Pharmaceutical and Toxicological Sciences, College of Pharmacy, University of Louisiana at Monroe, Monroe, LA 71209, USA; chamcheu@ulm.edu
* Correspondence: hang_ma@uri.edu; Tel.: +1-401-874-7654

Received: 20 June 2020; Accepted: 16 July 2020; Published: 19 July 2020

Abstract: Black cumin (*Nigella sativa*) seed extract has been shown to improve dermatological conditions, yet its beneficial effects for skin are not fully elucidated. Herein, Thymocid®, a chemically standardized black cumin seed extract, was investigated for its cosmeceutical potential including anti-aging properties associated with modulation of glycation, collagen cross-linking, and collagenase and elastase activities, as well as antimelanogenic effect in murine melanoma B16F10 cells. Thymocid® (50, 100, and 300 μg/mL) inhibited the formation of advanced glycation end-products (by 16.7–70.7%), collagen cross-linking (by 45.1–93.3%), collagenase activity (by 10.4–92.4%), and elastases activities (type I and III by 25.3–75.4% and 36.0–91.1%, respectively). In addition, Thymocid® (2.5–20 μg/mL) decreased melanin content in B16F10 cells by 42.5–61.6% and reduced cellular tyrosinase activity by 20.9% (at 20 μg/mL). Furthermore, Thymocid® (20 μg/mL for 72 h) markedly suppressed the mRNA expression levels of melanogenesis-related genes including microphthalmia-associated transcription factor (*MITF*), tyrosinase-related protein 1 (*TYRP1*), and *TYRP2* to 78.9%, 0.3%, and 0.2%, respectively. Thymocid® (10 μg/mL) also suppressed the protein expression levels of MITF (by 15.2%) and TYRP1 (by 97.7%). Findings from this study support the anti-aging and antimelanogenic potential of Thymocid® as a bioactive cosmeceutical ingredient for skin care products.

Keywords: black cumin; *Nigella sativa*; Thymocid®; skin aging; glycation; collagen; collagenase; elastase; melanogenesis; cosmeceutical

---

1. Introduction

Nigella sativa Linn. (family Ranunculaceae), commonly known as black seed, black cumin, or cumin noir, originated in regions of Eastern Europe, the Middle East, northern Africa, the Indian subcontinent, and the west and middle of Asia [1]. In these regions, it has a long history of use for culinary purpose as a spice, natural seasoning, or flavoring, as well as for medicinal purposes in traditional folk medicine systems to treat a variety of ailments [2]. Black cumin seeds are recognized for their great nutritional value as a source of nutrients including iron, copper, zinc, phosphorus, calcium, thiamin, niacin, pyridoxine, and folic acid [3]. In addition, black cumin seeds are a good source of plant-based proteins as they have been reported to have a high index of net protein utilization, protein efficiency ratio, and net dietary protein energy percent [3]. Moreover, phytochemical investigations of black cumin seeds revealed that thymoquinone (TQ) and its derivatives including thymohydroquinone, dithymoquinone, thymol, and carvacrol are the major chemicals in black cumin seed oil [4]. In addition, other phytochemicals, such as alkaloids including nigellicine, nigellimine, nigellidine, 17-*O*-(β-D-glucopyranosyl)-4-*O*-methylnigellidine, 4-*O*-methylnigellidine, nigelanoid, nigeglanine, and 4-*O*-methylnigeglanine have been identified as minor constituents of black cumin seed oil [4–6]. Apart from their nutritional values, black cumin seed extracts (BCSEs) have been reported to display diverse biological and pharmacological activities including antioxidant, anti-inflammatory, anticarcinogenic and antimutagenic, antidiabetic, antimicrobial, and immunological effects [7–10]. Moreover, pre-clinical and clinical studies have reported efficacious activities of BCSEs on dermatological conditions. Several in vitro studies demonstrated that protective effects of a BCSE on cutaneous disorders could be attributed to its antimicrobial effects including antibacterial, antiviral, antifungal, and antiparasitic activity [11]. BCSE's antimicrobial effects have also been studied in a clinical trial, where a treatment with 10% oil lotion of BCSE for 2 months exerted an anti-acne effect via reduction in the mean lesion count of papules and pustules [12]. However, only limited number of studies have reported BCSEs' cosmeceutical applications, such as its modulatory effects on the production of melanin in the melanophores from wall lizard (*Podarcis muralis*), which was attributed to the stimulation of cholinergic receptors [13]. However, to date, the cosmeceutical effects of BCSE as an antiwrinkle/-aging agent, have not yet been reported.

Skin wrinkling is a natural and observable index of the process of aging. It can be exacerbated by a combination of external oxidative stress (e.g., ultraviolet light radiation and pollutants) and endogenous factors (e.g., protein glycation and degradation). Skin wrinkles are formed when the physiological structure of skin tissue is impaired. Skin structure is collectively maintained by a group of connective and supportive proteins including collagen, elastin, claudin, laminin, nidogen, occludin, zonula occludens, and junctional adhesion molecule [14–18]. Amongst these proteins, collagen and elastin are major supportive molecules for the structure of skin tissue, and their structural damage and degradation are directly associated with the formation of skin wrinkles [16]. Collagen is a long-lasting protein with a half-life of over a decade and is subject to chronic internal stress such as glycation. The formation of advanced glycation end-products (AGEs) leads to the alteration of protein structure and the generation of free radicals, which further results in the impairment of the protein's physiological functions including its ability of maintaining the structure of skin tissue. Protein degradation is another crucial contributing factor to the formation of skin wrinkles [14]. Elastin, an elastic protein that maintains skin structure following stretching or contraction, can be degraded by a protease enzyme known as elastase [19]. Therefore, antiglycation agents and inhibitors of collagenase and elastase from dietary natural products have attracted immense research interest for the management of skin aging [20]. However, the protective effects of BCSEs against collagen and elastin degradation remain unclear.

Over the past decade, our laboratory has conducted phytochemical and biological investigations of dietary natural products including extracts of medicinal plants and functional foods. During the course of our studies, several botanical extracts exhibited promising anti-aging and beneficial effects for skin [21–23]. Among these herbal extracts, a BCSE was investigated for its phytochemical constituents and biological effects. This led to the identification of TQ (as the major phytochemical of this BCSE) and other bioactive compounds such as indazole-type alkaloids with antihyperglycemic effects [6]. A proprietary BCSE, namely, Thymocid®, is commercially produced by a cold compressed method without the use of extraction solvents. It contains omega-6 fatty acids such as linoleic acid and is chemically standardized to TQ content. As part of our group's continuing research efforts to study the beneficial effects to skin of bioactive dietary ingredients, the current study was designed to evaluate the cosmeceutical properties of Thymocid® with in vitro enzymatic and cell-based assays for

(1) effects on glycation of bovine serum albumin and collagen cross-linking; (2) anticollagenase and anti-elastase activities; (3) effects on melanogenesis in murine melanoma B16F10 cells; and, (4) effects on the expressions of melanogenesis-related genes and proteins in B16F10 cells.

2. Materials and Methods

2.1. Chemicals

Thymoquinone (TQ), 3-(4,5-dimethylthiazol-2-yl)-5-(3-carboxymethoxyphenyl)-2-(4-sulfenyl)-2-H-tetrazolium salt (MTS), methanol (analytical grade), trifluoroacetic acid, Tris-HCl buffer (pH 10), phosphate buffer saline (PBS, pH 7.4), kojic acid, aminoguanidine (AG), epigallocatechin gallate (EGCG), L-tyrosine, L-3,4-dihydroxyphenylalanine (L-DOPA), N-succinyl-Ala-Ala-ala-p-nitroanilide (AAAPVN), mushroom tyrosinase, elastase enzymes (type I and III) from porcine pancreas, bovine serum albumin (BSA), methylglyoxal (MGO), 3-(4,5-dimethylthiazol-2-yl)-5-(3-carboxymethoxyphenyl)-2-(4-sulfophenyl)-2H-tetrazolium (MTS) reagent, Triton X-100 agent, and sodium azide were purchased from Sigma-Aldrich Co. (St. Louis, MO, USA). Bovine type I collagen was obtained from Advanced BioMatrix Inc. (PureCol®, Catalog #5005; Carlsbad, CA, USA). A commercially available black cumin seed extract (Thymocid®) was kindly provided by Verdure Sciences (Noblesville, IN, USA).

2.2. Quantification of Thymoquinone (TQ) in Thymocid®

The level of TQ in Thymocid® was quantified by using the high-performance liquid chromatography (HPLC) method with an Hitachi HPLC instrument (Hitachi Instruments, Inc., San Jose, CA, USA), an Alltima C_{18} column (250 × 4.6 mm i.d., 5 µm), and a solvent system consisting of 0.1% trifluoroacetic acid in water (A) and methanol (B). A linear gradient eluting method was used as follows: 0–20 min, 50–100% B; 20–21 min, 100–50% B; 20–28 min, 50 % B with a total run time of 28 min, a flow rate of 0.75 mL/min, and an injection volume of 10 µL. Thymocid® was dissolved in dimethyl sulfoxide (DMSO) at various concentrations (0.1–4 mg/mL) and monitored at a range of wavelengths from 200 to 400 nm with a photodiode array detector (see Supplementary Materials Figure S1). A standard curve of TQ at various concentrations (1–200 µg/mL) monitored at the wavelength of 254 nm, which is the characteristic wavelength for TQ, was constructed for its quantification in Thymocid® (see Supplementary Materials Figure S2).

2.3. Bovine Serum Albumin (BSA)–Fructose Glycation Assay

The antiglycation assay was performed according to previously reported methods using a BSA–fructose model [24]. Briefly, a BSA–fructose reaction mixture containing BSA (10 mg/mL) and D-fructose (100 mM) was prepared in phosphate buffer saline (PBS; 0.2 M, pH 7.2). Next, different concentrations of Thymocid® (50, 100, and 300 µg/mL) were added to the BSA—fructose mixture and incubated at 37 °C on a shaking rack for a duration of 14 days. Aminoguanidine (AG) was used as a positive control. The formation of AGEs was monitored by the measurement of intrinsic fluorescence at excitation and emission wavelengths of 340 and 435 nm, respectively, using a plate reader.

2.4. Collagen Cross-Linking Assay

A reaction mixture consisting of bovine type I collagen (1.5 mg/mL), Thymocid® (at 50, 100, and 300 µg/mL), methylglyoxal (MGO; 5 mM), and sodium azide (10 mM) in PBS was incubated at 37 °C for 30 days. The reaction mixture of each sample (200 µL) was then transferred to a 96-well black fluorescence reading plate. The level of cross-linked collagen was monitored by measuring the fluorescent intensity of each well with excitation and emission wavelengths at 340 and 435 nm, respectively, using a plate reader.

2.5. Collagenase Inhibition Assay

A collagenase inhibition assay was conducted using a collagenase activity assay kit (Abcam Inc., Cambridge, MA, USA) following the manufacturer's instructions. Briefly, assay buffer (180 µL) and a mixture of test sample at various concentrations or a positive control, 1,10-phenanthroline (10 µL), and collagenase solution (10 µL) were incubated in a 96-well plate at 37 °C for 15 min. The substrate N-(3-[2-furyl]-acryloyl)-Leu-Gly-Pro-Ala (FALGPA) solution (100 µL) was then added to each well. The absorbance of each well was recorded at a wavelength of 345 nm with a kinetic mode for 30 min using a plate reader. The activity of collagenase was calculated as follow: collagenase activity = (ΔODc/ΔT) × 0.2/(0.53 × V), where ΔODc = difference of optical density (OD) reading from sample at different time points, 0.2 = reaction volume (mL), 0.53 = millimolar extinction coefficient of FALGPA, V = sample volume added into the reaction well (mL). The inhibition rate was calculated as inhibition% = 100 × (Activity$_{Enzyme}$ − Activity$_{Sample}$)/Activity$_{Enzyme}$ %.

2.6. Elastase Inhibition Assay

An elastase inhibition assay was conducted using previously reported methods with minor modifications [25]. Briefly, a mixture of test samples (10 µL), elastase solution (20 µL; 0.5 U/mL), and Tris-HCl buffer (140 µL; 2 mM; pH 8.0) were incubated in a 96-well plate at room temperature for 15 min. Then, substrate AAAPVN solution (50 µL; 1 mg/mL) in Tris-HCL buffer was added to each well, and the reaction mixtures were allowed to incubate at room temperature for 1 min. The absorbance of each well was recorded at a wavelength of 410 nm using a plate reader.

2.7. Tyrosinase Inhibition Assay

A tyrosinase inhibition assay was conducted as per previously reported methods with slight modifications [26]. In brief, a mixture of test samples (40 µL) and mushroom tyrosinase solution (100 U/mL) in PBS (0.1 M; 120 µL) was incubated in wells of a 96-well plate at room temperature for 15 min. Following incubation, L-tyrosine solution (2.5 mM; 40 µL) or L-DOPA solution (2.5 mM; 40 µL) in PBS was added to each well, and the reaction mixtures were incubated at 37 °C for 30 min. The absorbance of each well was recorded at a wavelength of 490 nm using a plate reader.

2.8. Cell Culture

Murine melanoma B16F10 cells obtained from American Type Culture Collection (ATCC, Rockville, MD, USA) were cultured as recommended by ATCC. Briefly, B16F10 cells were grown in Dulbecco's modified Eagle's medium (DMEM; Life Technologies, Gaithersburg, MD, USA) supplemented with 10% fetal bovine serum (Life Technologies) and 1% antibiotic solution (Sigma-Aldrich Co., St. Louis, MO, USA). Cells were maintained at 37 °C in the presence of 5% CO_2 and constant humidified atmosphere. Test samples were dissolved in DMSO as stock solution and then diluted with cell culture medium to the desired concentrations (DMSO < 0.1%).

2.9. Cell Viability Assay

The viability of B16F10 cells was determined by the MTS assay as described previously with minor modifications [27]. Briefly, cells were seeded in 96-well plates at a density of 5×10^3 cells per well and allowed to attach overnight. Next, the culture medium was replaced with fresh medium supplemented with various concentrations of Thymocid® (2.5, 5, 10, 20, and 40 µg/mL) for 72 h. After the incubation, freshly prepared MTS reagent (20 µL) was added to each well and incubated at 37 °C for 30 min, and optical density of each well was measured at a wavelength of 490 nm using a plate reader.

2.10. Melanogenesis Assay

The antimelanogenic effect of Thymocid® was evaluated by the measurement of melanin content in B16F10 cells following previously reported method with modifications [27]. Briefly, cells were seeded in 96-well plates at a density of 5×10^3 cells per well and allowed to attach for 24 h. Cells were then treated with Thymocid® at concentrations of 2.5, 5, 10, and 20 µg/mL for 72 h. Then cells were lysed by adding sodium hydroxide solution (0.1 M; 1 mL), and lysed cells were centrifuged (3000× g for 5 min). The supernatant was decanted, and the cell pellet was exposed to sodium hydroxide (200 µL), followed by placing in a water bath at 80 °C for 1 h. Samples were then briefly mixed by vortex, and melanin content was transferred to a 96-well plate. The absorbance of each well was measured at a wavelength of 405 nm using a plate reader.

2.11. Cellular Tyrosinase Activity Assay

The cell-based tyrosinase assay was performed using a previously reported method with modifications [28]. Briefly, B16F10 cells were seeded at 4×10^4 cells per well in a 24-well plate and allowed to grow for 24 h prior to being treated with Thymocid® (2.5, 5, 10, and 20 µg/mL) for 72 h. Next, cells were harvested and washed twice with ice-cold PBS followed by centrifugation at 12,000× g for 10 min, and cells pellets were re-suspended in PBS containing Triton X-100 (1%). The cells were lysed by a freeze-and-thaw cycle to release tyrosinase from the melanosome membrane. Cellular tyrosinase was collected by centrifugation at 10,000× g at 4 °C for 30 min. A reaction mixture consisting of cellular tyrosinase (20 µg in 200 µL of PBS) and L-DOPA solution (1.25 mM) was incubated at 37 °C for 30 min. The formation of dopachrome was quantified by measuring the optical density at a wavelength of 495 nm using a plate reader.

2.12. Real-Time Polymerase Chain Reaction (RT-PCR)

The mRNA expression level of melanin-synthesis-related genes including microphthalmia-associated transcription factor (*MITF*), tyrosinase (*TYR*), tyrosinase-related protein-1 (*TYRP1*), and tyrosinase-related protein-2 (*TYRP2*) were measured by a real-time polymerase chain reaction (RT-PCR) assay with a previously reported method [27]. Briefly, B16F10 cells were seeded in 6-well plates at a density of 1×10^5 cells per well and allowed to grow for 24 h. Then cells were treated with Thymocid® (2.5 or 10 µg/mL) for 24, 48, or 72 h. Total RNA was isolated from cells using TRIzol reagent (Invitrogen, Carlsbad, CA, USA) according to the manufacturer's instructions. The extracted genes were quantified by using a SYBR Green kit (Thermo Fisher Scientific, Grand Island, NY, USA) and compared to levels of *b2m* rRNA as a reference housekeeping gene.

2.13. Preparation of Cellular Protein Lysates and Western Blot Assay

B16F10 cells were seeded in 6-well plates at a density of 1.0×10^5 cells per well and allowed to grow for 24 h and were then treated with Thymocid® (2.5 and 10 µg/mL) and cultured for 72 h. After washing with PBS, B16F10 cells were harvested, whole-cell lysates were prepared and quantified, and the protein expressions of MITF, TYR, TYRP1, and TYRP2 were quantified by Western blot assay, as described previously [27]. Antibodies including anti-MITF (ab3201), anti-TYRP1 (ab178676), anti-TYRP2 (ab103463), and anti-β-Actin antibody (ab8227) were obtained from Abcam, Cambridge, MA, USA. The western blot (WB) bands were detected on X-ray films using an enhanced chemiluminescence (ECL) detection kit (GE Healthcare, Piscataway, NJ, USA) according to the manufacturer's protocol.

2.14. Statistical Analyses

Data are presented as mean ± standard deviation (S.D.) of at least three replicated experiments. Two-tailed unpaired Student's t test or ANOVA with Tukey post-test was used for statistical analysis of the data using the GraphPad Prism software 6.0 or Office Excel 2010 software. Significance for all tests was defined as $p \leq 0.05$ (*), $p \leq 0.01$ (**), $p \leq 0.001$ (***), and $p \leq 0.0001$ (****).

3. Results

3.1. Thymocid® Inhibits the Formation of Advanced Glycation End-Products (AGEs) and Collagen Cross-Linking

First, a commercially available BCSE, Thymocid®, was standardized to thymoquinone (TQ), as its major phytochemical marker. The level of TQ in Thymocid® was quantified using a standard curve based on the HPLC analysis (see Supplementary Materials Figure S2) and was determined to be 5.12% (Figure 1).

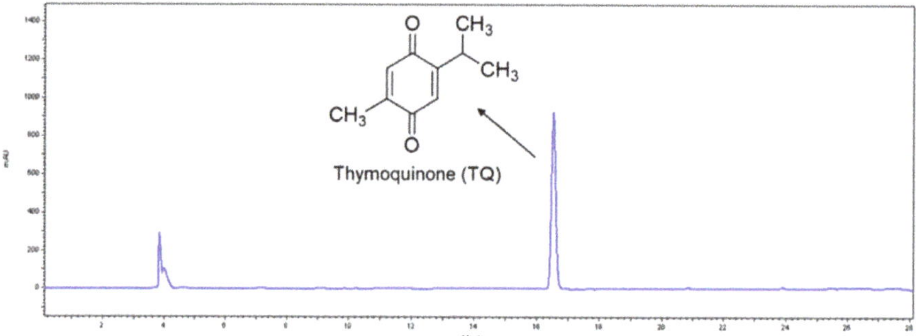

Figure 1. Chemical structure of thymoquinone (TQ) and HPLC profile of a black cumin seed extract (Thymocid®) standardized to TQ content.

Next, Thymocid® was assessed for the anti-skin-aging effects by measuring its inhibitory effects on the formation of advanced glycation end-products (AGEs) using a BSA model and protein cross-linking using a type I collagen model. Thymocid® (at 50, 100, and 300 µg/mL) reduced the fructose-induced formation of AGEs by 16.7%, 32.5%, and 70.7%, respectively (Figure 2A), in a concentration-dependent manner. Aminoguanidine (AG; employed as a positive control) had an inhibition rate of 59.6% at 100 µg/mL. In addition, Thymocid® (at 50, 100, and 300 µg/mL) inhibited methylglyoxal (MGO)-induced collagen cross-linking by 45.1%, 92.6%, and 93.3%, respectively (Figure 2B), whereas, AG (at 100 µg/mL) was less active with an inhibition rate of 12.5%.

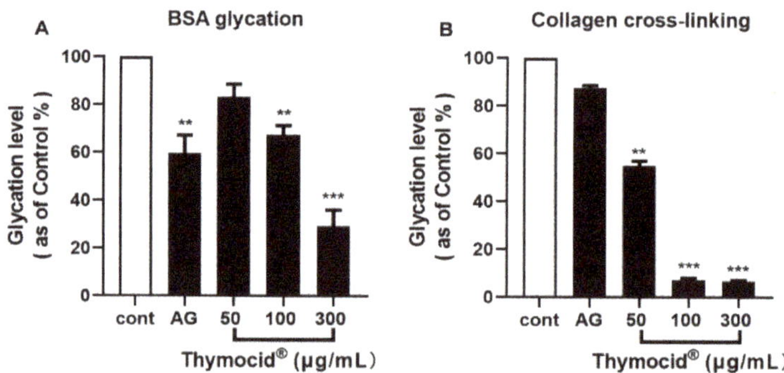

Figure 2. Effects of Thymocid® on the formation of advanced glycation end-products (AGEs) in two glycation models including bovine serum albumin (BSA)–fructose (**A**) and type I collagen cross-linking (**B**). The inhibitory effects of Thymocid® (50, 100, and 300 µg/mL) and aminoguanidine (AG) (a positive control) on the formation of AGEs and level of protein cross-linking were determined by fluorescent assays. Values are expressed in means ± standard deviation (S.D.) from three experiment replicates. Significance was defined as ** $p < 0.01$ and *** $p < 0.001$ when compared to the control group.

3.2. Thymocid® Inhibits Collagenase Activity

The effect of Thymocid® on collagen degradation was evaluated by assessing the inhibitory effects on collagenase enzyme activity. The OD_{345} values of enzymes treated with 1,10-phenanthroine (phen), a known inhibitor (employed as a positive control), was observed to decrease over 30 min, while Thymocid® (at 62.5–1000 µg/mL) decreased the OD_{345} values within 10–30 min, suggesting that both 1,10-phenanthroine and Thymocid® reduced the activity of collagenase (Figure 3A). Thymocid® (at 62.5–1000 µg/mL) reduced collagenase activity by 10.4–92.4%, respectively (Figure 3B), while phen (at 10 mM) showed an inhibition rate of 98.4%.

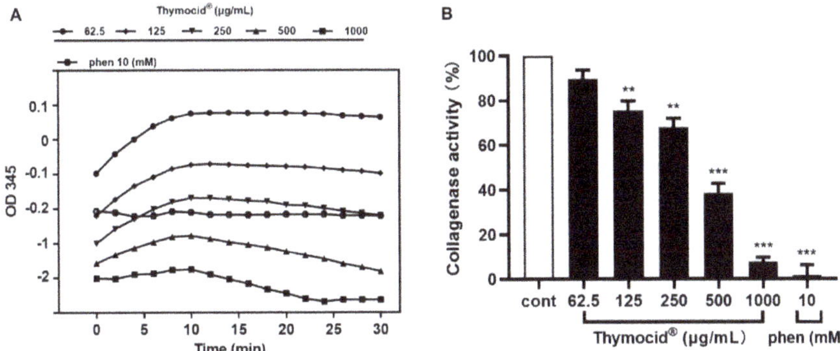

Figure 3. The optical density at a wavelength of 345 nm (OD_{345}) values of collagenase enzyme treated with Thymocid® or a positive control phen (1,10-phenanthroine), over 30 min in a kinetic mode (**A**). The inhibitory effect of Thymocid® (62.5–1000 µg/mL) and phen (10 mM) on collagenase activity was measured by a colorimetric assay (**B**). The inhibition rates are expressed as mean ± S.D. from three replicated experiments. Significance was defined as ** $p < 0.001$ and *** $p < 0.0001$ when compared to the control group.

3.3. Thymocid® Inhibits Elastase Activity

Thymocid® was further evaluated for its antiwrinkle property by measuring its inhibitory effects on the activity of elastases (type I and III). Thymocid® (at concentrations of 62.5–1000 µg/mL) inhibited elastase activities in a concentration-dependent manner as it reduced the activity of type I and III elastase by 25.3–75.4% and 36.0–91.1%, respectively (Table 1). Epigallocatechin gallate (EGCG; 92 µg/mL), used as a positive control, provided an inhibition rate of 73.0% and 75.2% on type I and III elastase, respectively. It should be noted that TQ, the major phytochemical in Thymocid®, also reduced the activities of collagenase and elastases (type I and III) by 18.0–30.0%, 16.1–30.2%, and 12.9–45.3%, at concentrations of 62.5, 125, 250, 500, and 1000 µg/mL, respectively (see Supplementary Materials Tables S1 and S2).

Table 1. Inhibitory activity of Thymocid® on elastase enzyme.

Sample	Concentration (µg/mL)	Inhibition Rate (%) [a]	
		Type-I	Type-III
Thymocid®	1000	75.4 ± 4.5	91.1 ± 1.2
	500	69.2 ± 1.3	88.6 ± 2.3
	250	56.2 ± 3.2	78.4 ± 3.2
	125	35.4 ± 6.9	61.3 ± 2.1
	62.5	25.3 ± 5.9	36.0 ± 2.7
EGCG [b]	92	73.0 ± 1.8	75.2 ± 4.0

[a] Values are expressed as mean ± S.D. from three replicated experiments. [b] Epigallocatechin gallate; positive control.

3.4. Thymocid® Increases Tyrosinase Activity

We further evaluated the effects of Thymocid® in melanin-biosynthesis-related bioassays. First, Thymocid® was evaluated for its modulatory effect on tyrosinase activity using two enzyme substrates including L-tyrosine and L-DOPA. As shown in Table 2, when tested with the different substrates, Thymocid® (62.5–1000 µg/mL) dose-dependently increased the activity of tyrosinase to 153.3–228.7% for L-tyrosine as substrate, and 113.7–146.3% for L-DOPA as substrate, respectively (Table 2). Kojic acid (at 10 µg/mL), a known tyrosinase inhibitor, was included as a positive control and it reduced tyrosinase activity to 49.5% and 66.1%, respectively. In addition, TQ (at 62.5, 125, 250, 500, and 1000 µg/mL) also increased the activity of tyrosinase to 120.7–110.8% and 100.8–117.7%, respectively, when assayed with substrates L-tyrosine and L-DOPA (see Supplementary Materials Table S3).

Table 2. Modulatory effect of Thymocid® on the activity of tyrosinase enzyme. Thymocid® was evaluated with mushroom tyrosinase with two substrates including L-tyrosine and L-DOPA.

Sample	Concentration (µg/mL)	Enzyme Activity (%) [a]	
		L-Tyrosine	**L-DOPA**
Thymocid®	1000	228.7 ± 9.6	146.3 ± 18.6
	500	192.0 ± 12.3	133.1 ± 6.1
	250	170.0 ± 6.8	133.9 ± 15.4
	125	162.5 ± 7.3	116.1 ± 6.7
	62.5	153.3 ± 6.6	113.7 ± 3.0
Kojic acid [b]	10	49.5 ± 3.0	66.2 ± 13.2

[a] Values are expressed as mean ± S.D. from three replicated experiments. [b] Positive control.

3.5. Thymocid® Reduces the Melanin Content in B16F10 Melanoma Cells

To further evaluate whether Thymocid® can modulate the production of melanin, cell-based assays were conducted in murine melanoma B16F10 cells. The cytotoxicity of Thymocid® (2.5, 5, 10, 20, and 40 µg/mL) on B16F10 cells was evaluated by measuring cell viability using the MTS assay. Thymocid® was nontoxic to B16F10 cells at concentrations ranging from 2.5 to 20 µg/mL, as it maintained the viability of B16F10 cells greater than 95.7% (Figure 4A), and these nontoxic concentrations were selected for further bioassays. Next, the antimelanogenic effect of Thymocid® in B16F10 cells was evaluated. We observed that Thymocid® (2.5, 5, 10, and 20 µg/mL) suppressed the production of melanin (Figure 4B) to 57.5%, 56.1%, 52.9%, and 38.4%, respectively, as compared to the control group without treatment of Thymocid® (Figure 4C). Furthermore, we assessed the effect of Thymocid® on cellular tyrosinase activity in B16F10 cells. Compared to the control group, Thymocid® reduced cellular tyrosinase activity by 20.9% at the highest tested concentration (20 µg/mL) (Figure 4D).

3.6. Thymocid® Suppresses the mRNA and Protein Expression Levels of Melanogenesis-Related Markers in B16F10 Cells

To further investigate the mechanism of Thymocid®'s suppression of melanogenesis (i.e., melanin production) in B16F10 cells, its effects on the expression of melanogenesis-related genes including *MITF, TYR, TYRP1*, and *TYRP2* were evaluated. As shown in Figure 5A, treatment of B16F10 cells with Thymocid® (at 20 µg/mL) for 24, 48, or 72 h, suppressed the mRNA expression of *MITF, TYR, TYRP1*, and *TYRP2* in a time-dependent manner. At 48 and 72 h, treatment with Thymocid® had a suppressive effect on the expression of *MITF, TYR, TYRP1*, and *TYRP2* as it reduced their mRNA expression to 42.0%, 76.5%, 62.2%, and 61.2%, and 3.3%, 83.6%, 0.3%, and 0.2%, respectively, whereas it only reduced expression of *MITF* to 78.9% within 24 h (Figure 5A). Moreover, we observed that treatment with Thymocid®, at 2.5 and 10 µg/mL, resulted in a concentration-dependent modulation of mRNA expression of *MITF, TYR, TYRP1*, and *TYRP2* to 80.1%, 98.6%, 93.4%, and 93.7%, and 73.0%, 104.9%, 76.6%, and 86.6%, respectively (Figure 5B).

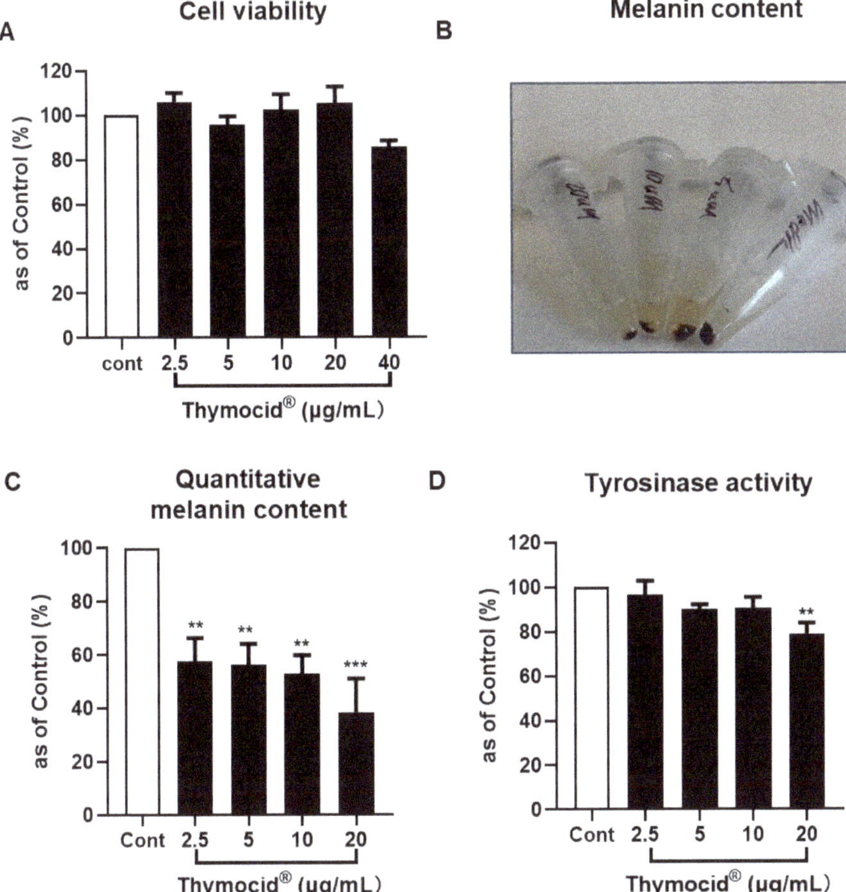

Figure 4. Effects of Thymocid® (2.5, 5, 10, and 20 µg/mL) on the cell viability of murine melanoma B16F10 cells (**A**), melanin content (**B**), and its quantification (**C**) in B16F10 cells cultured with Thymocid® for 72 h, and cellular tyrosinase activity in B16F10 cells treated with Thymocid® for 72 h (**D**). Significance was defined as ** $p < 0.01$ and *** $p < 0.001$ when compared to the control group and values are presented as the means ± S.D. from three experiment replicates.

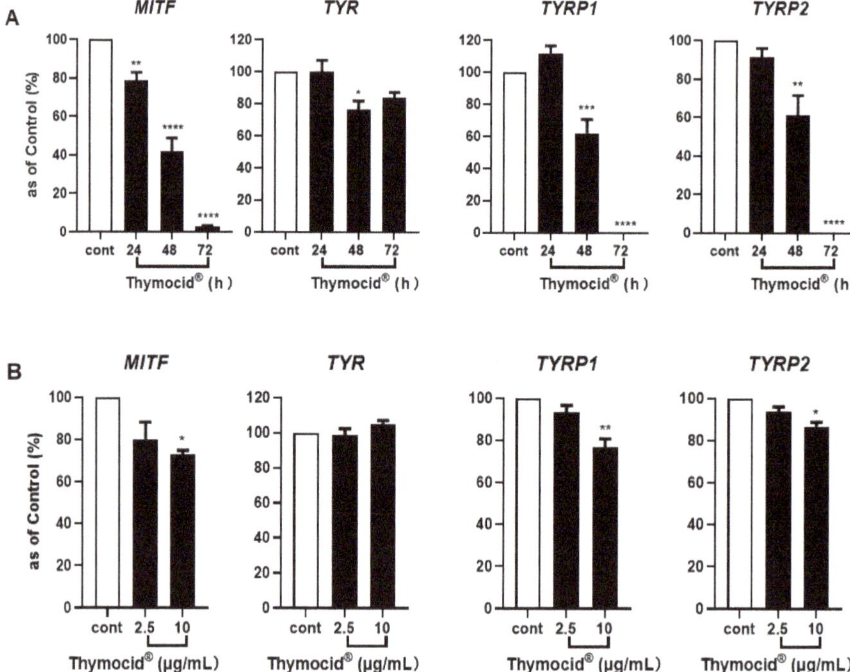

Figure 5. Effects of Thymocid® on the expression of melanogenesis-related genes and proteins in melanoma B16F10 cells. Cells were treated with Thymocid® (20 µg/mL) for 24, 48, and 72 h and the expression of genes including *MITF, TYR, TYRP1,* and *TYRP2* were determined by real-time qPCR assay (**A**). B16F10 cells were treated with Thymocid® (2.5 and 10 µg/mL) for 72 h, and the expression of genes including *MITF, TYR, TYRP1,* and *TYRP2* were determined by real-time qPCR (**B**). *B2M* was used as an internal control in real-time qPCR assay. Significance was defined as * $p < 0.05$, ** $p < 0.01$, *** $p < 0.005$, **** $p < 0.001$ when compared to the control group. Values are presented as the means ± S.D. from three experimental replicates.

Next, we evaluated the effect of Thymocid® on the expression levels of proteins related to melanogenesis in B16F10 cells by Western blotting assay. The densitometric data showed that treatment with Thymocid® reduced the expression of TYRP1 and TYRP2 in B16F10 cells (Figure 6A). Quantitative analysis of data from Western blotting assay revealed that Thymocid® (2.5 and 10 µg/mL) reduced suppressed the protein expression levels of TYRP1 (by 69.1–97.7%) and TYRP2 (by 9.4–9.3%), respectively (Figure 6B), whilst the expression of MITF was only slightly decreased by the treatment with Thymocid® by 18.8–15.2% at 2.5 and 10 µg/mL, respectively) (Figure 6B).

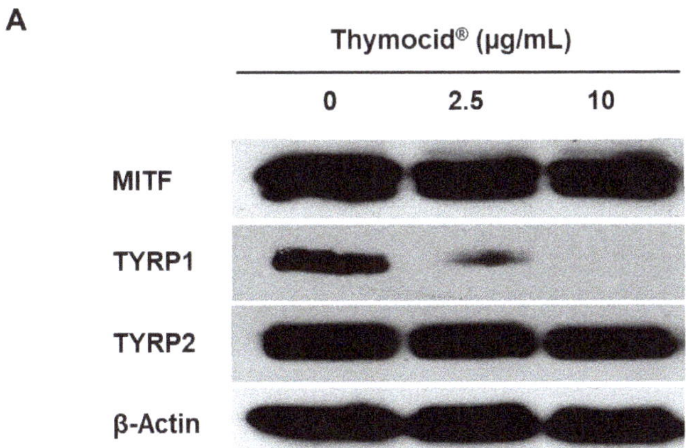

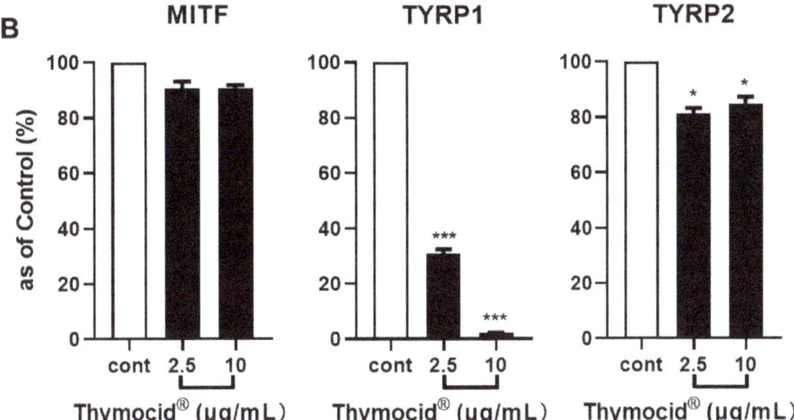

Figure 6. Effects of Thymocid® on the expression of the melanogenesis-related proteins in B16F10 cells. Cells were treated with Thymocid® (2.5 and 10 μg/mL) for 72 h and the expression of proteins including MITF, TYRP1, and TYRP2 were determined by Western blotting assay (densitometric data; **A**). Equal protein loading was confirmed by using protein β-actin as an internal household protein. The protein expression in B16F10 cells treated with Thymocid® (2.5 and 10 μg/mL) for 72 h were compared to the control group (**B**). Values are presented as the means ± S.D. from three experimental replicates. Significance was defined as: * $p < 0.05$ and *** $p < 0.001$ when compared to the control group.

4. Discussion

Thymoquinone (TQ), a major bioactive compound in BCSE [29], was used to standardize Thymocid®. TQ has been reported to show several skin beneficial effects such as chemoprevention of skin tumorigenesis [30] and anti-inflammatory effects against chemical-toxin-induced ear edema [30]. Although TQ may contribute to the overall skin beneficial effects of Thymocid®, other phytochemicals present in Thymocid® may also exert biological effects in an additive, complementary, and/or synergistic manner. In fact, TQ showed effects similar to those of Thymocid® in some bioassays, including anti-elastase and antityrosinase conducted in this study (Supplementary Materials Tables S1–S3); however, TQ at similar concentrations was less active than Thymocid®, suggesting that other phytochemicals present in Thymocid® also contributed to its overall biological effects. In addition, although only one major peak appeared in the HPLC chromatogram of Thymocid® monitored at the wavelength of 254 nm (retention time at 16.5 min; Figure 1), it is possible that other phytochemicals, including non-aromatic molecules, such as aliphatic compounds, may also be present in Thymocid® without being detectable by HPLC analysis. Future studies to identify these aliphatic compounds using proper analytical tools (e.g., gas chromatography–mass spectrometry) and evaluate their biological activities are warranted. Moreover, the presence of other TQ derivatives including thymol, carvacrol, thymohydroquinone, and dithymoquinone, as well as alkaloids including nigellicine, nigellimine, nigellidine, and nigelanoid are also reported in published studies on the chemical composition of BCSEs [4,6]. Although compounds including carvacrol and thymol were not identified and quantified in Thymocid® in the current study, it has been reported that carvacrol and thymol exhibited skin-protective effects including inhibitory effects on collagenase and elastase [31]. Therefore, it is possible that several phytochemicals including aliphatic compounds and alkaloids in Thymocid® may also exert biological effects that contribute to the overall skin beneficial effects of BCSE. Therefore, a "whole-food" approach was used to evaluate the biological effects of Thymocid®, rather than testing its individual compounds. Thymocid® showed anti-aging effects by maintaining protein structure against glycation (Figure 2). The inhibitory effect of Thymocid® on the formation of AGEs in a BSA–fructose model (Figure 2A) was in agreement with a published study showing that TQ reduced the formation of AGEs in a model with human serum albumin and glucose [32]. In addition, a BCSE has been reported to show preventive effects against glycation-induced DNA damage [33]. Furthermore, Thymocid® protected the structure of bovine type I collagen from glycation-induced cross-linking (Figure 2B). Given that the formation of AGEs is an oxidative process, it is possible that the antiglycation effect of Thymocid® may attribute to its antioxidant effects [34]. The protective effects of Thymocid® on the structure of skin's connective proteins were supported by its anti-collagenase (Figure 3) and anti-elastase activities (Table 1). This is in agreement with a published study showing that a BCSE was able to inhibit the activity of human neutrophil elastase [35]. However, due to the structural and functional difference between porcine pancreatic elastase and human neutrophil elastase [36], further studies on Thymocid®'s effects on elastase in human skin fibroblast cells are warranted to support its antiwrinkle effects.

Melanin, the end-product of melanogenesis, is produced in melanocytes within melanosomes. It is a critical factor for the color of various organs and tissues including eyes, hair, and skin [37–39]. Melanin's main role is related to skin protection against ultraviolet rays; however, excessive production and accumulation of melanin lead to cutaneous hyperpigmentation disorders, including freckles, skin discoloration, and pigmented age spots [38,40]. Over the years, this can lead to the enhancement of the degradation of cutaneous extracellular matrix proteins as observed in biological processes including skin aging, discoloration, and solar elastosis [41]. The process of melanogenesis is catalyzed by three major melanocyte-specific enzymes, namely, TYR, TYRP1, and TYRP2 (also known as dopachrome tautomerase) [40], with TYR being the rate-limiting enzyme for melanin biosynthesis [42–44]. These enzymes are appropriate targets for the improvement of skin conditions including skin tone, aging, whitening, or discoloration with plant-based nutrients [42,45]. Thymocid® promoted the activity of tyrosinase, which corroborates a previously reported study showing that BCSE was able to increase melanin production in melanophores isolated from wall lizard (*Podarcis muralis*) [29]. However, further evaluations on the melanin content in murine melanoma B16F10 cells indicated that Thymocid®, at nontoxic concentrations, suppressed the production of melanin (Figure 4B,C) while it decreased cellular tyrosinase activity (Figure 4D). This is similar to a previously reported study, in which a BCSE was able to reduce melanin production in B16F10 cells stimulated with alpha-melanocyte-stimulating hormone [46]. The BCSE's antimelanogenic effect (observed in this study and a study from another group) is contradictory with its melanogenic effect reported in a model of melanophores from wall lizard [30]. Several factors may account for this contradiction. First, BCSE's effects on melanogenesis were assessed with different models (murine melanoma cells vs. wall lizard melanophores), which may have distinct response to the treatment with BCSE. Next, levels of TQ in BCSE used in the wall lizard melanophores study and in Thymocid® are different (0.0356% vs. 5.12%). Therefore, further studies with more physiological relevant models including human melanocytes are warranted to confirm the modulatory effects of Thymocid® on melanogenesis. Mechanistically, our studies showed that Thymocid® inhibited cellular tyrosinase activity (Figure 4D) and suppressed the gene and protein expressions of melanogenesis-related markers in B16F10 cells (Figures 5 and 6), which may collectively contribute to its overall antimelanogenic effects. We noted that Thymocid® was able to downregulate both gene and protein expression of TYRP1 in B16F10 cells. TYRP1 is an enzyme catalyzing the oxidation of dihydroxyindole carboxylic acid during the process of melanin biosynthesis in murine melanoma cells [47]. In addition, TYRP1 is involved in maintaining the protein structure of tyrosinase and modulating its oxidative activity, which was in agreement with the observation of decreased cellular tyrosinase activity in B16F10 cells. Therefore, TYRP1 may be a valid molecular target for the antimelanogenic effects of Thymocid®. However, further studies with human melanocytes as well as in skin tissue architecture that simulate human skin are warranted to confirm this. In addition, further studies on Thymocid®'s cosmeceutical characterizations including skin permeability, bioavailability, and proper formulations are warranted.

5. Conclusions

In summary, Thymocid®, a chemically standardized and commercially available BCSE, protected the structure of BSA and type I collagen by inhibition of protein glycation and collagen cross-linking, respectively. Thymocid® also inhibited the activity of skin-wrinkle-related enzymes including collagenase and elastase. The inhibitory effects of Thymocid® on collagenase and elastase may contribute to its overall anti-aging effects. In addition, Thymocid® showed inhibitory effects on the production of melanin content in B16F10 cells. This antimelanogenic effect was associated with its modulation of cellular tyrosinase activity and expressions of melanogenesis-related genes and proteins including MITF, TYR, TYRP1, and TYRP2. Our findings highlight the potential of Thymocid® as a bioactive ingredient for cosmeceutical applications.

Supplementary Materials: The following are available online at http://www.mdpi.com/2072-6643/12/7/2146/s1, Figure S1. HPLC chromatograms of Thymocid®, Figure S2. Standard curve for quantification of TQ in Thymocid®, Table S1. Inhibitory effect of TQ on collagenase activity, Table S2. Inhibitory effect of TQ on elastase activity, Table S3. Inhibitory effect of TQ on tyrosinase activity.

Author Contributions: Conceptualization, N.P.S. and H.M.; methodology, H.L., N.A.D., W.L., J.X., G.W.D., J.C.C., and H.M.; data curation, H.L., N.A.D., J.X., and W.L.; writing—original draft preparation, H.L., N.A.D., and W.L.; writing—review and editing, J.C.C., N.P.S., and H.M.; supervision, J.A.D.; funding acquisition, D.L. All authors have read and agreed to the published version of the manuscript.

Funding: H.L. was supported by funding from the Department of Education of Guangdong Province (no: 2017KSYS010, 2019KZDZX2003) and the Jiangmen Program for Innovative Research Team (no: 2018630100180019806). J.C.C. is supported in part by a start-up fund from the University of Louisiana at Monroe (ULM) College of Pharmacy, an Institutional Development Award (IDeA), Networks of Biomedical Research Excellence (INBRE)-LBRN Administrative Supplement grant # 3P20GM103424-18S1, and an LBRN Pilot Awards grant # P2O GM103424-18 from National institute of Health (NIH)/NIGMS.

Acknowledgments: Spectrometric data were acquired from instruments located at the University of Rhode Island in the RI-INBRE core facility obtained from grant # P20GM103430 from the National Center for Research Resources (NCRR), a component of the National Institutes of Health (NIH). Black cumin seed extract Thymocid® was kindly provided by Verdure Sciences (Noblesville, IN, USA).

Conflicts of Interest: The authors declare no conflict of interest.

Abbreviations

AG	aminoguanidine
AGEs	advanced glycation end-products
AAAPVN	N-succinyl-Ala-Ala-Ala-p-nitroanilide
BCSEs	black cumin seed extracts
BSA	bovine serum albumin
EGCG	epigallocatechin gallate
FALGPA	N-(3-[2-furyl]-acryloyl)-Leu-Gly-Pro-Ala
L-DOPA	L-3,4-dihydroxyphenylalanine
MGO	methylglyoxal
MITF	microphthalmia-associated transcription factor
phen	1,10-phenanthroine
TQ	thymoquinone
TYR	tyrosinase
TYRP1	tyrosinase-related protein-1
TYRP2	tyrosinase-related protein-2

References

1. Bharti, U.; Hamal, I.A.; Jaiswal, A.; Patel, S. Black cumin (*Nigella sativa* L.)—A review. *J. Plant Dev.* **2009**, *4*, 1–43.
2. Hasan, R.; Javaid, A.; Fatima, S. The effects of short-term administration of weight reducing herbal drug (mehzileen) on serum enzymes in common rabbits. *J. Basic Appl. Sci.* **2012**, *8*. [CrossRef]
3. Takruri, H.R.H.; Dameh, M.A.F. Study of the nutritional value of black cumin seeds (*Nigella sativa* L.). *J. Sci. Food Agric.* **1998**, *76*, 404–410. [CrossRef]
4. Eid, A.M.; Elmarzugi, N.A.; Abu Ayyash, L.M.; Sawafta, M.N.; Daana, H.I. A review on the cosmeceutical and external applications of *Nigella sativa*. *J. Trop. Med.* **2017**, *2017*. [CrossRef]
5. Akram Khan, M.; Afzal, M. Chemical composition of *Nigella sativa* Linn: Part 2 Recent advances. *Inflammopharmacology* **2016**, *24*, 67–79. [CrossRef]
6. Yuan, T.; Nahar, P.; Sharma, M.; Liu, K.; Slitt, A.; Aisa, H.A.; Seeram, N.P. Indazole-type alkaloids from *Nigella sativa* seeds exhibit antihyperglycemic effects via AMPK activation in vitro. *J. Nat. Prod.* **2014**, *77*, 2316–2320. [CrossRef]
7. Agarwal, R.; Kharya, M.D.; Shrivastava, R. Antimicrobial & anthelmintic activities of the essential oil of *Nigella sativa* Linn. *Indian J. Exp. Biol.* **1979**, *17*, 1264–1265.
8. Mahmoud, M.R.; El-Abhar, H.S.; Saleh, S. The effect of *Nigella sativa* oil against the liver damage induced by *Schistosoma mansoni* infection in mice. *J. Ethnopharmacol.* **2002**, *79*, 1–11. [CrossRef]
9. Aljabre, S.H.M.; Randhawa, M.A.; Akhtar, N.; Alakloby, O.M.; Alqurashi, A.M.; Aldossary, A. Antidermatophyte activity of ether extract of *Nigella sativa* and its active principle, thymoquinone. *J. Ethnopharmacol.* **2005**, *101*, 116–119. [CrossRef]
10. Ali, B.H.; Blunden, G. Pharmacological and toxicological properties of *Nigella sativa*. *Phytother. Res.* **2003**, *17*, 299–305. [CrossRef]
11. Aljabre, S.H.M.; Alakloby, O.M.; Randhawa, M.A. Dermatological effects of *Nigella sativa*. *J. Dermatol. Dermatol. Surg.* **2015**, *19*, 92–98. [CrossRef]
12. Abdul-Ameer, N.; Al-Harchan, H. Treatment of acne vulgaris with *Nigella sativa* oil lotion. *Iraq. Postgrad. Med. J.* **2010**, *2*, 140–143.
13. Ali, S.A.; Meitei, K.V. Nigella sativa seed extract and its bioactive compound thymoquinone: The new melanogens causing hyperpigmentation in the wall lizard melanophores. *J. Pharm. Pharmacol.* **2011**, *63*, 741–746. [CrossRef] [PubMed]
14. Lee, J.; Ji, J.; Park, S.H. Antiwrinkle and antimelanogenesis activity of the ethanol extracts of *Lespedeza cuneata* G. Don for development of the cosmeceutical ingredients. *Food Sci. Nutr.* **2018**, *6*, 1307–1316. [CrossRef]
15. Wang, L.; Oh, J.Y.; Jayawardena, T.U.; Jeon, Y.J.; Ryu, B.M. Anti-inflammatory and anti-melanogenesis activities of sulfated polysaccharides isolated from *Hizikia fusiforme*: Short communication. *Int. J. Biol. Macromol.* **2020**, *142*, 545–550. [CrossRef]

16. Jung, H.J.; Kyoung Lee, A.; Park, Y.J.; Lee, S.; Kang, D.; Jung, Y.S.; Young Chung, H.; Ryong Moon, H. (2E,5E)-2,5-Bis(3-hydroxy-4-methoxybenzylidene) cyclopentanone exerts anti-melanogenesis and anti-wrinkle activities in B16F10 melanoma and hs27 fibroblast cells. *Molecules* **2018**, *23*, 1415. [CrossRef]

17. Lim, H.Y.; Jeong, D.; Park, S.H.; Shin, K.K.; Hong, Y.H.; Kim, E.; Yu, Y.G.; Kim, T.R.; Kim, H.; Lee, J.; et al. Antiwrinkle and antimelanogenesis effects of tyndallized *Lactobacillus acidophilus* KCCM12625P. *Int. J. Mol. Sci.* **2020**, *21*, 1620. [CrossRef]

18. Jeong, S.; Yoon, S.; Kim, S.; Jung, J.; Kor, M.; Shin, K.; Lim, C.; Han, H.S.; Lee, H.; Park, K.Y.; et al. Anti-wrinkle benefits of peptides complex stimulating skin basement membrane proteins expression. *Int. J. Mol. Sci.* **2020**, *21*, 73. [CrossRef]

19. Ganceviciene, R.; Liakou, A.I.; Theodoridis, A.; Makrantonaki, E.; Zouboulis, C.C. Skin anti-aging strategies. *Dermato Endocrinol.* **2012**, *4*, 308–319. [CrossRef]

20. Mukherjee, P.K.; Maity, N.; Nema, N.K.; Sarkar, B.K. Bioactive compounds from natural resources against skin aging. *Phytomedicine* **2011**, *19*, 64–73. [CrossRef]

21. Liu, C.; Guo, H.; DaSilva, N.A.; Li, D.; Zhang, K.; Wan, Y.; Gao, X.H.; Chen, H.D.; Seeram, N.P.; Ma, H. Pomegranate (*Punica granatum*) phenolics ameliorate hydrogen peroxide-induced oxidative stress and cytotoxicity in human keratinocytes. *J. Funct. Foods* **2019**, *54*, 559–567. [CrossRef]

22. Sheng, J.; Liu, C.; Petrovas, S.; Wan, Y.; Chen, H.D.; Seeram, N.P.; Ma, H. Phenolic-enriched maple syrup extract protects human keratinocytes against hydrogen peroxide and methylglyoxal induced cytotoxicity. *Dermatol. Ther.* **2020**, *33*. [CrossRef] [PubMed]

23. Liu, C.; Guo, H.; Dain, J.A.; Wan, Y.; Gao, X.-H.; Chen, H.-D.; Seeram, N.P.; Ma, H. Cytoprotective effects of a proprietary red maple leaf extract and its major polyphenol, ginnalin A, against hydrogen peroxide and methylglyoxal induced oxidative stress in human keratinocytes. *Food Funct.* **2020**. [CrossRef]

24. Zhang, Y.; Ma, H.; Liu, W.; Yuan, T.; Seeram, N.P. New antiglycative compounds from cumin (*Cuminum cyminum*) spice. *J. Agric. Food Chem.* **2015**, *63*, 10097–10102. [CrossRef]

25. Thring, T.S.A.; Hili, P.; Naughton, D.P. Anti-collagenase, anti-elastase and anti-oxidant activities of extracts from 21 plants. *Bmc Complement. Altern. Med.* **2009**, *9*. [CrossRef] [PubMed]

26. Brotzman, N.; Xu, Y.; Graybill, A.; Cocolas, A.; Ressler, A.; Seeram, N.P.; Ma, H.; Henry, G.E. Synthesis and tyrosinase inhibitory activities of 4-oxobutanoate derivatives of carvacrol and thymol. *Bioorg. Med. Chem. Lett.* **2019**, *29*, 56–58. [CrossRef] [PubMed]

27. Ma, H.; Xu, J.; DaSilva, N.A.; Wang, L.; Wei, Z.; Guo, L.; Johnson, S.L.; Lu, W.; Xu, J.; Gu, Q.; et al. Cosmetic applications of glucitol-core containing gallotannins from a proprietary phenolic-enriched red maple (*Acer rubrum*) leaves extract: Inhibition of melanogenesis via down-regulation of tyrosinase and melanogenic gene expression in B16F10 melanoma ce. *Arch. Dermatol. Res.* **2017**, *309*, 265–274. [CrossRef]

28. Chan, Y.Y.; Kim, K.H.; Cheah, S.H. Inhibitory effects of *Sargassum polycystum* on tyrosinase activity and melanin formation in B16F10 murine melanoma cells. *J. Ethnopharmacol.* **2011**, *137*, 1183–1188. [CrossRef]

29. Amin, B.; Hosseinzadeh, H. Black cumin (*Nigella sativa*) and its active constituent, thymoquinone: An overview on the analgesic and anti-inflammatory effects. *Planta Med.* **2016**, *82*, 8–16. [CrossRef]

30. Kundu, J.K.; Liu, L.; Shin, J.W.; Surh, Y.J. Thymoquinone inhibits phorbol ester-induced activation of NF-κB and expression of COX-2, and induces expression of cytoprotective enzymes in mouse skin in vivo. *Biochem. Biophys. Res. Commun.* **2013**, *438*, 721–727. [CrossRef]

31. Laothaweerungsawat, N.; Sirithunyalug, J.; Chaiyana, W. Chemical compositions and anti-skin-ageing activities of *Origanum vulgare* L. essential oil from tropical and mediterranean region. *Molecules* **2020**, *25*, 1101. [CrossRef] [PubMed]

32. Losso, J.N.; Bawadi, H.A.; Chintalapati, M. Inhibition of the formation of advanced glycation end products by thymoquinone. *Food Chem.* **2011**, *128*, 55–61. [CrossRef]

33. Pandey, R.; Kumar, D.; Ali, A. Nigella sativa seed extracts prevent the glycation of protein and DNA. *Curr. Perspect. Med. Aromat. Plants* **2018**, *1*, 1–7.

34. Burits, M.; Bucar, F. Antioxidant activity of *Nigella sativa* essential oil. *Phytother. Res.* **2000**, 323–328. [CrossRef]

35. Kacem, R.; Meraihi, Z. Effects of essential oil extracted from *Nigella sativa* (L.) seeds and its main components on human neutrophil elastase activity. *Yakugaku Zasshi* **2006**, *126*, 301–305. [CrossRef]

36. Schmelzer, C.E.H.; Jung, M.C.; Wohlrab, J.; Neubert, R.H.H.; Heinz, A. Does human leukocyte elastase degrade intact skin elastin? *FEBS J.* **2012**, *279*, 4191–4200. [CrossRef]

37. Regad, T. Molecular and cellular pathogenesis of melanoma initiation and progression. *Cell. Mol. Life Sci.* **2013**, *70*, 4055–4065. [CrossRef]

38. Osborne, R.; Hakozaki, T.; Laughlin, T.; Finlay, D.R. Application of genomics to breakthroughs in the cosmetic treatment of skin ageing and discoloration. *Br. J. Dermatol.* **2012**, *166*, 16–19. [CrossRef]
39. Videira, I.F.D.S.; Moura, D.F.L.; Magina, S. Mechanisms regulating melanogenesis. *An. Bras. Dermatol.* **2013**. [CrossRef]
40. Del Marmol, V.; Beermann, F. Tyrosinase and related proteins in mammalian pigmentation. *FEBS Lett.* **1996**, *381*, 165–168. [CrossRef]
41. Gilchrest, B.A. A review of skin ageing and its medical therapy. *Br. J. Dermatol.* **1996**, *135*, 867–875. [CrossRef] [PubMed]
42. Iwata, M.; Corn, T.; Iwata, S.; Everett, M.A.; Fuller, B.B. The relationship between tyrosinase activity and skin color in human foreskins. *J. Investig. Dermatol.* **1990**, *95*, 9–15. [CrossRef] [PubMed]
43. Kobayashi, T.; Urabe, K.; Winder, A.; Jiménez-Cervantes, C.; Imokawa, G.; Brewington, T.; Solano, F.; García-Borrón, J.C.; Hearing, V.J. Tyrosinase related protein 1 (TRP1) functions as a DHICA oxidase in melanin biosynthesis. *EMBO J.* **1994**, *13*, 5818–5825. [CrossRef] [PubMed]
44. Sulaimon, S.S.; Kitchell, B.E. Review article the biology of melanocytes. *Vet. Dermatol.* **2003**, *14*, 57–65. [CrossRef]
45. Kim, Y.J.; Uyama, H. Tyrosinase inhibitors from natural and synthetic sources: Structure, inhibition mechanism and perspective for the future. *Cell. Mol. Life Sci.* **2005**, *62*, 1707–1723. [CrossRef]
46. Mady, R.F.; El-Hadidy, W.; Elachy, S. Effect of Nigella sativa oil on experimental toxoplasmosis. *Parasitol. Res.* **2016**, *115*, 379–390. [CrossRef]
47. Sarangarajan, R.; Boissy, R.E. Tyrp1 and oculocutaneous albinism type 3. *Pigment Cell Res.* **2001**, *14*, 437–444. [CrossRef]

Article

Lactobacillus helveticus-Fermented Milk Whey Suppresses Melanin Production by Inhibiting Tyrosinase through Decreasing MITF Expression

Nobutomo Ikarashi [1,*,†], **Natsuko Fukuda** [1,†], **Makiba Ochiai** [1], **Mami Sasaki** [1], **Risako Kon** [1], **Hiroyasu Sakai** [1], **Misaki Hatanaka** [2] **and Junzo Kamei** [1]

[1] Department of Biomolecular Pharmacology, Hoshi University, 2-4-41 Ebara, Shinagawa-ku, Tokyo 142-8501, Japan; s151206@hoshi.ac.jp (N.F.); s161077@hoshi.ac.jp (M.O.); s161131@hoshi.ac.jp (M.S.); r-kon@hoshi.ac.jp (R.K.); sakai@hoshi.ac.jp (H.S.); kamei@hoshi.ac.jp (J.K.)
[2] Asahi Calpis Wellness Co., Ltd., 2-4-1 Ebisu-minami, Shibuya-ku, Tokyo 150-0022, Japan; misaki.hatanaka@asahicalpis-w.co.jp
* Correspondence: ikarashi@hoshi.ac.jp; Tel.: +81-3-5498-5918
† These authors contributed equally to this work.

Received: 17 June 2020; Accepted: 8 July 2020; Published: 14 July 2020

Abstract: Whey obtained from milk fermented by the *Lactobacillus helveticus* CM4 strain (LHMW) has been shown to improve skin barrier function and increase skin-moisturizing factors. In this study, we investigated the effects of LHMW on melanin production to explore the additional impacts of LHMW on the skin. We treated mouse B16 melanoma cells with α-melanocyte-stimulating hormone (α-MSH) alone or simultaneously with LHMW and measured the amount of melanin. The amount of melanin in B16 cells treated with α-MSH significantly increased by 2-fold compared with that in control cells, and tyrosinase activity was also elevated. Moreover, treatment with LHMW significantly suppressed the increase in melanin content and elevation of tyrosinase activity due to α-MSH. LHMW also suppressed the α-MSH-induced increased expression of tyrosinase, tyrosinase-related protein 1 (TRP1), and dopachrome tautomerase (DCT) at the protein and mRNA levels. Furthermore, the mRNA and protein microphthalmia-associated transcription factor (MITF) expression levels were significantly increased with treatment with α-MSH alone, which were also suppressed by LHMW addition. LHMW suppression of melanin production is suggested to involve inhibition of the expression of the tyrosinase gene family by lowering the MITF expression level. LHMW may have promise as a material for cosmetics with expected clinical application in humans.

Keywords: whey; *Lactobacillus helveticus*; melanin; α-melanocyte-stimulating hormone; tyrosinase; tyrosinase-related protein 1; dopachrome tautomerase; microphthalmia-associated transcription factor; cosmetics

---

1. Introduction

Melanin, which is the end product of melanogenesis, is generated in the melanosomes of melanocytes and is an important factor determining the color of human skin, hair, and eyes [1,2]. Melanin is associated with protection of the skin from ultraviolet rays; however, excess melanin production and accumulation on the skin cause pigmentation disorders, such as freckles, skin discoloration, and pigmented age spots [3].

In addition to ultraviolet irradiation, melanin production is induced by various hormones, including α-melanocyte-stimulating hormone (α-MSH) and estrogen [4,5], as well as by environmental stimulation from chemical substances such as theophylline [6,7] and extracellular stimuli, such as cytokines [8]. Melanogenesis is catalyzed by three types of melanocyte-specific enzymes: tyrosinase, tyrosinase-related protein 1 (TRP1), and dopachrome tautomerase (DCT) [9]. Tyrosinase is a rate-limiting enzyme of the melanin production pathway, in which l-tyrosine is converted to l-3,4-dihydroxyphenylalanine (l-DOPA) via hydroxylation. l-DOPA is in turn oxidized to become DOPA quinone [10]. DCT, which is also called TRP2, catalyzes the tautomerization of dopachrome to produce 5,6-dihydroxyindole-2-carboxylic acid (DHICA) [11]. TRP1 also oxidizes DHICA to produce carboxylate indole-quinone [12]. TRP1 and DCT function downstream from tyrosinase in the melanin biosynthesis pathway [2]. Therefore, these enzymes are suitable targets for the development of cosmetics aimed at skin whitening, and such novel materials have been actively explored [13,14].

Milk fermented by the *Lactobacillus helveticus* CM4 strain (LH-fermented milk), a lactic acid bacterium, has been reported to have a hypotensive effect [15,16] and a learning–memory improvement effect [17]. LH-fermented milk cofermented with yeast has been shown to have life-extension [18] and antitumor effects [19]. Moreover, whey obtained from LH-fermented milk (LH-fermented milk whey (LHMW)) was shown to strengthen epidermal barrier function when taken orally and was effective in the prevention of dermatitis [20]. Furthermore, the expression of profilaggrin, which is an important factor for skin moisturizing, was reported to increase when LHMW was added to normal human epidermal keratinocytes [21]. As such, it has been clarified that LHMW has useful effects on the skin, and its application has been gaining attention. To expand the effects of LHMW on the skin, in this study, we investigated the effect of LHMW on melanin production. Specifically, we treated B16 cells, a mouse melanocyte cell line, with α-MSH with and without LHMW to investigate the suppressive effect of LHMW upon induction of melanin production and to explore the underlying molecular mechanism.

2. Materials and Methods

2.1. Materials

Dulbecco's modified Eagle medium (DMEM) and 2-amino-2-hydroxymethyl-1,3-propanediol (Tris) were purchased from Fujifilm Wako Pure Chemical Co., Ltd. (Osaka, Japan). The cell proliferation reagent water-soluble tetrazolium salt (WST-1) was purchased from Roche (Mannheim, Germany). Bovine serum albumin (BSA) and TRI reagent were purchased from Sigma-Aldrich Corp. (St. Louis, MO, USA). Mouse antihuman tyrosinase (T311) antibody, mouse antihuman TRP1 (G-9) antibody, mouse antihuman TRP2/DCT (C-9) antibody, and mouse antihuman microphthalmia-associated transcription factor (MITF; D-9) antibody were purchased from Santa Cruz Biotechnology, Inc. (Santa Cruz, CA, USA). Rabbit anti-β-actin antibody was purchased from BioLegend (San Diego, CA, USA). Donkey antimouse IgG-HRP antibody, donkey antirabbit IgG-HRP antibody, and enhanced chemiluminescence (ECL) prime Western blotting detection reagents were purchased from GE Healthcare (Waukesha, WI, USA). A high-capacity cDNA synthesis kit was purchased from Applied Biosystems (Foster City, CA, USA). SsoAdvanced Universal SYBR Green Supermix was purchased from Bio-Rad Laboratories (Hercules, CA, USA).

2.2. Preparation of Fermented Milk Whey

Fermented milk was prepared as reported previously [21]. In brief, reconstituted, pasteurized 9% (*w/w*) skim milk solution was fermented with the *L. helveticus* CM4 strain at 35 °C for 24 h. This sample was separated into the whey fraction by ultrafiltration (MW < 5 kDa).

2.3. Cell Culture

B16 mouse melanoma (Riken Cell Bank, Ibaraki, Japan) cells were maintained in DMEM containing 100 U/mL penicillin G potassium, 100 μg/mL streptomycin, and 10% FBS. B16 cells were seeded in a plate at a density of 2.5×10^4 cells/cm^2 and incubated in a CO$_2$ incubator for 2 days. α-MSH (final concentration: 50 nM) alone or simultaneously with LHMW (final concentration: 1–5%) was added to B16 cells and cultured.

2.4. WST-1 Assay

B16 cells seeded in a 96-well plate were treated with LHMW and cultured for 48 h. After removing the medium and washing the cells with phosphate-buffered saline (PBS), WST-1 reagent was added to the cells. After incubation for 4 h, the absorbance at 450 nm and 620 nm was measured using a microplate reader (MTP-450 microplate reader, Corona Electric Co., Ltd., Ibaraki, Japan).

2.5. Measurement of Melanin Content

The amount of melanin was measured according to a previously described method, with slight modifications [22,23]. Briefly, α-MSH alone or simultaneously with LHMW was added to B16 cells seeded in a 6-well plate and cultured for 48 h. After B16 cells were washed with PBS, 500 μL of a 1 M NaOH solution was added to each well. The cells were collected and then incubated at 80 °C for 60 min. Absorbance at 415 nm was measured using a microplate reader. Melanin content was normalized by protein content and expressed as a percentage of the control.

2.6. Measurement of Tyrosinase Activity

Tyrosinase activity was measured according to a previously described method, with slight modifications [24,25]. α-MSH alone or simultaneously with LHMW was added to B16 cells seeded in a 6-well plate and cultured for 24 h. After B16 cells were washed with PBS, 200 μL of PBS containing 1% Triton X-100 was added to each well. The cells were collected and homogenized with an ultrasonic homogenizer (Handy Sonic, TOMY SEIKO Co., Ltd., Tokyo, Japan) on ice and then centrifuged (11,000× g, 20 min, 4 °C). Then, 100 μL of PBS containing 0.1% L-DOPA was added to 30 μL of the supernatant, and the mixture was incubated at 37 °C for 30 min. Absorbance at 492 nm was measured using a microplate reader. Tyrosinase activity was normalized by protein content and expressed as a percentage of the control.

2.7. Real-Time PCR

α-MSH alone or simultaneously with LHMW was added to B16 cells seeded in a 6-well plate and cultured for 3 h or 24 h. After washing B16 cells with PBS, total RNA was extracted using TRI reagent. Total RNA was used to calculate the RNA concentration and confirm the purity by measuring the absorbance at 260 nm and 280 nm with a microspectrophotometer (NanoDrop Lite Spectrophotometer, Thermo Fisher Scientific, Waltham, MA, USA). cDNA was synthesized from RNA using a high-capacity cDNA synthesis kit. Real-time PCR was performed to detect the expression of each gene using the specific primers shown in Table 1. The mRNA expression level was normalized using the housekeeping gene glyceraldehyde-3-phosphate dehydrogenase (GAPDH).

Table 1. Primer sequences used for real-time PCR.

Gene	Forward	Reverse
Tyrosinase	CAAAGGGGTGGATGACCGTG	AACTTACAGTTTCCGCAGTTGA
Trp1	ATGAAATCTTACAACGTCCTCCC	GCACACTCTCGTGGAAACTGA
Dct	TTCAACCGGACATGCAAATGC	GCTTCTTCCGATTACAGTCGGG
MITF	CAAATGGCAAATACGTTACCCG	CAATGCTCTTGCTTCAGACTCT
GAPDH	GGCAAATTCAACGGCACAGT	AGATGGTGATGGGCTTCCC

2.8. Real-Time PCR Preparation of Samples for Western Blotting

α-MSH alone or simultaneously with LHMW was added to B16 cells seeded in a 6-well plate and cultured for 3 h or 24 h. After washing B16 cells with PBS, 500 μL of RIPA buffer was added, and the cells were collected using a cell scraper. The cell suspension solution was placed on ice for 30 min, homogenized by an ultrasonic homogenizer (Handy Sonic, TOMY SEIKO Co., Ltd.), and centrifuged (15,000× g, 15 min, 4 °C). The obtained supernatant was used as a sample solution for Western blotting.

2.9. Western Blotting

The protein concentration was measured by the bicinchoninic acid (BCA) method. After the addition of an equal volume of loading buffer (100 mM Tris, 20% glycerol, 0.004% bromophenol blue, 4% sodium dodecyl sulfate, and 10% 2-mercaptoethanol; pH 6.8) to the sample solution, the samples were separated using SDS-PAGE. The protein was transferred to a polyvinylidene difluoride membrane and blocked with skim milk solution. The membrane was reacted with the following primary antibodies: mouse antihuman tyrosinase antibody, mouse antihuman TRP1 antibody, mouse antihuman TRP2/DCT antibody, mouse antihuman MITF antibody, or rabbit anti-β-actin antibody. After washing, the membrane was incubated with a secondary antibody, donkey antimouse IgG-HRP antibody or donkey antirabbit IgG-HRP antibody. The antibodies were detected with ECL prime Western blotting detection reagent. The protein signal was visualized using a CCD camera (ImageQuant LAS500, GE Healthcare).

2.10. Statistical Analyses

The experimental values are shown as the mean ± standard deviation (SD). Dunnett's test and Tukey's test were used for the statistical analyses.

3. Results

3.1. Effect of LHMW on Melanin Production Stimulated by α-MSH

We investigated the effect of LHMW on melanin production stimulated by α-MSH.

When B16 cells were treated with α-MSH and the condition of the cells was observed under a microscope, it was confirmed that melanin production was enhanced. In contrast, cotreatment with LHMW suppressed the amount of melanin stimulated by α-MSH in a concentration-dependent manner. In particular, treatment with 3% LHMW suppressed melanin production induced by α-MSH to approximately the same level as that in the control (Figure 1). The WST-1 assay further showed that treatment with up to 5% LHMW did not affect the cell survival rate of B16 cells (Figure S1).

These results confirmed that LHMW suppressed melanin production stimulated by α-MSH.

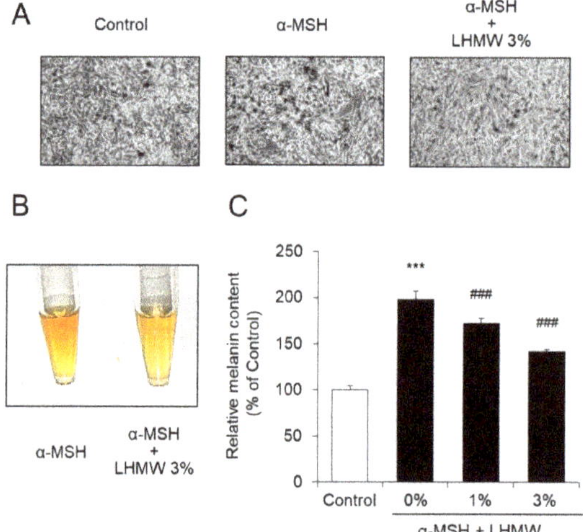

Figure 1. Effect of *Lactobacillus helveticus*-fermented milk whey (LHMW) on melanin production stimulated by α-melanocyte-stimulating hormone (α-MSH). α-MSH alone or simultaneously with LHMW was added to B16 cells and cultured for 48 h (**A**), and the intracellular melanin was eluted (**B**). The amount of melanin was calculated by measuring the absorbance, and the average value of the control was expressed as 100% (**C**) (mean ± SD, *n* = 6, ***; *p* < 0.001 vs. control, ###; *p* < 0.001 vs. vehicle).

3.2. Effect of LHMW on Tyrosinase Activity

Tyrosinase is a rate-limiting enzyme of melanogenesis, which is important in the determination of melanin content [13,14]. Therefore, we further investigated whether the suppressive effect of LHMW on melanin production was caused by inhibitory activity on tyrosinase.

B16 cells treated with α-MSH showed approximately 2-fold higher tyrosinase activity than the control cells, and LHMW suppressed the increase in tyrosinase activity stimulated by α-MSH in a concentration-dependent manner. Specifically, treatment with 3% LHMW resulted in tyrosinase activity similar to that of the control condition (Figure 2).

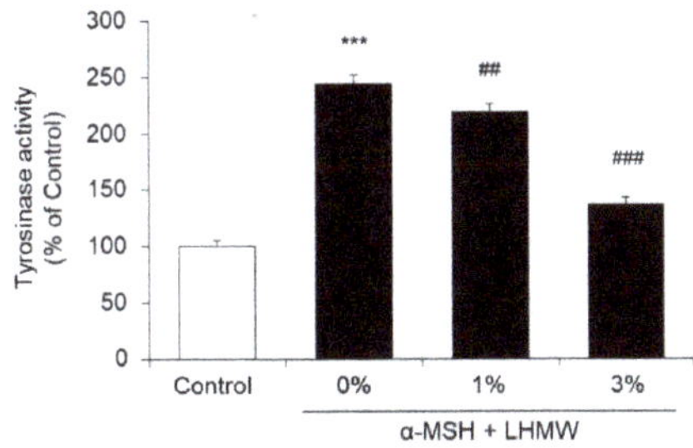

Figure 2. Effect of LHMW on tyrosinase activity. α-MSH alone or simultaneously with LHMW was added to B16 cells and cultured for 24 h. The tyrosinase activity was measured, and the average value of the control was expressed as 100% (mean ± SD, $n = 6$, ***; $p < 0.001$ vs. control, ##; $p < 0.01$, ###; $p < 0.001$ vs. vehicle).

Therefore, the suppression of melanin production by LHMW was considered to be caused by its inhibitory activity against tyrosinase.

3.3. Effects of LHMW on the Protein Expression of Tyrosinase, TRP1, and DCT

We investigated whether the inhibitory effect of LHMW on tyrosinase activity was due to suppression of tyrosinase protein expression. We also analyzed the effect of LHMW on other important proteins for melanin production, including TRP1 and DCT.

The tyrosinase expression level in the α-MSH-treated B16 cells was significantly higher than that in the control cells, and cotreatment with LHMW reduced the protein expression of tyrosinase to approximately the same level as that in the control cells. Similarly, the protein expression levels of TRP1 and DCT increased upon treatment with α-MSH, which were suppressed by LHMW cotreatment to the same level as that in the control cells (Figure 3).

These results indicated that suppressed tyrosinase protein expression was associated with the inhibitory effect of LHMW on tyrosinase activity. LHMW also reduced the levels of TRP1 and DCT expression to suppress melanin production.

3.4. Effects of LHMW on Tyrosinase, Trp1, and Dct mRNA Levels

We investigated whether the effect of LHMW in reducing the protein expression levels of tyrosinase, TRP1, and DCT occurred via transcriptional repression.

The expression of tyrosinase mRNA in B16 cells treated with α-MSH was higher than that in the control cells, and LHMW cotreatment reduced tyrosinase mRNA expression to the same level as that in the control cells. The same effects were observed for Trp1 and Dct mRNA levels (Figure 4).

These results clearly demonstrated that LHMW suppressed the transcription of the tyrosinase gene family induced by α-MSH to reduce the protein and mRNA expression levels of tyrosinase, Trp1, and Dct.

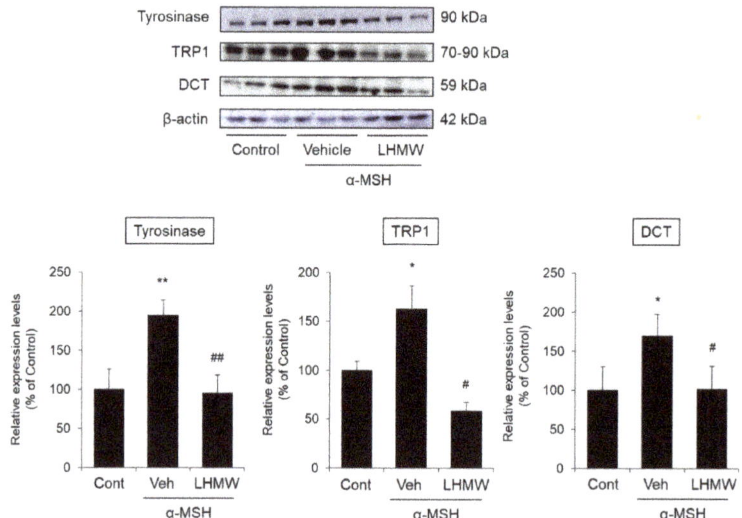

Figure 3. Effects of LHMW on the protein expression of tyrosinase, TRP1, and DCT. α-MSH alone or simultaneously with LHMW was added to B16 cells and cultured for 24 h. The protein expression levels of tyrosinase, TRP1, and DCT were analyzed by Western blotting and normalized to that of β-actin. The average value of the control was expressed as 100% (mean ± SD, $n = 6$, *; $p < 0.05$, **; $p < 0.01$ vs. control, #; $p < 0.05$, ##; $p < 0.01$ vs. vehicle).

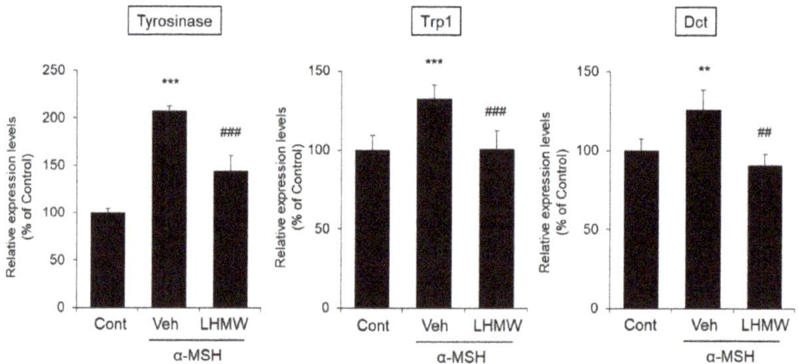

Figure 4. Effects of LHMW on the mRNA expression of tyrosinase, Trp1, and Dct. α-MSH alone or simultaneously with LHMW was added to B16 cells and cultured for 24 h. The mRNA expression levels of tyrosinase, Trp1, and Dct were analyzed by real-time PCR and normalized to that of GAPDH. The average value of the control was expressed as 100% (mean ± SD, $n = 6$, **; $p < 0.01$, ***; $p < 0.001$ vs. control, ##; $p < 0.01$, ###; $p < 0.001$ vs. vehicle).

3.5. Effect of LHMW on MITF Expression

MITF has been shown to regulate the transcription of tyrosinase, Trp1, and Dct [26,27]. Therefore, we investigated whether the reduction in the expression levels of these genes by LHMW was caused by decreasing MITF expression.

B16 cells treated with α-MSH showed significantly increased mRNA and protein levels of MITF compared with those of the control cells. However, upon the addition of LHMW, MITF expression was significantly decreased at both the protein and mRNA levels (Figure 5).

These results suggested that LHMW may have suppressed the transcription of the tyrosinase gene family by suppressing increased MITF expression.

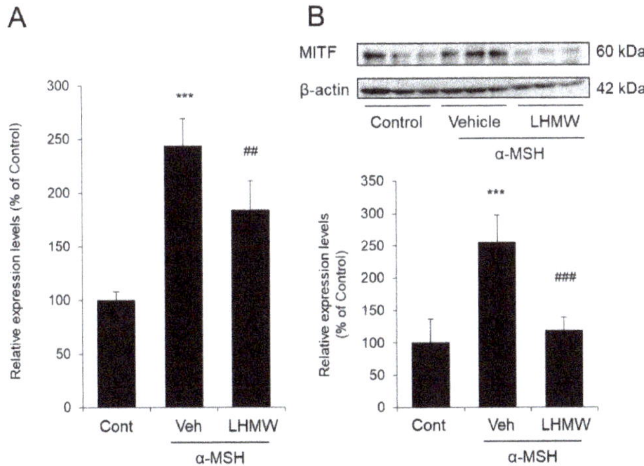

Figure 5. Effect of LHMW on MITF expression. α-MSH alone or simultaneously with LHMW was added to B16 cells and cultured for 3 h. The mRNA (**A**) and protein (**B**) expression levels of MITF were analyzed by real-time PCR and Western blotting and normalized to those of β-actin and GAPDH, respectively. The average value of the control was expressed as 100% (mean ± SD, $n = 6$, ***; $p < 0.001$ vs. control, ##; $p < 0.01$, ###; $p < 0.001$ vs. vehicle).

4. Discussion

Several recent reports have demonstrated the beneficial effects of probiotics such as lactic acid bacteria and bifidobacteria [28–32]. In addition, many findings on the beneficial effects of whey derived from fermented milk have attracted attention [33–36]. In this study, we examined the effects of LHMW on melanogenesis with the aim of exploring a potential novel function of LHMW on the skin.

We used mouse melanoma B16 cells, which have been widely applied in studies of melanogenesis. B16 cells are known to show enhanced melanin production under stimulation with α-MSH [37,38]. In this study, we confirmed that under α-MSH stimulation, the amount of intracellular melanin increased by approximately 2-fold of that in the control condition. In contrast, cotreatment with LHMW at a dose that did not induce cytotoxicity significantly suppressed the increase in melanin production stimulated by α-MSH in a concentration-dependent manner. Specifically, treatment with 3% LHMW reduced the intracellular melanin content to almost the same level as that of the control (Figure 1). These findings suggest that LHMW could be a useful cosmetic material for the development of a skin-whitening agent.

Melanogenesis is controlled by genes in the tyrosinase family, including tyrosinase, TRP1, and DCT [9]. Among them, melanin production in melanocytes is controlled mainly by activation and expression of the rate-limiting enzyme tyrosinase [39]. Upon addition of α-MSH to B16 cells, tyrosinase activity increased, which was suppressed by cotreatment with LHMW in a concentration-dependent manner (Figure 2). In addition, LHMW decreased the protein expression level of tyrosinase while simultaneously suppressing the increased protein levels of TRP1 and DCT under α-MSH stimulation (Figure 3). Based on these findings, the suppressive effect of LHMW on melanin production could have been caused by its ability to suppress the expression and activity of tyrosinase and reduce the protein expression levels of TRP1 and DCT.

The tyrosinase gene family, including tyrosinase, TRP1, and DCT, is associated with pigmentation, proliferation, and survival, and these effects are strictly controlled by MITF [26,27]. Specifically, MITF binds to the M box in the promoter region to control the expression of these tyrosinase genes [40]. Thus, MITF is a transcription factor that plays a very important role in melanogenesis [41,42]. In this study, LHMW suppressed the increase in mRNA expression levels of the tyrosinase, Trp1, and Dct genes stimulated by α-MSH in mouse B16 melanocytes (Figure 4). Therefore, suppression of the transcription of these genes appears to be associated with a reduction in the protein expression levels of the tyrosinase gene family by LHMW. We further investigated the effect of LHMW on MITF expression in B16 cells under α-MSH stimulation, which demonstrated that LHMW significantly suppressed MITF expression at both the mRNA and protein levels (Figure 5). Given these findings, the suppression of transcription and protein expression of the tyrosinase gene family by LHMW could have been caused by LHMW's ability to decrease MITF expression.

How can LHMW suppress the expression of MITF? It has been reported that α-MSH increases the expression level of MITF, as explained below [43,44]. First, when α-MSH binds to the melanocortin-1 receptor (Mc1R), adenylate cyclase is activated to increase the intracellular cyclic 3′,5′-adenosine monophosphate (cAMP) level. The increase in cAMP content in turn activates protein kinase A (PKA) and accelerates phosphorylation of the cAMP-response element-binding protein (CREB) to ultimately induce MITF expression. That is, phosphorylated CREB upregulates the transcription of MITF to ultimately increase the MITF protein expression level. To date, a material with an anti-melanin effect that targets these processes has been identified [45]. Therefore, LHMW may reduce MITF expression by suppressing adenylate cyclase, PKA, or phosphorylated CREB. We also found that LHMW reduced the mRNA level of Mc1R compared to that in cells treated with α-MSH alone (data not shown). Similar to the tyrosinase gene family, Mc1R expression is also controlled by MITF [46]. In other words, MITF controls the expression of both melanogenesis-related enzymes and their receptors. Therefore, LHMW can reduce the responsiveness to α-MSH stimulation by suppressing MITF in addition to decreasing the expression of the tyrosinase gene family.

As described above, ultraviolet irradiation enhances melanin synthesis, which is considered to be one of the causes of oxidative stress [47]. Therefore, it has been suggested that antioxidants exhibit anti-melanogenesis effects, and, in fact, the usefulness of various antioxidants as whitening agents has been clarified [48,49]. Relatedly, whey protein has been found to exhibit an antioxidant effect [50]. Based on these facts, there is a possibility that antioxidant effects may be one of the mechanisms involved in the anti-melanogenesis action of LHMW.

Although this study clarified that LHMW has a suppressive effect on melanin production, its active ingredient remains unclear. In general, whey is rich in peptides and proteins [51]. The *L. helveticus* CM4 strain used in this study has strong proteolytic activity and decomposes milk proteins to produce many peptides during fermentation [15,52]. In this study, the components contained in LHMW that reduced the expression of MITF could not be identified. Although it is still a matter of speculation, it is thought that the peptides exerted the action. In the future, it will be possible to clarify the usefulness of LHMW by searching for active ingredients and comparing with regular milk whey and other lactic acid bacteria-fermented milk whey.

In summary, our results show that LHMW suppresses melanin production, which is suggested to involve inhibition of the expression of the tyrosinase gene family by lowering the MITF expression level. LHMW may have promise as a material for cosmetics with expected clinical application in humans.

Supplementary Materials: The following are available online at http://www.mdpi.com/2072-6643/12/7/2082/s1, Figure S1: Cell viability.

Author Contributions: Conceptualization, N.I., N.F. and M.H.; methodology, N.I., N.F. and M.H.; formal analysis, N.I., N.F., M.O., M.S., R.K. and H.S.; writing—original draft preparation, N.I. and N.F.; writing—review and editing, J.K.; All authors have read and agreed to the published version of the manuscript.

Funding: This study was funded by Asahi Calpis Wellness, Co. Ltd.

Acknowledgments: We thank Takumi Togashi, Naoya Takayama, Ryotaro Yoshida, Ayuka Miyazawa, and Yui Shinozaki (Department of Biomolecular Pharmacology, Hoshi University) for their technical assistance.

Conflicts of Interest: M.H. is employees of Asahi Calpis Wellness Co., Ltd. The other authors declare that the research was conducted in the absence of any financial relationships that could be constructed as potential conflicts of interest.

References

1. Regad, T. Molecular and cellular pathogenesis of melanoma initiation and progression. *Cell. Mol. Life Sci. CMLS* **2013**, *70*, 4055–4065. [CrossRef]

2. Videira, I.F.; Moura, D.F.; Magina, S. Mechanisms regulating melanogenesis. *An. Bras. Dermatol.* **2013**, *88*, 76–83. [CrossRef] [PubMed]

3. Osborne, R.; Hakozaki, T.; Laughlin, T.; Finlay, D.R. Application of genomics to breakthroughs in the cosmetic treatment of skin ageing and discoloration. *Br. J. Dermatol.* **2012**, *166* (Suppl. 2), 16–19. [CrossRef]

4. Dika, E.; Patrizi, A.; Lambertini, M.; Manuelpillai, N.; Fiorentino, M.; Altimari, A.; Ferracin, M.; Lauriola, M.; Fabbri, E.; Campione, E.; et al. Estrogen receptors and melanoma: A Review. *Cells* **2019**, *8*, 1463. [CrossRef] [PubMed]

5. Natale, C.A.; Duperret, E.K.; Zhang, J.; Sadeghi, R.; Dahal, A.; O'Brien, K.T.; Cookson, R.; Winkler, J.D.; Ridky, T.W. Sex steroids regulate skin pigmentation through nonclassical membrane-bound receptors. *eLife* **2016**, *5*, e15104. [CrossRef]

6. Huang, H.C.; Yen, H.; Lu, J.Y.; Chang, T.M.; Hii, C.H. Theophylline enhances melanogenesis in B16F10 murine melanoma cells through the activation of the MEK 1/2, and Wnt/beta-catenin signaling pathways. *Food Chem. Toxicol.* **2020**, *137*, 111165. [CrossRef]

7. Martinez-Liarte, J.H.; Solano, F.; Garcia-Borron, J.C.; Jara, J.R.; Lozano, J.A. Alpha-MSH and other melanogenic activators mediate opposite effects on tyrosinase and dopachrome tautomerase in B16/F10 mouse melanoma cells. *J. Investig. Dermatol.* **1992**, *99*, 435–439. [CrossRef]

8. Imokawa, G.; Ishida, K. Inhibitors of intracellular signaling pathways that lead to stimulated epidermal pigmentation: Perspective of anti-pigmenting agents. *Int. J. Mol. Sci.* **2014**, *15*, 8293–8315. [CrossRef] [PubMed]

9. Del Marmol, V.; Beermann, F. Tyrosinase and related proteins in mammalian pigmentation. *FEBS Lett.* **1996**, *381*, 165–168. [CrossRef]

10. Sulaimon, S.S.; Kitchell, B.E. The biology of melanocytes. *Vet. Dermatol.* **2003**, *14*, 57–65. [CrossRef]

11. Yokoyama, K.; Yasumoto, K.; Suzuki, H.; Shibahara, S. Cloning of the human DOPAchrome tautomerase/tyrosinase-related protein 2 gene and identification of two regulatory regions required for its pigment cell-specific expression. *J. Biolog. Chem.* **1994**, *269*, 27080–27087.

12. Kobayashi, T.; Urabe, K.; Winder, A.; Jimenez-Cervantes, C.; Imokawa, G.; Brewington, T.; Solano, F.; Garcia-Borron, J.C.; Hearing, V.J. Tyrosinase related protein 1 (TRP1) functions as a DHICA oxidase in melanin biosynthesis. *EMBO J.* **1994**, *13*, 5818–5825. [CrossRef] [PubMed]

13. Iwata, M.; Corn, T.; Iwata, S.; Everett, M.A.; Fuller, B.B. The relationship between tyrosinase activity and skin color in human foreskins. *J. Investig. Dermatol.* **1990**, *95*, 9–15. [CrossRef] [PubMed]

14. Kim, Y.J.; Uyama, H. Tyrosinase inhibitors from natural and synthetic sources: Structure, inhibition mechanism and perspective for the future. *Cell. Mol. Life Sci. CMLS* **2005**, *62*, 1707–1723. [CrossRef] [PubMed]

15. Aihara, K.; Kajimoto, O.; Hirata, H.; Takahashi, R.; Nakamura, Y. Effect of powdered fermented milk with *Lactobacillus helveticus* on subjects with high-normal blood pressure or mild hypertension. *J. Am. Coll. Nutr.* **2005**, *24*, 257–265. [CrossRef]

16. Nakamura, Y.; Masuda, O.; Takano, T. Decrease of tissue angiotensin I-converting enzyme activity upon feeding sour milk in spontaneously hypertensive rats. *Biosci. Biotechnol. Biochem.* **1996**, *60*, 488–489. [CrossRef]

17. Ohsawa, K.; Uchida, N.; Ohki, K.; Nakamura, Y.; Yokogoshi, H. *Lactobacillus helveticus*-fermented milk improves learning and memory in mice. *Nutr. Neurosci.* **2015**, *18*, 232–240. [CrossRef]

18. Arai, K.; Murota, I.; Hayakawa, K.; Kataoka, M.; Mitsuoka, T. Effects of administration of pasteurized fermented milk to mice on the life-span and intestinal flora. *J. Jpn. Soc. Food Nutr.* **1980**, *23*, 219–223.

19. Takano, T.; Arai, K.; Murota, I.; Hayakawa, K.; Mizutani, T.; Mitsuoka, T. Effects of feeding sour milk on longevity and tumorigenesis in mice and rats. *Bifidobact. Microflora* **1985**, *4*, 31–37. [CrossRef]

20. Baba, H.; Masuyama, A.; Yoshimura, C.; Aoyama, Y.; Takano, T.; Ohki, K. Oral intake of *Lactobacillus helveticus*-fermented milk whey decreased transepidermal water loss and prevented the onset of sodium dodecylsulfate-induced dermatitis in mice. *Biosci. Biotechnol. Biochem.* **2010**, *74*, 18–23. [CrossRef]

21. Baba, H.; Masuyama, A.; Takano, T. Short communication: Effects of *Lactobacillus helveticus*-fermented milk on the differentiation of cultured normal human epidermal keratinocytes. *J. Dairy Sci.* **2006**, *89*, 2072–2075. [CrossRef]

22. Li, H.R.; Habasi, M.; Xie, L.Z.; Aisa, H.A. Effect of chlorogenic acid on melanogenesis of B16 melanoma cells. *Molecules* **2014**, *19*, 12940–12948. [CrossRef] [PubMed]

23. Wu, L.C.; Lin, Y.Y.; Yang, S.Y.; Weng, Y.T.; Tsai, Y.T. Antimelanogenic effect of c-phycocyanin through modulation of tyrosinase expression by upregulation of ERK and downregulation of p38 MAPK signaling pathways. *J. Biomed. Sci.* **2011**, *18*, 74. [CrossRef] [PubMed]

24. Park, S.Y.; Jin, M.L.; Kim, Y.H.; Kim, Y.; Lee, S.J. Aromatic-turmerone inhibits alpha-MSH and IBMX-induced melanogenesis by inactivating CREB and MITF signaling pathways. *Arch. Dermatol. Res.* **2011**, *303*, 737–744. [CrossRef] [PubMed]

25. Tu, C.X.; Lin, M.; Lu, S.S.; Qi, X.Y.; Zhang, R.X.; Zhang, Y.Y. Curcumin inhibits melanogenesis in human melanocytes. *Phytother. Res. PTR* **2012**, *26*, 174–179. [CrossRef]

26. Levy, C.; Khaled, M.; Fisher, D.E. MITF: Master regulator of melanocyte development and melanoma oncogene. *Trends Mol. Med.* **2006**, *12*, 406–414. [CrossRef]

27. Ye, Y.; Chu, J.H.; Wang, H.; Xu, H.; Chou, G.X.; Leung, A.K.; Fong, W.F.; Yu, Z.L. Involvement of p38 MAPK signaling pathway in the anti-melanogenic effect of San-bai-tang, a Chinese herbal formula, in B16 cells. *J. Ethnopharmacol.* **2010**, *132*, 533–535. [CrossRef]

28. Goto, H.; Sagitani, A.; Ashida, N.; Kato, S.; Hirota, T.; Shinoda, T.; Yamamoto, N. Anti-influenza virus effects of both live and non-live *Lactobacillus acidophilus* L-92 accompanied by the activation of innate immunity. *Br. J. Nutr.* **2013**, *110*, 1810–1818. [CrossRef]

29. Lee, E.S.; Song, E.J.; Nam, Y.D.; Lee, S.Y. Probiotics in human health and disease: From nutribiotics to pharmabiotics. *J. Microbiol.* **2018**, *56*, 773–782. [CrossRef]

30. Parisa, A.; Roya, G.; Mahdi, R.; Shabnam, R.; Maryam, E.; Malihe, T. Anti-cancer effects of Bifidobacterium species in colon cancer cells and a mouse model of carcinogenesis. *PLoS ONE* **2020**, *15*, e0232930. [CrossRef]

31. Saez-Lara, M.J.; Gomez-Llorente, C.; Plaza-Diaz, J.; Gil, A. The role of probiotic lactic acid bacteria and bifidobacteria in the prevention and treatment of inflammatory bowel disease and other related diseases: A systematic review of randomized human clinical trials. *BioMed Res. Int.* **2015**, *2015*, 505878. [CrossRef] [PubMed]

32. Sugawara, T.; Sawada, D.; Ishida, Y.; Aihara, K.; Aoki, Y.; Takehara, I.; Takano, K.; Fujiwara, S. Regulatory effect of paraprobiotic *Lactobacillus gasseri* CP2305 on gut environment and function. *Microb. Ecol. Health Dis.* **2016**, *27*, 30259.

33. Garcia, G.; Agosto, M.E.; Cavaglieri, L.; Dogi, C. Effect of fermented whey with a probiotic bacterium on gut immune system. *J. Dairy Res.* **2020**, *87*, 134–137. [CrossRef] [PubMed]

34. Imaizumi, K.; Hirata, K.; Zommara, M.; Sugano, M.; Suzuki, Y. Effects of cultured milk products by Lactobacillus and Bifidobacterium species on the secretion of bile acids in hepatocytes and in rats. *J. Nutr. Sci. Vitaminol.* **1992**, *38*, 343–351. [CrossRef] [PubMed]

35. Marshall, K. Therapeutic applications of whey protein. *Altern. Med. Rev.* **2004**, *9*, 136–156. [PubMed]

36. Yamamoto, N.; Akino, A.; Takano, T. Antihypertensive effect of the peptides derived from casein by an extracellular proteinase from *Lactobacillus helveticus* CP790. *J. Dairy Sci.* **1994**, *77*, 917–922. [CrossRef]

37. Chen, Y.J.; Chen, Y.Y.; Lin, Y.F.; Hu, H.Y.; Liao, H.F. Resveratrol inhibits alpha-melanocyte-stimulating hormone signaling, viability, and invasiveness in melanoma cells. *Evid. Based Complement. Altern. Med. eCAM* **2013**, *2013*, 632121. [CrossRef]

38. Zhou, J.; Ren, T.; Li, Y.; Cheng, A.; Xie, W.; Xu, L.; Peng, L.; Lin, J.; Lian, L.; Diao, Y.; et al. Oleoylethanolamide inhibits alpha-melanocyte stimulating hormone-stimulated melanogenesis via ERK, Akt and CREB signaling pathways in B16 melanoma cells. *Oncotarget* **2017**, *8*, 56868–56879. [CrossRef]

39. Pillaiyar, T.; Manickam, M.; Namasivayam, V. Skin whitening agents: Medicinal chemistry perspective of tyrosinase inhibitors. *J. Enzym. Inhibid. Med. Chem.* **2017**, *32*, 403–425. [CrossRef]

40. Bentley, N.J.; Eisen, T.; Goding, C.R. Melanocyte-specific expression of the human tyrosinase promoter: Activation by the microphthalmia gene product and role of the initiator. *Mol. Cell. Biol.* **1994**, *14*, 7996–8006. [CrossRef]

41. Busca, R.; Ballotti, R. Cyclic AMP a key messenger in the regulation of skin pigmentation. *Pigment Cell Res.* **2000**, *13*, 60–69. [CrossRef] [PubMed]

42. Tachibana, M. Cochlear melanocytes and MITF signaling. *J. Investig. Dermatol. Symp. Proc.* **2001**, *6*, 95–98. [CrossRef] [PubMed]

43. D'Mello, S.A.; Finlay, G.J.; Baguley, B.C.; Askarian-Amiri, M.E. Signaling pathways in melanogenesis. *Int. J. Mol. Sci.* **2016**, *17*, 1144. [CrossRef]

44. Gillbro, J.M.; Olsson, M.J. The melanogenesis and mechanisms of skin-lightening agents—Existing and new approaches. *Int. J. Cosmet. Sci.* **2011**, *33*, 210–221. [CrossRef]

45. Seo, G.Y.; Ha, Y.; Park, A.H.; Kwon, O.W.; Kim, Y.J. *Leathesia difformis* extract inhibits alpha-MSH-induced melanogenesis in B16F10 cells via down-regulation of CREB signaling pathway. *Int. J. Mol. Sci.* **2019**, *20*, 536. [CrossRef] [PubMed]

46. Aoki, H.; Moro, O. Involvement of microphthalmia-associated transcription factor (MITF) in expression of human melanocortin-1 receptor (MC1R). *Life Sci.* **2002**, *71*, 2171–2179. [CrossRef]

47. Brenner, M.; Hearing, V.J. The protective role of melanin against UV damage in human skin. *Photochem. Photobiol.* **2008**, *84*, 539–549. [CrossRef]

48. He, W.; Li, X.; Peng, Y.; He, X.; Pan, S. Anti-oxidant and anti-melanogenic properties of essential oil from peel of Pomelo cv. Guan Xi. *Molecules* **2019**, *24*, 242. [CrossRef]

49. Wang, J.J.; Wu, C.C.; Lee, C.L.; Hsieh, S.L.; Chen, J.B.; Lee, C.I. Antimelanogenic, antioxidant and antiproliferative effects of *Antrodia camphorata* fruiting bodies on B16-F0 melanoma cells. *PLoS ONE* **2017**, *12*, e0170924. [CrossRef]

50. Corrochano, A.R.; Buckin, V.; Kelly, P.M.; Giblin, L. Invited review: Whey proteins as antioxidants and promoters of cellular antioxidant pathways. *J. Dairy Sci.* **2018**, *101*, 4747–4761. [CrossRef]

51. Svanborg, S.; Johansen, A.G.; Abrahamsen, R.K.; Skeie, S.B. The composition and functional properties of whey protein concentrates produced from buttermilk are comparable with those of whey protein concentrates produced from skimmed milk. *J. Dairy Sci.* **2015**, *98*, 5829–5840. [CrossRef] [PubMed]

52. Wakai, T.; Shinoda, T.; Uchida, N.; Hattori, M.; Nakamura, Y.; Beresford, T.; Ross, R.P.; Yamamoto, N. Comparative analysis of proteolytic enzymes need for processing of antihypertensive peptides between *Lactobacillus helveticus* CM4 and DPC4571. *J. Biosci. Bioeng.* **2013**, *115*, 246–252. [CrossRef] [PubMed]

 *nutrients*

Article

A Glycosaminoglycan-Rich Fraction from Sea Cucumber *Isostichopus badionotus* Has Potent Anti-Inflammatory Properties In Vitro and In Vivo

Leticia Olivera-Castillo [1,*], George Grant [2,3], Nuvia Kantún-Moreno [1], Hirian A. Barrera-Pérez [4], Jorge Montero [1], Miguel A. Olvera-Novoa [1], Leydi M. Carrillo-Cocom [5], Juan J. Acevedo [6], Cesar Puerto-Castillo [1], Victor May Solís [1], Juan A. Pérez-Vega [1], Judit Gil-Zamorano [7], Enrique Hernández-Garibay [8], María A. Fernández-Herrera [9], Mayra Pérez-Tapia [10], Oscar Medina-Contreras [11], Jairo R. Villanueva-Toledo [12], Rossanna Rodriguez-Canul [1] and Alberto Dávalos [7,*]

[1] Departamento Recursos del Mar, Centro de Investigación y de Estudios Avanzados del IPN-Unidad Mérida, Antigua Carretera a Progreso Km. 6, Merida 97310, Yucatan, Mexico; nuviakm@gmail.com (N.K.-M.); lostrinosdeldiablo@gmail.com (J.M.); miguel.olvera@cinvestav.mx (M.A.O.-N.); cesarp@cinvestav.mx (C.P.-C.); victormay_29@hotmail.com (V.M.S.); juan.vega@cinvestav.mx (J.A.P.-V.); rossana.rodriguez@cinvestav.mx (R.R.-C.)

[2] School of Medicine, Medical Sciences and Nutrition, University of Aberdeen, Aberdeen AB25 2ZD, UK; g.grant@abdn.ac.uk

[3] Applied Science Research Foundation, A.C., Calle 26 No. 144 x 21 y 21A, Col. San Pedro Cholul, Merida 97138, Yucatan, Mexico

[4] Laboratorio de Anatomía Patológica (ANAPAT), Av. Yucatán 630, Fracc. Jardines de Mérida, Mérida 97135, Yucatan, Mexico; hbarrerap@gmail.com

[5] Facultad de Ingeniería Química, Universidad Autónoma de Yucatán, Calle 43 No. 613 x 90, Col. Inalámbrica, Mérida 97069, Yucatan, Mexico; leydi.carrillo@correo.uady.mx

[6] Departamento de Fisiología y Farmacología, Facultad de Medicina, Universidad Autónoma del Estado de Morelos, Calle Leñeros s/n, Col. Los Volcanes, Cuernavaca 62350, Morelos, Mexico; juan.acevedo@uaem.mx

[7] IMDEA Food Institute, CEI UAM+CSIC, Carretera de Cantoblanco 8, 28049 Madrid, Spain; judit.gil@imdea.org

[8] Centro Regional de Investigación Pesquera de Ensenada, Carretera Tijuana-Ensenada Km. 107.5, El Sauzal de Rodríguez, Ensenada 22860, Baja California, Mexico; enrique.garibay@inapesca.gob.mx

[9] Departamento de Física Aplicada, Centro de Investigación y de Estudios Avanzados del IPN-Unidad Mérida, Antigua Carretera a Progreso Km. 6, Mérida 97310, Yucatan, Mexico; marietafernandezh@gmail.com

[10] Instituto Politécnico Nacional, Unidad de Investigación, Desarrollo e Innovación Médica y Biotecnológica (UDIMEB), Prolongación de Carpio y Plan de Ayala s/n, Col. Santo Tomás, Del. Miguel Hidalgo, Ciudad de México 11340, Mexico; smpt.2011@hotmail.com

[11] Immunology and Proteomics Laboratory, Mexico Children's Hospital "Federico Gómez", Mexico City 06720, Mexico; omedina@himfg.edu.mx

[12] Cátedras CONACYT-Fundacion IMSS, A.C., CONACYT, Av. Insurgentes Sur 1582, Del. Benito Juárez, Col. Crédito Constructor, Ciudad de Mexico 03940, Mexico; jvillanuevat@conacyt.mx

* Correspondence: leticia.olivera@cinvestav.mx (L.O.-C.); alberto.davalos@imdea.org (A.D.); Tel.: +52-9999-42-94-00 (L.O.-C.); +34-912796985 (A.D.)

Received: 18 May 2020; Accepted: 3 June 2020; Published: 6 June 2020

Abstract: Sea cucumber body wall contains several naturally occurring bioactive components that possess health-promoting properties. *Isostichopus badionotus* from Yucatan, Mexico is heavily fished, but little is known about its bioactive constituents. We previously established that *I. badionotus* meal had potent anti-inflammatory properties in vivo. We have now screened some of its constituents for anti-inflammatory activity in vitro. Glycosaminoglycan and soluble protein preparations reduced 12-O-tetradecanoylphorbol-13-acetate (TPA)-induced inflammatory responses in HaCaT cells while an ethanol extract had a limited effect. The primary glycosaminoglycan (fucosylated chondroitin sulfate; FCS) was purified and tested for anti-inflammatory activity in vivo. FCS modulated the expression of critical genes, including NF-kB, TNFα, iNOS, and COX-2, and attenuated inflammation and tissue damage caused by TPA in a mouse ear inflammation model. It also mitigated colonic colitis caused in mice by dextran sodium sulfate. FCS from *I. badionotus* of the Yucatan Peninsula thus had strong anti-inflammatory properties in vivo.

Keywords: holothuroids; glycosaminoglycans; inflammation; ear-inflammation

1. Introduction

Sea cucumbers are used as food supplements or traditional medicines in many parts of the world. In Asia, they are ascribed a wide range of health-promoting actions, including antimicrobial, anti-thrombotic, anti-coagulant, neuroprotective, wound healing, antioxidant, anti-hypertensive, anti-tumor, anti-inflammatory, and immune-stimulating properties [1,2]. Harvested species belong to genera including *Apostichopus, Atinopyga, Cucumaria, Holothuria, Isostichopus, Parastichopus* and *Stichopus*. These are known to contain many potential therapeutic compounds, such as triterpene glycosides (saponins), fucosylated chondroitin sulfates, fucoidans, glycosaminoglycans, sulfated polysaccharides, sterols, phenolics, cerebrosides, lectins and proteins/peptides [1,2]. However, it remains unclear exactly which compounds or combinations thereof are required to reliably treat or manage specific diseases. Further, the profile and concentrations of the naturally occurring bioactive components in sea cucumbers is species-, locality-, and growth environment-dependent. Recent work has therefore concentrated on isolation and characterization of individual constituent bioactive compounds from sea cucumber and assessment of their specific health-modulating properties in vitro and in vivo [1–3].

Isostichopus badionotus is distributed along the coast of the Caribbean Sea and Western Atlantic Ocean. In recent years, the species has been intensively harvested in Mexican waters, particularly around the Yucatan Peninsula [4,5]. Remarkably, despite the high demand for this sea cucumber species, little is known of its nutritional, medicinal, or therapeutic properties in vivo. Two studies have established that *I. badionotus* from the Yucatan Peninsula has potential health-promoting effects. Diets containing the *I. badionotus* body wall were hypo-cholesterolaemic for rats [6]. In addition, they substantially altered gene expression in the rat intestine by down-regulating pro-inflammatory genes and up-regulating genes essential for gut barrier integrity and repair [7]. Further, studies in other in vitro and in vivo models confirmed the anti-inflammatory properties of *I. badionotus* body wall meal [7].

The present research aimed to identify anti-inflammatory factors from the body wall of *I. badionotus* from the Yucatan Peninsula and assess their efficacy in vivo.

2. Materials and Methods

2.1. Sea Cucumber Collection and Processing

Adult *Isostichopus badionotus* (Selenka, 1867) were collected from the seafloor off the coast of Sisal, Yucatan, Mexico (SAGARPA permit No. DGOPA/1009/210809/08761) and processed as before to obtain desalted, lyophilized sea cucumber meal [6].

2.2. Preliminary Extractions

A crude ethanol extract was obtained from desalted and lyophilized sea cucumber meal following Guo et al. [8], a soluble protein extract was prepared according to Ridzwan et al. [9] and a crude glycosaminoglycan preparation (GAGs) prepared using the basic technique of Vieira et al. [10]. The compositions of these preliminary extracts were monitored by thin-layer chromatography or polyacrylamide gel electrophoresis.

2.3. Large-Scale Extraction and Characterization of Glycosaminoglycans (GAGs)

GAGs were isolated essentially according to Vieira et al. [10]. Ten grams (10 g) desalted and lyophilized meal was added to 300 mL 0.1 M sodium acetate buffer (pH 6) containing 5 mM EDTA, 5 mM L-cysteine and 1 g papain. The mixture was incubated at 60 °C for 24 h, and the resulting enzymatic liquor centrifuged (2000× g 10 min at 10 °C). The supernatant was mixed with two volumes 95% ethanol, stored overnight at −10 °C and the precipitate recovered by centrifugation. This was resuspended and dialyzed against distilled water in a 12–14 kDa cut-off membrane (Spectra/Por). The preparation was further fractionated by preparative anion exchange chromatography on a QXL Hitrap column (GE Healthcare) fitted to an Akta Pryme plus (GE Healthcare). Elution was done with NaCl (up to 1.2 M made up in 20 mM Tris-HCl, pH 8.0) and absorbance (280 nm) and conductivity monitored.

2.4. Chemical Characterization of GAGs Fraction

Total carbohydrates were determined by UV spectrophotometry based on a method using sulfuric acid to form furfural derivatives [11]. Uronic acids were quantified using meta-hydroxy diphenyl with galacturonic acid as a standard [12]. Sulfates were measured using potassium sulfate as a standard [13]. Total proteins were determined with the bicinchoninic acid assay using a commercial kit (BCA Protein Assay kit, Thermo Scientific). Fucose was quantified using deoxy sugar and L-fucose as standards [14].

Gram-negative and Gram-positive bacteria, fungi and yeasts were screened for by culturing the GAGs preparations on appropriate agar plates or liquid media. Assessment of lipopolysaccharides (LPS) content was done using a commercial kit (Pierce LAL Chromogenic Endotoxin Quantitation, Thermo Scientific; detection limit <1 EU/mL) following manufacturer instructions. No viable microbes or LPS were detected in the GAGs samples.

2.5. Infrared Spectroscopy

Spectra were recorded on an Agilent Cary 630 FTIR spectrometer within the wavenumber range between 4000–600 cm^{-1}, using the Attenuated total reflection (ATR) technique. The standards employed for the IR analysis were L-fucose, D-glucuronic acid, D (+) galacturonic acid, and chondroitin sulfate from shark.

2.6. NMR Spectroscopy

The 1H NMR spectra of crude GAGs samples were measured at 25 °C and 600 MHz using an Agilent spectrometer. For the purified GAGs the spectra were measured at 400 MHz using a Bruker Advance 400 Ultrashield apparatus. Spectra processing was done with the MestReNova software (Mestrelab Research, Santiago de Compostela, Spain).

2.7. Polyacrylamide Gel Electrophoresis

PAGE was performed in a 6% polyacrylamide gel (37.5% acrylamide/1% bisacrylamide) using a mini-protean system (Bio-Rad, Mexico City, Mexico). Briefly, GAGs in Tris-HCL buffer (pH 6.8) were mixed with non-reducing sample buffer, boiled for 5 min, loaded onto the gel and electrophoresed at 100 V for 110 min. The gel was stained with 0.5% Alcian Blue solution in 3% acetic acid/25% isopropanol for two hours and de-stained with a 10% acetic acid/40% ethanol solution overnight. Finally, the gel was digitized. A broad range molecular weight standard (Bio-Rad, Mexico City, Mexico) was used.

2.8. Action of GAGs on TPA-Stimulated HaCaT (Human Keratinocyte) Cells and Mouse Splenocytes In Vitro

Spontaneously immortalized human keratinocytes (HaCaT) were cultured in Dulbecco's Modified Eagle Medium (DMEM) supplemented with 10% fetal bovine serum (FBS). 3×10^5 HaCat cells were seeded in 96-well plates (Corning Costar Sigma Mexico) and cultured overnight until 80% confluence. Media was replaced and cells were stimulated with 3.2 mg/mL GAGs sample in the presence of 2.5 ng/mL TPA [15]. After 24 h media were collected, and the cells were re-stimulated with 2.5 ng/mL 12-O-tetradecanoylphorbol-13-acetate (TPA) and 5 µg/mL brefeldin for 3 h. The cells were harvested with 0.25% trypsin/1 mM EDTA and stained for flow cytometry analysis.

Isolated HaCaT/splenocyte cells were resuspended in phosphate buffered saline (PBS) containing 5% FBS. Unspecific staining was blocked with PBS-10% FBS for 15 min at 4 °C. Cells were permeabilized using fixation/permeabilization buffer (eBioscience) for 15 min. Samples were then stained at 4 °C for 30 min with AlexaFluor 647-labeled IFNγ antibodies, then washed two times in permeabilization/wash buffer. An amount of 3×10^5 events were acquired on a CytoFlex (Beckman Coulter, Brea, CA, USA) flow cytometer, and analyzed using the CytExpert software (Beckman Coulter) [16].

2.9. Action of GAGs on the Viability of Fibroblasts In Vitro

A thiazolyl blue tetrazolium blue (MTT) assay with human fibroblast cell line (hFB) was used to evaluate the viability of GAGs activity [17]. Cells were cultivated routinely in DMEM/F-12 medium without phenol red and supplemented with 10% FBS at 37 °C in a 5% CO^2 and humidified atmosphere. Cells were seeded at 2×10^4 cells/well in 100 μL medium into 96-well microplates and incubated for 24 h. Fresh culture medium (100 μL) containing GAGs was then added to the wells to final concentrations of 0.039, 0.078, 0.156, 0.313, 0.625 and 1.25 mg/mL. Cells exposed only to culture medium were used as a control, and cells exposed only to medium without cells were used as a background. Cells with dimethyl sulfoxide (DMSO) were used as a toxicity control, and cells with recombinant human fibroblast growth factor (rFGF) were used as a proliferation control. After 48 h incubation, 100 μL MTT (5 mg/mL) were added to each well and left for 4 h. The medium was then removed and 200 μL DMSO was added to resuspend the formazan crystals. Optical density (OD) was measured at 570 nm by a spectrophotometric plate reader (Multiskan FC Thermo Scientific). The percentage of relative cell viability was calculated as:

$$[(\text{OD treated} - \text{OD background})/(\text{OD control} - \text{OD background})] \times 100\%$$

Data were expressed as the mean ± standard deviation and analysed with a one-way analysis of variance (ANOVA) test followed by Tukey's post hoc test ($p < 0.05$). Analysis was run on the GraphPad Prism 7 statistical program.

2.10. Experimental Animals

Male CD1 and C57Bl/6 mice were obtained from the Centre for Research and Advanced Studies of the National Polytechnic Institute (CIVNESTAV-IPN) at Zacatenco, Mexico. Animal experiments were performed at CINVESTAV-IPN Mérida. Animals were kept under standard conditions (12 h light/dark cycle, 23 ± 2 °C, 65% humidity). Food and water were freely available. Animals were acclimatized for seven days prior to the experiment. All experimental protocols were approved by the Institutional Animal Care and Use Committee of CINVESTAV-IPN (No. 0126-15) and complied with the applicable Mexican Official Norm (NOM-062-ZOO-1999), "Technical Specifications for the Care and Use of Laboratory Animals", as well as all applicable federal and institutional regulations.

2.11. Mouse Ear Anti-Inflammatory Activity

Anti-inflammatory activity was evaluated using the mouse ear model, as described previously [7,18]. Under anesthesia (pentobarbital 25 mg/kg animal weight), acute inflammation was induced by topical application of 12-O-tetradecanoylphorbol-13-acetate (TPA; phorbol myristate acetate; Sigma Cat. P8139; 2.5 μg in 10 μL acetone) to both surfaces of the right ear of CD1 mice. Ten microliters (10 μL) vehicle were topically applied to each surface of the left ear. GAGs extract (125 μg/20 μL water/per ear, each surface) or the anti-inflammatory dexamethasone (DXA) were applied (250 μg/10 μL water/per ear, each surface) 15 min after treatment with TPA. Six hours later, the mice were re-anesthetized, and an electronic micrometer used to assess ear edema and thickness. The animals were then euthanized by cervical dislocation, and each ear removed and weighed (total ear weight). A sterile biopsy punch (Integra Miltex®, Integra LifeSciences Corp, PA.USA.) was used to remove two representative samples (5 mm diameter discs) from each ear, and the discs weighed. Inflammation was quantified as the difference between the average weight of the right ear discs (TPA alone or TPA+dexamethasone or TPA+extract) and that of the left ear discs (vehicle: acetone). Samples were taken from the ear discs for analysis of gene expression (immersed in 400 μL RNAlater [Ambion] and stored at −80 °C) and for histological evaluation (fixed in 10% formalin).

2.12. Histological Analysis of Ear Samples

Fixed samples were mounted in paraffin blocks (Richard-Allan Scientific Paraffin Type 6®, ThermoFisher Scientific, Mexico City, Mexico, cut into sections (2 μm thickness) with a ThermoScientific Microm HM 325® rotary microtome (Mexico City, Mexico) and processed using an AutotechniconDuo® 2A System(Technicon Instruments Corp, New York, NY, USA). Sections were stained simultaneously with haematoxylin-eosin (H&E), examined with a conventional optical microscope and images taken with a digital camera (Evolution™ LC Color; Olympus, Mexico City, Mexico). A representative area was selected for qualitative light microscopic analysis. Histology scores for ear samples were generated by two independent pathologists and averaged for analysis. Briefly, samples were evaluated for the presence of neutrophil infiltration and edema, and each factor deemed to be either null, slight, moderate or intense [6,7]. Numerical values were assigned to each intensity: null (0); slight (1); moderate (2); and intense (3). A sample's total score was calculated by adding the values assigned each factor.

2.13. Dextran Sodium Sulfate (DSS)-Colitis

Sixteen C57Bl/6 mice (8 weeks old, 25 g) were given free access to water containing Dextran Sodium Sulfate (MP Biomedicals, USA, DSS, 36–50 kDa) [25 g/liter] for 4 days and water for a further 2 days. In addition, half of the mice were dosed orally with GAGs solution (2 mg/mouse/d; ~80 mg/kg body weight (BW)/d) on days −1, 0, and +1 to +5 while the remainder were daily administered a vehicle substance. The mice were euthanized by cervical dislocation on day 6 and dissected. The small intestine and colon were removed, their length measured, and representative samples of colon collected for analysis. Untreated control mice had free access to water throughout the study.

2.14. Gene Expression

RNA was isolated from tissue using the Animal Tissue RNA Purification Kit (Norgen), following manufacturer instructions. Genomic DNA (gDNA) was removed using the TURBO DNA-free™ Kit (Ambion). cDNA was obtained using the RevertAid H Minus First Strand cDNA kit (Thermo Scientific), according to manufacturer instructions. Single-strand DNA (ssDNA) was diluted at a 1:20 ratio for further use in real-time PCR (qRT-PCR). Quantitative real-time PCR (qRT-PCR) assays were run using specific oligos in a Rotor Gene Q (2-plex) real-time PCR detection system with QuantiNova SYBR Green PCR master mix (Qiagen). Reactions (including non-template as negative controls) were performed in quadruplicate. Thermo-cycling conditions were 5 min at 95 °C, followed by 40 cycles of 15 s at 94 °C, and 45 s at 62 °C (annealing-extension). β-actin (ACTB) [ear assay] and elongation factor 2 (EEF2) [DSS-colitis] were used as housekeeping genes. Relative expression was calculated using the $2^{-\Delta\Delta Ct}$ method, as previously described [19], and results expressed relative to the controls.

2.15. Statistical Analysis

Statistical analysis was applied using generalized linear modelling (GLM) to estimate the effect of the factor levels (i.e., ear-weight results), as well as the variable of gene expression (fold change) for genes. Over-dispersion was detected (from residual Poisson regression model) and corrected using Poisson GLM. The standard errors were multiplied by the square root of the dispersion parameter using a quasi-Poisson GLM model. The variance; $Var [Yi] = \varphi\mu$, where μ is the mean and φ the dispersion parameter, was calculated with the GLM model. Significant difference was set at p-value < 0.05. A customized R (www.r-project.org) function and GraphPad Prism were used for statistical analyses.

3. Results

3.1. Extracts from the Body-Wall of I. badionotus Inhibit TPA-Induced Proliferation of Keratinocyte and Splenocyte Cells in Culture

Previous data from our group indicated that sea cucumber *I. badionotus* meal had potent anti-inflammatory activity in vitro and in vivo [7]. To ascertain the general nature of the naturally occurring bioactive components in *I. badionotus* responsible for this ameliorative action, an initial study was conducted in which three extracts of desalted and lyophilized sea cucumber meal (an ethanolic extract [8], a soluble protein extract [9] and a crude glycosaminoglycan (GAGs) preparation [10]) were prepared and assayed for their effects on TPA-stimulated human keratinocyte (HaCaT) cells and mouse splenocytes in vitro.

The GAGs and soluble protein fractions prepared from *I. badionotus* body wall inhibited TPA-mediated proliferation of HaCat cells (Figure 1a) and of mouse splenocytes (Supplemental Figure S1a). In contrast, while the ethanol extract, expected to contain saponins, partially reduced TPA-induced proliferation of HaCaT cells (Figure 1a), it had limited activity against TPA-stimulated splenocytes (Supplemental Figure S1a). Analysis of the preliminary extracts revealed that the GAGs fraction was comprised of primarily one major component whereas both the protein and ethanol extracts were complex mixtures (data not shown). The present studies therefore concentrated on the GAGs fraction.

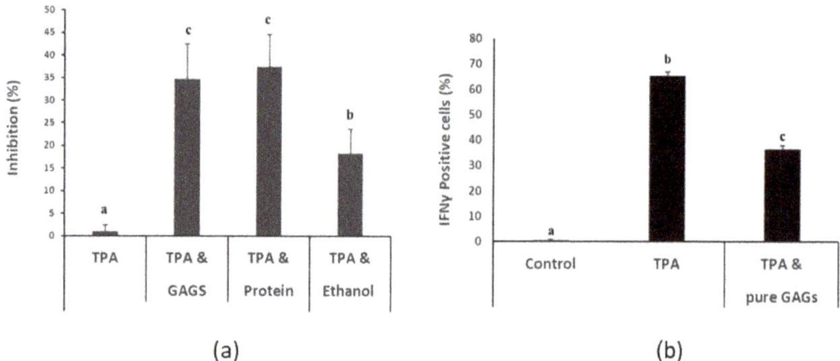

Figure 1. Extracts (crude GAGs, soluble proteins, ethanol soluble) from the body wall of *I. badionotus* inhibit 12-O-tetradecanoylphorbol-13-acetate (TPA)-induced proliferation of HaCat [human keratinocyte] cells in culture (**a**). Purified GAGs (3 μg/mL) limit TPA-mediated production of Interferon gamma (IFNγ)-positive HaCaT cells (**b**). N ≥ 4 per group and values with distinct superscripts differ significantly from each other ($p \leq 0.05$).

3.2. Large-Scale Preparation and Characterization of GAGs

The preliminary GAGs fraction prepared using the basic method of Vieira et al. [10] was comprised mainly of fucosylated chondroitin sulfate (FCS) but also contained small amounts of impurities, including fucoidan. An additional ion-exchange chromatography was applied to the isolation procedure. This method produced a glycosaminoglycan preparation comprised of fucose, N-acetyl-galactosamine, uronic acid and sulfates (Table 1). Polyacrylamide gel electrophoresis (Figure 2a), 1H NMR (Figure 2b; Supplemental Figure S2a) and infrared spectral analysis (Figure 2c,d, Supplemental Figure S2b) indicated the GAGS preparation to be fucosylated chondroitin sulfate (FCS). The sample appeared to be free of fucoidan and other contaminants. However, their presence in trace amounts in the GAGs preparation cannot be completely discounted.

Table 1. Chemical composition (*w/w*, %) of GAGs extracted from *Isostichopus badionotus* from the Yucatan Peninsula compared to a non-fucosylated chondroitin sulfate from shark cartilage (CSS).

	GAGs	CSS
Total CHO	24.77 ± 2.5	25.36 ± 0.06
Fucose	6.40 ± 0.54	ND
Uronic acid	8.55 ± 0.48	20.6 ± 0.04
Sulfates	11.12 ± 0.60	15.06 ± 0.001
Soluble proteins	4.15 ± 0.01	5.01 ± 0.001

Results are the average of three independent determinations ± SD. CHO = carbohydrates; ND = not detected.

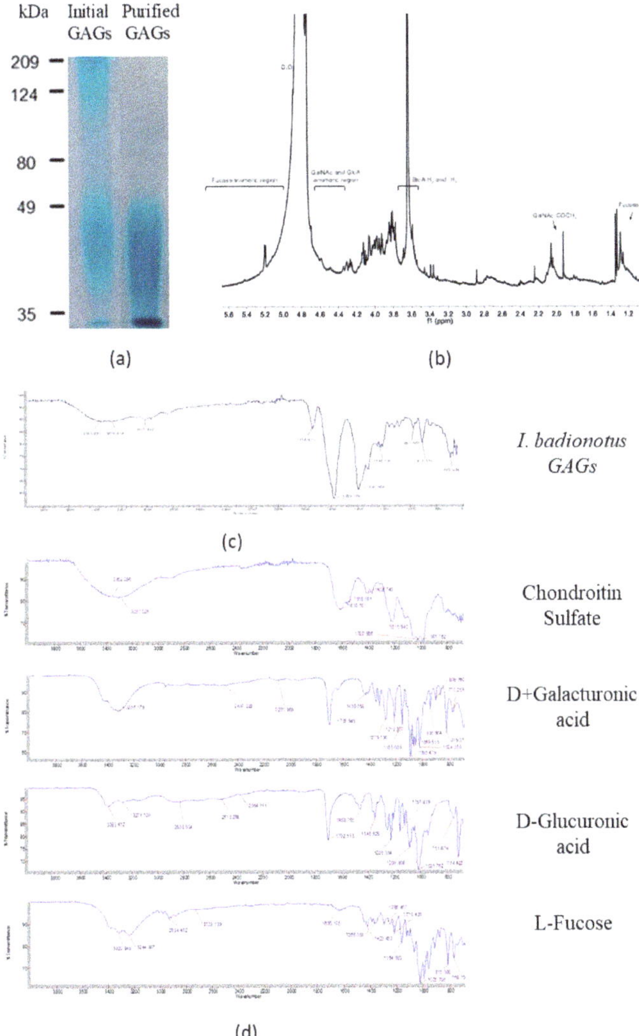

Figure 2. GAGs preparation isolated from *I. badionotus* body wall. (**a**) Alcian Blue-stained polyacrylamide gel electrophoretogram of initial and purified GAGs from sea cucumber (*I. badionotus*). (**b**) 1H NMR spectra at 600 MHz of fucosylated chondroitin sulfate of sea cucumber (*I. badionatus*). (**c**) Infrared spectra of GAGs from sea cucumber (*I. badionotus*). (**d**) Overlay comparing GAGs to standards.

The bioactivity in vitro of the purified FCS was confirmed since it was found to inhibit a TPA-mediated expansion in Interferon gamma (IFNγ)-positive HaCaT cells at levels of around 3 µg

FCS/mL (Figure 1b). At much higher concentrations, FCS was cytotoxic (IC50, 80 µg/mL) in a human breast-derived fibroblast cell line (Supplemental Figure S1b).

3.3. FCS Exerts Anti-Inflammatory Activity In Vivo

Given their potent anti-inflammatory activity against keratinocytes in vitro we evaluated the biological activity of the purified FCS in a mouse ear inflammation model. As reported previously [7], topical administration of 12-O-tetradecanoylphorbol-13-acetate [TPA] to mouse ear triggered inflammatory cell infiltration, edema and major histological disruption within six hours (Figure 3a–c). Co-treatment of the ear with FCS attenuated these inflammatory responses, as did dexamethasone (Figure 3a–c). In contrast, chondroitin sulfate from shark cartilage had little or no effect on TPA-induced inflammation (Figure 3d).

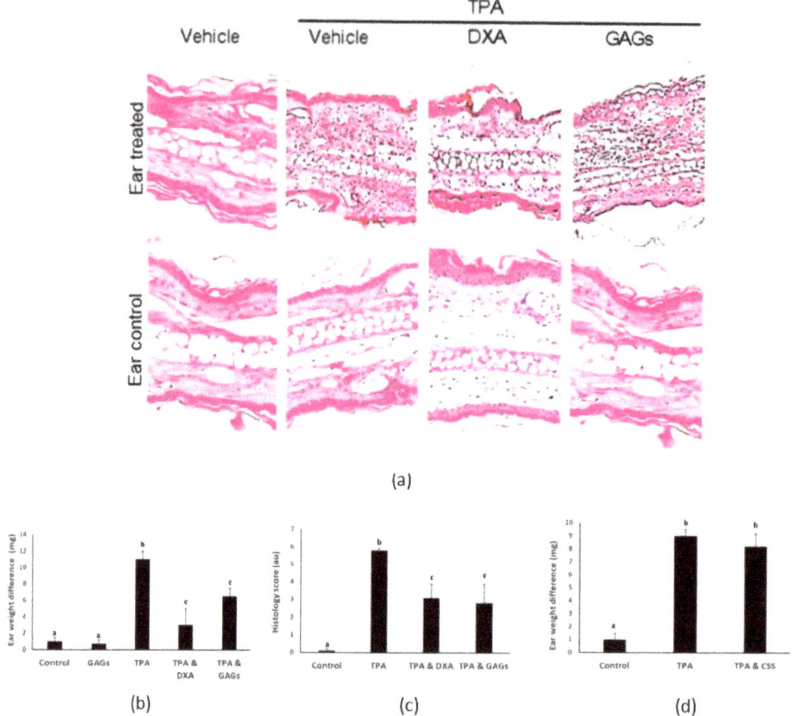

Figure 3. GAGs from sea cucumber (*I. badionotus*) body wall exert anti-inflammatory effects in a mouse ear inflammation model. Mouse ears exposed to vehicle (control, left ear), or inflammatory agent TPA alone, or TPA followed by dexamethasone (DXA) or GAGs (right ear). (**a**) Representative photomicrographs of transverse section of mouse ears sensitized with TPA or TPA and test substance, stained with haematoxylin-eosin (magnification 10X). Images are representative of five mice with similar results. (**b**) Differences in ear weights (right–left) after treatment with TPA or TPA and test substance; $n \geq 5$ per group. (**c**) Histological score for ear micrographs in (**a**); values are the mean ± SEM, $n = 3$ per group. (**d**) Differences in ear weights (right–left) after treatment with TPA or TPA and shark cartilage (CSS); $n \geq 3$ per group. Values with distinct superscripts differ significantly from each other ($p \leq 0.05$).

Expression of the TNFα, IL-6, NF-kB, iNOS, IL-10, IL-11, COX-2 and STAT3 genes in mouse ear tissue increased following application of TPA (Figure 4). Co-administration of FCS suppressed these TPA-mediated responses, except for IL-6, which was increased further, and IL-10, which was unaffected (Figure 4). By contrast, TPA-mediated expression of all these genes was inhibited by co-treatment with dexamethasone (Figure 4). FCS alone had little or no effect on expression of inflammation-associated genes, except for NF-kB, which was slightly up-regulated and IL-6, which was marginally suppressed (Figure 4).

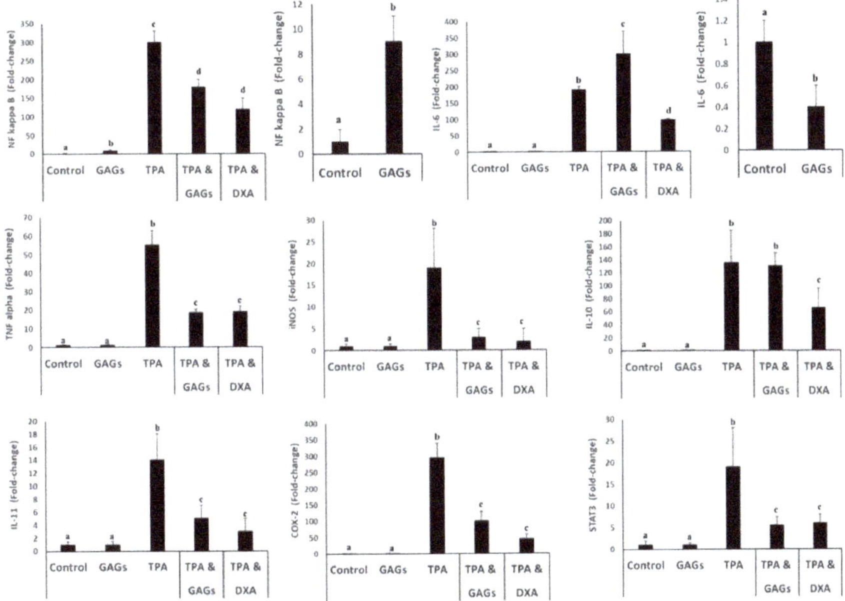

Figure 4. GAGs from sea cucumber (*I. badionotus*) target principal inflammation-related genes. Mouse ears were exposed to vehicle alone (left ear), GAGs, inflammatory agent TPA followed by GAGs, or TPA followed by dexamethasone (DXA) as an anti-inflammatory control (right ear). Genes NF kappa B (NF-kB), IL-6, TNF alpha (TNFα), iNOS, IL-10, IL-11, COX-2 and STAT3 were evaluated by qRT-PCR. Values (fold change) are the mean ± SD. Values with distinct superscripts differ significantly from each other ($p < 0.05$).

3.4. FCS from I. badionotus with Sulfate Levels of 4.5% or Above Have Similar Anti-Inflammatory Properties

During our studies we noted that sulfation of the purified FCS from *I. badionotus* steadily reduced when the preparation was repeatedly reconstituted and lyophilized (Figure 5a). Freshly prepared GAGs were tested alongside a sulfate-depleted sample in the mouse ear inflammation assay. Surprisingly, we found that the sulfate-depleted GAGs samples exhibited the same potent anti-inflammatory activity as did the freshly prepared ones (Figure 5b). This may indicate that sulfate levels of approximately 4.5% (*w/w*) may be adequate to maximize the anti-inflammatory actions of FCS from *I. badionotus*.

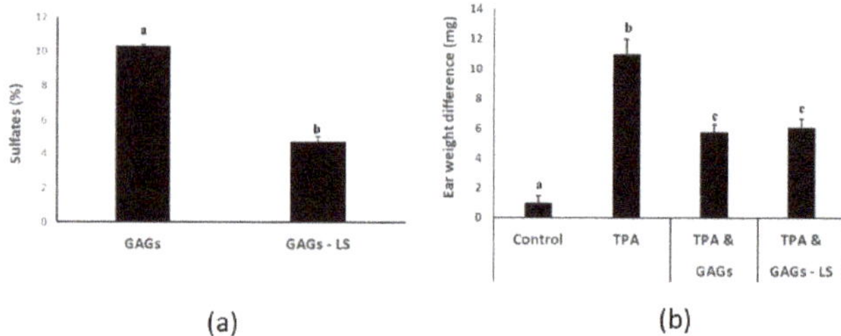

(a)   (b)

Figure 5. GAGs from sea cucumber (*I. badionotus*) with sulfate levels of 4.5% or above have similar anti-inflammatory properties. (**a**) The sulfate content of freshly-isolated GAGs used in analyses was quantified. The GAGs—LS were derived from the same sample but were reconstituted and lyophilized at least 3 times prior to sulfate analysis. (**b**) Differences in ear weights (right-left) in mouse ears exposed to vehicle alone (left ear), or inflammatory agent TPA followed by GAGs or GAGs—LS [low sulfate] (right ear). For each figure, $n \geq 4$ per group and values with distinct superscripts differ significantly from each other ($p \leq 0.05$).

3.5. Orally-Administered FCS from I. badionotus Ameliorate DSS-Colitis in Mice

Having established that FCS was anti-inflammatory in the ear inflammation model, we then did a pilot study evaluating its ability to protect against acute intestinal inflammation in a DSS-colitis mouse model [20,21]. As expected, mice administered 2.5% DSS in their drinking water for four days followed by water alone for two days lost body weight (Figure 6a), the expression of TNFα gene in their colonic tissue was greatly up-regulated (Figure 6b) and their colons were significantly shortened (Figure 6c, Supplemental Figure S2). The actions of DSS on the intestine were confined to the lower bowel, meaning the small intestine was unaffected by this treatment (Figure 6d).

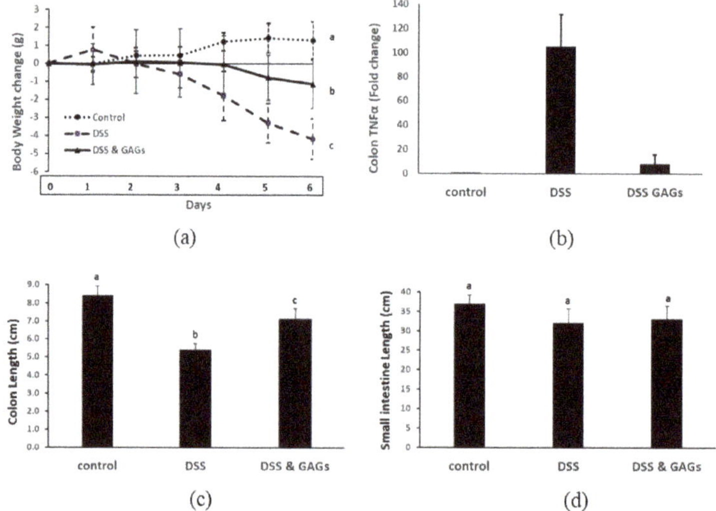

(a)   (b)

(c)   (d)

Figure 6. Orally-administered GAGs from sea cucumber (*I. badionotus*) body wall exert anti-inflammatory effects on Dextran sodium Sulfate (DSS)-colitis in mice. DSS (25 g/L in drinking water) was offered mice for four days, followed by water alone for two days. A proportion of the mice were dosed daily with GAGS (80 mg/kg body weight (BW)/d) for one day prior to and during the experimental period. (**a**) Weight change in mice dosed with DSS ± co-treatment with GAGs. (**b**) Colonic TNFα gene expression in mice dosed with DSS ± co-treatment with GAGs. (**c**) Colon lengths in mice dosed with DSS ± co-treatment with GAGs. (**d**) Small intestine lengths in mice dosed with DSS ± co-treatment with GAGs. For each figure, $n \geq 4$ per group and values with distinct superscripts differ significantly from each other ($p \leq 0.05$).

Oral administration of FCS immediately prior to and during the experimental period counteracted the effects of DSS on mice (Figure 6a–c; Supplemental Figure S3). In other words, co-treatment with FCS attenuated the body weight loss, expression of colonic TNFα gene and colon shortening caused by DSS.

4. Discussion

Previous studies established that lyophilized sea cucumber (*Isostichopus badionotus*) from the Yucatan Peninsula have potent anti-inflammatory actions when tested in various animal models [7]. We have now shown that these protective effects are multi-factorial, and are largely mediated by fucosylated chondroitin sulfate (FCS), a principal component of the body wall. This natural occurring bioactive component limited the responses of HaCat (keratinocyte) cells to 12-O-tetradecanoylphorbol-13-acetate (TPA), modulated expression of critical genes and attenuated inflammation and tissue damage caused by TPA in mouse ear and mitigated colonic colitis caused by dextran sodium sulfate (DSS) in a pilot study in mice.

FCS from *I. badionotus* ameliorates hyperlipidaemia and metabolic syndromes in mice [22,23]. Given our original finding that *I. badionotus* meal was both hypolipidemic and anti-inflammatory in animal models [6,7], the present observations that *I. badionotus* FCS suppressed skin and intestinal inflammation is not inconsistent with the findings of Li et al. [22,23], and indicates that *I. badionotus* FCS has a broad spectrum of health-promoting actions in vivo.

Fucoidans from sea cucumber, including those of *I. badionotus*, are also known to have anti-inflammatory, hypolipidemic and health-promoting properties in vivo [24–26]. As we found in our preliminary studies, small amounts of fucoidan can co-extract with glycosaminoglycans in the Vieira et al. [10] isolation procedure. However, introduction of an ion exchange chromatography step in the purification process significantly reduced the residual amounts of fucoidan in the final FCS preparation. The fucoidan doses necessary to ameliorate experimental hyperlipidaemia and metabolic syndrome in vivo are similar [24–26] to those required for FCS to treat the same experimental disorders [22,23]. Given that our FCS preparation contained at most only trace amounts of fucoidan, it is unlikely that fucoidan made a significant contribution to the potent anti-inflammatory actions of our *I. badionotus* FCS preparation.

The cytotoxicity of *I. badionotus* FCS for cells in culture was higher (IC50, 80 μg/mL) than previously reported [27]. However, only low levels (3 μg/mL) of FCS were needed to suppress TPA-induced responses in HaCaT cells, and mouse splenocytes in vitro; indeed these were at least thirteen-to fifteen-fold less than needed to cause any significant adverse effects on human fibroblast viability in vitro. Therefore, cytotoxicity did not influence the anti-inflammatory properties of *I. badionotus* FCS.

Sulfation levels, sulfate pattern (2,4-O-di-sulphation) and alpha-L-fucosyl branched units are considered essential factors in FCS bioactivity, particularly its anticoagulative and anti-thrombotic activities [28–31]. However, we found that sulfate concentrations over 4.5% were of no apparent advantage for the anti-inflammatory actions of *I. badionotus* FCS. This level of sulfation may therefore be the threshold required for maximal bioactivity.

The composition, structure and bioactivity of sea cucumber FCS vary considerably between species and even within the same species depending on the exact location and environment of the capture site [27,28,32]. It is, therefore, essential to establish that FCS derived from a new source has the required therapeutic properties for practical clinical use. The results of this study demonstrate that FCS from *I. badionotus* captured off the Yucatan Peninsula has excellent potential for use in clinical treatment or management of inflammatory disorders.

TPA activates the nuclear factor-B molecular signaling pathway. This, in turn, initiates the expression of pro-inflammatory genes, such as TNFα and COX-2, as well as a range of cell survival factors essential for the preservation of cellular integrity and subsequent down-regulation of acute inflammatory responses [28,33–35]. FCS limited inflammation and tissue damage in TPA-treated mouse ears by down-regulation of NF-kB and down-stream genes such as COX-2 and TNFα. *I. badionotus* FCS's mode of action is unknown, but reports for some sea cucumbers indicate that their constituent glycosaminoglycans can act on the NF-kB-pathway by blocking the translocation of active NF-kB [RelA(p65)/p50] from the cytoplasm to the nucleus, thereby preventing activation of downstream genes [36,37]. This action would be consistent with the gene changes observed in TPA/FCS-treated ear samples except for the slight upregulation of NF-kB caused by FCS alone. However, if FCS prevented even basal levels of active NF-kB from reaching the nucleus, the small FCS-linked increase in NF-kB expression may have been a compensatory response aimed at overcoming the loss of molecular signaling. IL-6 expression is highly dependent on NF-kB [38] so the reduced expression of IL-6 in FCS-treated ear may also be a result of the failure of active NF-kB to reach or accumulate in the nucleus.

DSS activates NF-kB in colonic epithelium and up-regulates an array of pro-inflammatory genes. These actions cause chronic colonic inflammation, tissue disruption, a characteristic shortening of the colon and rapid loss of body weight [20,39]. Oral administration of *I. badionotus* FCS (80 mg/kg BW/d) attenuated the harmful effects of DSS on the colon and limited the loss of body weight caused by DSS. As in the ear inflammation assay, this amelioration of DSS colitis may have been due to the direct inhibitory effects of FCS upon NF-kB in epithelial cells leading to a diminution in colonic inflammation.

FCS may also have had indirect protective actions. DSS disrupts the intestine microbiota, leading to loss of protective and short-chain fatty acid (SCFA)-producing bacteria and a concomitant expansion in the populations of opportunistic pathobionts [40–42]. In previous studies, FCS partially ameliorated intestinal dysbiosis associated with experimental metabolic syndromes [22,41]. *I. badionotus* FCS could therefore have similar effects on DSS-induced dysbiosis. It may reduce the levels of opportunistic pathogens and promote an expansion in the numbers of SCFA-producers or protective bacteria which produce compounds that inhibit activation, release, or translocation of NF-kB [20,39,41,43]. These actions would limit the severity of epithelial inflammation and gut damage caused by DSS.

I. badionotus body wall has potent anti-inflammatory actions in vivo [6,7]. Here we show that these properties are due, at least in part, to FCS, one of its major extractable components. However, in agreement with reports for other species [1,2], our in vitro data also indicates that other body wall constituents have anti-inflammatory properties. It remains to be established whether the naturally occurring bioactive components in *I. badionotus* body wall act additively or synergistically to provide its overall anti-inflammatory effect.

5. Conclusions

I. badionotus body wall contains several components with anti-inflammatory properties. They may act additively or synergistically to give maximal activity. Nonetheless, the present pre-clinical studies have established that *I. badionotus* FCS has pharmacologically promising anti-inflammatory activity on its own. Given the immense need for novel anti-inflammatory products of a known and predictable mode of action, robust clinical trials are needed to establish if *I. badionotus* FCS has similar demonstrable and repeatable health benefits for humans.

Supplementary Materials: The following are available online at http://www.mdpi.com/2072-6643/12/6/1698/s1, Figure S1: (a) Crude GAGs from the body wall of *I. badionotus* inhibit 12-O-tetradecanoylphorbol-13-acetate (TPA)-induced proliferation of mouse splenocytes. (b) High concentrations of purified GAGs reduce the viability of a human breast-derived fibroblast cell line. N ≥ 4 per group and values with distinct superscripts differ significantly from each other ($p \leq 0.05$). Figure S2: (a) 1H NMR spectra at 600 MHz of fucosylated chondroitin sulfate of sea cucumber (*I. badionatus*). (b) Infrared spectra of GAGs from sea cucumber (*I. badionotus*). Figure S3: Representative colons from mice treated with DSS and *I. badionotus* GAGs (a), or DSS (b) and untreated controls (c).

Author Contributions: Conceptualization, L.O.-C.; G.G. and A.D.; methodology, L.O.-C., G.G., A.D., N.K.-M., O.M.-C.; validation, O.M.-C., L.M.C.-C., M.P.-T., J.J.A. and E.H.-G.; formal analysis, L.O.-C., N.K.-M., V.M.-S., H.A.B.-P., M.A.F.-H., L.M.C.-C., J.M., O.M.-C., C.P.-C., J.A.P.-V., J.R.V.-T.; investigation, L.O.-C., G.G., A.D.; resources, L.O.-C., R.R.-C., M.A.O.-N., M.P.-T.; data curation; L.O.-C., N.K.-M., V.M.S., G.G., A.D., H.A.B.-P., O.M.-C., M.A.F.-H., J.G.-Z.; original draft preparation, L.O.-C., G.G., A.D.; writing—review and editing, G.G., A.D., L.O.-C., M.A.O.-N.; visualization, L.O.-C., G.G., A.D.; supervision, L.O.-C., G.G., A.D.; project administration, R.R.-C.; funding acquisition, L.O.-C., R.R.-C., A.D., G.G. All authors have read and agreed to the published version of the manuscript.

Funding: This study was supported by grants and funds: CONACyT fondos mixtos Yucatán (M0023-A1) "El pepino de mar como un alimento funcional: obtención de sus principios bióactivos, caracterización biológica, efectos sobre el metabolismo y sistema inmune utilizando un modelo murino"; CONACyT basic science grant (I0017) "Actividad antiinflamatoria y cicatrizante del pepino de mar (*Isostichopus badionotus*) en un modelo murino: caracterización de la actividad farmacológica y los mecanismos moleculares involucrados.

Conflicts of Interest: The authors declare no conflicts of interest. The funders had no role in the design of the study, in the collection, analyses, or interpretation of data; in the writing of the manuscript, or in the decision to publish the results.

References

1.   Khotimchenko, Y. Pharmacological Potential of Sea Cucumbers. *Int. J. Mol. Sci.* **2018**, *19*, 1342. [CrossRef]

2.   Pangestuti, R.; Arifin, Z. Medicinal and health benefit effects of functional sea cucumbers. *J. Tradit. Complement. Med.* **2018**, *8*, 341–351. [CrossRef]

3.   Zhu, Z.; Zhu, B.; Sun, Y.; Ai, C.; Wu, S.; Wang, L.; Song, S.; Liu, X. Sulfated polysaccharide from sea cucumber modulates the gut microbiota and its metabolites in normal mice. *Int. J. Boil. Macromol.* **2018**, *120*, 502–512. [CrossRef]

4.   Gamboa-Álvarez, M.Á.; López-Rocha, J.A.; Poot-López, G.R.; Aguilar-Perera, A.; Villegas-Hernández, H. Rise and decline of the sea cucumber fishery in Campeche Bank, Mexico. *Ocean Coast. Manag.* **2020**, *184*, 105011. [CrossRef]

5.   López-Rocha, J.A.; Velázquez-Abunader, I. Fast decline of the sea cucumber Isostichopus badionotus as a consequence of high exploitation in Yucatan, Mexico. *Reg. Stud. Mar. Sci.* **2019**, *27*, 100547. [CrossRef]

6.   Olivera-Castillo, L.; Grant, G.; Kantún-Moreno, N.; Acevedo, J.J.; Puc-Sosa, M.; Montero, J.; Olvera-Novoa, M.A.; Negrete-León, E.; Santa-Olalla, J.; Ceballos-Zapata, J.; et al. Sea cucumber (Isostichopus badionotus) body-wall preparations exert anti-inflammatory activity in vivo. *PharmaNutrition* **2018**, *6*, 74–80. [CrossRef]

7.   Olivera-Castillo, L.; Davalos, A.; Grant, G.; Valadez-González, N.; Montero, J.; Barrera-Pérez, H.A.M.; Chi, Y.A.C.; Olvera-Novoa, M.A.; Ceja-Moreno, V.; Acereto-Escoffié, P.; et al. Diets Containing Sea Cucumber (Isostichopus badionotus) Meals Are Hypocholesterolemic in Young Rats. *PLoS ONE* **2013**, *8*, e79446. [CrossRef] [PubMed]

8.   Guo, L.; Gao, Z.; Zhang, L.; Guo, F.; Chen, Y.; Li, Y.; Huang, C. Saponin-enriched sea cucumber extracts exhibit an antiobesity effect through inhibition of pancreatic lipase activity and upregulation of LXR-β signaling. *Pharm. Boil.* **2015**, *54*, 1–14. [CrossRef] [PubMed]

9.   Ridzwan, B.H.; Farah Hanis, Z.; Daud, J.M.; Althunibat, O.Y. Protein profiles of three species of Malaysian sea cucumber; Holothuria edulis Lesson, H. scabra Jaeger and Stichopus horrens Selenka. *Eur. J. Sci. Res.* **2012**, *75*, 255–261.

10. Vieira, R.P.; A Mourão, P. Occurrence of a unique fucose-branched chondroitin sulfate in the body wall of a sea cucumber. *J. Boil. Chem.* **1988**, *263*, 18176–18183.

11. Albalasmeh, A.A.; Berhe, A.A.; Ghezzehei, T. A new method for rapid determination of carbohydrate and total carbon concentrations using UV spectrophotometry. *Carbohydr. Polym.* **2013**, *97*, 253–261. [CrossRef] [PubMed]

12. Bitter, T.; Muir, H. A modified uronic acid carbazole reaction. *Anal. Biochem.* **1962**, *4*, 330–334. [CrossRef]

13. Craigie, J.S.; Wen, Z.C.; Van Der Meer, J.P.; Wen, J.S.C.Z.C. Interspecific, Intraspecific and Nutritionally-Determined Variations in the Composition of. Agars from *Gracilaria* spp. *Bot. Mar.* **1984**, *27*, 55–62. [CrossRef]

14. Dische, Z. New color reactions for determination of sugars in polysaccharides. *Methods Biochem. Anal.* **1955**, *2*, 313–358. [PubMed]

15. Magcwebeba, T.U.; Swart, A.C.; Swanvelder, S.; Joubert, E.; Gelderblom, W. Anti-Inflammatory Effects of Aspalathus linearis and Cyclopia spp. Extracts in a UVB/Keratinocyte (HaCaT) Model Utilising Interleukin-1α Accumulation as Biomarker. *Molecules* **2016**, *21*, 1323. [CrossRef]

16. Medina-Contreras, O.; Harusato, A.; Nishio, H.; Flannigan, K.L.; Ngo, V.L.; Leoni, G.; Neumann, P.-A.; Geem, D.; Lili, L.N.; Ramadas, R.A.; et al. Cutting Edge: IL-36 Receptor Promotes Resolution of Intestinal Damage. *J. Immunol.* **2015**, *196*, 34–38. [CrossRef]

17. Caamal-Herrera, I.O.; Carrillo-Cocom, L.M.; Escalante-Réndiz, D.Y.; Araiz-Hernández, D.; Azamar-Barrios, J.A. Antimicrobial and antiproliferative activity of essential oil, aqueous and ethanolic extracts of Ocimum micranthum Willd leaves. *BMC Complement. Altern. Med.* **2018**, *18*, 55. [CrossRef]

18. Gábor, M. Models of Acute Inflammation in the Ear. *Inflamm. Protoc.* **2003**, *225*, 129–138.

19. Livak, K.J.; Schmittgen, T.D. Analysis of relative gene expression data using real-time quantitative PCR and the $2^{-\Delta\Delta CT}$ method. *Methods* **2001**, *25*, 402–408. [CrossRef]

20. Delday, M.; Mulder, I.; Logan, E.T.; Grant, G. Bacteroides thetaiotaomicronAmeliorates Colon Inflammation in Preclinical Models of Crohn's Disease. *Inflamm. Bowel Dis.* **2018**, *25*, 85–96. [CrossRef]

21. Wirtz, S.; Popp, V.; Kindermann, M.; Gerlach, K.; Weigmann, B.; Fichtner-Feigl, S.; Neurath, M.F. Chemically induced mouse models of acute and chronic intestinal inflammation. *Nat. Protoc.* **2017**, *12*, 1295–1309. [CrossRef] [PubMed]

22. Li, S.; Li, J.; Mao, G.; Wu, T.; Lin, D.; Hu, Y.; Ye, X.; Tian, D.; Chai, W.; Linhardt, R.J.; et al. Fucosylated chondroitin sulfate from Isostichopus badionotus alleviates metabolic syndromes and gut microbiota dysbiosis induced by high-fat and high-fructose diet. *Int. J. Boil. Macromol.* **2018**, *124*, 377–388. [CrossRef] [PubMed]

23. Li, S.; Li, J.; Mao, G.; Hu, Y.; Ye, X.; Tian, D.; Linhardt, R.J.; Chen, S. Fucosylated chondroitin sulfate oligosaccharides from Isostichopus badionotus regulates lipid disorder in C57BL/6 mice fed a high-fat diet. *Carbohydr. Polym.* **2018**, *201*, 634–642. [CrossRef] [PubMed]

24. Li, S.; Li, J.; Mao, G.; Yan, L.; Hu, Y.; Ye, X.; Tian, D.; Linhardt, R.J.; Chen, S. Effect of the sulfation pattern of sea cucumber-derived fucoidan oligosaccharides on modulating metabolic syndromes and gut microbiota dysbiosis caused by HFD in mice. *J. Funct. Foods* **2019**, *55*, 193–210. [CrossRef]

25. Yin, J.; Yang, X.; Xia, B.; Yang, Z.; Wang, Z.; Wang, J.; Li, T.; Lin, P.; Song, X.; Guo, S. The fucoidan from sea cucumber Apostichopus japonicus attenuates lipopolysaccharide-challenged liver injury in C57BL/6J mice. *J. Funct. Foods* **2019**, *61*, 103493. [CrossRef]

26. Xu, X.; Chang, Y.; Xue, C.; Wang, J.; Shen, J. Gastric Protective Activities of Sea Cucumber Fucoidans with Different Molecular Weight and Chain Conformations: A Structure–Activity Relationship Investigation. *J. Agric. Food Chem.* **2018**, *66*, 8615–8622. [CrossRef]

27. Panagos, C.G.; Thomson, D.S.; Moss, C.; Hughes, A.D.; Kelly, M.S.; Liu, Y.; Chai, W.; Venkatasamy, R.; Spina, D.; Page, C.P.; et al. Fucosylated chondroitin sulfates from the body wall of the sea cucumber Holothuria forskali: Conformation, selectin binding, and biological activity. *J. Boil. Chem.* **2014**, *289*, 28284–28298. [CrossRef]

28. Li, S.; Li, J.; Zhi, Z.; Wei, C.; Wang, W.; Ding, T.; Ye, X.; Hu, Y.; Linhardt, R.J.; Chen, S. Macromolecular properties and hypolipidemic effects of four sulfated polysaccharides from sea cucumbers. *Carbohydr. Polym.* **2017**, *173*, 330–337. [CrossRef]

29. Pomin, V.H.; Mourão, P.A.D.S. Specific sulfation and glycosylation—A structural combination for the anticoagulation of marine carbohydrates. *Front. Microbiol.* **2014**, *4*, 33. [CrossRef]

30. Luo, L.; Wu, M.; Xu, L.; Lian, W.; Xiang, J.; Lu, F.; Zhao, J. Comparison of physicochemical characteristics and anti-coagulant activities of polysaccharides from three sea cucumbers. *Mar. Drugs* **2013**, *11*, 399–417. [CrossRef]

31. Wu, N.; Zhang, Y.; Ye, X.; Hu, Y.; Ding, T.; Chen, S. Sulfation pattern of fucose branches affects the anti-hyperlipidemic activities of fucosylated chondroitin sulfate. *Carbohydr. Polym.* **2016**, *147*, 1–7. [CrossRef] [PubMed]

32. Ustyuzhanina, N.E.; Bilan, M.I.; Nifantiev, N.E.; Usov, A.I. New insight on the structural diversity of holothurian fucosylated chondroitin sulfates. *Pure Appl. Chem.* **2019**, *91*, 1065–1071. [CrossRef]

33. Lee, J.; Kang, U.; Seo, E.K.; Kim, Y.S. Heme oxygenase-1-mediated anti-inflammatory effects of tussilagonone on macrophages and 12-O-tetradecanoylphorbol-13-acetate-induced skin inflammation in mice. *Int. Immunopharmacol.* **2016**, *34*, 155–164. [CrossRef] [PubMed]

34. Napetschnig, J.; Wu, H. Molecular basis of NF-κB signaling. *Annu. Rev. Biophys.* **2013**, *42*, 443–468. [CrossRef]

35. Chang, M.S.; Chen, B.C.; Yu, M.T.; Sheu, J.R.; Chen, T.F.; Lin, C.H. Phorbol 12-myristate 13-acetate up-regulates cyclooxygenase-2 expression in human pulmonary epithelial cells via Ras, Raf-1, ERK, and NF-κB, but not p38 MAPK, pathways. *Cell. Signal.* **2005**, *17*, 299–310. [CrossRef] [PubMed]

36. Zhao, Y.; Zhang, D.; Wang, S.; Tao, L.; Wang, A.; Chen, W.; Zhu, Z.; Zheng, S.; Gao, X.; Lu, Y. Holothurian Glycosaminoglycan Inhibits Metastasis and Thrombosis via Targeting of Nuclear Factor-κB/Tissue Factor/Factor Xa Pathway in Melanoma B16F10 Cells. *PLoS ONE* **2013**, *8*, e56557. [CrossRef]

37. Li, S.; Jiang, W.; Hu, S.; Song, W.; Ji, L.; Wang, Y.; Cai, L. Fucosylated chondroitin sulphate from Cusumaria frondosa mitigates hepatic endoplasmic reticulum stress and inflammation in insulin resistant mice. *Food Funct.* **2015**, *6*, 1547–1556. [CrossRef]

38. Luo, Y.; Zheng, S. Hall of Fame among Pro-inflammatory Cytokines: Interleukin-6 Gene and Its Transcriptional Regulation Mechanisms. *Front. Immunol.* **2016**, *7*, 503. [CrossRef]
39. Zhang, R.; Yuan, S.; Ye, J.; Wang, X.; Zhang, X.; Shen, J.; Yuan, M.; Liao, W. Polysaccharide from flammuliana velutipes improves colitis via regulation of colonic microbial dysbiosis and inflammatory responses. *Int. J. Boil. Macromol.* **2020**, *149*, 1252–1261. [CrossRef]
40. Wu, H.; Rao, Q.; Ma, G.-C.; Yu, X.-H.; Zhang, C.-E.; Ma, Z.-J. Effect of Triptolide on Dextran Sodium Sulfate-Induced Ulcerative Colitis and Gut Microbiota in Mice. *Front. Pharmacol.* **2020**, *10*, 1652. [CrossRef]
41. Hu, S.; Wang, J.; Xu, Y.; Yang, H.; Wang, J.; Xue, C.; Yan, X.; Su, L. Anti-inflammation effects of fucosylated chondroitin sulphate from Acaudina molpadioides by altering gut microbiota in obese mice. *Food Funct.* **2019**, *10*, 1736–1746. [CrossRef] [PubMed]
42. Wang, J.; Zhang, C.; Guo, C.; Li, X. Chitosan Ameliorates DSS-Induced Ulcerative Colitis Mice by Enhancing Intestinal Barrier Function and Improving Microflora. *Int. J. Mol. Sci.* **2019**, *20*, 5751. [CrossRef] [PubMed]
43. Xia, Y.; Chen, Y.; Wang, G.; Yang, Y.; Song, X.; Xiong, Z.; Zhang, H.; Lai, P.; Wang, S.; Ai, L. Lactobacillus plantarum AR113 alleviates DSS-induced colitis by regulating the TLR4/MyD88/NF-κB pathway and gut microbiota composition. *J. Funct. Foods* **2020**, *67*, 103854. [CrossRef]

Article

Treatment with Modified Extracts of the Microalga *Planktochlorella nurekis* Attenuates the Development of Stress-Induced Senescence in Human Skin Cells

Jagoda Adamczyk-Grochala [1], Maciej Wnuk [1], Magdalena Duda [2], Janusz Zuczek [2] and Anna Lewinska [1,*]

[1] Department of Biotechnology, University of Rzeszow, Pigonia 1, 35-310 Rzeszow, Poland; ag1jagoda@gmail.com (J.A.-G.); mawnuk@gmail.com (M.W.)
[2] Bioorganic Technologies sp. z o.o., Sedziszow Malopolski, Sielec 1A, 39-120 Sielec, Poland; mduda@bioorganictechnologies.pl (M.D.); janusz@bioorganictechnologies.pl (J.Z.)
* Correspondence: alewinska@o2.pl or lewinska@ur.edu.pl; Tel.: +48-17-851-86-09

Received: 22 February 2020; Accepted: 3 April 2020; Published: 6 April 2020

Abstract: More recently, we have proposed a safe non-vector approach to modifying the biochemical profiles of the microalga *Planktochlorella nurekis* and obtained twelve clones with improved content of lipids and selected pigments and B vitamins and antioxidant activity compared to unaffected cells. In the present study, the biological activity of water and ethanolic extracts of modified clones is investigated in the context of their applications in the cosmetic industry and regenerative medicine. Extract-mediated effects on cell cycle progression, proliferation, migration, mitogenic response, apoptosis induction, and oxidative and nitrosative stress promotion were analyzed in normal human fibroblasts and keratinocytes in vitro. Microalgal extracts did not promote cell proliferation and were relatively non-cytotoxic when short-term treatment was considered. Long-term stimulation with selected microalgal extracts attenuated the development of oxidative stress-induced senescence in skin cells that, at least in part, was correlated with nitric oxide signaling and increased niacin and biotin levels compared to an unmodified microalgal clone. We postulate that selected microalgal extracts of *Planktochlorella nurekis* can be considered to be used in skin anti-aging therapy.

Keywords: microalgae; *Planktochlorella nurekis*; skin cells; proliferation; senescence

---

1. Introduction

The skin is a natural barrier that protects the human body against a number of environmental stressors, such as physical, chemical, or biological agents. Environmental factors such as sun radiation (ultraviolet radiation, visible light, and infra-red radiation), air pollutants (polycyclic aromatic hydrocarbons (PAH), volatile organic compounds (VOCs), particulate matters (PMs), and ozone) and tobacco smoke may affect and/or promote the progression of inflammatory skin diseases such as atopic dermatitis (AD), acne and psoriasis, skin aging, and cancerogenesis (e.g., photo-aging and photo-carcinogenesis) [1,2]. Moreover, the skin may be subjected to photo-pollution stress (combined action of UV radiation and air pollutants, such as PAH and PMs) that may result in synergistic photo-toxic damage [2,3].

Skin aging, a complex biological process that can be modulated by several intrinsic (chronological or intrinsic skin aging) and environmental factors (e.g., UV radiation, extrinsic skin aging), is accompanied by both phenotypic changes in cutaneous cells and structural and functional changes of extracellular matrix (ECM) components, namely collagens, elastin and proteoglycans that are needed to maintain the tensile strength, elasticity and hydration of the skin [4,5]. Mitochondrial DNA damage, increased reactive oxygen species (ROS) production, and telomere shortening are considered to be major players during intrinsic skin aging [4,5]. Moreover, the accumulation of replicative old senescent skin cells may also promote the pathological remodeling of ECM [4]. Cumulative exposure to UV radiation can result in the damage of chromophore-rich cutaneous biomolecules and oxidative stress, affecting skin cells and matrix components, and stimulate the expression of ECM proteases (e.g., matrix metalloproteinases, MMPs) via transcription factor activator protein-1 (AP-1) signaling, leading to collagen degradation and wrinkle formation [4].

Macroalgae (multicellular algae) and microalgae (unicellular algae) are a rich source of natural bioactive compounds and secondary metabolites with antioxidant, anti-inflammatory, anti-microbial and anticancer activities that may have some beneficial effects during skin anti-aging therapies, de-pigmentation, and acne, and chronic wound and melanoma treatments [6–8]. As algae are exposed to UV-radiation-mediated oxidative stress in their natural environments and adapted to cope with oxidative biomolecule damage by the means of elevated production of antioxidants and UV-absorbing compounds, algal extracts and isolated chemicals may provide similar UV protection as sunscreens currently used in the market [8]. Photo-protective activity of selected algal compounds, namely, shinorine, porphyra-334, palythene, eckstolonol, eckol, mycosporine-glycine, mycosporine methylamine-serine, sargachromenol, fucoxanthin, tetraprenyltoluquinol chromane meroterpenoid, scytonemin, and sargaquinoic acid, has been previously reported [7,9–12]. A number of algal bioactive compounds may protect against photo-aging by decreasing the levels of MMPs and maintaining the pools of skin collagen [6–8]. Extracts of the green microalga *Chlorella vulgaris* may also promote collagen synthesis in the skin, stimulate tissue regeneration and limit wrinkle formation, and thus, may have applications in skin anti-aging therapies [6,8].

Although algae and related groups are widely studied, only a limited number of microalgal and cyanobacterial species have been comprehensively characterized and commercially used, namely, *Spirulina*, *Chlorella*, *Haematococcus*, *Dunaliella*, *Botryococcus*, *Phaeodactylum*, and *Porphyridium* [8,13,14]. In 2014, a new microalgal species was reported, namely, *Planktochlorella nurekis* [15]. More recently, we provided a comprehensive biochemical characterization of the microalga *Planktochlorella nurekis* and selected twelve clones with elevated levels of lipids and several pigments and B vitamins, and increased antioxidant activity [16]. However, *Planktochlorella nurekis*-mediated effects in cellular models in vitro have never been addressed, e.g., the ability to stimulate cell proliferation and migration and to promote some anti-aging effects. In the present study, we use human fibroblasts and keratinocytes to investigate mitogenic, promigratory and anti-senescence activity of water and ethanolic extracts obtained from twelve modified clones of *Planktochlorella nurekis* and their corresponding unaffected cells. Extract-mediated ability to block the development of oxidative stress-induced senescence in human skin cells is documented and discussed.

2. Materials and Methods

2.1. Cell Culture and Extract Preparation and Treatment

Human foreskin fibroblasts BJ (lot 62341989, catalog number ATCC® CRL-2522™) were obtained from American Type Culture Collection (ATCC, Manassas, VA, USA) and human epidermal keratinocytes HEK (lot 3057, catalog number 102-05F and lot 1831898, catalog number C0015C) were obtained from Cell Application Inc. (San Diego, CA, USA) and Thermo Fisher Scientific (Waltham, MA, USA), respectively. Population doubling levels were monitored as described previously [17] and only replicative young cells with high proliferative potential were used. Cells were cultured at 37 °C in a cell culture incubator in the presence of 5% CO_2. BJ fibroblasts were cultured in Dulbecco's modified Eagle's medium (DMEM) containing 10% fetal calf serum (FCS), 100 U/mL penicillin, 0.1 mg/mL streptomycin, and 0.25 µg/mL amphotericin B (Corning, Tewksbury, MA, USA). HEK cells were cultured in EpiLife™ basal medium with the Human Keratinocyte Growth Supplement Kit (HKGS Kit, Thermo Fisher Scientific, Waltham, MA, USA) containing bovine pituitary extract (BPE, 0.2% *v/v*), recombinant human insulin-like growth factor-I (0.01 µg/mL), hydrocortisone (0.18 µg/mL), bovine transferrin (5 µg/mL), human epidermal growth factor (0.2 ng/mL) and gentamicin/amphotericin. BJ cells were passaged using 0.25% trypsin/2.21 mM EDTA (Corning, Tewksbury, MA, USA) and HEK cells were passaged using 0.025% trypsin/0.01% EDTA (Thermo Fisher Scientific, Waltham, MA,

USA) and trypsin was then inactivated using Trypsin Neutralizer Solution (Thermo Fisher Scientific, Waltham, MA, USA). Typically, cells were seeded at 10^4 cells per cm^2 of a culture flask.

Twelve clones of the microalga *Planktochlorella nurekis* with improved biochemical features were used [16]. A detailed description of biochemical profiles of modified microalgal clones can be found elsewhere [16]. To prepare microalgal extracts of twelve clones of the microalga *Planktochlorella nurekis* and one control clone, two solvents were considered, namely, water and 80% ethanol. To obtain water extracts (WE), 100 mg of microalgal dry weight were added to sterile ultra pure water to give the stock concentration of 100 mg/mL. The samples were then boiled at 100 °C for 10 min and centrifuged (13,000 rpm, RT, 10 min). Supernatants were collected and stored until use at −20 °C. To obtain ethanolic extracts (EE), 100 mg of microalgal dry weight were added to 80% ethanol to give the stock concentration of 100 mg/mL. The samples were then incubated at 37 °C for 24 h with shaking (1000 rpm) and centrifuged (13,000 rpm, RT, 10 min). Supernatants were collected and stored until use at −20 °C. Fibroblasts and keratinocytes were treated with microalgal extracts for 24 h (the majority of experiments), up to 72 h (wound healing assay) or up to 7 days (senescence-associated beta-galactosidase activity). The solvent action (water, 80% ethanol) alone was also considered and the solvents used had no effect on cells.

2.2. MTT Assay

To study the extract-mediated changes in metabolic activity (thiazolyl blue tetrazolium bromide (MTT) assay), BJ and HEK cells were seeded at the concentration of 5000 cells per a well of a 96-well plate and cultured overnight. Microalgal extracts were then added (water extracts at the concentrations ranging from 1 ng/mL to 1000 µg/mL and ethanolic extracts at the concentrations ranging from 1 ng/mL to 500 µg/mL) for 24 h. After the removal of microalgal extracts, cells were incubated with MTT solution (0.5 mg/mL, 4 h). After incubation, the MTT solution was removed, and 200 µL DMSO was added to dissolve the formazan crystals. Absorbance was read at 570 and 630 nm using a microplate reader. Metabolic activity was calculated as a ΔA (A570-A630). Metabolic activity at control conditions (untreated cells) was considered as 100%. According to MTT results, the concentrations of 100 µg/mL water and 100 µg/mL ethanolic extracts and the concentrations of 100 µg/mL water and 1 µg/mL ethanolic extracts were selected for further analysis (BJ and HEK cells, respectively).

2.3. Cell Cycle

Extract-mediated changes in the phases of the cell cycle were analyzed using a Muse® Cell Analyzer and a Muse® Cell Cycle Assay Kit according to the manufacturer's instructions (Merck KGaA, Darmstadt, Germany). Briefly, BJ and HEK cells were treated with microalgal extracts for 24 h and then fixed using ice-cold 70% ethanol at −20 °C. Cells were then stained using kit reagent and G0/G1 (gap 1), S (synthesis), and G2 (gap 2)/M (mitosis) phases were analyzed [%]. Representative histograms are also shown.

2.4. Cell Proliferation

Extract-mediated changes in cell proliferation were investigated as cell number analysis and Ki67 (proliferation marker) immunostaining. The cell number was automatically calculated using TC10™ Automated Cell Counter (Bio-Rad, Hercules, CA, USA). The subpopulations of Ki67-positive and Ki67-negative cells (%) were analyzed using a Muse® Cell Analyzer and a Muse® Ki67 Proliferation Kit (Merck KGaA, Darmstadt, Germany). Briefly, BJ and HEK cells were treated with microalgal extracts for 24 h and then fixed and permeabilized using kit reagents for fixation and permeabilization and then incubated with the Ki67 antibody. Control Ki67-non-specific antibody was used to calculate the subpopulation of Ki67-negative cells (grey histogram). Representative histograms are also shown.

2.5. Mitogenic Activity

Extract-induced mitogenic activity was revealed using a Muse® Cell Analyzer and a Muse® MAPK Activation Dual Detection Kit. Briefly, BJ and HEK cells were treated with microalgal extracts for

24 h, fixed, permeabilized and extracellular signal-regulated kinase 1/2 (ERK1/2) activity was measured using two directly conjugated antibodies, a phospho-specific anti-phospho-ERK1/2 (Thr202/Tyr204, Thr185/Tyr187)-phycoerythrin, and anti-ERK1/2-PECy5 conjugated antibody (Merck KGaA, Darmstadt, Germany) according to the manufacturer's instructions. Three subpopulations were considered, namely, ERK1/2-positive (inactivated fraction), phospho-ERK1/2-positive (activated fraction), and ERK1/2-negative (non-expressing fraction). Representative dot-plots are also shown.

2.6. Wound Healing Assay

Extract-mediated cell migration was evaluated using a wound healing assay [18]. Briefly, BJ and HEK cells were cultured until almost 100% confluency was achieved, and a fresh medium containing 100 µg/mL water extracts was added. A sterile pipette tip was applied to make a cross-shaped wound by streaking across a monolayer of BJ and HEK cells. The migration of cells into the wound was documented at time 0, 24, 48, and 72 h after wounding using an inverted microscope. Every 24 h, a fresh medium containing 100 µg/mL water extract was added.

2.7. Apoptosis versus Necrosis

Extract-induced apoptosis was evaluated using a Muse® Cell Analyzer and a Muse® Annexin V and Dead Cell Assay Kit (Merck KGaA, Darmstadt, Germany). Briefly, BJ and HEK cells were stained according to manufacturer's instructions and four-cell subpopulations were revealed, namely, (1) non-apoptotic cells (live cells): Annexin V (−) and 7-AAD (−), (2) early apoptotic cells: Annexin V (+) and 7-AAD (−), (3) late apoptotic and dead cells: Annexin V (+) and 7-AAD (+), and (4) mostly nuclear debris (necrotic cells): Annexin V (−) and 7-AAD (+) (%). Representative dot-plots are also shown. Necrosis was also evaluated using a trypan blue exclusion assay. BJ and HEK cells were incubated with 0.4% trypan blue, and then dead cells with porous cell membranes (blue-stained cells) were automatically calculated (%) using a TC10™ Automated Cell Counter (Bio-Rad, Hercules, CA, USA).

2.8. Superoxide and Nitric Oxide Levels

Extract-mediated changes in the levels of superoxide and nitric oxide were evaluated using a Muse® Cell Analyzer and a Muse® Oxidative Stress Kit and Muse® Nitric Oxide Kit, respectively, according to the manufacturer's instructions. Briefly, for evaluation of oxidative stress, BJ and HEK cells were incubated with superoxide specific fluorogenic probe dihydroethidium, and, for evaluation of nitrosative stress, human fibroblasts and keratinocytes were incubated with nitric oxide specific fluorogenic probe DAX-J2™ Orange and a death marker, 7-AAD. Two subpopulations were revealed, namely, superoxide-positive and superoxide-negative and four subpopulations were documented, namely, (1) nitric oxide-negative, 7-AAD-negative, (2) nitric oxide-positive, 7-AAD-negative, (3) nitric oxide-positive, 7-AAD-positive, and (4) nitric oxide-negative, 7-AAD-positive, respectively. Representative histograms or dot-plots are also shown.

2.9. Senescence-Associated Beta-Galactosidase Activity

A cellular model of hydrogen peroxide-induced senescence was considered, as described previously [17,19]. Briefly, BJ and HEK cells were treated with 100 µM hydrogen peroxide for 2 h, and then water and ethanolic extracts were added for up to 7 days to analyze their protective effects. Every 48 h, a fresh medium containing microalgal extracts was added. After 7 days of hydrogen peroxide removal, senescence-associated beta-galactosidase (SA-beta-gal) activity was assayed as described previously [17,19]. The effect of water and ethanolic extracts alone (pro-senescence activity) was also investigated. BJ and HEK cells were treated with water and ethanolic extracts for 24 h, and, 7 days after microalgal extracts removal, SA-beta-gal activity was analyzed. SA-beta-gal-positive and SA-beta-gal-negative cells were scored (%).

2.10. Preliminary Analysis of Anticancer Activity

Water and ethanolic extracts were also initially analyzed in terms of their anticancer activity. Extract-mediated changes in metabolic activity (MTT assay) of normal human fibroblasts (BJ cells) were compared with extract-induced changes in metabolic activity of selected human cancer cells, namely, MDA-MB-231 breast cancer cells (ATCC® HTB-26™, ATCC, Manassas, VA, USA), U-2 OS osteosarcoma cells (92022711, ECACC, Porton Down, Salisbury, UK), and U-251 MG glioblastoma cells (09063001, ECACC, Porton Down, Salisbury, UK). Briefly, cells were seeded at 5000 cells per well of a 96-well plate and cultured overnight. Cells were then treated with water and ethanolic extracts (100 µg/mL) for 24 h, and, after microalgal extracts removal, cells were incubated with MTT reagent as described above.

2.11. Statistical Analysis

The results represent the mean ± standard deviation (SD) from at least three independent experiments. Statistical significance was assessed by 1-way analysis of variance (ANOVA) using GraphPad Prism 5 and with Dunnett's multiple comparison test.

3. Results and Discussion

3.1. The Effect of Water and Ethanolic Extracts of the Microalga Planktochlorella Nurekis on Cell Cycle Progression and Proliferation of Human Skin Cells

In 2014, a new species of microalgae was reported, namely, *Planktochlorella nurekis* [15]. More recently, we comprehensively characterized its biochemical features and also selected twelve microalgal clones with improved biochemical profiles based on colchicine and cytochalasin B-mediated changes in ploidy state and DNA content [16]. However, the biological activity of water and ethanolic extracts of the microalga *Planktochlorella nurekis* has not been addressed. In the present study, we focus on the evaluation of extract-mediated effects on cell proliferation, migration, mitogenic activity, overall cytotoxicity and the ability to attenuate the development of oxidative stress-induced senescence in human skin cells that may have potential applications in the cosmetic industry and regenerative medicine. We also asked if the modified and improved clones of the microalga *Planktochlorella nurekis* may have some advantages over control unaffected microalgal cells.

Two types of skin cells were selected for the analysis, namely, normal human fibroblasts (BJ cells) and normal human keratinocytes (HEK cells). As extracts of different *Chlorella* species may act on human and rodent cells at a wide range of concentrations and the effects may also rely on solvents used [20–24], we decided to study both water and ethanolic extracts and the concentrations ranging from 1 to 1000 µg/mL (Figure 1).

We noticed that 100 µg/mL water extracts of seven modified clones (WE1, WE2, WE3, WE5, WE6, WE8, WE12) promoted metabolic activity (MTT assay) of BJ cells compared to control conditions (Figure 1a, red arrows). Control water extract (CWE) also caused a slight increase in metabolic activity (Figure 1a, a red arrow). A similar effect was not observed when BJ cells were treated with ethanolic extracts (Figure 1b). Moreover, we have established that ethanolic extracts can be applied at the concentrations up to 100 µg/mL without promoting adverse changes in metabolic activity (Figure 1b). At the concentration of 500 µg/mL, all ethanolic extracts, both modified and control, caused a statistically significant decrease in metabolic activity of BJ cells (Figure 1b, $p < 0.001$).

More recently, it has been observed that a hot water extract of *Chlorella vulgaris* at the concentrations of 400 and 600 µg/mL increased the number of viable cells of young human dermal fibroblasts (HDFs) [24]. Interestingly, the extract-mediated effect depended on replicative age of HDFs, namely, stimulatory effects in pre-senescent and senescent HDFs were noticed when 200 and 100 µg/mL water extract was used, respectively, and the concentrations ranging from 500 to 800 µg/mL decreased cell viability of senescent HDFs [24]. Water and 80% ethanolic extracts of *Chlorella sorokiniana* did not affect the cell viability of Chinese hamster lung fibroblasts (V79 cells) when used at the concentrations up to 80 and 20 µg/mL, respectively [23]. However, treatment with 12.5, 25, and 50 µg/mL hot water extract of *Chlorella vulgaris* increased cell viability by 124%, 135.4%, and 155.0% in IEC-6 rat small intestine epithelial cells compared to untreated cells [21]. A slight cytotoxic effect of 50 and 100 µg/mL ethanolic extract of *Chlorella vulgaris* was also reported using 3T3-L1 mouse preadipocytes as a cellular model in vitro [22].

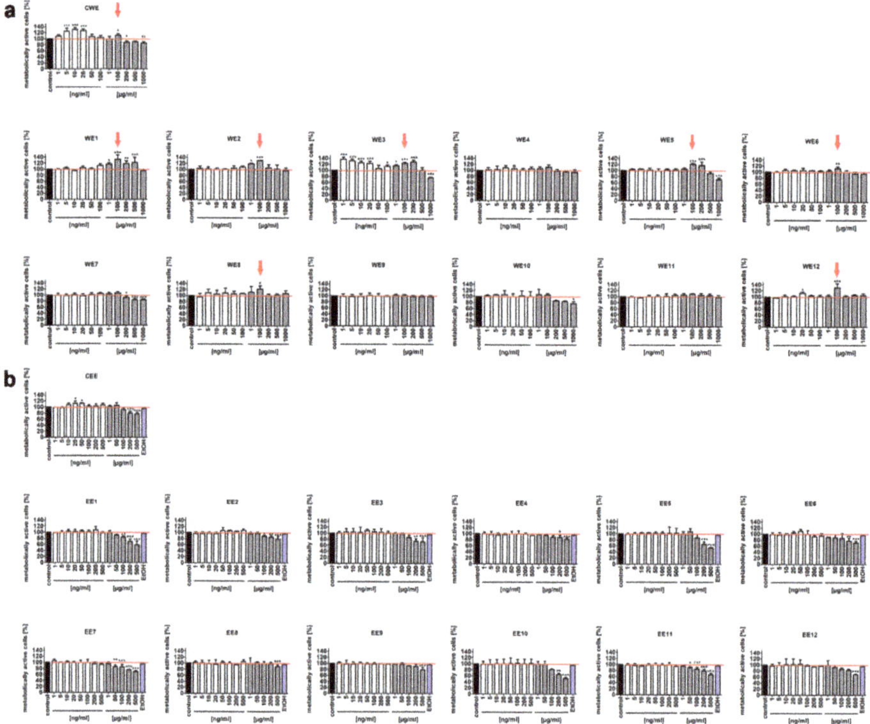

Figure 1. Extract-mediated changes in metabolic activity (MTT assay) of normal human fibroblasts (BJ). Metabolic activity at standard growth conditions (control, a black bar) is considered as 100%. (**a**) The effects of water extracts (WE, twelve modified clones from WE1 to WE12) are shown. Control clone water extract is denoted as CWE. (**b**) The effects of ethanolic extracts (EE, twelve modified clones from EE1 to EE12) are shown. Control clone ethanolic extract is denoted as CEE. To emphasize extract action, a red horizontal line is added. The effect of 80% ethanol (a solvent of ethanolic extracts, a violet bar) is also shown. White bars indicate the extract concentrations in ng/mL, whereas grey bars indicate the extract concentrations in µg/mL. Red arrows indicate the concentration of water extract (100 µg/mL) that was selected for further analysis based on the stimulatory effect on the metabolic activity of the majority of modified clone water extracts used. Bars indicate SD, $n = 5$, *** $p < 0.001$, ** $p < 0.01$, * $p < 0.05$ compared to the control (ANOVA and Dunnett's a posteriori test).

Three selected concentrations (0.1, 1, and 100 µg/mL) were also considered when studying extract-mediated changes in the metabolic activity of human keratinocytes (Figure 2). In general, HEK cells were found to be more sensitive to microalgal extract treatment than BJ cells, especially in terms of ethanolic extract at the concentration of 100 µg/mL (Figures 1 and 2b). Moreover, a pro-stimulatory effect on the metabolic activity of HEK cells was limited to 100 µg/mL WE2 (Figure 2a, a red arrow). There are no literature data on the effects of *Chlorella* sp. extracts on normal keratinocytes in vitro. The effect of two extracts obtained from two different microalgal species derived from the Blue Lagoon, namely coccoid and filamentous algae, on the vitality of primary human epidermal keratinocytes was studied [25]. Coccoid algae extract (157 µg/mL, $p < 0.05$) promoted cell viability, whereas filamentous algae extract, when used from the concentration of 12 µg/mL, stimulated cytotoxic effect ($p < 0.05$) [25].

However, the authors did not provide information about the microalgal species used [25]. It has also been reported that 0.1% (1 mg/mL) aqueous crude extract of the cyanobacterium *Spirulina platensis* caused a 2-fold increase in cell viability of the HS2 human keratinocyte cell line [26].

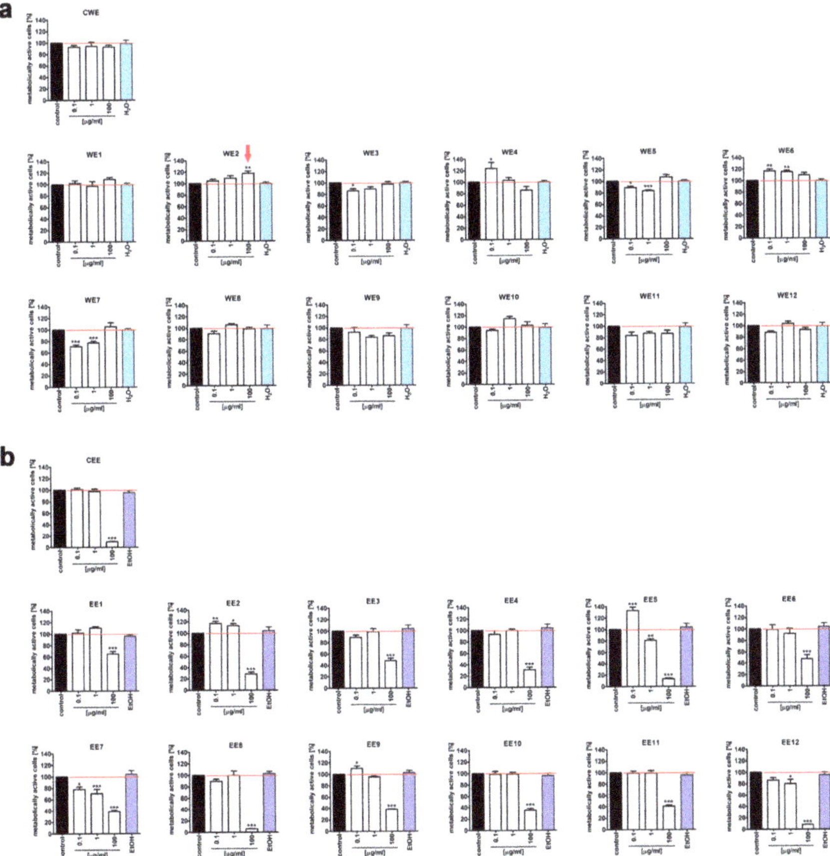

Figure 2. Extract-mediated changes in metabolic activity (MTT assay) of normal human keratinocytes (HEK). Metabolic activity at standard growth conditions (control, a black bar) is considered as 100%. (**a**) The effects of water extracts (WE, twelve modified clones from WE1 to WE12, white bars) are shown. Control clone water extract is denoted as CWE. (**b**) The effects of ethanolic extracts (EE, twelve modified clones from EE1 to EE12, white bars) are shown. Control clone ethanolic extract is denoted as CEE. To emphasize extract action, a red horizontal line is added. The effect of water (a blue bar) and 80% ethanol (a violet bar) (solvents used) is also shown. Red arrow indicates the concentration of water extract (100 µg/mL) that was selected for further analysis based on the stimulatory effect on the metabolic activity of BJ cells of the majority of modified clone water extracts used. Bars indicate SD, $n = 5$, *** $p < 0.001$, ** $p < 0.01$, * $p < 0.05$ compared to the control (ANOVA and Dunnett's a posteriori test).

Based on obtained results, 100 µg/mL water extracts (from WE1 to WE12 and control CWE) and 100 µg/mL ethanolic extracts (from EE1 to EE12 and control CEE) were selected for further analysis using BJ cells, and 100 µg/mL water extracts (from WE1 to WE12 and control CWE) and 1 µg/mL ethanolic extract (from EE1 to EE12 and control CEE) were selected for further analysis using HEK cells. As the MTT test measures direct change in metabolic activity (the activity of mitochondrial dehydrogenases), one cannot discriminate between extract-mediated changes in cell proliferation, cell number, or cytostatic or cytotoxic effects.

Thus, we then decided to study extract-mediated pro-stimulatory effects more comprehensively (Figures 3–5). Firstly, we analyzed extract-mediated changes in the phases of cell cycle of both BJ cells and HEK cells (Figure 3). No significant changes in the cell cycle progression were revealed upon stimulation of BJ and HEK cells with water and ethanolic extracts (Figure 3). It has been previously reported that the hot water extract of *Chlorella vulgaris* promoted the accumulation of cells at S and G2/M phases of the cell cycle in young human dermal fibroblasts [27]. However, the authors considered a four-times higher extract concentration (400 µg/mL) than used in the present study (100 µg/mL) [27]. Secondly, two cell proliferation parameters were considered, namely, Ki67 immunostaining (Figure 4) and cell number count (Figure 5a). In general, no significant changes in the levels of Ki67-positive cells were revealed after treatments with water and ethanolic extracts in HEK cells (Figure 4c,d). Ethanolic extracts EE1, EE5, and EE6 caused a slight decrease in the levels of Ki67-positive cells in BJ cells (Figure 4b, $p < 0.05$). Microalgal extracts also did not affect the cell number of BJ and HEK cells (Figure 5a).

Thus, one can conclude that extract-mediated stimulatory effects on metabolic activity (MTT assay) in BJ and HEK cells were not associated with increased cell proliferation (this study). In contrast, stimulation of IEC-6 rat small intestine epithelial cells with hot water extract of *Chlorella vulgaris* (up to 50 µg/mL) resulted in both an increase in metabolic activity and cell proliferation that was mediated by the activation of MAPK (ERK1/2) and PI3K/Akt pathways, pivotal regulators of cell survival, proliferation and growth [21]. Moreover, *Spirulina* crude protein also promoted the proliferation and migration of IEC-6 cells by activating the EGFR/MAPK signaling pathway [28]. Thus, we have then asked a question of whether microalgal extracts may modulate mitogenic response and cell migration of BJ cells and HEK cells. The mitogenic response was studied by analyzing the changes in the phosphorylation status of two mitogen-activated protein kinases, namely ERK1/2 (Figure S1). However, no ERK1/2 activation was revealed upon microalgal extract stimulation both in BJ cells and HEK cells (Figure S1). Finally, we have studied the effect of microalgal extracts on cell migration using wound healing assay (Figure S2). No extract-mediated stimulatory effects were observed in BJ cells (Figure S2a). In contrast, water extracts WE2, WE6, WE7, WE8, WE9, WE10, and WE11 promoted cell migration upon 48 h treatment after wounding in HEK cells (Figure S2b). There is a limited number of studies reporting the promigratory and wound healing effects of *Chlorella* sp. extracts in vitro and in vivo [29–31]. *Chlorella vulgaris* dressing promoted wound closure and minimized the formation of scar tissue during the healing process compared to the control wounds created on the dorsal surface of rats [29]. Moreover, oral (500 mg/kg) and topical (10%) administration of ROQUETTE *Chlorella* sp. had beneficial effects on skin lesions in mice [30]. The wound healing potential of *Arthrospira* (*Spirulina*) *platensis* extract has also been compared to the wound healing potential of *Chlorella vulgaris* extract using human fibroblasts as a cellular model [31]. It has been concluded that *Arthrospira* extract stimulated wound closure more efficiently than *Chlorella* extract [31]. The effect of solvent used has been also analyzed [32]. Treatment with 50 µg/mL water extract of *Spirulina platensis* promoted cell proliferation and migration of human fibroblasts in vitro, whereas the effects of methanolic and ethanolic extracts, when used at the same concentration, were mild to moderate [32]. The authors claimed that the promigratory and wound healing effects of water extracts were achieved by the presence of cinnamic acid, naringenin, kaempferol, temsirolimus, phosphatidylserine isomeric derivatives, and sulphoquinovosyl diacylglycerol [32]. Moreover, skin cream containing 1.125% *Spirulina platensis* crude extract showed an enhanced wound healing effect compared to control formulation using HS2 human keratinocytes as a cellular model [26].

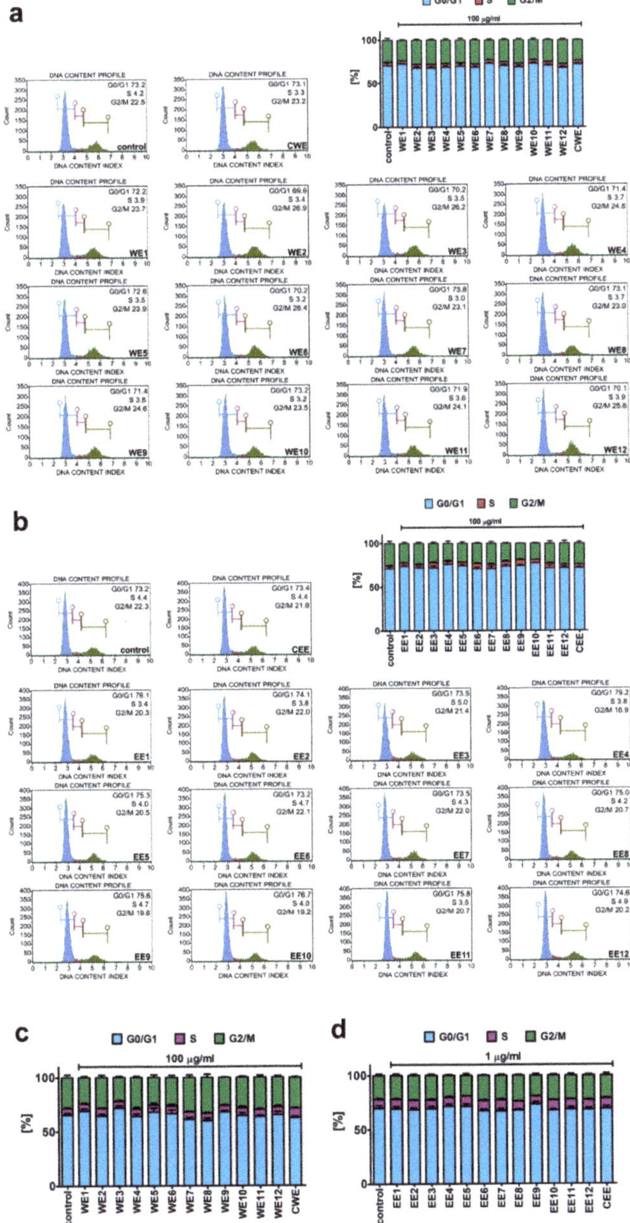

Figure 3. Extract-mediated changes in the cell cycle of BJ cells ((**a**), 100 µg/mL water extracts; (**b**), 100 µg/mL ethanolic extracts) and HEK cells ((**c**), 100 µg/mL water extracts; (**d**), 1 µg/mL ethanolic extracts). (**a,c**) The effects of water extracts (WE, twelve modified clones from WE1 to WE12) are shown. Control clone water extract is denoted as CWE. (**b,d**) The effects of ethanolic extracts (EE, twelve modified clones from EE1 to EE12) are shown. Control clone ethanolic extract is denoted as CEE. The percentage of cells in the G0/G1, S, and G2/M phases of the cell cycle was assessed using a Muse® Cell Analyzer and a Muse® Cell Cycle Kit. Representative histograms are shown (BJ cells). Bars indicate SD, *n* = 3.

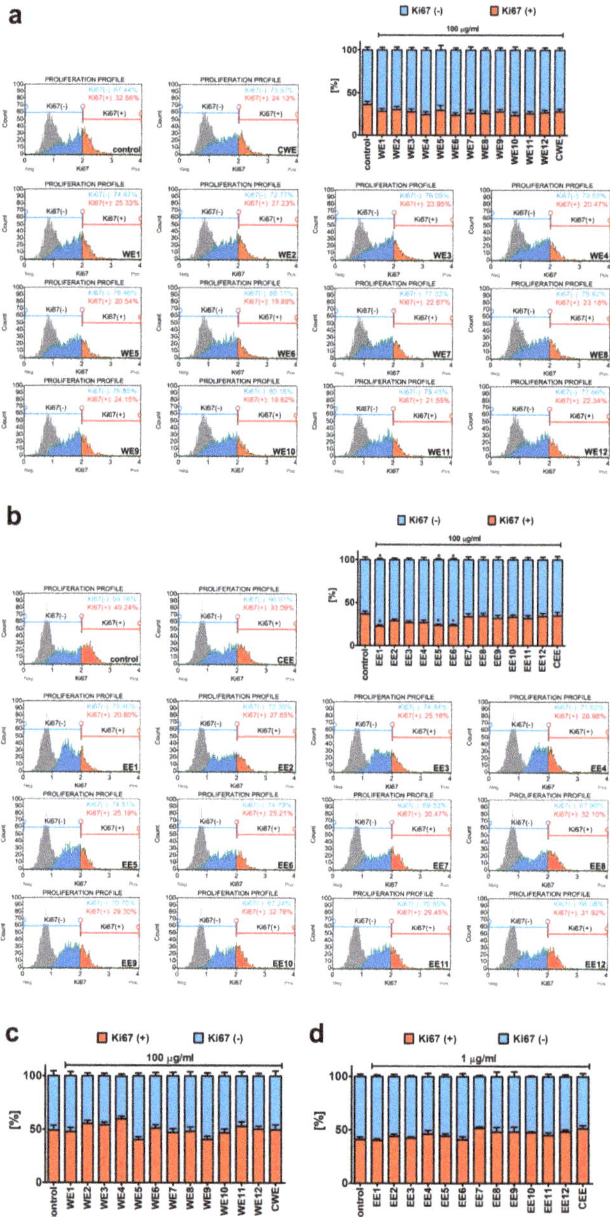

Figure 4. Extract-mediated changes in cell proliferation of BJ cells ((**a**), 100 μg/mL water extracts; (**b**), 100 μg/mL ethanolic extracts) and HEK cells ((**c**), 100 μg/mL water extracts; (**d**), 1 μg/mL ethanolic extracts). (**a**,**c**) The effects of water extracts (WE, twelve modified clones from WE1 to WE12) are shown. Control clone water extract is denoted as CWE. (**b**,**d**) The effects of ethanolic extracts (EE, twelve modified clones from EE1 to EE12) are shown. Control clone ethanolic extract is denoted as CEE. Cell proliferation was assayed using a Muse® Cell Analyzer and a Muse® Ki67 Proliferation Kit. Representative histograms are presented (BJ cells). A negative control without incubation with Ki67 specific antibody is denoted as a grey histogram in each analyzed sample. Bars indicate SD, *n* = 3. * *p* < 0.05 compared to the control (ANOVA and Dunnett's a posteriori test).

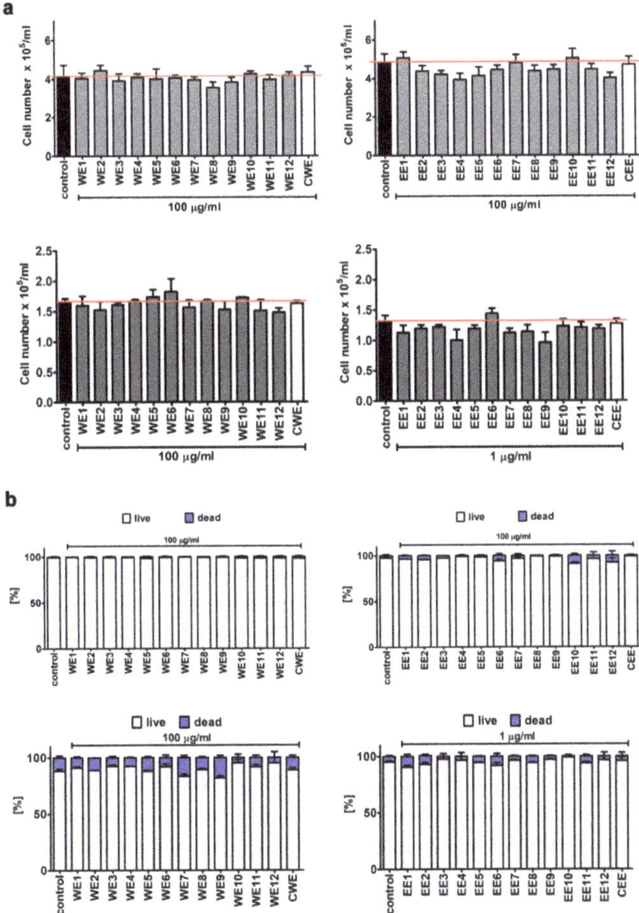

Figure 5. Extract-mediated changes in BJ and HEK cell numbers (**a**) and the levels of necrotic cells (trypan blue exclusion assay) (**b**). (**a**) Cell number was analyzed using TC10™ automated cell counter. To emphasize extract action, a red horizontal line is added. Black bars indicate control conditions, white bars indicate treatments with control extracts (CWE or CEE) and grey bars indicate treatments with modified water or ethanolic extracts. (**b**) BJ and HEK cells were incubated with 0.4% trypan blue and dead cells with porous cell membranes (blue-stained cells) were automatically calculated (%) using TC10™ automated cell counter. (**a,b**) Upper panel: BJ cells; lower panel: HEK cells. The effects of water extracts (WE, twelve modified clones from WE1 to WE12) and ethanolic extracts (EE, twelve modified clones from EE1 to EE12) are shown. Control clone water extract is denoted as CWE and control clone ethanolic extract is denoted as CEE. Bars indicate SD, *n* = 3.

3.2. Extract-Mediated Cytotoxicity, Oxidative and Nitrosative Stress

We have then considered if water and ethanolic extracts may also exert some adverse effects, namely, extract-mediated cytotoxicity, oxidative, and nitrosative stress were investigated (Figures 5b, 6, 7 and 8).

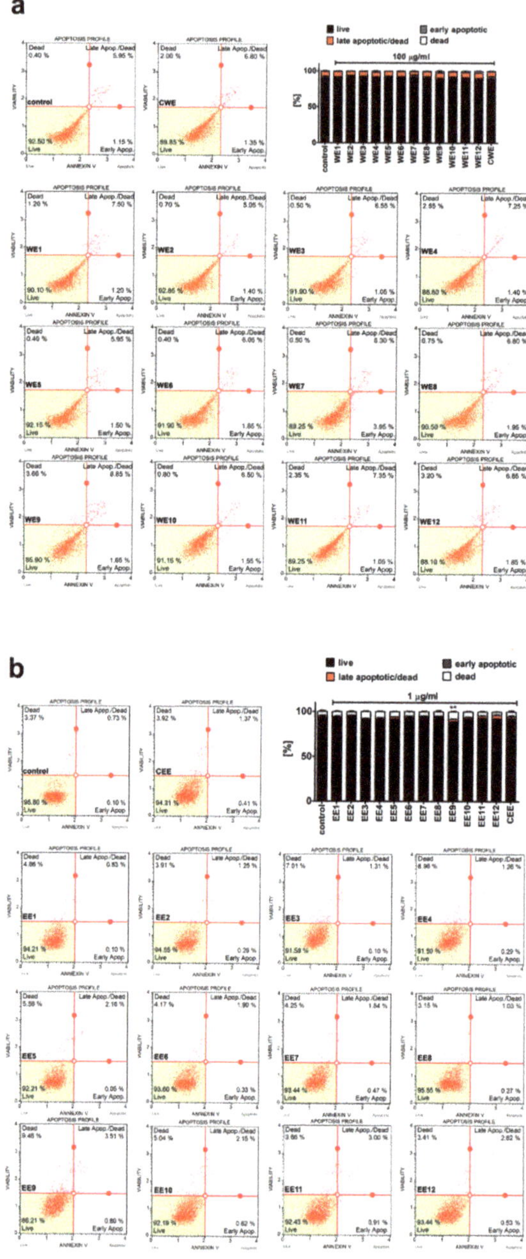

Figure 6. Extract-induced apoptosis in BJ cells ((**a**) 100 μg/mL water extracts) and HEK cells ((**b**) 1 μg/mL ethanolic extracts). (**a**) The effects of water extracts (WE, twelve modified clones from WE1 to WE12) are shown. Control clone water extract is denoted as CWE. (**b**) The effects of ethanolic extracts (EE, twelve modified clones from EE1 to EE12) are shown. Control clone ethanolic extract is denoted as CEE. Phosphatidylserine externalization was analyzed using a Muse® Cell Analyzer and a Muse® Annexin V and Dead Cell Assay Kit. Representative dot-plots are shown. Bars indicate SD, $n = 3$. ** $p < 0.01$ compared to the control (ANOVA and Dunnett's a posteriori test).

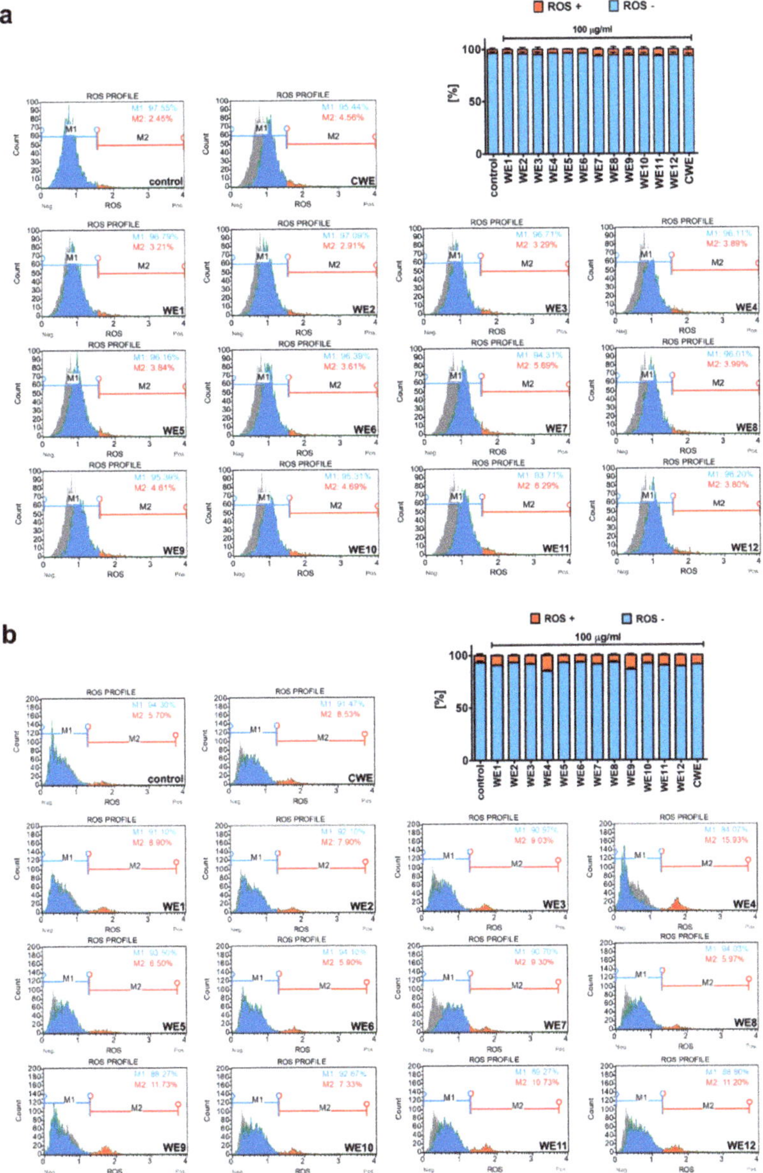

Figure 7. Extract-mediated changes in the levels of superoxide in BJ cells ((**a**) 100 µg/mL water extracts) and HEK cells ((**b**) 100 µg/mL water extracts). (**a**,**b**) The effects of water extracts (WE, twelve modified clones from WE1 to WE12) are shown. Control clone water extract is denoted as CWE. Superoxide levels were measured using a Muse® Cell Analyzer and a Muse® Oxidative Stress Kit. Representative histograms are presented. The control sample is denoted as a grey histogram in each sample analyzed. Bars indicate SD, *n* = 3.

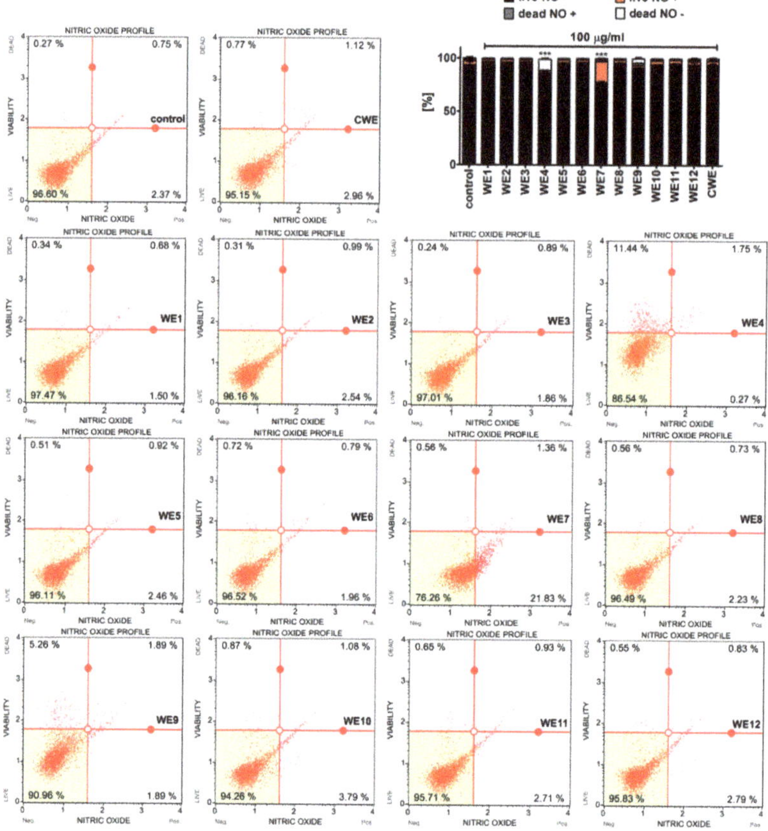

Figure 8. Extract-mediated changes in the levels of nitric oxide in BJ cells (100 µg/mL water extracts). The effects of water extracts (WE, twelve modified clones from WE1 to WE12) are shown. Control clone water extract is denoted as CWE. Nitric oxide levels were measured using a Muse® Cell Analyzer and a Muse® Nitric Oxide Kit. Representative dot-plots are presented. Bars indicate SD, $n = 3$. *** $p < 0.001$ compared to the control (ANOVA and Dunnett's a posteriori test).

No significant increase in the levels of necrotic cells was noticed upon microalgal extract stimulation in BJ cells and HEK cells, as judged by trypan blue exclusion assay (Figure 5b). However, a very slight increase in necrotic cell fraction in HEK cells was observed after treatment with ethanolic extract EE9, as revealed using a 7-AAD death marker (Figure 6b, $p < 0.01$). In contrast, no signs of an apoptotic mode of cell death were documented after treatment with both water and ethanolic extracts in BJ and HEK cells (Figure 6).

Thus, one can conclude that selected and analyzed concentrations of microalgal extracts were relatively non-cytotoxic when short-term treatment (24 h) was considered (this study). It has also been reported that a water extract of *Chlorella vulgaris* (400 µg/mL) lowered the levels of early apoptotic cells compared to control conditions in young human fibroblasts [27]. However, the levels of apoptotic cells at standard growth conditions were established to be more than 35%, which suggest the effect of some stress stimuli [27]. The solvent used and type of cells considered may also account for the cytotoxicity results. Briefly, 100 and 250 µg/mL methanolic extract of *Chlorella* elevated the levels of late apoptotic cells to 60.25% and 82.66% compared to untreated 3T3-L1 preadipocytes (25.88% of late apoptotic cells), respectively [20].

It is widely accepted that oxidative stress and nitrosative stress-promoting conditions can be detrimental to the cellular components causing, e.g., DNA and protein oxidative damage [33]. However, slight to moderate changes in the levels of reactive oxygen species (ROS) and reactive nitrogen species (RNS) may also have regulatory roles in the cell being a part of signal transduction pathways [33–36]. Thus, we then decided to analyze the effect of microalgal extracts on the levels of superoxide (Figure 7) and nitric oxide (Figure 8).

Microalgal extracts did not induce oxidative stress in BJ cells (Figure 7a). In HEK cells, a minor increase in the levels of superoxide was observed after treatment with water extracts WE4 and WE9 (Figure 7b). However, these effects were of no statistical significance (Figure 7b). Except for a statistically significant increase in the levels of nitric oxide of about 20% in WE7-treated BJ cells compared to control conditions (Figure 8, $p < 0.001$), microalgal extracts did not stimulate the production of nitric oxide in BJ cells (Figure 8) and HEK cells (data not shown). This moderate increase in the levels of nitric oxide after WE7 treatment may have a regulatory and signal transduction role. Indeed, the protective function of nitric oxide against UVB radiation-mediated oxidative stress and copper toxicity has been established in different *Chlorella* species [37,38]. The addition of nitric oxide to *Chlorella pyrenoidosa* suspensions irradiated by UVB promoted the activity of catalase and superoxide dismutase [37]. Moreover, exogenous nitric oxide limited copper-induced oxidative burst in *Chlorella vulgaris* [38]. Thus, nitric oxide may have an antioxidative role, at least in part, during oxidative stress conditions [37,38].

3.3. Attenuation of the Development of Stress-Induced Senescence in Skin Cells by Microalgal Extracts

Cellular senescence may be due to telomere shortening-based limited proliferative ability of human cells in vitro (replicative senescence) or may be promoted by diverse stress stimuli, namely, physical, chemical, or biological factors (stress-induced premature senescence, SIPS) [39]. As cellular senescence may contribute to aging and age-related diseases, it has been postulated that interventions based on the suppression of induction of cellular senescence, delay the onset of cellular senescence and/or removal of senescent cells from the body may be therapeutically important [40,41]. We then analyzed if microalgal extract may promote protective effects against oxidative stress-mediated stimulation of senescence in fibroblasts and keratinocytes (Figure 9).

We used a cellular model of stress-induced premature senescence that is based on the incubation of cells with 100 µM hydrogen peroxide for 2 h, followed by a 7-day culture without an oxidant [17,19]. After treatment, about 90% population of BJ cells and 95% population of HEK cells were characterized by senescence-associated beta-galactosidase (SA-beta-gal) activity (Figure 9). We have then tested if microalgal extracts may affect the levels of SA-beta-gal-positive cells when added after hydrogen peroxide treatment (Figure 9). Interestingly, selected microalgal extracts were found to be protective against the development of oxidative stress-induced senescence in fibroblasts and keratinocytes (Figure 9). The following ranking of extract-mediated anti-senescence activity was established, namely, for the effects of water extracts in hydrogen peroxide-treated BJ and HEK cells: WE7 (2% of SA-beta-gal-positive cells) > WE9 (15% of SA-beta-gal-positive cells) > WE10 (36% of SA-beta-gal-positive cells) > WE5 (40% of SA-beta-gal-positive cells) > WE11 (78% of SA-beta-gal-positive cells) > WE4 (79% of SA-beta-gal-positive cells) and WE9 (62% of SA-beta-gal-positive cells) > WE7 (70% of SA-beta-gal-positive cells), respectively (Figure 9). Ethanolic extracts were less effective, but post-treatment with selected extracts also resulted in a decrease in the levels of SA-beta-gal-positive cells, namely EE9 (49% of SA-beta-gal-positive cells), EE4 (52% of SA-beta-gal-positive cells), and EE2 (57% of SA-beta-gal-positive cells) in hydrogen peroxide-treated BJ cells and EE8 (31% of SA-beta-gal-positive cells) in hydrogen peroxide-treated HEK cells (Figure 9). To conclude, WE7 and WE9 promoted the most pronounced anti-senescence activity in both cell lines (Figure 9). In the case of WE7, the protective effect may rely on WE7-mediated nitric oxide signaling, at least in BJ cells (Figure 8) and increased pools of niacin and biotin compared to unmodified control microalgal clone (CWE) [16]. A plethora of biological processes can be modulated by niacin-derived nucleotides (e.g., nicotinamide adenine dinucleotide, NADH, and its phosphorylated form NADPH), namely, redox reactions, antioxidative protection, DNA repair, and the activity of signaling pathways, which may stimulate some protective effects during stress conditions [42]. However, more studies are needed to establish the molecular mechanism of microalgal extract-mediated attenuation of the development of oxidative stress-induced senescence in fibroblasts and keratinocytes.

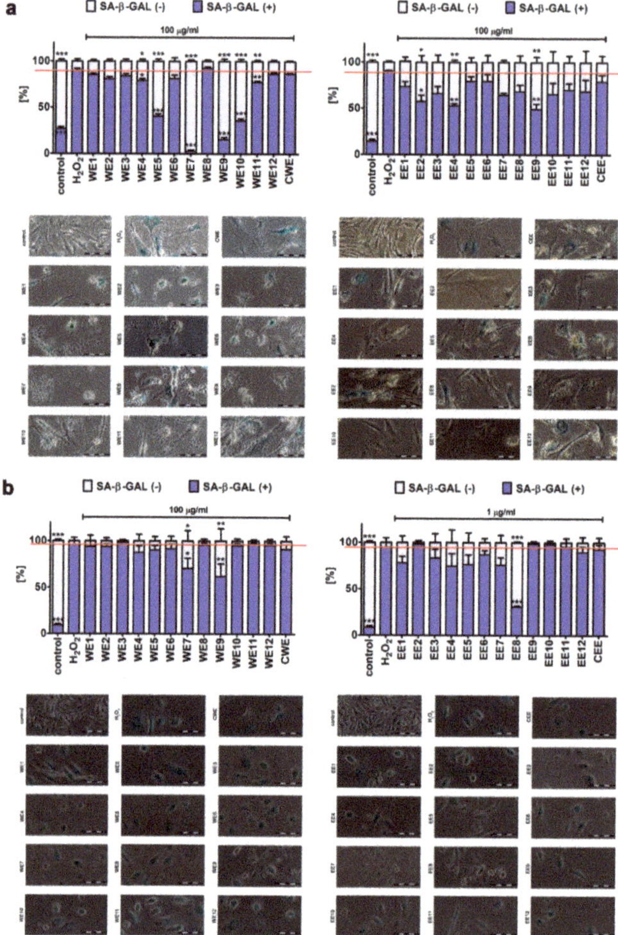

Figure 9. Extract-mediated changes in the levels of hydrogen peroxide-induced senescent BJ cells ((**a**), 100 µg/mL water extracts and 100 µg/mL ethanolic extracts) and HEK cells ((**b**), 100 µg/mL water extracts and 1 µg/mL ethanolic extracts). The effects of water extracts (WE, twelve modified clones from WE1 to WE12) and ethanolic extracts (EE, twelve modified clones from EE1 to EE12) are shown. Control clone water extract is denoted as CWE and control clone ethanolic extract is denoted as CEE. Senescence-associated β-galactosidase activity. Representative microphotographs are shown. Scale bars 100 µm, objective 20x. To emphasize extract action, a red horizontal line is added. Bars indicate SD, $n = 3$, *** $p < 0.001$, ** $p < 0.01$, * $p < 0.05$ compared to hydrogen peroxide treatment (ANOVA and Dunnett's a posteriori test).

Of course, we have also tested if the treatments with microalgal extracts alone may stimulate some pro-senescence effects (Figure S3). The pro-senescence activity of microalgal extracts was limited (Figure S3). In HEK cells, only water extract WE3 promoted an increase in the levels of SA-beta-gal-positive cells compared to untreated cells, whereas treatment with ethanolic extract EE10 resulted in the most pronounced pro-senescence activity compared to other microalgal extracts in BJ cells (Figure S3).

We also observed that treatment with selected microalgal extracts for 7 days after 2 h pre-treatment with hydrogen peroxide (long-term stimulation) may result in cell loss, as judged by the analysis of microphotographs (Figure 9). As we did not reveal any signs of cytotoxicity after short-term treatments (24 h, majority of assays when cells were plated at the concentration of 10,000 cells per cm^2) and after stimulation with microalgal extracts for 3 days (wound healing assay based on high cell density protocol; Figure S2), one can conclude that perhaps the observed decrease in cell numbers (SA-β-gal assay when cells are typically seeded at the density of 2000 cells per cm^2 to avoid cell culture overcrowding when the test is terminated after 7 days; Figure 9) may be due to extract-mediated cytostatic effects and various cell responses based on the number of cells plated into the culture vessel. Moreover, prolonged antiproliferative effects of selected microalgal extracts after 24 h stimulation and 7 days of the removal of microalgal extracts were also noticed (Figure S3). Thus, we then decided to investigate this phenomenon more comprehensively, namely, the analysis of absolute numbers of senescent and non-senescent cells (cell count test) was performed after 2 h stimulation with hydrogen peroxide and subsequent cell culture with microalgal extracts for 7 days or 24 h treatment with microalgal extracts and subsequent cell culture without microalgal extracts for 7 days (Figure S4). In the case of 2 h pre-treatment with hydrogen peroxide followed by long-term stimulation with microalgal extracts for 7 days (Figure S4a,b), we analyzed the action of microalgal extracts compared to hydrogen peroxide treatment (Figure S4a,b) as we expected that hydrogen peroxide would promote the cytostatic effect itself. Thus, hydrogen peroxide treatment is assumed as a reference (Figure 4a,b). Cytostatic effects of microalgal extracts were limited to BJ cells treated with WE7 (a decrease in cell count of about 46% compared to hydrogen peroxide treatment, $p < 0.001$; Figure S4a). WE7-treated cells were non-senescent cells (non-stained cells; Figure 9a; WE7 treatment). In contrast, pro-stimulatory effects were observed in the case of HEK cells treated with WE9, EE7, and EE8 (an increase in cell number of about 30% compared to hydrogen peroxide treatment, $p < 0.05$, Figure S4b). These cells were also non-senescent cells (non-stained cells; Figure 9b; WE9, EE7, and EE8 treatments). More pronounced cytostatic effects of microalgal extracts were noticed when BJ cells were treated with selected ethanolic extracts for 24 h and then cultured for up to 7 days without the extracts (Figure S4c). In this case, control conditions were considered as a reference (100%; Figure S4c,d). EE1, EE3, and EE4 decreased the number of BJ cells by about 50% compared to standard growth conditions ($p < 0.001$, Figure S4c). According to microphotographs, these cells were mainly non-senescent cells (non-stained cells; Figure S3a; EE1, EE3, and EE4 treatments). Similar effects were not observed in the case of HEK cells treated with ethanolic extracts (Figure S4d). Minor to moderate cytostatic action of microalgal extracts was also noticed in the case of BJ cells treated with WE11, EE2, and EE11, and HEK cells treated with WE4 (a decrease in cell count ranging from 20% to 30% compared to the control; $p < 0.001$, $p < 0.01$ and $p < 0.05$; Figure S4c,d). Thus, one can conclude that the cytostatic action of microalgal extracts is limited, and, in some cases, extract-mediated pro-stimulatory effects are also observed (Figure S4b).

Data on *Chlorella* extract-associated anti-aging action are limited to selected models of replicative senescence, namely, human fibroblasts and myocytes at different passage numbers (young, pre-senescent, and senescent cells) [24,27,43–45] and UV-induced cytotoxicity and senescence [46,47]. Hot water extract of *Chlorella vulgaris* lowered the levels of DNA damage in young, pre-senescent, and senescent fibroblasts in control conditions [27] and during oxidative stress [43]. However, the antioxidant activity of hot water extract of *Chlorella vulgaris* was limited to pre-senescent fibroblasts as judged by the extract-mediated increase in catalase (CAT), superoxide dismutase (SOD), and glutathione peroxidase (GPx) activity [44]. More recently, *Chlorella* extract-mediated modulation of the expression of genes involved in oxidative stress response and DNA damage response and insulin/insulin-like growth factor-1 signaling, cell differentiation, and cell proliferation pathways was analyzed using young, pre-senescent, and senescent fibroblasts [24]. The authors concluded that *Chlorella* extract decreased the expression of *SOD1*, *CAT*, and copper chaperone *CCS* in young fibroblasts and increased the expression of *SOD2* in young, pre-senescent, and senescent fibroblasts [24]. Moreover, *Chlorella* extract promoted a decrease in mRNA levels of two regulators of cell cycle progression, namely, *TP53* (p53) and *CDKN2A* (p16) in replicative old fibroblasts that may suggest some anti-aging effects. Indeed, *Chlorella*-derived peptide (CDP) inhibited UVB-induced matrix metalloproteinase-1 (MMP-1) expression in skin fibroblasts by suppressing the expression of transcription factor AP-1 and cysteine-rich 61 (CYR61) and monocyte chemoattractant protein-1 (MCP-1) production [46]. Thus, CDP may attenuate MMP-1-stimulated UV-associated premature skin aging [46]. Similar protective effects and related mechanisms were observed against UV-induced photo-damage in *Arthrospira platensis* extract-treated human dermal fibroblasts [48].

3.4. Preliminary Analysis of Extract-Mediated Anticancer Activity

As *Chlorella* sp. water and ethanolic extracts have been repeatedly reported to possess pro-apoptotic and growth inhibitory activity against cancer cells of different origin both in vitro and in vivo [49–51], we also initially evaluated the anticancer activity of *Planktochlorella nurekis* water and ethanolic extracts (100 µg/mL) against selected cellular models of cancer in vitro, namely, MDA-MB-231 breast cancer cells, U-2 OS osteosarcoma cells, and U-251 MG glioblastoma cells (Figure S5). An MTT assay was used, and extract-mediated changes in metabolic activity of cancer cells were compared to extract-mediated changes in metabolic activity of normal human fibroblasts (BJ cells; Figure S5). Water extracts at the concentration of 100 µg/mL did not affect metabolic activity of U-2 OS and U-251 MG cells, whereas water extracts WE2, WE3, and WE5 stimulated metabolic activity of about 25% to 30% in MDA-MB-231 cells (Figure S5a; $p < 0.05$). Thus, *Planktochlorella nurekis* water extracts cannot be considered to be used in anticancer therapies, at least not those involving osteosarcoma, glioblastoma, or breast cancers (Figure S5a). In contrast, selected ethanolic extracts acted against cancer cells, especially U-2 OS and U-251 MG cells (Figure S5b). The most pronounced anticancer effects were observed when U-2 OS and U-251 MG cells were treated with EE8, EE7, EE6, and EE2 compared to similar treatments of BJ fibroblasts (Figure S5b, $p < 0.001$). However, the anticancer activity of *Planktochlorella nurekis* ethanolic extracts requires further comprehensive studies.

4. Conclusions

We analyzed for the first time the biological activity of twelve clones of the microalga *Planktochlorella nurekis* cells with improved biochemical features compared to their unmodified counterpart using human skin cells, namely, fibroblasts and keratinocytes, as two cellular models in vitro. As selected microalgal extracts blocked the development of oxidative stress-induced senescence in human skin cells, we strongly believe that this approach may have potential applications in such clinical contexts where protection against the occurrence of old damaged cells has great importance, e.g., in skin anti-aging therapies. Perhaps microalgal extracts may also be active against stress-induced senescence in cells derived from tissues and organs other than skin. More studies involving other cell types are needed to confirm such assumptions and reveal the underlying mechanism(s).

Supplementary Materials: The following are available online at http://www.mdpi.com/2072-6643/12/4/1005/s1. Figure S1: Extract-mediated ERK1/2 activity in BJ cells ((a), 100 µg/mL water extracts) and HEK cells ((b), 100 µg/mL water extracts); Figure S2: Extract-mediated effects on cell migration. Scratch wound healing assay; Figure S3: Pro-senescence activity of microalgal extracts in BJ cells ((a), 100 µg/mL water extracts and 100 µg/mL ethanolic extracts) and HEK cells ((b), 100 µg/mL water extracts and 1 µg/mL ethanolic extracts); Figure S4: Extract-mediated changes in BJ (a) and HEK cell number (b) after 2 h stimulation with hydrogen peroxide and subsequent cell culture for 7 days in the presence of water (left) and ethanolic extracts (right), and the effect of 24 h treatment with water (left) and ethanolic (right) extracts on BJ (c) and HEK cell number (d) after subsequent cell culture for 7 days without microalgal extracts; Figure S5: Preliminary analysis of anticancer activity of water ((a), 100 µg/mL) and ethanolic ((b), 100 µg/mL) extracts against MDA-MB-231 breast cancer, U-2 OS osteosarcoma and U-251 MG glioblastoma cells.

Author Contributions: Conceptualization, A.L. and J.Z.; methodology, A.L. and M.W.; software, A.L. and M.W.; validation, A.L., M.W., and J.A.-G.; formal analysis, A.L.; investigation, J.A.-G., A.L., and M.W.; resources, J.Z.; data curation, M.D. and J.Z.; writing—original draft preparation, A.L.; writing—review and editing, A.L. and M.W.; visualization, A.L.; supervision, J.Z.; project administration, M.D. and J.Z.; funding acquisition, J.Z. All authors have read and agreed to the published version of the manuscript.

Funding: The study was supported by the project no. POIR.01.01.01-00-0405/17 from National Centre for Research and Development. The project was co-financed from European Union funds as a part of the Smart Growth Operational Programme—measure: R&D projects of enterprises; sub-measure: industrial research and development work implemented by enterprises.

Conflicts of Interest: The authors declare no conflict of interest. The funders had no role in the design of the study; in the collection, analyses, or interpretation of data; in the writing of the manuscript, or in the decision to publish the results.

References

1. Kim, K.E.; Cho, D.; Park, H.J. Air pollution and skin diseases: Adverse effects of airborne particulate matter on various skin diseases. *Life Sci.* **2016**, *152*, 126–134. [CrossRef] [PubMed]
2. Parrado, C.; Mercado-Saenz, S.; Perez-Davo, A.; Gilaberte, Y.; Gonzalez, S.; Juarranz, A. Environmental stressors on skin aging. Mechanistic insights. *Front. Pharmacol.* **2019**, *10*, 759. [CrossRef] [PubMed]
3. Soeur, J.; Belaidi, J.P.; Chollet, C.; Denat, L.; Dimitrov, A.; Jones, C.; Perez, P.; Zanini, M.; Zobiri, O.; Mezzache, S.; et al. Photo-pollution stress in skin: Traces of pollutants (PAH and particulate matter) impair redox homeostasis in keratinocytes exposed to UVA1. *J. Dermatol. Sci.* **2017**, *86*, 162–169. [CrossRef]
4. Naylor, E.C.; Watson, R.E.; Sherratt, M.J. Molecular aspects of skin ageing. *Maturitas* **2011**, *69*, 249–256. [CrossRef] [PubMed]
5. Zhang, S.; Duan, E. Fighting against skin aging: The way from bench to bedside. *Cell Transplant.* **2018**, *27*, 729–738. [CrossRef] [PubMed]
6. Wang, H.D.; Chen, C.C.; Huynh, P.; Chang, J.S. Exploring the potential of using algae in cosmetics. *Bioresour. Technol.* **2015**, *184*, 355–362. [CrossRef]
7. Wang, H.D.; Li, X.C.; Lee, D.J.; Chang, J.S. Potential biomedical applications of marine algae. *Bioresour. Technol.* **2017**, *244*, 1407–1415. [CrossRef]
8. Ariede, M.B.; Candido, T.M.; Jacome, A.L.M.; Velasco, M.V.R.; de Carvalho, J.C.M.; Baby, A.R. Cosmetic attributes of algae—A review. *Algal Res.* **2017**, *25*, 483–487. [CrossRef]
9. Urikura, I.; Sugawara, T.; Hirata, T. Protective effect of fucoxanthin against UVB-induced skin photoaging in hairless mice. *Biosci. Biotechnol. Biochem.* **2011**, *75*, 757–760. [CrossRef]
10. Kim, M.S.; Oh, G.H.; Kim, M.J.; Hwang, J.K. Fucosterol inhibits matrix metalloproteinase expression and promotes type-1 procollagen production in UVB-induced HaCaT cells. *Photochem. Photobiol.* **2013**, *89*, 911–918. [CrossRef]
11. Ryu, J.; Park, S.J.; Kim, I.H.; Choi, Y.H.; Nam, T.J. Protective effect of porphyra-334 on UVA-induced photoaging in human skin fibroblasts. *Int. J. Mol. Med.* **2014**, *34*, 796–803. [CrossRef] [PubMed]
12. Balboa, E.M.; Li, Y.X.; Ahn, B.N.; Eom, S.H.; Dominguez, H.; Jimenez, C.; Rodriguez, J. Photodamage attenuation effect by a tetraprenyltoluquinol chromane meroterpenoid isolated from *Sargassum muticum*. *J. Photochem. Photobiol. B* **2015**, *148*, 51–58. [CrossRef] [PubMed]
13. Rosenberg, J.N.; Oyler, G.A.; Wilkinson, L.; Betenbaugh, M.J. A green light for engineered algae: Redirecting metabolism to fuel a biotechnology revolution. *Curr. Opin. Biotechnol.* **2008**, *19*, 430–436. [CrossRef] [PubMed]
14. Borowitzka, M.A. High-value products from microalgae—Their development and commercialisation. *J. Appl. Phycol.* **2013**, *25*, 743–756. [CrossRef]
15. Škaloud, P.; Němcová, Y.; Pytela, J.; Bogdanov, N.I.; Bock, C.; Pickinpaugh, S.H. *Planktochlorella nurekis* gen. et sp. nov. (*Trebouxiophyceae, Chlorophyta*), a novel coccoid green alga carrying significant biotechnological potential. *Fottea* **2014**, *14*, 53–62. [CrossRef]
16. Szpyrka, E.; Broda, D.; Oklejewicz, B.; Podbielska, M.; Slowik-Borowiec, M.; Jagusztyn, B.; Chrzanowski, G.; Kus-Liskiewicz, M.; Duda, M.; Zuczek, J.; et al. A non-vector approach to increase lipid levels in the microalga *Planktochlorella nurekis*. *Molecules* **2020**, *25*, 270. [CrossRef]
17. Lewinska, A.; Adamczyk-Grochala, J.; Kwasniewicz, E.; Deregowska, A.; Semik, E.; Zabek, T.; Wnuk, M. Reduced levels of methyltransferase DNMT2 sensitize human fibroblasts to oxidative stress and DNA damage that is accompanied by changes in proliferation-related miRNA expression. *Redox Biol.* **2018**, *14*, 20–34. [CrossRef]
18. Lewinska, A.; Bocian, A.; Petrilla, V.; Adamczyk-Grochala, J.; Szymura, K.; Hendzel, W.; Kaleniuk, E.; Hus, K.K.; Petrillova, M.; Wnuk, M. Snake venoms promote stress-induced senescence in human fibroblasts. *J. Cell. Physiol.* **2019**, *234*, 6147–6160. [CrossRef]
19. Lewinska, A.; Adamczyk-Grochala, J.; Bloniarz, D.; Olszowka, J.; Kulpa-Greszta, M.; Litwinienko, G.; Tomaszewska, A.; Wnuk, M.; Pazik, R. AMPK-mediated senolytic and senostatic activity of quercetin surface functionalized Fe_3O_4 nanoparticles during oxidant-induced senescence in human fibroblasts. *Redox Biol.* **2020**, *28*, 101337. [CrossRef]
20. Chon, J.W.; Sung, J.H.; Hwang, E.J.; Park, Y.K. *Chlorella* methanol extract reduces lipid accumulation in and increases the number of apoptotic 3T3-L1 cells. *Ann. N. Y. Acad. Sci.* **2009**, *1171*, 183–189. [CrossRef]
21. Song, S.H.; Kim, I.H.; Nam, T.J. Effect of a hot water extract of *Chlorella vulgaris* on proliferation of IEC-6 cells. *Int. J. Mol. Med.* **2012**, *29*, 741–746. [PubMed]

22. Ilavenil, S.; Kim, D.H.; Vijayakumar, M.; Srigopalram, S.; Roh, S.G.; Arasu, M.V.; Lee, J.S.; Choi, K.C. Potential role of marine algae extract on 3T3-L1 cell proliferation and differentiation: An in vitro approach. *Biol. Res.* **2016**, *49*, 38. [CrossRef] [PubMed]

23. Himuro, S.; Ueno, S.; Noguchi, N.; Uchikawa, T.; Kanno, T.; Yasutake, A. Safety evaluation of *Chlorella sorokiniana* strain CK-22 based on an in vitro cytotoxicity assay and a 13-week subchronic toxicity trial in rats. *Food Chem. Toxicol.* **2017**, *106*, 1–7. [CrossRef] [PubMed]

24. Jaafar, F.; Durani, L.W.; Makpol, S. *Chlorella vulgaris* modulates the expression of senescence-associated genes in replicative senescence of human diploid fibroblasts. *Mol. Biol. Rep.* **2020**, *47*, 369–379. [CrossRef] [PubMed]

25. Grether-Beck, S.; Muhlberg, K.; Brenden, H.; Felsner, I.; Brynjolfsdottir, A.; Einarsson, S.; Krutmann, J. Bioactive molecules from the blue lagoon: In vitro and in vivo assessment of silica mud and microalgae extracts for their effects on skin barrier function and prevention of skin ageing. *Exp. Dermatol.* **2008**, *17*, 771–779. [CrossRef]

26. Gunes, S.; Tamburaci, S.; Dalay, M.C.; Deliloglu Gurhan, I. In vitro evaluation of *Spirulina platensis* extract incorporated skin cream with its wound healing and antioxidant activities. *Pharm. Biol.* **2017**, *55*, 1824–1832. [CrossRef]

27. Saberbaghi, T.; Abbasian, F.; Mohd Yusof, Y.A.; Makpol, S. Modulation of cell cycle profile by *Chlorella vulgaris* prevents replicative senescence of human diploid fibroblasts. *Evid. Based Complement. Altern. Med.* **2013**, *2013*, 780504. [CrossRef]

28. Jeong, S.J.; Choi, J.W.; Lee, M.K.; Choi, Y.H.; Nam, T.J. *Spirulina* crude protein promotes the migration and proliferation in IEC-6 cells by activating EGFR/MAPK signaling pathway. *Mar. Drugs* **2019**, *17*, 205. [CrossRef]

29. Zailan, N.; Abdul Rashid, A.H.; Das, S.; Abdul Mokti, N.A.; Hassan Basri, J.; Teoh, S.L.; Wan Ngah, W.Z.; Mohd Yusof, Y.A. Comparison of *Chlorella vulgaris* dressing and sodium alginate dressing: An experimental study in rats. *Clin. Ter.* **2010**, *161*, 515–521.

30. Hidalgo-Lucas, S.; Bisson, J.F.; Duffaud, A.; Nejdi, A.; Guerin-Deremaux, L.; Baert, B.; Saniez-Degrave, M.H.; Rozan, P. Benefits of oral and topical administration of roquette *Chlorella* sp. On skin inflammation and wound healing in mice. *Anti-Inflamm. Anti-Allergy Agents Med. Chem.* **2014**, *13*, 93–102. [CrossRef]

31. Bari, E.; Arciola, C.R.; Vigani, B.; Crivelli, B.; Moro, P.; Marrubini, G.; Sorrenti, M.; Catenacci, L.; Bruni, G.; Chlapanidas, T.; et al. In vitro effectiveness of microspheres based on silk sericin and *Chlorella vulgaris* or *Arthrospira platensis* for wound healing applications. *Materials* **2017**, *10*, 983. [CrossRef] [PubMed]

32. Syarina, P.N.; Karthivashan, G.; Abas, F.; Arulselvan, P.; Fakurazi, S. Wound healing potential of *Spirulina platensis* extracts on human dermal fibroblast cells. *EXCLI J.* **2015**, *14*, 385–393. [PubMed]

33. Schieber, M.; Chandel, N.S. ROS function in redox signaling and oxidative stress. *Curr. Biol.* **2014**, *24*, R453–R462. [CrossRef] [PubMed]

34. Finkel, T. Signal transduction by reactive oxygen species. *J. Cell Biol.* **2011**, *194*, 7–15. [CrossRef]

35. Russell, E.G.; Cotter, T.G. New insight into the role of reactive oxygen species (ROS) in cellular signal-transduction processes. *Int. Rev. Cell Mol. Biol.* **2015**, *319*, 221–254.

36. Labunskyy, V.M.; Gladyshev, V.N. Role of reactive oxygen species-mediated signaling in aging. *Antioxid. Redox Signal.* **2013**, *19*, 1362–1372. [CrossRef]

37. Chen, K.; Feng, H.; Zhang, M.; Wang, X. Nitric oxide alleviates oxidative damage in the green alga *Chlorella pyrenoidosa* caused by UV-B radiation. *Folia Microbiol.* **2003**, *48*, 389–393. [CrossRef]

38. Singh, A.; Sharma, L.; Mallick, N. Antioxidative role of nitric oxide on copper toxicity to a chlorophycean alga, *Chlorella. Ecotoxicol. Environ. Saf.* **2004**, *59*, 223–227. [CrossRef]

39. De Magalhaes, J.P.; Passos, J.F. Stress, cell senescence and organismal ageing. *Mech. Ageing Dev.* **2018**, *170*, 2–9. [CrossRef]

40. Naylor, R.M.; Baker, D.J.; van Deursen, J.M. Senescent cells: A novel therapeutic target for aging and age-related diseases. *Clin. Pharmacol. Ther.* **2013**, *93*, 105–116. [CrossRef]

41. Childs, B.G.; Gluscevic, M.; Baker, D.J.; Laberge, R.M.; Marquess, D.; Dananberg, J.; van Deursen, J.M. Senescent cells: An emerging target for diseases of ageing. *Nat. Rev. Drug Discov.* **2017**, *16*, 718–735. [CrossRef] [PubMed]

42. Kennedy, D.O. B vitamins and the brain: Mechanisms, dose and efficacy—A review. *Nutrients* **2016**, *8*, 68. [CrossRef] [PubMed]

43. Makpol, S.; Yaacob, N.; Zainuddin, A.; Yusof, Y.A.; Ngah, W.Z. *Chlorella vulgaris* modulates hydrogen peroxide-induced DNA damage and telomere shortening of human fibroblasts derived from different aged individuals. *Afr. J. Tradit. Complement. Altern. Med.* **2009**, *6*, 560–572. [CrossRef] [PubMed]

44. Makpol, S.; Yeoh, T.W.; Ruslam, F.A.; Arifin, K.T.; Yusof, Y.A. Comparative effect of *Piper betle*, *Chlorella vulgaris* and tocotrienol-rich fraction on antioxidant enzymes activity in cellular ageing of human diploid fibroblasts. *BMC Complement. Altern. Med.* **2013**, *13*, 210. [CrossRef] [PubMed]

45. Zainul Azlan, N.; Mohd Yusof, Y.A.; Alias, E.; Makpol, S. *Chlorella vulgaris* improves the regenerative capacity of young and senescent myoblasts and promotes muscle regeneration. *Oxid. Med. Cell. Longev.* **2019**, *2019*, 3520789. [CrossRef] [PubMed]

46. Chen, C.L.; Liou, S.F.; Chen, S.J.; Shih, M.F. Protective effects of *Chlorella*-derived peptide on UVB-induced production of MMP-1 and degradation of procollagen genes in human skin fibroblasts. *Regul. Toxicol. Pharmacol.* **2011**, *60*, 112–119. [CrossRef]

47. Shih, M.F.; Cherng, J.Y. Protective effects of *Chlorella*-derived peptide against UVC-induced cytotoxicity through inhibition of caspase-3 activity and reduction of the expression of phosphorylated FADD and cleaved PARP-1 in skin fibroblasts. *Molecules* **2012**, *17*, 9116–9128. [CrossRef]

48. Lee, J.J.; Kim, K.B.; Heo, J.; Cho, D.H.; Kim, H.S.; Han, S.H.; Ahn, K.J.; An, I.S.; An, S.; Bae, S. Protective effect of *Arthrospira platensis* extracts against ultraviolet B-induced cellular senescence through inhibition of DNA damage and matrix metalloproteinase-1 expression in human dermal fibroblasts. *J. Photochem. Photobiol. B* **2017**, *173*, 196–203. [CrossRef]

49. Chung, J.-G.; Peng, H.-Y.; Chu, Y.-C.; Hsieh, Y.-M.; Wang, S.-D.; Chou, S.-T. Anti-invasion and apoptosis induction of *Chlorella* (*Chlorella sorokiniana*) in Hep G2 human hepatocellular carcinoma cells. *J. Funct. Foods* **2012**, *4*, 302–310. [CrossRef]

50. Lin, P.Y.; Tsai, C.T.; Chuang, W.L.; Chao, Y.H.; Pan, I.H.; Chen, Y.K.; Lin, C.C.; Wang, B.Y. *Chlorella sorokiniana* induces mitochondrial-mediated apoptosis in human non-small cell lung cancer cells and inhibits xenograft tumor growth in vivo. *BMC Complement. Altern. Med.* **2017**, *17*, 88. [CrossRef]

51. Zhang, Z.D.; Liang, K.; Li, K.; Wang, G.Q.; Zhang, K.W.; Cai, L.; Zhai, S.T.; Chou, K.C. *Chlorella vulgaris* induces apoptosis of human non-small cell lung carcinoma (NSCLC) cells. *Med. Chem.* **2017**, *13*, 560–568. [CrossRef] [PubMed]

Article

Expression Profiles of Genes Encoding Cornified Envelope Proteins in Atopic Dermatitis and Cutaneous T-Cell Lymphomas

Magdalena Trzeciak [1], Berenika Olszewska [1,*], Monika Sakowicz-Burkiewicz [2], Małgorzata Sokołowska-Wojdyło [1], Jerzy Jankau [3], Roman Janusz Nowicki [1] and Tadeusz Pawełczyk [2]

[1] Department of Dermatology, Venereology and Allergology, Medical University of Gdansk, 80-214 Gdansk, Poland; mtrzeciak@gumed.edu.pl (M.T.); mwojd@gumed.edu.pl (M.S.-W.); rnowicki@gumed.edu.pl (R.J.N.)
[2] Department of Molecular Medicine, Medical University of Gdansk, 80-214 Gdansk, Poland; ssak@gumed.edu.pl (M.S.-B.); tkpaw@gumed.edu.pl (T.P.)
[3] Department of Plastic Surgery, Medical University of Gdańsk, 80-214 Gdańsk, Poland; jjankau@gumed.edu.pl
* Correspondence: berenika.olszewska@o2.pl; Tel.: +48-58-584-40-10

Received: 30 January 2020; Accepted: 22 March 2020; Published: 24 March 2020

Abstract: The skin barrier defect in cutaneous T-cell lymphomas (CTCL) was recently confirmed to be similar to the one observed in atopic dermatitis (AD). We have examined the expression level of cornified envelope (CE) proteins in CTCL, AD and healthy skin, to search for the differences and their relation to the courses of both diseases. The levels of *FLG, FLG2, RPTN, HRNR, SPRR1A, SPRR1B, SPRR3* and *LELP-1* mRNA were determined by qRT-PCR, while protein levels were examined using the ELISA method in skin samples. We have found that mRNA levels *of FLG, FLG2, LOR, CRNN* and *SPRR3v1* were decreased ($p \leq 0.04$), whereas mRNA levels of *RPTN, HRNR* and *SPRR1Av1* were increased in lesional and nonlesional AD skin compared to the healthy control group ($p \leq 0.04$). The levels of *FLG, FLG2, CRNN, SPRR3v1* mRNA increased ($p \leq 0.02$) and *RPTN, HRNR* and *SPRR1Av1* mRNA decreased ($p \leq 0.005$) in CTCL skin compared to the lesional AD skin. There was a strong correlation between the stage of CTCL and increased *SPRR1Av1* gene expression at both mRNA (R = 0.89; $p \leq 0.05$) and protein levels (R = 0.94; $p \leq 0.05$). FLG, FLG2, RPTN, HRNR and SPRR1A seem to play a key role in skin barrier dysfunction in CTCL and could be considered a biomarker for differential diagnosis of AD and CTCL. *SPRR1Av1* transcript levels seem to be a possible marker of CTCL stage, however, further studies on a larger study group are needed to confirm our findings.

Keywords: mycosis fungoides; atopic dermatitis; cutaneous lymphomas; cornified envelope proteins; FLG

---

1. Introduction

Cutaneous T-cell lymphomas (CTCLs) represent a rare heterogeneous group of extranodal non-Hodgkin lymphomas. CTCLs are characterized by an infiltration of the skin with neoplastic CD4 + CD45RO + skin-homing T-cells [1,2]. The pathogenesis of CTCL remains unexplained. There are a few hypotheses, including chronic antigen stimulation, viral or bacterial, leading to a loss of immune-surveillance and, therefore, the proliferation of neoplastic T-cells. Moreover, chromosomal instability and abnormal expression of genes involved in the cell cycle and a complex series of interactions between different cells in skin microenvironment have been suggested [3–6]. The inflammatory microenvironment of the skin seems to play a crucial role in CTCL pathogenesis, as the predominance of Th2 over Th1 cells in inflammatory microenvironment seems to be responsible for the suppression of antitumor response, proliferation of malignant cells and escape from immunosurveillance [7–9]. The clinical spectrum of CTCL varies widely; the predominant subtype is mycosis fungoides (MF), accounting for 60% [1]. The clinical presentation of MF follows some clinical stages, ranging from patch-like lesions in the early stage, through plaque stage, erythroderma and tumors; usually, all stages are accompanied by severe pruritus [1]. MF is also a significant diagnostic challenge in histopathological examination as it may mimic many benign dermatoses such as atopic dermatitis (AD) [1,2]. CTCL shares many clinical and histological characteristics with AD. Not only the clinical presentation but also immunological features are overlapping. Both MF and AD show infiltration of the skin by skin-homing T-cells [10,11]. The skin of MF patients similar to AD shows an impaired skin barrier with decreased expression of filaggrin, loricrin and antimicrobial peptides in the skin, which results in increased susceptibility towards *Staphylococcus aureus* colonization and infections [11,12]. The malignant T-cells were reported to be responsible for changes in the epidermis in CTCL and impaired epidermal barrier function [13]. Therefore, skin infections have been a significant clinical problem both in AD and in CTCL as a result of compromised skin barrier and a progressive immunodeficiency in CTCL. In fact, patients with progressive CTCL die more frequently from infection rather than from the disease itself [14]. Changes in the expression of cornified envelope proteins in the skin were shown to impact the development and course of AD [15–17]. In the face of so many similarities between CTCL and AD, evaluation of expression of genes encoding the cornified envelope (CE) proteins such as the late cornified envelope-like proline-rich protein-1 (LELP-1), small proline-rich proteins (SPRR1A, SPRR1B, SPRR3), repetin (RPTN), cornulin (CRNN), hornerin (HRNR), loricrin (LOR) and filaggrin (FLG, FLG2) might provide accurate information and findings to differentiate those diseases and determine their relation to the course and development of both diseases.

2. Materials and Methods

2.1. Patients and Samples

The study group included: 11 patients with AD (seven males, four females, median age 30, age range 14–59) and nine CTCL patients (eight males 8, females 1, median age 65, age range 26–75) treated at the Department of Dermatology, Venerology and Allergology of the Medical University of Gdańsk. CTCL patients were diagnosed according to the International Society of Cutaneous Lymphoma (ISCL) and the European Organization of Research and Treatment of Cancer (EORTC) criteria [18], while the diagnosis of AD was based on Hanifin and Rajka criteria [19]. The control group comprised 11 healthy volunteers (one male, 10 females, median age 34, age range 20–54), without a medical history of immunological diseases, allergies or malignancies. Punch biopsies for gene expression and protein level analysis were obtained from CTCL patients (nine from lesional skin), AD patients (11 from lesional skin and 11 from nonlesional skin) and healthy controls (11 from healthy skin).

The study was approved by the Ethics Committee of the Medical University of Gdańsk (NKBBN/317/2018) and was conducted according to the principles of the Declaration of Helsinki. All participants signed an informed consent prior to any study procedure.

2.2. Isolation of Total RNA from Skin Fragments

Isolation of total RNA was carried out in accordance with the Chomczyński procedure, with our own modifications [20]. The skin fragments (4-mm punch biopsies) were homogenized in a sterile tube with 1 mL of TRI reagent (Sigma-Aldrich, Poznań, Poland). Next, chloroform (250 µL) was added, and the samples were vigorously shaken, incubated at 4 °C for 15 min and centrifuged (10,000× g for 15 min at 4 °C). The upper aqueous phase was removed into a new Eppendorf tube where an equal volume of isopropanol was added, and RNA was precipitated by overnight incubation at −20 °C, followed by centrifugation (10,000× g for 15 min at 4 °C). RNA pellets were washed first with 96% and then with 75% (*v/v*) ethanol, air-dried, resolved in diethyl-pyrocarbonate-treated thermo-sterilized water (30 µL) and stored at −20 °C until further analysis.

2.3. Relative Quantitative Real-Time RT-PCR Analysis

FLG, FLG2, RPTN, HRNR, LELP-1, SPRR 1A, SPRR1B, SPRR3, LOR and *HRNR* mRNA levels were analyzed by real-time PCR (RT-PCR) with TaqMan primer–probe sets using the Path-ID Multiplex

One-Step RT-PCR kit (Path-ID Multiplex One-Step RT-PCR Kit, Applied Biosystems, Foster City, CA, USA). The reference transcript (*ACTB* or *TBP* or *G6PD*) was used as an internal standard and was amplified together with each target gene transcript in the same way using primers and probes (Table 1). Data analysis was performed using LightCycler 480 II software (Roche Diagnostics International Ltd., Rotkreuz, Switzerland).

Table 1. Gene transcript, primers and TaqMan probes used for real-time polymerase chain reaction (RT-PCR).

Gene Transcript	Primers	TaqMan Probe	Transcript of Reference Gene
FLG NM_002016.1	(F) ggactctgagaggcgatctg (R) tgctcccgagaagatccat	Universal ProbeLibrary Probe #38 (Roche)	Universal ProbeLibrary Reference Gene Assay Roche, Human *TBP* Gene Assay
FLG2 NM_001014342.2	(F) tgactatggcctgcaacaa (R) ctttgaccctgaagctttgc	Universal ProbeLibrary Probe #73 (Roche)	Universal ProbeLibrary Reference Gene Assay Roche, Human *G6PD* Gene Assay
LELP1 NM_001010857.1	(F) cccaagtgtgaacaaaagtg (R) ttcgaaacagcgttgcag	Universal ProbeLibrary Probe #26	Universal ProbeLibrary Reference Gene Assay Roche, Human *ACTB* Gene Assay
SPRR1Avar.1 NM_001199828.1	(F) tcgggtgcatttgaggat (R) aaggaagactagggatggttca	Universal ProbeLibrary Probe #60 (Roche)	Universal ProbeLibrary Reference Gene Assay Roche, Human *ACTB* Gene Assay
SPRR1B NM_003125.2	(F) gagagacttaagatgaaagcaatga (R) tgaaagtgaatttaatgggggta	Universal ProbeLibrary Probe #24 (Roche)	Universal ProbeLibrary Reference Gene Assay Roche, Human *ACTB* Gene Assay
SPRR3var.1 NM_005416.2	(F) tcaggagcttagaggattcttca (R) ttctgctggtaagaactcatgc	Universal ProbeLibrary Probe #89 (Roche)	Universal ProbeLibrary Reference Gene Assay Roche, Human *ACTB* Gene Assay
RPTN NM_001122965.1	(F) gctcttggctgagtttggag (R) aggttcaagatggtttccaca	Universal ProbeLibrary Probe #65 (Roche)	Universal ProbeLibrary Reference Gene Assay Roche, Human *TBP* Gene
HRNR NM_001009931.2	(F) caggggcaagatgggtattc (R) ccagaacttccccctcat	Universal ProbeLibrary Probe #69 (Roche)	Universal ProbeLibrary Reference Gene Assay Roche, Human *TBP* Gene Assay
CRNN NM_016190.2	(F) tggtcaaagatggatgcaag (R) cctgtcctcccggtactgt	Universal ProbeLibrary Probe #51 (Roche)	Universal ProbeLibrary Reference Gene Assay Roche, Human *G6PD* Gene
LOR NM_000427.2	(F) ctcacccttcctggtgctt (R) gaggtcttcacgcagtcca	Universal ProbeLibrary Probe #12 (Roche)	Universal ProbeLibrary Reference Gene Assay Roche, Human *ACTB* Gene Assay

2.4. Protein Level Determination

Protein levels of FLG, FLG2, LELP-1, SPRR1A and SPRR1B were determined using an enzyme-linked immunoabsorbent assay (ELISA) standard kit (Antibody-Protein ELISA kit, MyBioSource Inc., San Diego, CA, USA). We analyzed 4 CTCL, 6 AD and 11 control group samples with ELISA. Skin specimens were weighed, rinsed with cold phosphate-buffered saline (PBS) and homogenized in PBS containing 0.5% Igepol CA-630 (1:2) on ice. The homogenates were subjected to three freeze–thaw cycles to break the cell membranes, and the homogenates were centrifuged for 5 min at 5000× *g*. The supernatants obtained were used for protein determination by ELISA kit according to the manufacturer's protocol. The quantity of investigated protein in the sample was interpolated from the standard curve and corrected for dilution.

2.5. Statistical Analysis

Statistical analysis was carried out using the Mann–Whitney U test. A comparison between lesional and nonlesional skin from the same patient was performed with the Wilcoxon matched-pairs test (marked in figures by dashed line). Associations between the two variables were assessed based on Spearman's rank correlation. The results were considered statistically significant when $p < 0.05$. Statistical analyses were performed using Statistica 13.3.

3. Results

3.1. mRNA Level of Cornified Envelop Proteins

Expression levels of almost all genes encoding the CE proteins differed significantly between the AD and control groups, except for *SPRR1B* and *LELP1*. We have found significantly lower levels of *FLG*, *FLG2*, *LOR* and *CRNN* mRNA in lesional and nonlesional AD skin compared to the healthy control group ($p \leq 0.02$). mRNA level of *SPRR3v1* was also decreased in AD samples, however, only in lesional skin ($p \leq 0.00008$). Moreover, mRNA levels of those genes were significantly decreased in lesional skin in comparison to the nonlesional skin of AD patients ($p \leq 0.04$), except for CRNN. On the other hand, mRNA levels of *LELP1, RPTN, HRNR and SPRR1Av1* were significantly higher in AD skin than healthy controls ($p \leq 0.05$). The lesional skin of AD patients demonstrated significantly higher expression of *HRNR* and *SPRR1Av1* mRNA in comparison to nonlesional AD skin samples ($p \leq 0.04$).

Differences in expression levels of genes encoding the CE proteins between CTCL skin and control group were observed only in a few genes. The mRNA levels of *FLG* and *FLG2* were significantly lower in CTCL skin compared to control skin ($p \leq 0.02$), whereas *RPTN* and *SPRR1Av1* mRNA expression were significantly higher ($p \leq 0.04$). All CE genes mRNA levels in skin of CTCL, AD and healthy subjects are demonstrated in detail in Figure 1.

Except for *LELP1, SPRR1B* and *SPRR3*, all other CE genes showed significant differences in mRNA expression between CTCL and both lesional and nonlesional AD skin samples. We observed that the levels of *LOR* and *CRNN* mRNA were significantly increased and *HRNR* mRNA was decreased in CTCL skin compared to both lesional and nonlesional AD skin. The levels of *FLG, FLG2* and *SPRR3v1* mRNA were significantly higher in CTCL than lesional AD skin. Moreover, *RPTN, HRNR* and *SPRR1Av1* mRNA expression were significantly lower in CTCL skin compared to the lesional AD skin.

3.2. Cornified Envelope Proteins Levels

The levels of FLG and FLG2 proteins were significantly lower in CTCL and both lesional and nonlesional AD skin compared to healthy controls skin. Both FLG and FLG2 levels were significantly lower in lesional than nonlesional AD skin samples. The level of FLG2 protein was increased in CTCL compared to AD skin samples ($p \leq 0.02$). SPRR1A protein level both in CTCL ($p \leq 0.01$) and AD skin ($p \leq 0.001$) was significantly increased in comparison with healthy skin. Moreover, the level of SPRR1A in CTCL was lower than in lesional ($p \leq 0.02$) and nonlesional AD skin ($p \leq 0.025$). On the other hand, the level of SPRR1B was distinctly higher in AD lesional and nonlesional skin samples compared with the skin of healthy subjects. Levels of FLG, FLG2, LELP1, SPRR1A and SPRR1B proteins in skin of CTCL, AD and healthy subjects are demonstrated in Figure 2. We have also found a significant correlation between increased *SPRR1Av1* gene expression at both mRNA (R = 0.89; $p \leq 0.05$) and protein levels (R = 0.94; $p \leq 0.05$) and stage of disease (Figure 3).

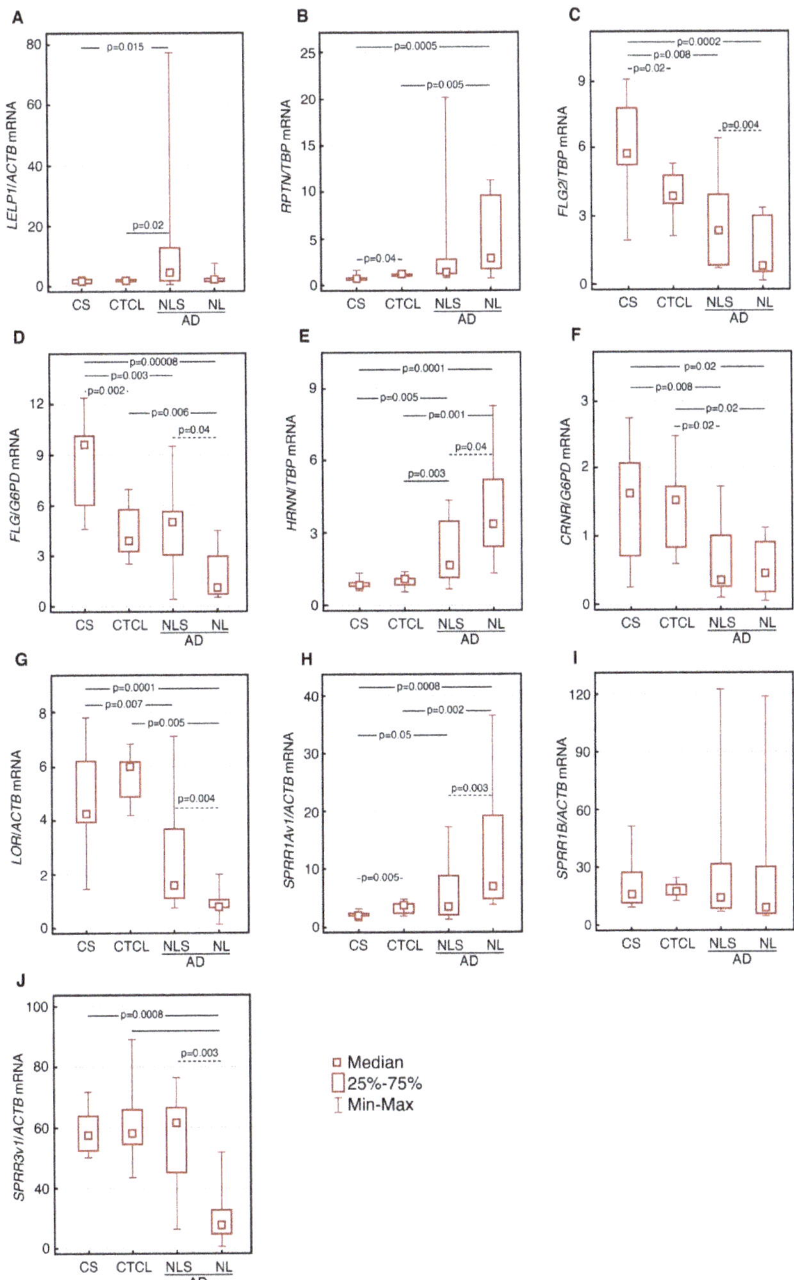

Figure 1. Levels of *LELP1* (**A**), *RPTN* (**B**), *FLG2* (**C**), *FLG* (**D**), *HRNR* (**E**), *CRNN* (**F**), *LOR* (**G**), *SPRR1Av1* (**H**), *SPRR1B* (**I**) and *SPRR3v1* (**J**) transcripts in skin of healthy subjects (CS), cutaneous T-cell lymphomas (CTCL), lesional atopic dermatitis (LS AD) and nonlesional AD (NLS AD).

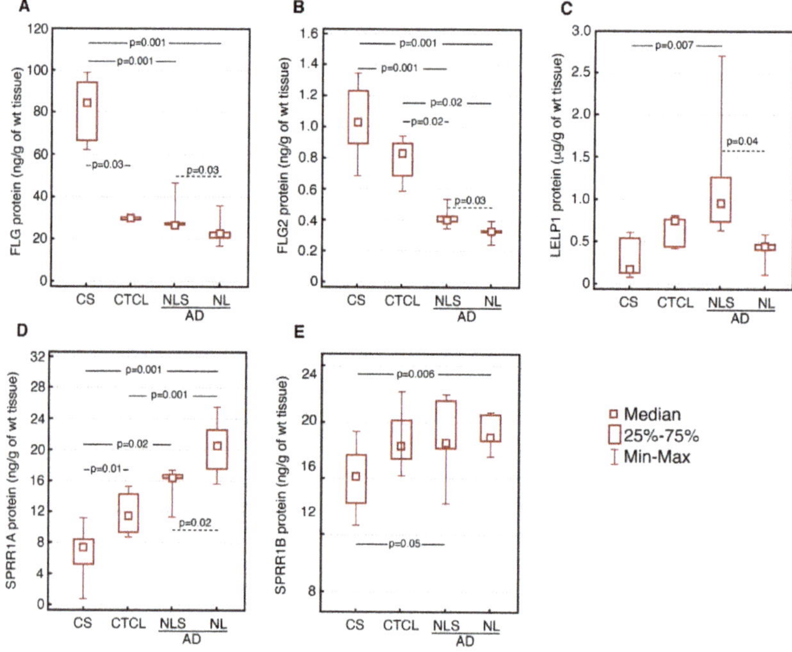

Figure 2. Levels of FLG (**A**), FLG2 (**B**), LELP1 (**C**), SPRR1A (**D**) and SPRR1B (**E**) proteins in skin of healthy subjects (CS) and patients with CTCL, lesional AD (LS AD) and nonlesional AD (NLS AD).

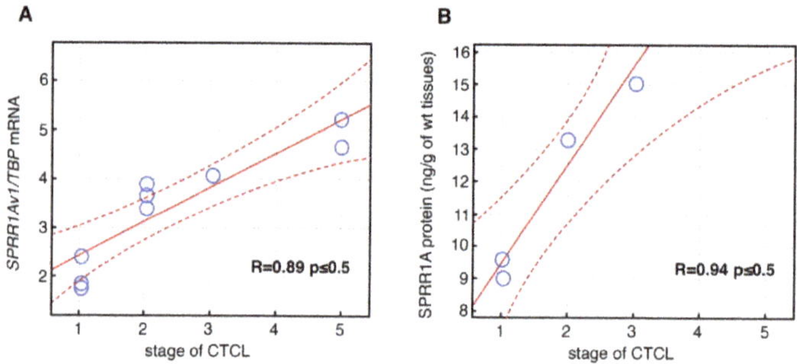

Figure 3. Correlation of *SPRR1Av1* gene expression at both mRNA (**A**) and protein levels (**B**) with stage of disease.

4. Limitations of This Study

We found some discrepancies between our finding and those of and other studies [11] that might be related to our relatively small study group with only a few cases of advanced-stage CTCL, which is a limitation of our study. Moreover, the major limitation is the disparity in the number of samples from female and male volunteers between the control, CTCL and AD groups. Therefore, it would be beneficial to perform further studies that would include a larger study group and a suitable number of female and male patients.

5. Discussion

Factors such as cornified envelope proteins that contribute to skin barrier dysfunction have been extensively studied, and their role in AD skin barrier defects has been proven [15–17,21–24]. However, reports concerning their role in CTCL are rather scarce. There is only one study, by Suga et al., that has analyzed skin barrier dysfunction in CTCL [11]. Decreased levels of *FLG* and *LOR* mRNA have already been demonstrated in the skin of CTCL and psoriasis patients, which confirms that skin barrier dysfunction is not specific to AD pathogenesis [11]. Due to some similarities between CTCL and AD, CE proteins appear to be a promising target for markers helpful in differential diagnosis of those diseases.

Our study has revealed decreased mRNA expression of *FLG, FLG2, LOR, CRNN* and *SPRR3v1* concomitant with increased *LELP1, RPTN, HRNR* and *SPRR1Av1* mRNA levels in AD compared to healthy controls, which is in line with previous findings [15–17,21–24]. Moreover, the majority of examined CE proteins were decreased in comparison to healthy skin, which confirms that pathological changes also occur in nonlesional AD skin. We have noticed similar upward and downward tendencies in the expression of genes encoding some of the CE proteins in AD and CTCL skin in comparison to healthy controls. However, alterations in *HRNR, SPRR3v1, LOR* and *CRNN* mRNA levels in skin are more specific for AD than CTCL. Despite similarities in CE protein expression, the same significant differences between CTCL and AD have also been demonstrated.

We observed that trends of mRNA expression of *LELP1, HRNR, CRNN, LOR, SPRR1B* and *SPRR3* in CTCL were consistent with expression in the control group rather than the AD group, indicating that those CE proteins do not participate in the CTCL skin barrier defect. It seems that FLG, FLG2, RPTN and SPRR1A are the true key players in CTCL skin barrier deficiency, as levels of their transcripts differed significantly in comparison to healthy controls. Moreover, mentioned levels of mRNA expression in skin and levels of FLG2 and SPRR1A proteins differed significantly between both AD and healthy subjects. FLG, FLG2, RPTN and SPRR1A together might be considered potential candidate biomarkers for differential diagnosis in AD and CTCL using either PCR or ELISA method.

Our findings concerning CTCL are partially in line with previous study results [11]. Filaggrin mRNA expression levels in skin of CTCL patients were significantly lower than in the control group, which is in accordance with the findings of Suga et al. [11]. However, *LOR* mRNA expression did not differ between CTCL and healthy skin, which is contrary to the previous report [11].

Recently, Kim et al. demonstrated that Th2 cytokines such as IL-4 and IL-13 inhibit expression of filaggrin and loricrin by primary human keratinocytes [25]. It seems that the immunity against infections is downregulated not only by a predominance of Th2 cytokine pattern in CTCL skin microenvironment but also by the skin barrier dysfunction. It would be worth examining whether expression of other CE proteins is also downregulated by Th2 cytokines. Furthermore, the hypothesis of CTCL pathogenesis, concerning chronic antigen stimulation with bacterial superantigens such as *S. aureus* leading to the development and clonal proliferation of malignant T-cells seems to be up to date [4,26–29]. In the resent study, Lindahl and coworkers reported inhibition of malignant T-cells and clinical improvement in patients with advanced-stage CTCL after antibiotic treatment [29]. It appears that CTCL progression is powered by bacteria. Therefore, the skin barrier seems to be essential in disease activity.

SPRR proteins were demonstrated to have protective functions, and their overexpression was associated with exposure to factors such as ultraviolet radiation [30]. Moreover, SPRR1 was demonstrated to be overexpressed in severe rather than moderate AD, suggesting compensation of other CE protein deficiencies [15,21]. We have noted a positive correlation between *SPRR1Av1* mRNA expression and the stage of disease, which is in line with previous findings indicating its remedial functions [15]. We have also observed significantly higher expression of *RPTN* mRNA in CTCL than in the control group. It could be speculated that overexpression of RPTN as of SPRR1Av1 compensates filaggrin deficiency; however, this needs further study.

According to the presented results, it seems that certain cornified envelope proteins play a key role in skin barrier dysfunction both in CTCL and AD. Moreover, differences in protein expression indicate that cornified envelope proteins might be considered biomarkers for differential diagnosis of AD and CTCL, especially in early-stage MF, when the differentiation even in histopathological examination is challenging. Skin barrier dysfunction is a well-known reason for AD development and exacerbation of the disease. Our results provide evidence that the skin barrier is also essential in the course of CTCL, probably contributing to the progression of the disease. The basic treatment for atopic dermatitis is the restoration of the skin barrier with emollients, which not only decreases transepidermal water loss and repairs the epidermal barrier but also prevents AD exacerbation, skin infections and AD development in infant AD risk groups [31,32]. Moreover, restoration of skin barrier functions indirectly decreases inflammation [33]. Based on our results, it seems that the approach to the CTCL treatment should be revised and complemented with the restoration of epidermal barrier functions by the usage of emollients.

In conclusion, we have demonstrated that skin barrier deficiency is not unique to AD, but it is also present in CTCL. Skin barrier dysfunction underlines AD development and course of disease, whereas in CTCL it rather contributes to the exacerbation of immunological processes due to skin infections. Only FLG, FLG2, RPTN, HRNR and SPRR1A proteins appear to be significant in skin barrier dysfunction. Moreover, they might be considered potential markers in difficult differential diagnosis of AD and CTCL.

Author Contributions: Conceptualization, M.T., M.S.-B. and B.O.; methodology, M.S.-B.; software, M.S.-B.; resources, B.O., J.J., M.T. and M.S.-W.; data curation, M.T., B.O. and M.S.-B.; writing—original draft preparation, B.O.; writing—review and editing, M.T, M.S.-W., R.J.N., J.J. and T.P.; visualization, M.S.-B.; supervision, M.S.-W., R.J.N, J.J. and T.P.; project administration, M.T, M.S.-B. and B.O.; funding acquisition, M.T. All authors have read and agreed to the published version of the manuscript.

Funding: The publiaction is a result of a research project No 2011/01/B/NZ1/00001 financed by the National Science Centre, Poland. The study was supported by the Polish Ministry of Science and Higher Education Grant. Project No. ST 02-0066.

Acknowledgments: The authors would like to thank all patients for their participation in the project.

Conflicts of Interest: The authors declare no conflict of interest.

Abbreviations

CTCL	cutaneous T-cell lymphomas
AD	atopic dermatitis
CE	cornified envelope
mRNA	messenger ribonucleic acid
LELP-1	late cornified envelope-like proline-rich protein-1
SPRR1A	small proline-rich proteins 1A
SPRR1B	small proline-rich proteins 1B
SPRR3	small proline-rich proteins 3
RPTN	repetin
CRNN	cornulin
HRNR	hornerin
LOR	loricrin
FLG	filaggrin
FLG2	filaggrin 2
RT-PCR	real-time polymerase chain reaction
ELISA	enzyme-linked immunoabsorbent assay
PBS	phosphate buffered saline

References

1. Willemze, R.; Cerroni, L.; Kempf, W.; Berti, E.; Facchetti, F.; Swerdlow, S.H.; Jaffe, E.S. The 2018 update of the WHO-EORTC classification for primary cutaneous lymphomas. *Blood* **2019**, *133*, 1703–1714. [CrossRef] [PubMed]
2. Wong, H.K.; Mishra, A.; Hake, T.; Porcu, P. Evolving Insights in the pathogenesis and therapy of Cutaneous T-cell lymphoma (Mycosis Fungoides and Sézary Syndrome). *Br. J. Haematol.* **2011**, *155*, 150–166. [CrossRef] [PubMed]
3. Siakantari, M. Classification and molecular pathogenesis of CTCL. *Melanoma Res.* **2010**, *20*, e20–e21. [CrossRef]
4. Vonderheid, E.C.; Bigler, R.D.; Hou, J.S. On the possible relationship between staphylococcal superantigens and in- creased Vbeta5.1 usage in cutaneous T-cell lymphoma. *Br. J. Haematol.* **2005**, *152*, 825–826.

5. McGirt, L.Y.; Jia, P.; Baerenwald, D.A.; Duszynski, R.J.; Dahlman, K.B.; Zic, J.A.; Zwerner, J.P.; Hucks, N.; Dave, U.; Zhao, Z.; et al. Whole-genome sequencing reveals oncogenic mutations in mycosis fungoides. *Blood* **2015**, *126*, 508–519. [CrossRef]

6. Almeida, A.C.D.S.; Abate, F.; Khiabanian, H.; Martínez-Escala, E.; Guitart, J.; Tensen, C.P.; Vermeer, M.H.; Rabadan, R.; Ferrando, A.; Palomero, T. The mutational landscape of cutaneous T cell lymphoma and Sézary syndrome. *Nat. Genet.* **2015**, *47*, 1465–1470. [CrossRef]

7. Hoppe, R.T.; Medeiros, L.; A Warnke, R.; Wood, G.S. CD8-positive tumor-infiltrating lymphocytes influence the long-term survival of patients with mycosis fungoides. *J. Am. Acad. Dermatol.* **1995**, *32*, 448–453. [CrossRef]

8. Guenova, E.; Watanabe, R.; Teague, J.E.; DeSimone, J.A.; Jiang, Y.; Dowlatshahi, M.; Schlapbach, C.; Schaekel, K.; Rook, A.H.; Tawa, M.; et al. TH2 cytokines from malignant cells suppress TH1 responses and enforce a global TH2 bias in leukemic cutaneous T-cell lymphoma. *Clin. Cancer Res.* **2013**, *19*, 3755–3763. [CrossRef]

9. Chong, B.F.; Wilson, A.J.; Gibson, H.M.; Hafner, M.S.; Luo, Y.; Hedgcock, C.J.; Wong, H.K. Immune function abnormalities in peripheral blood mononuclear cell cytokine expression differentiates stages of cutaneous T-cell lymphoma/mycosis fungoides. *Clin. Cancer Res.* **2008**, *14*, 646–653. [CrossRef]

10. Miyagaki, T.; Sugaya, M. Erythrodermic cutaneous T-cell lymphoma: How to differentiate this rare disease from atopic dermatitis. *J. Dermatol. Sci.* **2011**, *64*, 1–6. [CrossRef]

11. Suga, H.; Sugaya, M.; Miyagaki, T.; Ohmatsu, H.; Kawaguchi, M.; Takahashi, N.; Fujita, H.; Asano, Y.; Tada, Y.; Kadono, T.; et al. Skin Barrier Dysfunction and Low Antimicrobial Peptide Expression in Cutaneous T-cell Lymphoma. *Clin. Cancer Res.* **2014**, *20*, 4339–4348. [CrossRef] [PubMed]

12. Talpur, R.; Bassett, R.; Duvic, M. Prevalence and treatment of Staphylococcus aureus colonization in patients with mycosis fungoides and Sézary syndrome. *Br. J. Dermatol.* **2008**, *159*, 105–112. [CrossRef] [PubMed]

13. Thode, C.; Woetmann, A.; Wandall, H.H.; Carlsson, M.C.; Qvortrup, K.; Kauczok, C.S.; Wobser, M.; Printzlau, A.; Ødum, N.; Dabelsteen, S. Malignant T cells secrete galectins and induce epidermal hyperproliferation and disorga-nized stratification in a skin model of cutane- ous T-cell lymphoma. *J. Investig. Dermatol.* **2015**, *135*, 238–246. [CrossRef] [PubMed]

14. Mirvish, E.D.; Pomerantz, R.G.; Geskin, L.J. Infectious agents in cutaneous T-cell lymphoma. *J. Am. Acad. Dermatol.* **2010**, *64*, 423–431. [CrossRef]

15. Trzeciak, M.; Sakowicz-Burkiewicz, M.; Wesserling, M.; Dobaczewska, D.; Gleń, J.; Nowicki, R.; Pawelczyk, T. Expression of Cornified Envelope Proteins in Skin and Its Relationship with Atopic Dermatitis Phenotype. *Acta Derm. Venereol.* **2017**, *97*, 36–41. [CrossRef]

16. Trzeciak, M.; Wesserling, M.; Bandurski, T.; Gleń, J.; Nowicki, R.; Pawelczyk, T. Association of a Single Nucleotide Polymorphism in a Late Cornified Envelope-like Proline-rich 1 Gene with Atopic Dermatitis. *Acta Derm. Venereol.* **2016**, *96*, 459–463. [CrossRef]

17. Trzeciak, M.; Sakowicz-Burkiewicz, M.; Wesserling, M.; Dobaczewska, D.; Bandurski, T.; Nowicki, R.; Pawelczyk, T.; Gleń, J. Altered Expression of Genes Encoding Cornulin and Repetin in Atopic Dermatitis. *Int. Arch. Allergy Immunol.* **2017**, *172*, 11–19. [CrossRef]

18. Olsen, E.A.; Vonderheid, E.; Pimpinelli, N.; Willemze, R.; Kim, Y.; Knobler, R.; Zackheim, H.; Duvic, M.; Estrach, T.; Lamberg, S.; et al. Revisions to the staging and classification of mycosis fungoides and Sézary syndrome: A proposal of the International Society for Cutaneous Lymphomas (ISCL) and the cutaneous lymphoma task force of the European Organization of Research and Treatment of Cancer (EORTC). *Blood* **2007**, *110*, 1713–1722.

19. Hanifin, J.M.; Rajka, G. Diagnostic features of atopic dermatitis. *Acta Derm Venereol* **1980**, *59*, 44–47.

20. Chomczynski, P.; Sacchi, N. Single-step method of RNA isolation by acid guanidinium thiocyanate-phenol-chloroform extraction. *Anal. Biochem.* **1987**, *162*, 156–159. [CrossRef]

21. Sugiura, H.; Ebise, H.; Tazawa, T.; Tanaka, K.; Sugiura, Y.; Uehara, M.; Kikuchi, K.; Kimura, T. Large-scale DNA microarray analysis of atopic skin lesions shows overexpression of an epidermal differentiation gene cluster in the alternative pathway and lack of protective gene expression in the cornified envelope. *Br. J. Dermatol.* **2005**, *152*, 146–149. [CrossRef] [PubMed]

22. Pellerin, L.; Henry, J.; Hsu, C.-Y.; Balica, S.; Jean-Decoster, C.; Méchin, M.-C.; Hansmann, B.; Rodríguez, E.; Weindinger, S.; Schmitt, A.-M.; et al. Defects of filaggrin-like proteins in both lesional and nonlesional atopic skin. *J. Allergy Clin. Immunol.* **2013**, *131*, 1094–1102. [CrossRef] [PubMed]

23. Batista, D.I.S.; Perez, L.; Orfali, R.L.; Zaniboni, M.C.; Samorano, L.P.; Pereira, N.V.; Sotto, M.N.; Ishizaki, A.S.; Oliveira, L.M.S.; Sato, M.N.; et al. Profile of skin barrier proteins (filaggrin, claudins 1 and 4) and Th1/Th2/Th17 cytokines in adults with atopic dermatitis. *J. Eur. Acad. Dermatol. Venereol.* **2015**, *29*, 1091–1095. [CrossRef] [PubMed]
24. De Benedetto, A.; Rafaels, N.M.; McGirt, L.Y.; Ivanov, A.I.; Georas, S.N.; Cheadle, C.; Berger, A.E.; Zhang, K.; Vidyasagar, S.; Yoshida, T.; et al. Tight junction defects in patients with atopic dermatitis. *J. Allergy Clin. Immunol.* **2010**, *127*, 773–786. [CrossRef]
25. Kim, B.; Leung, D.; Boguniewicz, M.; Howell, M. Loricrin and Involucrin Expression is Down-Regulated by Th2 Cytokines through STAT-6. *J. Allergy Clin. Immunol.* **2008**, *121*, S272. [CrossRef]
26. Jackow, C.M.; Cather, J.C.; Hearne, V.; Asano, A.T.; Musser, J.M.; Duvic, M. Association of erythrodermic cutaneous T-cell lymphoma, superantigen-positive Staphylococcus aureus, and oligoclonal T-cell receptor V beta gene expansion. *Blood* **1997**, *89*, 32–40. [CrossRef]
27. Fanok, M.H.; Sun, A.; Fogli, L.K.; Narendran, V.; Eckstein, M.; Kannan, K.; Dolgalev, I.; Lazaris, C.; Heguy, A.; Laird, M.E.; et al. Role of Dysregulated Cytokine Signaling and Bacterial Triggers in the Pathogenesis of Cutaneous T-Cell Lymphoma. *J. Investig. Dermatol.* **2018**, *138*, 1116–1125. [CrossRef]
28. Willerslew-Olsen, A.; Krejsgaard, T.; Lindahl, L.M.; Bonefeld, C.M.; A Wasik, M.; Koralov, S.B.; Geisler, C.; Kilian, M.; Iversen, L.; Woetmann, A.; et al. Bacterial Toxins Fuel Disease Progression in Cutaneous T-Cell Lymphoma. *Toxins* **2013**, *5*, 1402–1421. [CrossRef]
29. Lindahl, L.M.; Willerslev-Olsen, A.; Gjerdrum, L.M.R.; Nielsen, P.R.; Blümel, E.; Rittig, A.H.; Celis, P.; Herpers, B.; Becker, J.C.; Stausbøl-Grøn, B.; et al. Antibiotics inhibit tumor and disease activity in cutaneous T-cell lymphoma. *Blood* **2019**, *134*, 1072–1083. [CrossRef]
30. Kelsell, D.; Byrne, C. SNPing at the Epidermal Barrier. *J. Investig. Dermatol.* **2011**, *131*, 1593–1595. [CrossRef]
31. Simpson, E.L.; Chalmers, J.; Hanifin, J.M.; Thomas, K.S.; Cork, M.; McLean, W.I.; Brown, S.; Chen, Z.; Chen, Y.; Williams, H.C. Emollient enhancement of the skin barrier from birth offers effective atopic dermatitis prevention. *J. Allergy Clin. Immunol.* **2014**, *134*, 818–823. [CrossRef] [PubMed]
32. Narbutt, J.; Lesiak, A. I-modulia®w badaniach klinicznych–zastosowanie emolientów u chorych na atopowe zapalenie skóry. *Forum Dermatol.* **2016**, *2*, 139–143.
33. Elias, P.M. Barrier-repair therapy for atopic dermatitis: Corrective lipid biochemical therapy. *Expert Rev. Dermatol.* **2008**, *3*, 441–452. [CrossRef]

Article

Anti-Melanogenic Effect of Ethanolic Extract of *Sorghum bicolor* on IBMX–Induced Melanogenesis in B16/F10 Melanoma Cells

Hye Ju Han, Seon Kyeong Park, Jin Yong Kang, Jong Min Kim, Seul Ki Yoo and Ho Jin Heo *

Division of Applied Life Science (BK21 plus), Institute of Agriculture and Life Science Gyeongsang National University, Jinju 52828, Korea; gksgpwn2527@naver.com (H.J.H); tjsrud2510@naver.com (S.K.P.); kangjy2132@naver.com (J.Y.K.); myrock201@naver.com (J.M.K.); ysyk9412@naver.com (S.K.Y.)
* Correspondence: hjher@gnu.ac.kr; Tel.: +82-55-772-1907

Received: 19 February 2020; Accepted: 18 March 2020; Published: 20 March 2020

Abstract: To evaluate possibility as a skin whitening agent of *Sorghum bicolor* (*S. bicolor*), its antioxidant activity and anti-melanogenic effect on 3-isobutyl-1-methylxanthine (IBMX)-induced melanogenesis in B16/F10 melanoma cells were investigated. The result of total phenolic contents (TPC) indicated that 60% ethanol extract of *S. bicolor* (ESB) has the highest contents than other ethanol extracts. Antioxidant activity was evaluated using the 2,2′-azino-bis-(3-ethylbenzothiazolin-6-sulfonic acid) diammonium salt (ABTS)/1,1-diphenyl-2-picryl-hydrazyl (DPPH) radical scavenging activities and malondialdehyde (MDA) inhibitory effect. These results showed ESB has significant antioxidant activities. Inhibitory effect against tyrosinase was also assessed using L-tyrosine (IC_{50} value = 89.25 µg/mL) and 3,4-dihydroxy-L-phenylalanine (L-DOPA) as substrates. In addition, ESB treatment effectively inhibited melanin production in IBMX-induced B16/F10 melanoma cells. To confirm the mechanism on anti-melanogenic effect of ESB, we examined melanogenesis-related proteins. ESB downregulated melanogenesis by decreasing expression of microphthalmia-associated transcription factor (MITF), tyrosinase and tyrosinase-related protein (TRP)-1. Finally, 9-hydroxyoctadecadienoic acid (9-HODE), 1,3-*O*-dicaffeoylglycerol and tricin as the main compounds of ESB were analyzed using the ultra-performance liquid chromatography-ion mobility separation-quadrupole time of flight/tandem mass spectrometry (UPLC-IMS-QTOF/MS^2). These findings suggest that ESB may have physiological potential to be used skin whitening material.

Keywords: anti-melanogenesis; B16/F10 melanoma cell; hydroxyoctadecadienoic acid; Sorghum bicolor; 3-isobutyl-1-methylxanthine

1. Introduction

Melanin is generated through a process, called as melanogenesis within the intracellular melanosomes of melanocytes, and the production and distribution of melanin determines the skin, eyes and hair color. In addition, melanin plays a vital role in protecting skin from various molecules and stimuli such as ultraviolet (UV) radiation, reactive oxygen species (ROS), α-melanocyte-stimulating hormone (α-MSH) and cyclic adenosine monophosphate (cAMP) elevating agents including forskolin and IBMX [1]. In particular, UV radiation directly and indirectly damages skin cells via ROS generation such as hydroxyl radicals, superoxide anions, and hydrogen peroxide [2]. The generated ROS causes single- and double-strand DNA breaks, and attacks important cellular structural molecules such as proteins and lipids [2]. Thus, melanin and antioxidant enzymes such as glutathione peroxidase, superoxide dismutase and catalase maintain a constant level by removing the ROS produced. However, excessive ROS production as a result of various factors activates melanogenesis in skin, and the abnormal melanin accumulation induces negative hyperpigmentation effects, causing age or liver spots, blotches and freckles [3,4]. Currently, natural compounds such as kojic acid, hydroquinone, arbutin and ascorbic acid, which are known as melanin synthesis inhibitors, are being utilized as skin whitening agent additives. However, these ingredients not only have poor penetration into skin, but also cause cytotoxicity and inflammation with long-term administration [5]. In this regard, it is necessary to develop a nonirritating and effective whitening agent to treat human skin hyperpigmentatzu, as well as to study natural resource materials for this.

Human skin cells commonly generate melanin as a photoprotection reaction. Melanogenesis is a complex process that is regulated via a variety of signaling pathways, and all signals ultimately upregulate MITF. As activated MITF promotes the tyrosinase gene family (tyrosinase and TRPs) expression, melanogenesis begins in earnest [6]. Generally, the first step in melanogenesis is that tyrosinase catalyzes the oxidation of L-tyrosine to 3,4-dihydroxy-L-phenylalanine (L-DOPA), and subsequently oxidizes L-DOPA into DOPA quinone. And then TRP-2 converts DOPA chrome to 5,6-dihydroxyindole-2-carboxylic acid (DHICA) or 5,6-dihydroxyindole (DHI). Finally, DHICA and DHI are oxidized by TRP-1, eventually leading to melanin production [6].

IBMX and other methylxanthines stimulate melanogenesis by inhibiting phosphodiesterase and thereby increasing levels of cAMP [7]. cAMP activated by IBMX phosphorylates extracellular-signal-regulated kinase (ERK) and phosphoinositide 3-kinases/protein kinase B (PI3K/Akt) signaling pathways, and ultimately leads to the activation of MITF in melanogenesis processes [7]. Therefore, the downregulation of these proteins can be regarded as an important factor for the anti-whitening effect.

Sorghum bicolor (*S. bicolor*), which belongs to the family Poaceae, is considered an important crop in tropical regions, including Africa, Central America, and arid regions, and is the fourth most important cereal crop of the world after wheat, rice and maize [8]. Recent studies have shown that *S. bicolor* is beneficial to humans because it contains phytochemicals such as phenolic compounds, tannins and anthocyanins [9]. These phytochemicals have gained increased attention due to its antioxidant, anti-obesity, anti-diabetes and anti-inflammatory effects [10–14]. While there are various research results, studies on the skin whitening effect of *S. bicolor* have not yet been conducted. Therefore, the present study aimed to investigate the antioxidant activity and anti-melanogenic effects of *S. bicolor*, and to examine their effects on IBMX-induced melanogenesis in B16/F10 melanoma cells.

2. Materials and Methods

2.1. Chemicals

Folin–Ciocalteu's reagent, L-DOPA, L-tyrosine, L-ascorbic acid, arbutin, acarbose, dimethyl sulfoxide (DMSO), bovine serum albumin, α-glucosidase, mushroom tyrosinase, bovine serum albumin, 3-(4,5-dimethyl-2-thiazolyl)-2,5-diphenyltetrazolium bromide (MTT) and IBMX were purchased from Sigma-Aldrich Chemical Co. (St. Louis, MO, USA). MITF (bsm-51339M) and tyrosinase (bs-0819R) were purchased from Bioss (Beijing, China). TRP-1 (sc-166857) and β-actin (sc-69879) were purchased from Santa Cruz Biotechnology (Dallas, TX, USA). Secondary antibodies (anti-rabbit and anti-mouse) were obtained from Cell Signaling Technology (Danvers, MA, USA).

2.2. Sample Preparation

S. bicolor (20 g) was extracted with various ethanol concentration (0, 20, 40, 60, 80 and 95%; 1L) at 40°C for 2 h using a reflux condenser. The extracts were filtered using a filter paper (Whatman International Limited, Kent, UK), and concentrated using a vacuum evaporator (N-1000; EYELA Co., Tokyo, Japan). After, the extracts were lyophilized using a vacuum freeze dryer (Il Shin Lab Co., Ltd., Yangju, Korea), and the dried extracts were stored at -20°C until used.

2.3. In Vitro Antioxidant Activity

2.3.1. Total Phenolic Contents (TPC)

Total phenolic contents were examined based on the principle that Folin–Ciocalteu's reagent is reduced to blue reaction product under alkaline conditions. A sample (1 mL) mixed with Folin–Ciocalteu's reagent and 7% sodium carbonate. The mixture was activated for 2 h, and then the absorbance was measured at 760 nm using a spectrophotometer (UV-1201; Shimadzu, Kyoto,

Japan). TPC was calculated from the standard curve of gallic acid and the results were expressed as mg GAE g^{-1}.

2.3.2. Radical Scavenging Activity

ABTS radical cation solution was produced by mixing 2.45 mM potassium persulfate and 7 mM ABTS with 100 mM potassium phosphate buffer (pH 7.4) containing 150 mM and allowing them to react for 24 h at room temperature. The ABTS solution was then diluted with distilled water to obtain an absorbance of 0.700 ± 0.020 at 734 nm. The sample was allowed to react with 980 μL the ABTS solution for 10 min at 37°C and then absorbance at 734 nm was measured using a spectrophotometer (UV-1201; Shimadzu, Kyoto, Japan).

DPPH radical solution was prepared by dissolving 0.1 mM DPPH in 80% methanol. The DPPH solution was diluted to an absorbance of 1.000 ± 0.020 at 517 nm. 50 μL of the sample was mixed with 1.45 mL of the DPPH solution and reacted for 30 min in the dark. After reacting, the mixture was determined at 517 nm.

2.3.3. Inhibitory Effect on Lipid Peroxidation

To measure the inhibitory effect on lipid peroxidation in brains tissue, the thiobarbituric acid (TBA) reactive substance method was used. Brain tissue was homogenated in 20 mM Tris-HCl buffer (pH 7.4), and centrifuged at 6000× *g* for 20 min. The supernatant was added to 0.1 mM L-ascorbic acid and 10 μM ferrous sulfate 37°C for 1 h incubation. Next, 30% trichloroacetic acid and 1% TBA were added to the mixture, which was then incubated in a water bath at 80°C for 20 min. Then, the TBA-MDA complex was measured using a spectrophotometer (UV-1201; Shimadzu, Kyoto, Japan) at 532 nm.

2.4. Tyrosinase Inhibitory Effect

The tyrosinase inhibitory effect was determined using L-tyrosine as a substrate. A sample was added to a 96-well plate and mixed with 0.1 M sodium phosphate buffer, tyrosinase and 0.1 mM L-tyrosine substrate to react at 37°C. After incubating, enzyme activity was measured using a microplate reader (EPOCH2; BioTek, Winooski, VT, USA) at 490 nm. Also, L-DOPA was used as a substrate to measure tyrosinase inhibitory activity. 67 mM sodium phosphate buffer, tyrosinase and 10 mM L-DOPA substrate were added to the sample to react at 37°C for 10 min. Tyrosinase activity was measured at 415 nm.

2.5. α-Glucosidase Inhibitory Effect

The α-glucosidase inhibitory effect was measured by mixing 0.1 M sodium phosphate buffer and α-glucosidase at 37°Cfor 10 min. After activating, the mixture was added to 5 mM 4-nitrophenyl-α-D-glucopyranoside in buffer at 37°C for 5 min, and then the absorbance was measured at 405 nm using a microplate reader (EPOCH2; BioTek, Winooski, VT, USA).

2.6. Cell Viability Assay

The cytotoxicity of the sample was estimated using an MTT assay. B16/F10 melanoma cells were cultured at 1×10^4 cells/well in a 96-well plate for 24 h. Thereafter, samples were treated to 1, 2, 5 and 10 μg/mL concentrations. Samples were treated for 48 h, after which the cells were treated with 10 μg/mL of MTT stock solution for 3 h. Finally, media was removed, and DMSO was added to solubilize the formazan. The amount of formazan was quantified using a microplate reader (EPOCH2; BioTek, Winooski, VT, USA) at 570 nm (excitation wavelength) and 655 nm (emission wavelength).

2.7. Measurement of Cellular Melanin Contents

To measure the amount of melanin, B16/F10 melanoma cells were cultured in a 24-well plate at a density of 4×10^4 cells/well above for 24 h. Then, the cells were treated with 100 mM IBMX and ESB or positive control (arbutin) for an additional 48 h. After treatment, the cells were washed and harvested using phosphate buffer saline (PBS). The harvested cells were then centrifuged at 16,000× *g* for 10 min, and the obtained pellet was dissolved with 1 N sodium hydroxide containing 10% DMSO at 90 °C for 20 min. The melanin contents were measured at 490 nm using a microplate reader (EPOCH2; BioTek, Winooski, VT, USA).

2.8. Western Blot Analysis

Pretreated B16/F10 melanoma cells were washed with PBS and lysed using RIPA buffer containing 1% protease inhibitors on ice. Protein concentrations were determined by Bradford protein assay, and the equal proteins were denatured with a Laemmli buffer for 10 min at 95°C. The denatured proteins were separated by 10% sodium dodecyl sulfate polyacrylamide gel electrophoresis (SDS-PAGE), and then were transferred onto a polyvinylidene difluoride (PVDF) membranes (Millipore, Billerica, MA, USA). Next, membranes were blocked with 5% skim milk in Tris-buffered saline with 0.1% Tween 20 (TBST) for 1 h, and then reacted with primary antibodies overnight at 4°C. The membranes were reacted with the secondary antibodies for 1 h and the immune complexes were visualized using an enhanced chemiluminescence reagent (Bionote, Hwaseong, Korea). The band images were detected and analyzed by Chemi-doc (iBright Imager, Thermo-Fisher Scientific, Waltham, MA, USA).

2.9. UPLC-IMS-QTOF/MS^2 Analysis

The main compounds in the 60% ethanolic extracts of *S. bicolor* were analyzed by UPLC-IMS-QTOF/MS^2 (Vion, Waters Corp., Milford, MA, USA). Chromatographic separation was carried out using an Acquity UPLC BEH C18 column (2.1 × 100 mm, 1.7 μm particle size; Waters Corp.) at a flow rate of 0.35 mL/min. The mobile phases consisted of solvent A (0.1% formic acid in distilled water) and B (acetonitrile), and analysis conditions included the following time gradient: 1% B at 0–1 min, 1–100% B at 1–7 min, 100% B at 7–8 min, 100–1% B at 8–8.2 min, 1% B at 8.2–10 min. Mass spectrometry was performed in the electrospray ionization (ESI) negative mode under the following conditions: source temperature, 100 °C; desolvation line temperature, 400 °C; ramp collision energy, 10–30 V; capillary voltage, 2.5 kV. Phenolic compounds were detected by a full scan ranging from *m/z* 50 to 1500.

2.10. Statistical Analysis

All results were presented as the means ± standard deviation (SD). The statistically difference between groups were determined by A one-way analysis of variance (ANOVA) using Duncan's new multiple-range test ($p < 0.05$) with SAS software version 9.4 (SAS Institute, Cary, NC, USA). And the interrelationship between the antioxidant effects (ABTS/DPPH radical scavenging activity and MDA inhibitory effect) and the inhibitory effect of melanogenesis-mediated enzymes (tyrosinase and α-glucosidase) was evaluated Pearson correlation analysis.

3. Results

3.1. Total Phenolic Contents

To compare antioxidant activity of each ethanolic extract from *S. bicolor*, TPC was investigated. As shown in Figure 1, TPC were different among the ethanolic extracts; the 60% ethanolic extract (150.08 ± 1.13 mg GAE/g) was significantly higher than the other extractions (0%; 70.83 ± 1.18, 20%; 100.42 ± 0.72, 40%; 114.67 ± 1.46, 80%; 118.33 ± 3.17 and 95%; 73.67 ± 1.46 mg GAE/g). Therefore, the 60% ethanolic extract was used for further experimental analyses.

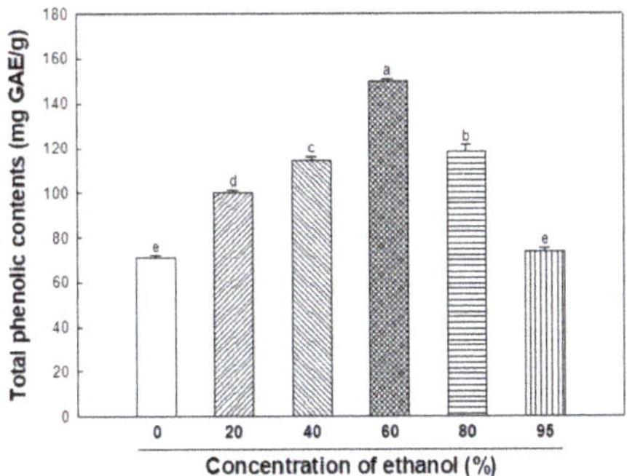

Figure 1. Total phenolic contents of various ethanolic extracts from *Sorghum bicolor*. Data were expressed as the means ± SD (*n* = 3). Each small letter represented statistical difference, and was statistically considered at *p* < 0.05.

3.2. Radical Scavenging Activity

The antioxidant activity of the 60% ethanolic extract from *S. bicolor* (ESB) was measured using the ABTS/DPPH assay and MDA assay, and the results were shown in Figure 2. The ABTS radical scavenging activity of ESB was exhibited 97.98% at 1000 μg/mL concentration, and vitamin C (99.82%), which is used as the positive control, and also had similar radical scavenging activity at same concentration (Figure 2A). And, ESB showed a 50% inhibition effects (IC_{50} value) of the ABTS radical at 409.71 μg/mL concentration. As shown in Figure 2B, the DPPH radical scavenging activity of ESB was showed 79.44% at 1000 μg/mL concentration, and the IC_{50} value was indicated at 612.53 μg/mL. The inhibitory effect of MDA production of ESB showed an inhibitory effect of 89.54% at 50 μg/mL concentration, and showed a higher inhibitory effect than the catechin (71.03%), which is the positive control, at the same concentration (Figure 2C). In addition, IC_{50} value of MDA was observed at 16.56 μg/mL concentrations of ESB.

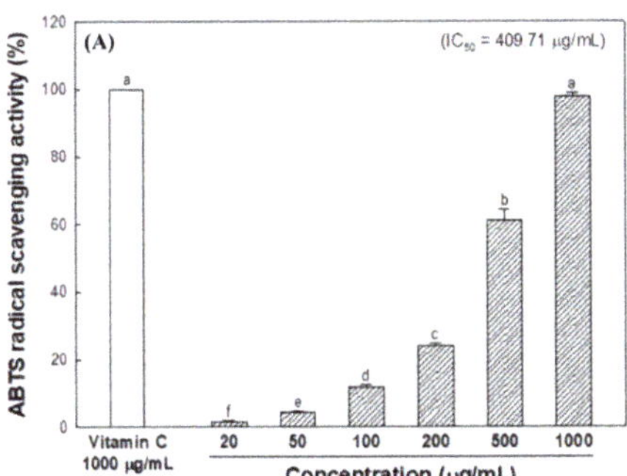
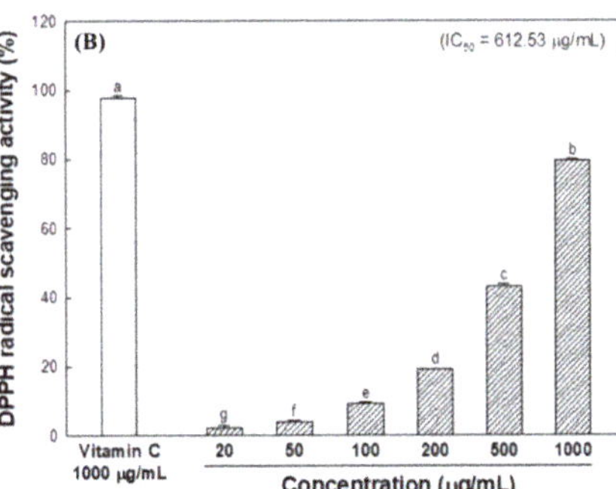

Figure 2. *Cont.*

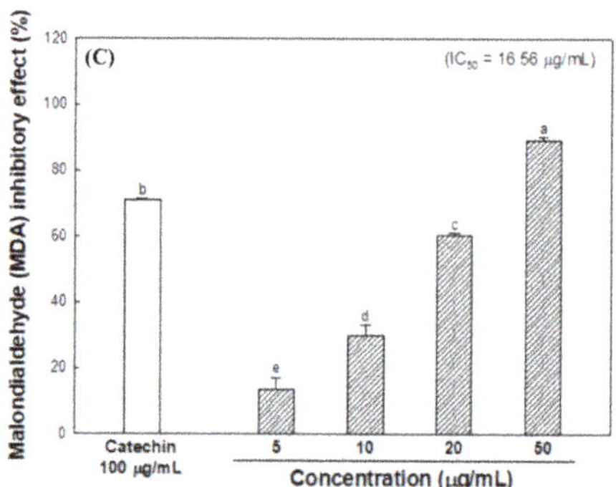

Figure 2. 2,2′-azino-bis-(3-ethylbenzothiazolin-6-sulfonic acid) diammonium salt (ABTS) radical scavenging activity (**A**), 1,1-diphenyl-2-picryl-hydrazyl (DPPH) radical scavenging activity (**B**) and malondialdehyde (MDA) inhibitory effect (**C**) of the 60% ethanolic extract from *Sorghum bicolor* (ESB). Data were expressed as the means ± SD (*n* = 3). Each small letter represented statistical difference, and was statistically considered at *p* < 0.05.

3.3. Inhibitory Effect of Melanogenesis-Mediated Enzymes

The inhibitory effect of melanogenesis-mediated enzymes was evaluated using tyrosinase and α-glucosidase (Figure 3). To measure inhibitory activity of tyrosinase against melanin synthesis, experiments were performed using L-tyrosine and L-DOPA as a substrate under cell-free conditions. As a result, the IC_{50} value was 89.25 μg/mL when L-tyrosine was used as a substrate (Figure 3A), but L-DOPA could not that more than 50% of inhibitory effect was observed (Figure 3B). At this time, the IC_{50} value of the positive control (arbutin) for L-tyrosine was 74.35 μg/mL, which was higher than ESB. ESB was found to have a better inhibitory effect on tyrosinase using L-tyrosine than using L-DOPA as a substrate. Therefore, based on the results of assay using two substrates, the tyrosinase inhibitory effect of ESB is considered to be related to the oxidation of L-tyrosine to L-DOPA rather than the oxidation of L-DOPA to DOPA quinone in melanogenesis. Tyrosinase inhibitory effect using L-DOPA as substrate and radical scavenging activity (ABTS; 0.943, *p* < 0.05, DPPH; 0.952, *p* < 0.05) indicated the positive strong correlations in Table 1.

To measure the inhibitory effect of melanogenesis-mediated enzyme, the inhibitory activity against α-glucosidase of ESB was measured. As shown in Figure 3C, ESB inhibited α-glucosidase by 33.72% at 200 μg/mL concentration. And the IC_{50} of ESB was 46.29 μg/mL, which is about five times higher than acarbose (216.05 μg/mL). These results indicate that ESB has better α-glucosidase inhibitory activity than acarbose as a positive control. And α-glucosidase inhibitory effect revealed strong positive correlations with MDA inhibitory effect (0.966, *p* < 0.05) in Table 1.

Table 1. Correlation between antioxidant activity and inhibitory effect of melanogenesis- mediated enzyme

	Pearson Correlation		
	Tyrosinase Inhibitory Effect		**α-Glucosidase Inhibitory Effect**
	L-tyrosine	**L-DOPA**	
ABTS radical scavenging activity	0.900	0.943[*]	0.792
DPPH radical scavenging activity	0.718	0.952[*]	0.937
MDA inhibitory effect	0.988	0.995	0.966[*]

*Correlation is significant (*p* < 0.05).

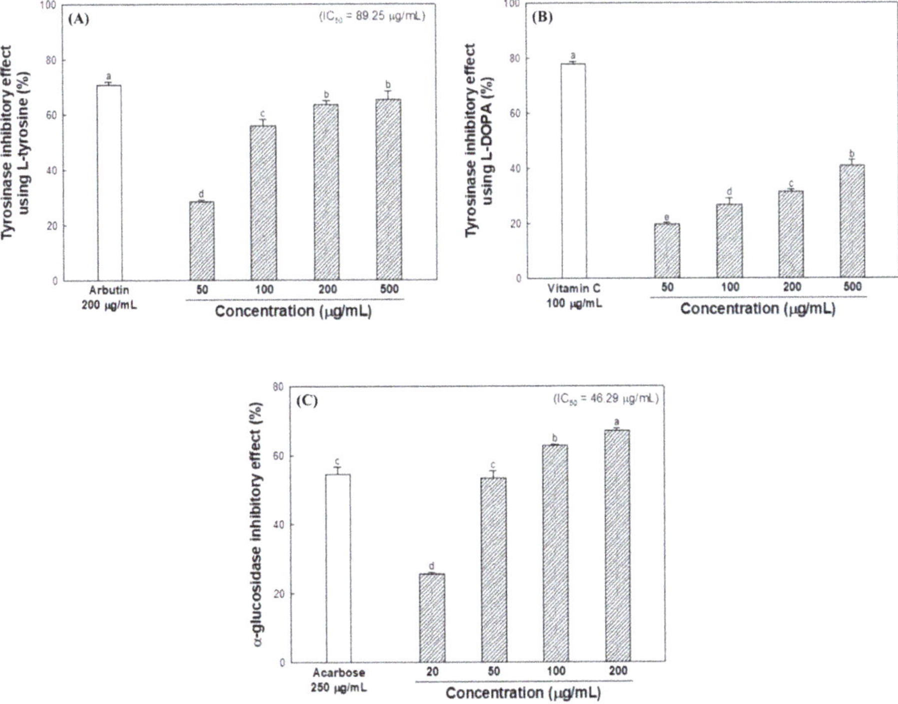

Figure 3. Inhibitory effect of the 60% ethanolic extract of *Sorghum bicolor* (ESB) on melanogenesis-mediated enzymes. Tyrosinase inhibitory effect using L-tyrosine (**A**) and L-DOPA (**B**) as substrates, and α-glucosidase inhibitory effect (C). Data were expressed as the means ± SD (*n* = 5). Each small letter represented statistical difference, and was statistically considered at *p* < 0.05.

3.4. Cell Viability and Cellular Melanin Synthesis

To confirm the cytotoxicity of ESB, B16/F10 melanoma cells were treated with ESB for 24 h at 1, 2, 5, 10 and 20 µg/mL concentrations. The results showed that ESB did not affect cell viability at the indicated concentrations when compared to the control (Figure 4A). Therefore, experiments were carried out by selecting 2, 5 and 10 µg/mL concentrations. Subsequently, B16/F10 melanoma cells were treated with IBMX to analyze the effect of ESB on melanin production. As shown in Figure 4B, the level of melanin content increased to 316.85% when treated with IBMX. On the other hand, the ESB treated group (108.60%) at 10 µg/mL concentration showed significantly decreased melanin contents compared to the IBMX treated group, and similar melanin contents compared to the arbutin group (101.79%) as a positive control.

3.5. Melanogenesis Pathway in B16/F10 Melanoma Cells

Melanogenesis is an essential process to protect human skin from external stimuli, and this process is regulated by various mechanisms. In this study, we measured the expression level of melanogenesis-regulating molecules to confirm the potential inhibitory mechanism of ESB on IBMX-induced melanin production in B16/F10 melanoma cells (Figure 5). The results showed that ESB effectively decreased IBMX-induced MITF expression (Figure 5A,B). Thus, the expression of tyrosinase (Figure 5A,C) and TRP-1 (Figure 5A,D) was downregulated by reduced MITF with ESB treatment. As a result, ESB treatment downregulated the expression of related proteins in IBMX-induced melanogenesis in B16/F10 melanoma cells.

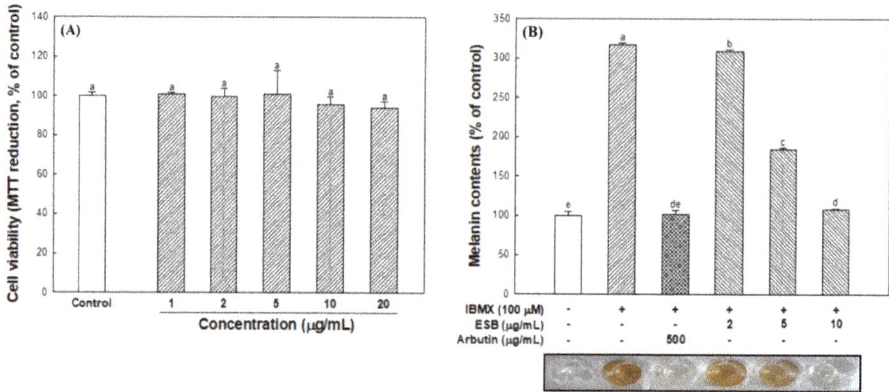

Figure 4. Cell viability (**A**) and melanin contents (**B**) in B16/F10 melanoma cells of the 60% ethanolic extract of *Sorghum bicolor* (ESB). Data were expressed as the means ± SD (n = 3). Each small letter represented statistical difference, and was statistically considered at $p < 0.05$.

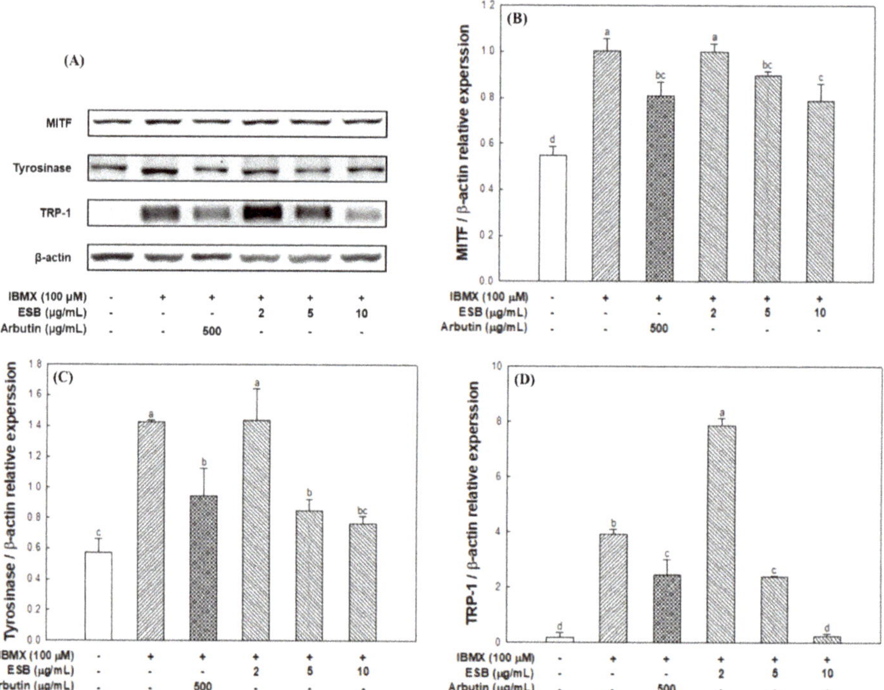

Figure 5. Anti-melanogenic effect of the 60% ethanolic extract of *Sorghum bicolor* (ESB) on 3-isobutyl-1-methylxanthine (IBMX)-induced melanogenesis in B16/F10 melanoma cells. Representative band images of proteins (**A**), relative expression of microphthalmia-associated transcription factor (MITF) (**B**), tyrosinase (**C**) and tyrosinase and tyrosinase-related protein (TRP)-1 (**D**). Data were expressed as the means ± SD (n = 3). Each small letter represented statistical difference, and was statistically considered at $p < 0.05$.

3.6. UPLC-IMS-QTOF/MS^2 Analysis

A profile of the main compounds of ESB was scanned by LC/MS in a range of *m/z* 50–1500, and the base peak chromatograms are shown in Figure 6. Further identification and characterization of the compounds was performed in comparison with literature data using MS^2 fragmentation data (Table 2). Based on the fragments in previous literature, peak 2 was 1-*O*-caffeoylglycerol (*m/z* 253.07, 179.03, 161.02 and 135.04) [15]; peak 3, dicaffeoylglycerides (*m/z* 415.10, 253.07, 179.03 and 135.04) [16]; peak 4, 1,3-*O*-dicaffeoylglycerol (*m/z* 415.10, 253.07, 179.03, 161.02 and 135.04) [16]; peak 5, *p*-coumaroyl-caffeoylglycerol (*m/z* 399.10, 253.07, 179.03, 161.02 and 135.04) [17]; peak 6, feruloyl-caffeoylglycerol (*m/z* 429.12, 253.07, 193.05, 161.02 and 134.03) [18]; peak 7, tricin (*m/z* 329.23, 314.04, 299.01 and 271.02) [19]; and peak 9, 9-hydroxyoctadecadienoic acid (9-HODE) (*m/z* 295.22, 277.21 and 171.10) [20], respectively. Among them, 1,3-*O*-dicaffeoylglycerol (Figure 6B), tricin (Figure 6C) and 9-HODE (Figure 6D) were identified as major compounds.

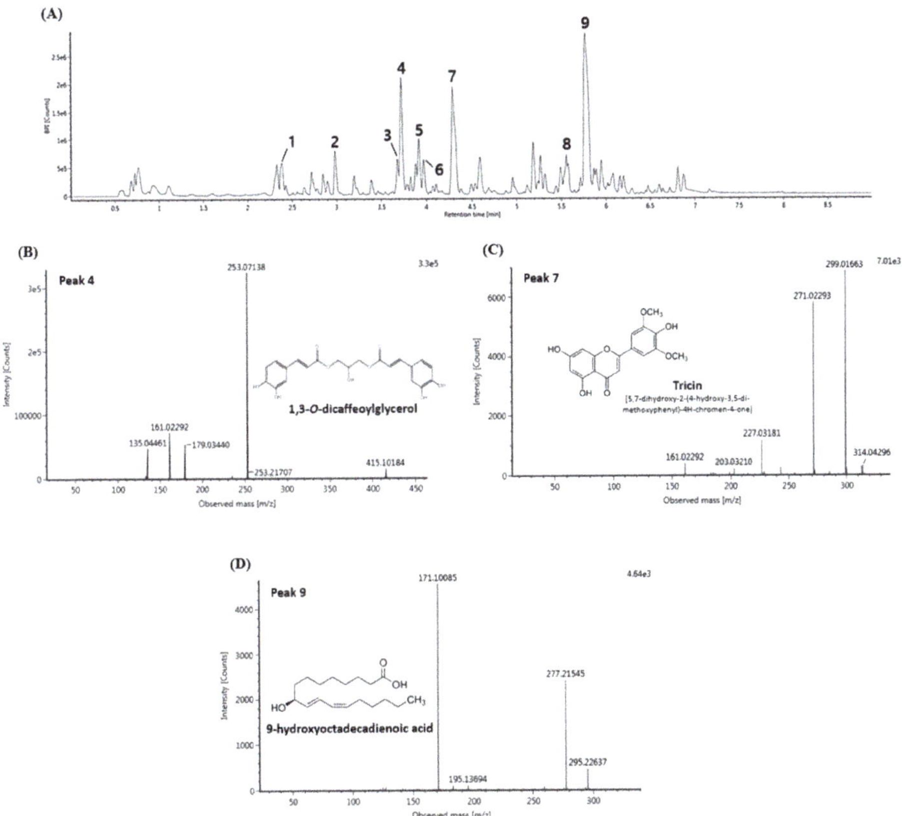

Figure 6. UPLC IMS-QTOF/MS spectra in negative ion mode (**A**) and MS^2 spectra and chemical structure of 1,3-*O*-dicaffeoylglycerol (**B**), Tricin (**C**) and 9-hydroxyoctadecadienoic acid (9-HODE) (**D**) from the 60% ethanolic extract of *Sorghum bicolor* (ESB).

Table 2. Identification of main compounds in the 60% ethanolic extract of *Sorghum bicolor* (ESB).

No.	RT (min)	[M-H]⁻ (m/z)	MS² Fragments (m/z)	Proposed Compound
1	2.38	625.19	407.11, 383.11, 221.06, 125.02, 89.02	Unknown
2	2.98	253.07	179.03, 161.02, 135.04	1-*O*-caffeoylglycerol
3	3.68	415.10	253.07, 179.03, 135.04	Dicaffeoylglycerides
4	3.72	415.10	253.07, 179.03, 161.02, 135.04	1,3-*O*-dicaffeoylglycerol
5	3.91	399.10	253.07, 179.03, 161.02, 135.04	*p*-coumaroyl-caffeoylglycerol
6	3.97	429.12	253.07, 193.05, 161.02, 134.03	Feruloyl-caffeoylglycerol
7	4.29	329.23	314.04, 299.01, 271.02	Tricin
8	5.55	315.25	297.24, 279.23	Unknown
9	5.77	295.22	277.21, 171.10	9-hydroxyoctadecadienoic acid (9-HODE)

4. Discussion

Melanogenesis is reported to be caused by increased oxidative stress from external stimuli. Oxidative stress causes oxidation of DNA and proteins, lipid peroxidation, and increased unsaturated fatty acids. This stress also leads to increase unnecessarily melanin synthesis in melanocytes on the skin. Thus, excessive melanin generated can cause pigmentation, and it can lead to skin cancer. IBMX stimulate melanogenesis by inhibiting phosphodiesterase and thereby increasing levels of cAMP [7]. As the level of cAMP in cells increases, it causes activation of the ERK and PI3K/Akt signaling pathways that promote the generation of proteins associated with melanogenesis. Therefore, we investigated the whitening effect of ESB on IBMX-induced melanogenesis in B16/F10 melanoma cells.

Phenolic compounds as the secondary metabolites of plants have the property of being stabilized by a resonance structure when they react with radicals [21]. Also, phenolic compounds act as important factors for the antioxidant activity such as ABTS/DPPH radical scavenging activity, and act as inhibitors of pigment formation because they have a chemical structure like that of L-tyrosine [22,23]. An ABTS/DPPH assay is the most common and simplest method for estimating in vitro antioxidant activity. In this study, ESB shown a high TPC (Figure 1) and ABTS/DPPH radical scavenging activity (Figure 2A,B). Especially, ESB was higher in the ABTS radicals scavenging activity than the DPPH radical scavenging activity. The DPPH assay is used to measure the radical scavenging activity of hydrophobic compounds, whereas the ABTS assay makes it possible to measure the radical scavenging activity against hydrophobic as well as hydrophilic compounds [24]. Therefore, it was confirmed that the radical scavenging activity of ESB was more affected by the hydrophilic compound.

UV radiation exposure causes ROS production and has direct and indirect effects on the skin. In particular, ROS induces lipid peroxidation by attacking unsaturated fatty acids, which are components of cell membranes [2]. The accumulation of lipid peroxides leads to the destruction of the antioxidant system of skin, causing pigmentation, inflammation and aging [25]. In this study, ESB showed an excellent inhibit lipid peroxidation (Figure 2C). The phenolic compounds contained in natural plants were reported to be associated with radical scavenging activity [26]. Also, Maresca et al. [27] reported that removal of ROS with antioxidants is effective in inhibiting melanin biosynthesis. A recent study reported that the ethyl acetate and aglycone fractions of *Dendropanax morbifera* leaf can act as skin cell protectants and natural antioxidants by protecting HaCaT cells against oxidative stress [28]. Moreover, treatment with the same fractions showed a better anti-melanogenic effect than arbutin, used as a positive control, against α-MSH-induced excessive melanin generation in B16/F10 melanoma cells [28].

Several mechanisms are associated with melanogenesis, and the common goal for skin whitening agents is the regulation of tyrosinase, which is an important enzyme in melanogenesis. Tyrosinase, a copper-containing enzyme, catalyzes two reactions in this pathway: the hydroxylation of L-tyrosine to L-DOPA and the oxidation of L-DOPA to DOPA quinone [6]. When exposed to large amounts of UV or ROS, the activity of tyrosinase is increased and melanin synthesis is also increased. Therefore, the results of measuring the inhibitory effect of tyrosinase in relation to melanin synthesis showed that ESB was found to have a better inhibitory effect on tyrosinase using L-tyrosine than using L-DOPA as a substrate (Figure 3A,B). A recent study reported the correlation between TPC and tyrosinase inhibition because phenolic compounds have a chemical structure similar to that of L-tyrosine, a substrate of tyrosinase. According to Choi et al. [29], it was observed that the efficiency of free radical scavenging activity and tyrosinase inhibitory effect increased in proportion to the fermentation time of *Cheonggukjang*, which was due to the amount of TPC increasing with fermentation time. Im & Lee [30] observed that the ethyl acetate fraction of 75% ethanolic extract from *Taraxacum coreanum* had abundant TPC and antioxidant activity, and the tyrosinase inhibitory effect was superior to other fractions (hexane, chloroform, butanol and aqueous).

Tyrosinase, one of the glycoproteins, is glycosylated when polypeptide chains translocate into the endoplasmic reticulum. Various enzymes are involved in this process, and among them, inhibition of α-glucosidase could inhibit the folding of tyrosinase to form an inactive structure without copper. As a result, tyrosinase is unable to be transported to the melanosome and inhibit melanin production. The polyphenol present in plants is capable not only of reducing oxidative stress, but also inhibits carbohydrate hydrolyzing enzymes such as α-glucosidase by binding to proteins [31]. A previous study indicated that B16/F10 melanoma cells in the presence of glycosylation inhibitors results in the reduction of melanogenesis by decreasing the total content of tyrosinase [32]. And Choi et al. [33] reported that deoxynojirimycin, one of the α-glucosidase inhibitors, blocked glycosylation of tyrosinase and inhibited tyrosinase migration to melanosomes. Ando et al. [34] have shown that it is possible to inhibit the melanin synthesis in melanoma cells by controlling the glycosylation of tyrosinase by α-glucosidase inhibitors such glutathione, ferritin and feldamycin. Kim et al. [35] reported the good correlations for Bee Pollen as natural antioxidant between TPC and tyrosinase inhibitory effect (0.973, $p < 0.05$), and also TPC gave the positive correlation with DPPH radical scavenging activity (0.897, $p < 0.05$). And the results suggest that polyphenols as natural antioxidants could be potential effective skin whitening based on their excellent antioxidant activity. In our study, antioxidant effect of ESB in the process of inhibiting the action of tyrosinase was indicated to be highly correlated through inhibition the oxidation of L-DOPA to DOPA quinone and the glycosylation of tyrosinase (Table 1). For this reason, the antioxidant effect of ESB has an effect on skin whitening effect by affecting the inhibition of melanogenesis-mediated enzyme.

The PKA signaling pathway is the main process involved in melanogenesis, and this is activated by IBMX, which is a cAMP elevating agent. The PKA signaling pathway can regulate transcription factors such as MITF and cAMP-responsive element-binding protein, which are involved in the expression of melanogenesis regulators such as tyrosinase, TRP-1 and TRP-2. Consequently, IBMX treatment increases the expression of MITF and the tyrosinase gene family (tyrosinase and TRP-1) through cAMP activation [7]. Park et al. [36] showed that the amount of melanin synthesis from *Papenfusiella kuromo* ethanolic extract (65.17 ± 13.40%) at a concentration of 40 μg/mL was similar to kojic acid (72.30 ± 3.92%) as a positive control, and reported that it showed potential as a natural whitening material. In this study, ESB showed an effective inhibitory effect on melanin synthesis similar to that of the positive control at low concentrations (Figure 4). After that, to assesses detailed anti-melanogenic effect of ESB, western blot analysis was performed on melanogenesis-mediated protein using B16/F10 melanoma cell.

Akt, which is involved in various mechanisms, plays a key role in multiple cellular processes such as apoptosis, glucose and skin metabolism. In particular, the increase of cAMP by IBMX treatment decreases Akt phosphorylation and its activation. Also, inactivated Akt induces the activation of GSK-3β, which means that phosphorylation is not achieved, that is, the level of p-GSK-3β is reduced [7]. According to previous literature, the aerial part of *Pueraria thunbergiana* downgrades MITF by activating the Akt/GSK-3β signaling pathway in B16/F10 melanoma cells [37]. Also, Huang et al. [38] showed that [6]-shogaol promoted p-Akt expression and inhibited melanogenesis signaling in B16/F10 melanoma cells. Activated GSK-3β upregulates the tyrosinase gene family by phosphorylating Ser289 of MITF, consequently promoting melanogenesis. MITF is a major transcription factor that binds to a tyrosinase gene promoter, and it subsequently increases the expression of enzymes related to melanocyte proliferation and melanogenesis [7]. When the expression of tyrosinase is promoted by MITF, it subsequently leads to an increase in the levels of TRP-1 and TRP-2, and as a result, these melanogenic enzymes produce melanin [6]. Oxidative stress caused by UV light stimulates melanin pigmentation in the skin. Therefore, the scavenging activity of radicals or ROS is effective to suppress melanin production, and polyphenols with antioxidant activity may have excellent whitening effects [38]. Choi et al. [39] showed that Korean bamboo stems (*Phyllostachys nigra* variety henosis) downregulated PKA/CREB-mediated MITF via scavenging of ABTS/DPPH radicals. And a 70% ethanolic extract of *Nelumbo nucifera* G. Leaf inhibited the expression of MITF, tyrosinase and TRPs with antioxidant activities such as electron donation ability and inhibition of xanthine oxidase [40]. In this experiment, it was shown that ESB treatment decreased the expression of MITF and tyrosinase gene family (tyrosinase and TRP-1), which promote the oxidation of DHICA to DHI in the melanogenesis pathway (Figure 5). Based on these results, ESB has been shown to downregulate IBMX-mediated melanogenesis based on its superior antioxidant effects.

Based on the previous literature, the main phenolic compounds in ESB were identified that 1,3-*O*-dicaffeoylglycerol (Figure 6B), tricin (Figure 6C) and 9-HODE (Figure 6D) [16,19,20]. Among them, 9-HODE, which is the most abundant compound in ESB, is one of the metabolites of linoleic acid, and shows a whitening effect on irritated skin by inhibiting tyrosinase [41]. Also, unlike other aliphatic compounds, 9-HODE is reported to be stable in all oxidation experiments and to have a whitening effect on ROS damage [41]. Meanwhile, tricin (5,7-dihydroxy- 2-(4-hydroxy-3,5-di-methoxyphenyl)-4H-chromen-4-one) has long been recognized for its beneficial health effects, such as antioxidant, antiviral, and antitumor/anticancer activities [42–44]. Mu et al. [45] reported that tricin exhibited a better inhibitory effect on tyrosinase activity compared with arbutin as a positive control. Moreover, Meng et al. [46] reported that tricin isolated from the methanol extract of young green barley (*Hordeum vulgare* L.) inhibited melanin biosynthesis and tyrosinase activity in B16/F10 melanoma cells through the hydroxyl group at the C-4′ position and methoxy groups at the C-3′,5′ positions of the tricin skeleton. Phenylpropanoids such as 1-*O*-caffeoylglycerol, dicaffeoylglycerides, and 1,3-*O*-dicaffeoylglycerol are organic compounds that are synthesized by plants from phenylalanine and tyrosine. Phenylpropanoid and their glycosylated forms were reported to be potent antioxidants either by directly scavenging ROS, or by acting as chain-breaking peroxyl radical scavengers [47].

p-Coumaric acid, which has a chemical structure similar to L-tyrosine, competes with L-tyrosine for active sites on tyrosinase [48]. It is known as a strong tyrosinase inhibitor, and it was reported that its anti-melanogenesis effect is stronger than structurally similar compounds such as cinnamic acid [48]. In addition, An et al. [48] reported that *p*-coumaric acid, a constituent of *Sasa quelpaertensis* Nakai, inhibited tyrosinase activity using L-DOPA as a substrate, and reduced melanin production in B16/F10 melanoma cells. Therefore, ESB may be considered to have an excellent whitening effect through *p*-coumaric acid. Furthermore, the MS spectra of 1-*O*-caffeoylglycerol and 1-3-*O*-dicaffeoyl glycerol showed similar fragmentation patterns at m/z 179, 161 and 135, and these fragments were reported as caffeic acid residue. They have a UV spectrum similar to that of a caffeic acid derivative with a λ_{max} at about 326 nm, and so are considered to be a caffeic acid derivative [48]. It was reported that caffeic acid, a derivative of various compounds, is a potent antioxidant *in vitro*/*in vivo*, and reduces tyrosinase activity and melanogenesis in B16/F10 melanoma cells. Consequently, the anti-melanogenic effect of ESB might be considered to be due to antioxidants and skin whitening substances such as aliphatic and phenolic compounds.

5. Conclusions

To evaluate possibility as a skin whitening agent of *Sorghum bicolor*, antioxidant, and anti-melanogenic effects of ESB on IBMX-induced melanogenesis were evaluated in B16/F10 melanoma cells. ESB exhibited significant antioxidant activities in ABTS/DPPH radical scavenging activity and MDA inhibitory effect. ESB prevented the first step of melanogenesis by inhibiting α-glucosidase and tyrosinase using L-tyrosine and L-DOPA as substrates. The anti-melanogenic effects of ESB on IBMX-induced melanogenesis were evaluated in B16/F10 melanoma cells, and the results showed that ESB inhibited IBMX-induced melanogenesis in B16/F10 melanoma cells. Melanin accumulation was inhibited by downregulation of the expression of MITF and the tyrosinase gene family (tyrosinase and TRP-1) in IBMX-induced melanogenesis. The major compounds of ESB were analyzed as 1,3-*O*-dicaffeoylglycerol, tricin and 9-HODE. Consequently, ESB showed *in vitro* antioxidant activity and an anti-melanogenic effect in B16/F10 melanoma cells, and thus, could be used as a skin whitening agent in the cosmetics market.

Author Contributions: Writing—original draft preparation, investigation, H.J.H. (Hye Ju Han) and S.K.P.; conceptualization, S.K.P and H.J.H. (Ho Jin Heo); methodology, J.Y.K.; validation, J.M.K.; formal analysis, S.K.Y.; writing—review and editing and supervision, H.J.H. (Ho Jin Heo). All authors have read and agreed to the published version of the manuscript.

Funding: This research received no external funding.

Acknowledgments: This study was supported by the Basic Science Research Program through the National Research Foundation (NRF) of Korea (NRF 2018R1D1A3B07043398) funded by the Ministry of Education, Republic of Korea. Hye Ju Han, Seon Kyeong Park, Jin Yong Kang, Jong Min Kim, and Seul Ki Yoo were supported by the BK21 Plus program, Ministry of Education, Republic of Korea.

Conflicts of Interest: The authors declare no conflict of interest.

References

1. Lin, Y.S.; Wu, W.C.; Lin, S.Y.; Hou, W.C. Glycine hydroxamate inhibits tyrosinase activity and melanin contents through downregulating cAMP/PKA signaling pathways. *Amino Acids* **2015**, *47*, 617–625. [CrossRef] [PubMed]

2. Fernandez-Garcia, E. Skin protection against UV light by dietary antioxidants. *Food Funct.* **2014**, *5*, 1994–2003. [CrossRef] [PubMed]

3. Agar, N.; Young, A.R. Melanogenesis: A photoprotective response to DNA damage? *Mutat. Res.-Fundam. Mol. Mech. Mutagen.* **2005**, *571*, 121–132. [CrossRef] [PubMed]

4. Schallreuter, K.U.; Wazir, U.; Kothari, S.; Gibbons, N.C.; Moore, J.; Wood, J.M. Human phenylalanine hydroxylase is activated by H_2O_2: A novel mechanism for increasing the L-tyrosine supply for melanogenesis in melanocytes. *Biochem. Biophys. Res. Commun.* **2004**, *322*, 88–92. [CrossRef]

5. Draelos, Z.D. Skin lightening preparations and the hydroquinone controversy. *Dermatol. Ther.* **2007**, *20*, 308–313. [CrossRef]

6. Slominski, A.; Tobin, D.J.; Shibahara, S.; Wortsman, J. Melanin pigmentation in mammalian skin and its hormonal regulation. *Physiol. Rev.* **2004**, *84*, 1155–1228. [CrossRef]

7. Byun, E.B.; Song, H.Y.; Mushtaq, S.; Kim, H.M.; Kang, J.A.; Yang, M.S.; Sung, N.Y.; Jang, B.S.; Byun, E.H. Gamma-Irradiated luteolin inhibits 3-Isobutyl-1-Methylxanthine-induced melanogenesis through the regulation of CREB/MITF, PI3K/Akt, and ERK pathways in B16BL6 melanoma cells. *J. Med. Food.* **2017**, *20*, 812–819. [CrossRef]

8. Dillon, S.L.; Shapter, F.M.; Henry, R.J.; Cordeiro, G.; Izquierdo, L.; Lee, L.S. Domestication to crop improvement: Genetic resources for *Sorghum* and *Saccharum* (Andropogoneae). *Ann. Bot.* **2007**, *100*, 975–989. [CrossRef]

9. Svensson, L.; Sekwati-Monang, B.; Lutz, D.L.; Schieber, A.; Ganzle, M.G. Phenolic acids and flavonoids in nonfermented and fermented red sorghum (*Sorghum bicolor* (L.) Moench). *J. Agric. Food. Chem.* **2010**, *58*, 9214–9220. [CrossRef]

10. Sikwese, F.E.; Duodu, K.G. Antioxidant effect of a crude phenolic extract from sorghum bran in sunflower oil in the presence of ferric ions. *Food Chem.* **2007**, *104*, 324–331. [CrossRef]

11. Shen, R.L.; Zhang, W.L.; Dong, J.L.; Ren, G.X.; Chen, M. Sorghum resistant starch reduces adiposity in high-fat diet-induced overweight and obese rats via mechanisms involving adipokines and intestinal flora. *Food Agric. Immunol.* **2015**, *26*, 120–130. [CrossRef]

12. Kim, J.M.; Park, Y.S. Anti-diabetic effect of sorghum extract on hepatic gluconeogenesis of streptozotocin-induced diabetic rats. *Nutr. Metab.* **2012**, *9*, 106. [CrossRef] [PubMed]

13. Park, J.H.; Lee, S.H.; Chung, I.M.; Park, Y.S. Sorghum extract exerts an anti-diabetic effect by improving insulin sensitivity via PPAR-γ in mice fed a high-fat diet. *Nutr. Res. Pract.* **2012**, *6*, 322–327. [CrossRef] [PubMed]

14. Burdette, A.; Garner, P.L.; Mayer, E.P.; Hargrove, J.L.; Hartle, D.K.; Greenspan, P. Anti-inflammatory activity of select sorghum (*Sorghum bicolor*) brans. *J. Med. Food* **2010**, *13*, 879–887. [CrossRef]

15. Wu, G.; Johnson, S.K.; Bornman, J.F.; Bennett, S.J.; Clarke, M.W.; Singh, V.; Fang, Z. Growth temperature and genotype both play important roles in sorghum grain phenolic composition. *Sci. Rep.* **2016**, *6*, 21835. [CrossRef]

16. Ma, C.; Xiao, S.Y.; Li, Z.G.; Wang, W.; Du, L.J. Characterization of active phenolic components in the ethanolic extract of *Ananas comosus* L. leaves using high-performance liquid chromatography with diode array detection and tandem mass spectrometry. *J. Chromatogr. A* **2007**, *1165*, 39–44. [CrossRef]

17. Steingass, C.B.; Glock, M.P.; Schweiggert, R.M.; Carle, R. Studies into the phenolic patterns of different tissues of pineapple (*Ananas comosus* [L.] Merr.) infructescence by HPLC-DAD-ESI-MS^n and GC-MS analysis. *Anal. Bioanal. Chem.* **2015**, *407*, 6463–6479. [CrossRef]

18. Wu, G.; Bennett, S.J.; Bornman, J.F.; Clarke, M.W.; Fang, Z.; Johnson, S.K. Phenolic profile and content of sorghum grains under different irrigation managements. *Food Res. Int.* **2017**, *97*, 347–355. [CrossRef]

19. Kang, J.; Price, W.E.; Ashton, J.; Tapsell, L.C.; Johnson, S. Identification and characterization of phenolic compounds in hydromethanolic extracts of sorghum wholegrains by LC-ESI-MS^n. *Food Chem.* **2016**, *211*, 215–226. [CrossRef]

20. Kloos, D.; Lingeman, H.; Mayboroda, O.A.; Deelder, A.M.; Niessen, W.M.A.; Giera, M. Analysis of biologically-active, endogenous carboxylic acids based on chromatography-mass spectrometry. *Trac-Trends Anal. Chem.* **2014**, *61*, 17–28. [CrossRef]

21. Soobrattee, M.A.; Neergheen, V.S.; Luximon-Ramma, A.; Aruoma, O.I.; Bahorun, T. Phenolics as potential antioxidant therapeutic agents: Mechanism and actions. *Mutat. Res.-Fundam. Mol. Mech. Mutagen.* **2005**, *579*, 200–213. [CrossRef]

22. Jeong, C.H.; Choi, S.G.; Heo, H.J. Analysis of nutritional components and evaluation of functional activities of *Sasa borealis* leaf tea. *Korean J. Food Sci. Technol.* **2008**, *40*, 586–592.

23. Wang, K.H.; Lin, R.D.; Hsu, F.L.; Huang, Y.H.; Chang, H.C.; Huang, C.Y.; Lee, M.H. Cosmetic applications of selected traditional Chinese herbal medicines. *J. Ethnopharmacol.* **2006**, *106*, 353–359. [CrossRef]

24. Floegel, A.; Kim, D.O.; Chung, S.J.; Koo, S.I.; Chun, O.K. Comparison of ABTS/DPPH assays to measure antioxidant capacity in popular antioxidant-rich US foods. *J. Food Compos. Anal.* **2011**, *24*, 1043–1048. [CrossRef]

25. Birben, E.; Sahiner, U.M.; Sackesen, C.; Erzurum, S.; Kalayci, O. Oxidative stress and antioxidant defense. *World Allergy Organ. J.* **2012**, *5*, 9. [CrossRef] [PubMed]

26. Woo, K.S.; Seo, H.I.; Lee, Y.H.; Kim, H.Y.; Ko, J.Y.; Song, S.B.; Lee, J.S.; Jung, K.Y.; Nam, M.H.; Oh, I.S.; et al. Antioxidant compounds and antioxidant activities of sweet potatoes with cultivated conditions. *J. Korean Soc. Food Sci. Nutr.* **2012**, *41*, 519–525. [CrossRef]

27. Maresca, V.; Flori, E.; Briganti, S.; Mastrofrancesco, A.; Fabbri, C.; Mileo, A.M.; Paggi, M.G.; Picardo, M. Correlation between melanogenic and catalase activity in in vitro human melanocytes: A synergic strategy against oxidative stress. *Pigment Cell Melanoma Res.* **2008**, *21*, 200–205. [CrossRef] [PubMed]

28. Park, S.A.; Park, J.; Park, C.I.; Jie, Y.J.; Hwang, Y.C.; Kim, Y.H.; Jeon, S.H.; Lee, H.M.; Ha, J.H.; Kim, K.J.; et al. Cellular antioxidant activity and whitening effects of *Dendropanax morbifera* leaf extracts. *Microbiol. Biotechnol. Lett.* **2013**, *41*, 407–415. [CrossRef]

29. Choi, H.K.; Lim, Y.S.; Kim, Y.S.; Park, S.Y.; Lee, C.H.; Hwang, K.W.; Kwon, D.Y. Free-radical-scavenging and tyrosinase-inhibition activities of Cheonggukjang samples fermented for various times. *Food Chem.* **2008**, *106*, 564–568. [CrossRef]

30. Im, D.Y.; Lee, K.I. Antioxidative and antibacterial activity and tyrosinase inhibitory activity of the extract and fractions from *Taraxacum coreanum Nakai*. *Korean J. Med. Crop Sci.* **2011**, *19*, 238–245. [CrossRef]

31. Mai, T.T.; Thu, N.N.; Tien, P.G.; Van Chuyen, N. Alpha-glucosidase inhibitory and antioxidant activities of Vietnamese edible plants and their relationships with polyphenol contents. *J. Nutr. Sci. Vitaminol.* **2007**, *53*, 267–276. [CrossRef] [PubMed]

32. Gillbro, J.M.; Olsson, M.J. The melanogenesis and mechanisms of skin-lightening agents–existing and new approaches. *Int. J. Cosmet. Sci.* **2011**, *33*, 210–221. [CrossRef] [PubMed]

33. Choi, H.J.; Ahn, S.M.; Chang, H.K.; Cho, N.S.; Joo, K.M.; Lee, B.G.; Chang, I.S.; Hwang, J.S. Influence of N-glycan processing disruption on tyrosinase and melanin synthesis in HM3KO melanoma cells. *Exp. Dermatol.* **2007**, *16*, 110–117. [CrossRef]

34. Ando, H.; Kondoh, H.; Ichihashi, M.; Hearing, V.J. Approaches to identify inhibitors of melanin biosynthesis *via* the quality control of tyrosinase. *J. Invest. Dermatol.* **2007**, *127*, 751–761. [CrossRef] [PubMed]

35. Kim, S.B.; Jo, Y.H.; Liu, Q.; Ahn, J.H.; Hong, I.P.; Han, S.M.; Hwan, B.Y.; Lee, M.K. Optimization of extraction condition of bee pollen using response surface methodology: Correlation between anti-melanogenesis, antioxidant activity, and phenolic content. *Molecules* **2015**, *20*, 19764–19774. [CrossRef] [PubMed]

36. Park, S.K.; Chung, Y.H.; Park, J.K.; Kim, S.J.; Jung, D.H.; Choi, H.G. Inhibitory effects of *Papenfusiella kuromo* ethanol extracts on melanin formation in B16F10 murine melanoma cells. *J. Korean Beauty Soc.* **2019**, *25*, 156–161.

37. Han, E.B.; Chang, B.Y.; Kim, D.S.; Cho, H.K.; Kim, S.Y. Melanogenesis inhibitory effect of aerial part of *Pueraria thunbergiana in vitro* and *in vivo*. *Arch. Dermatol. Res.* **2015**, *307*, 57–72. [CrossRef]

38. Huang, H.C.; Chang, S.J.; Wu, C.Y.; Ke, H.J.; Chang, T.M. [6]-Shogaol inhibits α-MSH-induced melanogenesis through the acceleration of ERK and PI3K/Akt-mediated MITF degradation. *Biomed. Res. Int.* **2014**, *2014*, 842569. [CrossRef]

39. Choi, M.H.; Jo, H.G.; Yang, J.; Ki, S.; Shin, H.J. Antioxidative and anti-melanogenic activities of bamboo stems (*Phyllostachys nigra* variety henosis) *via* PKA/CREB-mediated MITF downregulation in B16F10 melanoma cells. *Int. J. Mol. Sci.* **2018**, *19*, 409. [CrossRef]

40. Yoo, D.H.; Joo, D.H.; Lee, S.Y.; Lee, J.Y. Antioxidant effect of *Nelumbo nucifera* G. leaf extract and inhibition of MITF, TRP-1, TRP-2, and tyrosinase expression in a B16F10 melanoma cell line. *Korean J. Life Sci.* **2015**, *25*, 1115–1123. [CrossRef]

41. Weindl, G.; Schäfer-Korting, M.; Schaller, M.; Korting, H.C. Peroxisome proliferator-activated receptors and their ligands. *Drugs* **2005**, *65*, 1919–1934. [CrossRef] [PubMed]

42. Kwon, Y.S.; Kim, E.Y.; Kim, W.J.; Kim, W.K.; Kim, C.M. Antioxidant constituents from *Setaria viridis*. *Arch. Pharm. Res.* **2002**, *25*, 300–305. [CrossRef] [PubMed]

43. Sakai, A.; Watanabe, K.; Koketsu, M.; Akuzawa, K.; Yamada, R.; Li, Z.; Sadanari, H.; Matsubara, K.; Murayama, T. Anti-human cytomegalovirus activity of constituents from *Sasa albo-marginata* (Kumazasa in Japan). *Antivir. Chem. Chemother.* **2008**, *19*, 125–132. [CrossRef] [PubMed]

44. Cai, H.; Sale, S.; Schmid, R.; Britton, R.G.; Brown, K.; Steward, W.P.; Gescher, A.J. Flavones as colorectal cancer chemopreventive agents—phenol-O-methylation enhances efficacy. *Cancer Prev. Res.* **2009**, *2*, 743–750. [CrossRef] [PubMed]

45. Mu, Y.; Li, L.; Hu, S.Q. Molecular inhibitory mechanism of tricin on tyrosinase. *Spectroc. Acta Pt A-Molec. Biomolec. Spectr.* **2013**, *107*, 235–240. [CrossRef] [PubMed]

46. Meng, T.X.; Irino, N.; Kondo, R. Melanin biosynthesis inhibitory activity of a compound isolated from young green barley (*Hordeum vulgare* L.) in B16 melanoma cells. *J. Nat. Med.* **2015**, *69*, 427–431. [CrossRef]

47. Korkina, L.G. Phenylpropanoids as naturally occurring antioxidants: From plant defense to human health. *Cell Mol. Biol.* **2007**, *53*, 15–25.

48. An, S.M.; Lee, S.I.; Choi, S.W.; Moon, S.W.; Boo, Y.C. *p*-Coumaric acid, a constituent of *Sasa quelpaertensis* Nakai, inhibits cellular melanogenesis stimulated by α-melanocyte stimulating hormone. *Br. J. Dermatol.* **2008**, *159*, 292–299. [CrossRef]

Communication

Enhanced Triacylglycerol Content and Gene Expression for Triacylglycerol Metabolism, Acyl-Ceramide Synthesis, and Corneocyte Lipid Formation in the Epidermis of Borage Oil Fed Guinea Pigs

Ju-Young Lee [1], Kwang-Hyeon Liu [2], Yunhi Cho [1,*] and Kun-Pyo Kim [1,*]

[1] Department of Medical Nutrition, Graduate School of East-West Medical Science, Kyung Hee University, Gyeonggi-do 17104, Korea; bell1924@naver.com

[2] BK21 Plus KNU Multi-Omics Based Creative Drug Research Team, College of Pharmacy and Research Institute of Pharmaceutical Sciences, Kyungpook National University, Daegu 41566, Korea; dstlkh@knu.ac.kr

* Correspondence: choyunhi@khu.ac.kr (Y.C.); kunpyokim@khu.ac.kr (K.-P.K.); Tel.: +82-31-201-3817 (Y.C.); +82-31-201-3819 (K.-P.K.)

Received: 14 October 2019; Accepted: 13 November 2019; Published: 18 November 2019

Abstract: Triacylglycerol (TAG) metabolism is related to the acyl-ceramide (Cer) synthesis and corneocyte lipid envelope (CLE) formation involved in maintaining the epidermal barrier. Prompted by the recovery of a disrupted epidermal barrier with dietary borage oil (BO: 40.9% linoleic acid (LNA) and 24.0% γ-linolenic acid (GLA)) in essential fatty acid (EFA) deficiency, lipidomic and transcriptome analyses and subsequent quantitative RT-PCR were performed to determine the effects of borage oil (BO) on TAG content and species, and the gene expression related to overall lipid metabolism. Dietary BO for 2 weeks in EFA-deficient guinea pigs increased the total TAG content, including the TAG species esterified LNA, GLA, and their C20 metabolized fatty acids. Moreover, the expression levels of genes in the monoacylglycerol and glycerol-3-phosphate pathways, two major pathways of TAG synthesis, increased, along with those of TAG lipase, acyl-Cer synthesis, and CLE formation. Dietary BO enhanced TAG content, the gene expression of TAG metabolism, acyl-Cer synthesis, and CLE formation.

Keywords: borage oil; triacylglycerol metabolism; acyl-ceramide; corneocyte lipid envelope; epidermis

---

1. Introduction

Triacylglycerol (TAG) is one of the most abundant lipids in the epidermis, and TAG synthesis and hydrolysis are reported to be related to acyl-ceramide (Cer) synthesis and corneocyte lipid envelope (CLE) formation [1–3]. The systemic lack of diacylglycerol O-acyltransferase 2 (*Dgat2*), the gene for TAG synthesis, leads to an impaired permeability barrier in the epidermis [4]. In addition, mutation of *CGI-58*, the gene that facilitates TAG hydrolysis, has been reported to induce severe ichthyosis along with impaired CLE formation [3]. In parallel, the epidermal levels of linoleic acid (LNA; 18:2n-6) esterified acyl-Cer, a key lipid involved in CLE formation, were decreased in cases of both impaired permeability barriers and ichthyosis [2,3], suggesting that LNA, which is hydrolyzed from TAG, could be required for acyl-Cer synthesis and CLE formation in the differentiated epidermis.

In *Dgat2*-deficient mice, the level of LNA esterified to TAG is reduced, and a skin phenotype is exhibited similar to that seen in essential fatty acid (EFA) deficiency, which is characterized by a scaly skin condition and increased trans-epidermal water loss (TEWL) [4]. Interestingly, levels of TAG are reduced in the epidermis of EFA-deficient mice [5]. In addition, γ-linolenic acid (GLA; 18:3n-6), another ω-6 EFA, has been reported to exert greater efficacy than LNA in recovering EFA deficiency in the epidermis [6]. Furthermore, our recent studies demonstrated that GLA is also esterified to acyl-Cer in the epidermis [7]. Therefore, we hypothesized that there is a relationship among LNA or GLA esterified TAG, CLE formation, and epidermal barrier homeostasis. To test this hypothesis, we performed lipidomic and transcriptome analyses to determine the effect of dietary BO containing high concentrations of GLA and LNA on TAG content and TAG metabolism related gene expression levels.

2. Materials and Methods

2.1. Animals

Male albino Hartley guinea pigs at 3 weeks of age were purchased from the Samtako Laboratory (Osan, Republic of Korea) and housed as described previously [8]. After 1 week adaptation period, EFA deficiency was induced by a diet containing 40 g/kg hydrogenated coconut oil (HCO) (400950; Dyets, Bethlehem, PA, USA) supplemented with 20 g/kg triolein (T7752) (Sigma-Aldrich, St Louis, MO, USA) for 8 weeks. Subsequently, EFA deficient guinea pigs were fed a diet containing 60 g/kg BO (Midlands Seed, Ashburton, New Zealand) (group HCO + BO; $n = 5$) or continued on the HCO diet (group HCO: EFA deficient control; $n = 5$) for 2 weeks. Animal care and handling was approved by the Animal Care and Use Review Committee of Kyung Hee University, Yongin, Republic of Korea (KHUAGC-17-020).

2.2. Lipidomic Analysis

The epidermal lipid was extracted using Folch solution (chloroform/methanol, 2:1, *V/V*) and high-performance thin-layer chromatography (HPTLC) was used for analysis of the total epidermal TAG content as described previously [9]. Lipidomic analysis was performed using direct infusion electrospray mass spectrometry (ES-MS) as described previously [10]. Specifically, epidermal lipid samples with chloroform/methanol (1:9, *V/V*) containing 7.5 mM ammonium acetate were infused into a Thermo LTQ XL ion trap mass spectrometer (Thermo Fisher Scientific, West Palm Beach, FL, USA) through a nanoelectrospray infusion system (TriVersa NanoMate, Advion Biosciences, Ithaca, NY, USA). The spectral data were recorded using the Thermo Xcalibur software (version 2.1, Thermo Fisher Scientific) and loaded into the Genedata Expressionist MSX module (Genedata AG, Basel, Switzerland) for data processing. After subtracting background noise, spectral data below an intensity of 300 were removed, and data alignment was performed using nonlinear transformation to map the original *m/z* data onto the common universal data. After spectrum peak detection, isotopic clustering of individual detected peaks was conducted using spectrum isotope clustering activity. After normalizing the spectral data, each compound was identified using commercially available standards or the Lipidomics Gateway of the LIPID MAPS (http://www.lipidmaps.org).

2.3. Transcriptome Analysis

The total RNA was extracted using TRIzol reagent (Gibco, New York, NY, USA) from the epidermis of a randomly selected subpopulation of 3 guinea pigs from each group, HCO and HCO + BO. The RNA was amplified, labelled with Cyanine 3 using the Quick AMP Labeling Kit (Agilent Technologies, Santa Clara, CA, USA), and purified. The labeled complementary RNA was hybridized to an Agilent 4 × 180K Custom Gene Expression Microarray. After washing, microarrays were scanned, and data were extracted with Feature Extraction 9.1 Software (Agilent Technologies, Santa Clara, CA, USA). Subsequently, the gene expression data were quantile-normalized and log2-transformed.

To identify the alteration of epidermal transcriptomes between groups HCO and HCO + BO, we extracted gene ontology (GO) terms from the genes (n = 1243) whose expression in group HCO + BO was 1.5-fold higher than that in group HCO, using a web-based gene set analysis toolkit (www.webgestalt.org) [11]. Next, to identify the alteration of TAG metabolism, we screened genes associated with the TAG biosynthetic process with a 2-fold change (FC) threshold. The expression levels of the selected genes were confirmed by quantitative reverse transcription polymerase chain reaction (qRT-PCR) using specific primers (Table 1) as described previously [8]. Lastly, we screened genes associated with TAG hydrolysis, acyl-Cer synthesis, and CLE formation to gain insight into the meaning of the altered TAG species.

Table 1. qRT-PCR primer sequences.

Gene [1]	Forward	Reverse
Gapdh	5′-AGAACATCATCCCCGCATCC-3′	5′-TCCACAACCGACACATTAGGT-3′
Mogat2	5′-TGCTCTACCTTTTGCTTATGGG-3′	5′-TGGCTTGTCTCGGTCCA-3′
Agpat4	5′-GCTGATTGTTATGTTAGGCGGA-3′	5′-GACTTTGGGGGTTTCTGGGA-3′
Elovl7	5′-GGACAGAGTTCCAGCGAGTA-3′	5′-ACAAGTGAGAGTCAAAAGCCTG-3′
Dgat2	5′-CTCCTCTGTCAAATCTCAGGC-3′	5′-TTACTCCAACAACACGCAGG-3′

[1] *Gapdh*: glyceraldehyde 3-phosphate dehydrogenase (NCBI Accession: NM_001172951), *Mogat2*: monoacylglycerol O-acyltransferase 2 (NCBI Accession: XM_003468553), *Agpat4*: 1-acylglycerol-3-phosphate O-acyltransferase 4 (NCBI Accession: XM_003466380), *Elovl7*: elongation of very long-chain fatty acids 7 (NCBI Accession: XM_003470212) and *Dgat2*: diacylglycerol O-acyltransferase 2 (NCBI Accession: XM_003468552).

2.4. General Statistical Analysis

All data are expressed as means ± standard deviation, and differences between groups HCO and HCO+BO were tested by unpaired Student's *t*-test. *p*-values < 0.05 were considered significant.

3. Results and Discussion

3.1. Altered TAG Content and Species

First, we assessed the total TAG content in the epidermis of groups HCO and HCO + BO using HPTLC analysis (Figure 1a). The total TAG content was significantly increased in group HCO + BO compared with group HCO (Figure 1b), as reported previously [5]. To identify which TAG species were altered between groups HCO and HCO + BO, we used a lipidomic analysis approach (Figure 1c and Table S1). TAG 52:4 and TAG 54:6 containing LNA, TAG 54:4 and TAG 56:7 containing GLA and dihomo-γ-linolenic acid (DGLA; 20:3*n*-6), and TAG 54:5 and TAG 56:6 containing arachidonic acid (ARA; 20:4*n*-6) were significantly increased in group HCO + BO compared with group HCO. Other TAG species were not altered in either group (data not shown). These results indicate that the increased total TAG content in group HCO + BO may have resulted from an increase in TAG species containing EFA (i.e., LNA and GLA and their C20 metabolized fatty acids (FA) such as DGLA and ARA).

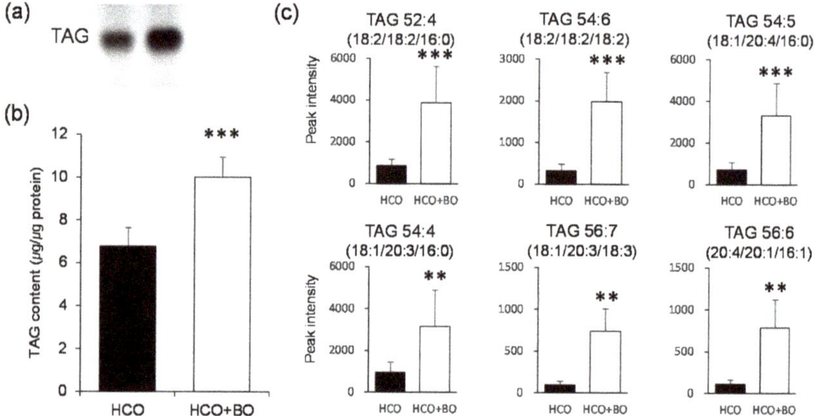

Figure 1. Triacylglycerol (TAG) content in the epidermis of guinea pigs fed a hydrogenated coconut oil (HCO) diet for 10 weeks (group HCO) or 8 weeks followed by 2 weeks of borage oil (BO) diet (group HCO + BO). Total TAG content was analyzed by high-performance thin-layer chromatography (HPTLC). (**a**) Representative band and (**b**) densitometric analysis. (**c**) Analysis of TAG species was performed by direct-infusion electrospray mass spectrometry (ES-MS). Values are means ± SD (n = 5). ** p < 0.01, *** p < 0.001 vs. group HCO by the unpaired Student's t test.

3.2. Altered Expression of Genes Related to TAG Synthesis

To identify the genes related to altered TAG content, we compared epidermal transcriptomes between groups HCO and HCO+BO using microarray analysis. First, we evaluated the overall meaning of up-regulated genes (FC > 1.5, n = 1243) in group HCO + BO (Table S2) by GO enrichment analysis (Figure 2a and Table S3). The up-regulated genes in group HCO + BO were involved in biological processes involved in the lipid metabolic process, keratinocyte differentiation, and epidermis development. Among the up-regulated genes associated with the lipid metabolic process (Table S4), we further checked genes related to TAG metabolism, especially with the two major pathways of TAG synthesis, the glycerophosphate pathway and monoacylglycerol (MAG) pathway [12]. As shown in Figure 2b, gene expression levels of monoacylglycerol O-acyltransferase 2 (*Mogat2*), 1-acylglycerol-3-phosphate O-acyltransferase 4 (*Agpat4*) and the elongation of very long-chain fatty acids 7 (*Elovl7*) and Dgat2 increased over two to thirteen-fold in group HCO + BO compared with group HCO; these results were confirmed by qRT-PCR (Figure 2c). Although there is little information about the MAG pathway in the epidermis, our results suggest that the MAG pathway seems to play a role in TAG synthesis in the epidermis, at least in guinea pigs. In the glycerophosphate pathway, *Agpat4* seems to play a control step in TAG synthesis in the guinea pig epidermis. Because *Agpat4*, an acyltransferase that transfers an acyl-CoA onto the sn-2 position of glycerol [1], has been reported to have a preference for a long-chain polyunsaturated fatty acyl-CoA (including 18:2 and 20:4-CoA) [13], the increase of the TAG species esterified LNA and ARA at the sn-2 position in group HCO + BO seems to be largely attributed to *Agpat4*. In addition, the increase in the TAG species esterified DGLA (the elongated C20 metabolized FA of GLA) in group HCO + BO might be attributed to the increased expression of *Elovl7*, which is known to have the highest activity toward 18:3-CoA [14]. *Dgat2*, which is responsible for the last step of TAG synthesis, is known to play a crucial role in skin barrier homeostasis [15]. Overexpression of *Dgat2* causes acyl-Cer synthesis [16], while *Dgat2*-deficient mice show decreased content of acyl-Cer [4], suggesting that the decreased CLE content in EFA-deficient epidermis [17] could be related to decreased *Dgat2* expression.

(a)

Biological process	*p*-value	FDR
lipid metabolic process	0	0
cellular lipid metabolic process	0	0
cellular lipid catabolic process	5.24E-11	1.59E-07
lipid catabolic process	7.51E-11	1.71E-07
keratinization	1.16E-10	1.93E-07
membrane lipid metabolic process	1.27E-10	1.93E-07
keratinocyte differentiation	4.83E-10	6.27E-07
epidermis development	9.54E-10	1.08E-06
skin development	1.82E-09	1.84E-06
epidermal cell differentiation	2.05E-09	1.86E-06

(b)

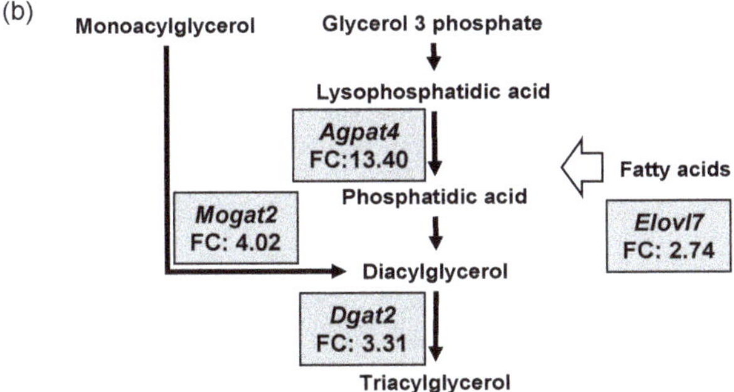

(c)

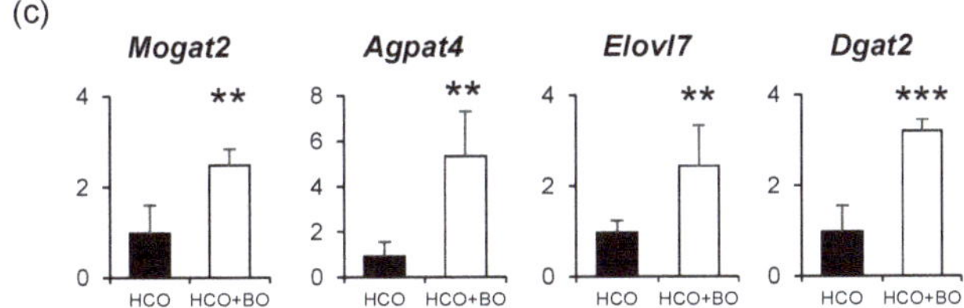

Figure 2. Gene expression levels in the epidermis of guinea pigs fed a hydrogenated coconut oil (HCO) diet for 0 weeks (group HCO) or 8 weeks followed by 2 weeks of borage oil (BO) diet (group HCO + BO). Transcriptomes were obtained by microarray analysis ($n = 3$). (**a**) The gene ontology enrichment analysis of up-regulated genes in group HCO + BO (Fold change; FC > 1.5) was performed using a web-based gene set analysis toolkit. Results are listed by false discovery rate (FDR) and the adjusted *p*-value is corrected for multiple comparisons. (**b**) Among genes related to triacylglycerol synthesis, up-regulated genes in group HCO + BO were selected (FC > 2). (**c**) The gene expression levels of monoacylglycerol O-acyltransferase 2 (*Mogat2*), 1-acylglycerol-3-phosphate O-acyltransferase 4 (*Agpat4*), and the elongation of very long-chain fatty acids 7 (*Elovl7*) and diacylglycerol O-acyltransferase 2 (*Dgat2*) were validated by a quantitative reverse transcription polymerase chain reaction (qRT-PCR, $n = 5$). Values are FC or means ± SD. ** $p < 0.01$, *** $p < 0.001$ vs. group HCO by the unpaired Student's *t* test.

3.3. Altered Expression of Genes Related to TAG Hydrolysis and CLE Formation

To understand the implication of altered TAG content and gene expression related to TAG metabolism such as *Dgat2*, we further evaluated the genes involved in TAG hydrolysis, acyl-Cer synthesis, and CLE formation (Table 2). Among the various candidate genes that could hydrolyze TAG [15], the expression levels of lipase family member N (*Lipn*, FC = 3.66) and K (*Lipk*, FC = 3.65) and the patatin-like phospholipase domain containing 5 (*Pnpla5*, FC = 2.35) were increased in group HCO + BO. In addition, the expression levels of genes that induce the hydroxylation of ultra-long-chain FA (*Cyp4f22*), acylation of FA into sphingoid bases (*Cers3*), transportation of lamellar granules (*Abca12*), and oxidation of LNA for CLE formation (*Alox12b* and *Aloxe3*) [18] were also increased approximately 1.5–3-fold (Table 2). Collectively, these results suggest that the EFA esterified into TAG is likely to be hydrolyzed by these lipases and esterified for acyl-Cer synthesis, which is subsequently utilized for CLE formation by these genes. Furthermore, because *Alox12b*, the dioxygenase required for the formation of covalent bond between acyl-Cer and corneocytes, has been reported to have a preference for GLA as well as LNA [19], it could be speculated that GLA-esterified acyl-Cer, which was reported in our previous studies, might be involved in maintaining the integrity of CLE. Therefore, the genes listed in Table 2 seem to be related to restoring the impaired skin barrier in EFA-deficient guinea pigs by BO supplementation [6].

Table 2. Up-regulated genes associated with epidermal lipase, acyl-Cer synthesis and CLE formation in group HCO + BO [1].

Gene [2]	FC [3]	Function
Lipn	3.66	Lipases
Lipk	3.65	Lipases
Pnpla5	2.35	Lipases
Elovl4	1.72	Elongation of fatty acids
Cyp4f22	1.57	Omega-hydroxylation of ultra-long-chain fatty acids
Cers3	1.61	Ceramide synthesis
Ugcg	1.53	Glucosylation of ceramide
Abca12	3.07	Transport via lamellar granules
Alox12b	2.13	Oxidation of linoleic acid in ceramide
Aloxe3	1.65	Oxidation of linoleic acid in ceramide

[1] Cer: ceramide, CLE: corneocyte lipid envelope, group HCO+BO: hydrogenated coconut oil (HCO) diet for 8 weeks followed by borage oil (BO) diet for 2 weeks. [2] *Lipn*: lipase family member N, *Lipk*: lipase family member K, *Pnpla5*: patatin-like phospholipase domain containing 5, *Elovl4*: elongation of very long-chain fatty acids 4, *Cyp4f22*: cytochrome P450 family 4 subfamily F member 22, *Cers3*: ceramide synthase 3, *Ugcg*: UDP-glucose ceramide glucosyltransferase, *Abca12*: ATP binding cassette subfamily A, *Alox12b*: arachidonate 12-lipoxygenase 12R type, *Aloxe3*: arachidonate lipoxygenase 3. [3] 3-fold change (FC) of group HCO + BO with respect to group HCO (HCO diet for 10 weeks).

4. Conclusions

Dietary BO induced the increase of TAG content and TAG species esterified LNA, GLA, DGLA, and ARA in parallel with the up-regulation of the expression of genes involved in TAG synthesis including *Mogat2*, *Agpat4*, *Elovl7*, and *Dgat2* as well as TAG hydrolysis, acyl-Cer synthesis, and CLE formation in the epidermis of EFA-deficient guinea pigs. These results suggest that TAG could serve as an important EFA-donor required for acyl-Cer synthesis.

Supplementary Materials: The following are available online at http://www.mdpi.com/2072-6643/11/11/2818/s1, Table S1: Experimental *m/z* and signal intensities of each triacylglycerol species quantitated, Table S2: Up-regulated gene list in HCO + BO (fold change; FC > 1.5), Table S3: Over-representation analysis with up-regulated genes in HCO + BO, Table S4: Up-regulated genes in HCO + BO related to GO: 0006629: lipid metabolic process.

Author Contributions: Conceptualization and methodology, Y.C. and K.-P.K.; formal analysis and investigation, J.-Y.L. and K.-H.L.; writing—original draft preparation, K.-P.K.; writing—review & editing, Y.C.; funding acquisition, Y.C. All authors read and approved the final manuscript.

Funding: This study was supported by the Mid-Career Research Program of the National Research Foundation (NRF) funded by the Ministry of Science, ICT & Future Planning (Grant No.: 2017R1A2B2007081), Republic of Korea.

Conflicts of Interest: The authors declare that they have no conflicts of interest.

References

1. Lu, B.; Jiang, Y.J.; Man, M.Q.; Brown, B.; Elias, P.M.; Feingold, K.R. Expression and regulation of 1-acyl-sn -glycerol- 3-phosphate acyltransferases in the epidermis. *J. Lipid Res.* **2005**, *46*, 2448–2457. [CrossRef] [PubMed]

2. Radner, F.P.W.; Streith, I.E.; Schoiswohl, G.; Schweiger, M.; Kumari, M.; Eichmann, T.O.; Rechberger, G.; Koefeler, H.C.; Eder, S.; Schauer, S.; et al. Growth Retardation, Impaired Triacylglycerol Catabolism, Hepatic Steatosis, and Lethal Skin Barrier Defect in Mice Lacking Comparative Gene Identification-58 (CGI-58). *J. Biol. Chem.* **2010**, *285*, 7300–7311. [CrossRef] [PubMed]

3. Ujihara, M.; Nakajima, K.; Yamamoto, M.; Teraishi, M.; Uchida, Y.; Akiyama, M.; Shimizu, H.; Sano, S. Epidermal triglyceride levels are correlated with severity of ichthyosis in Dorfman–Chanarin syndrome. *J. Dermatol. Sci.* **2010**, *57*, 102–107. [CrossRef] [PubMed]

4. Stone, S.J.; Myers, H.M.; Watkins, S.M.; Brown, B.E.; Feingold, K.R.; Elias, P.M.; Farese, R. V Lipopenia and skin barrier abnormalities in DGAT2-deficient mice. *J. Biol. Chem.* **2004**, *279*, 11767–11776. [CrossRef] [PubMed]

5. Bibel, D.J.; Miller, S.J.; Brown, B.E.; Pandey, B.B.; Elias, P.M.; Shinefield, H.R.; Aly, R. Antimicrobial activity of stratum corneum lipids from normal and essential fatty acid-deficient mice. *J. Invest. Dermatol.* **1989**, *92*, 632–638. [CrossRef] [PubMed]

6. Chung, S.; Kong, S.; Seong, K.; Cho, Y. γ-Linolenic Acid in Borage Oil Reverses Epidermal Hyperproliferation in Guinea Pigs. *J. Nutr.* **2002**, *132*, 3090–3097. [CrossRef] [PubMed]

7. Shin, K.-O.; Kim, K.; Jeon, S.; Seo, C.-H.; Lee, Y.-M.; Cho, Y. Mass Spectrometric Confirmation of γ-Linolenic Acid Ester-Linked Ceramide 1 in the Epidermis of Borage Oil Fed Guinea Pigs. *Lipids* **2015**, *50*, 1051–1056. [CrossRef]

8. Kim, K.-P.; Jeon, S.; Kim, M.-J.; Cho, Y. Borage oil restores acidic skin pH by up-regulating the activity or expression of filaggrin and enzymes involved in epidermal lactate, free fatty acid, and acidic free amino acid metabolism in essential fatty acid-deficient Guinea pigs. *Nutr. Res.* **2018**, *58*, 26–35. [CrossRef]

9. Shin, J.; Kim, Y.J.; Kwon, O.; Kim, N.-I.; Cho, Y. Associations among plasma vitamin C, epidermal ceramide and clinical severity of atopic dermatitis. *Nutr. Res. Pract.* **2016**, *10*, 398. [CrossRef] [PubMed]

10. Shon, J.C.; Shin, H.-S.; Seo, Y.K.; Yoon, Y.-R.; Shin, H.; Liu, K.-H. Direct infusion MS-based lipid profiling reveals the pharmacological effects of compound K-reinforced ginsenosides in high-fat diet induced obese mice. *J. Agric. Food Chem.* **2015**, *63*, 2919–2929. [CrossRef] [PubMed]

11. Liao, Y.; Wang, J.; Jaehnig, E.J.; Shi, Z.; Zhang, B. WebGestalt 2019: gene set analysis toolkit with revamped UIs and APIs. *Nucleic Acids Res.* **2019**, *47*, 199–205. [CrossRef] [PubMed]

12. Jiang, Y.J.; Feingold, K.R. The expression and regulation of enzymes mediating the biosynthesis of triglycerides and phospholipids in keratinocytes/epidermis. *Dermatoendocrinol.* **2011**, *3*, 70–76. [CrossRef] [PubMed]

13. Eto, M.; Shindou, H.; Shimizu, T. A novel lysophosphatidic acid acyltransferase enzyme (LPAAT4) with a possible role for incorporating docosahexaenoic acid into brain glycerophospholipids. *Biochem. Biophys. Res. Commun.* **2014**, *443*, 718–724. [CrossRef] [PubMed]

14. Naganuma, T.; Sato, Y.; Sassa, T.; Ohno, Y.; Kihara, A. Biochemical characterization of the very long-chain fatty acid elongase ELOVL7. *FEBS Lett.* **2011**, *585*, 3337–3341. [CrossRef] [PubMed]

15. Radner, F.P.W.; Fischer, J. The important role of epidermal triacylglycerol metabolism for maintenance of the skin permeability barrier function. Biochim. Biophys. *Acta - Mol. Cell Biol. Lipids* **2014**, *1841*, 409–415. [CrossRef] [PubMed]

16. Ohno, Y.; Kamiyama, N.; Nakamichi, S.; Kihara, A. PNPLA1 is a transacylase essential for the generation of the skin barrier lipid ω-O-acylceramide. *Nat. Commun.* **2017**, *8*, 14610. [CrossRef] [PubMed]

17. Elias, P.M.; Gruber, R.; Crumrine, D.; Menon, G.; Williams, M.L.; Wakefield, J.S.; Holleran, W.M.; Uchida, Y. Formation and functions of the corneocyte lipid envelope (CLE). Biochim. Biophys. *Acta - Mol. Cell Biol. Lipids* **2014**, *1841*, 314–318. [CrossRef] [PubMed]

18. Akiyama, M. Corneocyte lipid envelope (CLE), the key structure for skin barrier function and ichthyosis pathogenesis. *J. Dermatol. Sci.* **2017**, *88*, 3–9. [CrossRef] [PubMed]

19. Sun, D.; McDonnell, M.; Chen, X.-S.; Lakkis, M.M.; Li, H.; Isaacs, S.N.; Elsea, S.H.; Patel, P.I.; Funk, C.D. Human 12(R)-Lipoxygenase and the Mouse Ortholog. *J. Biol. Chem.* **1998**, *273*, 33540–33547. [CrossRef] [PubMed]

Article

Synergistic Cytoprotective Effects of Rutin and Ascorbic Acid on the Proteomic Profile of 3D-Cultured Keratinocytes Exposed to UVA or UVB Radiation

Agnieszka Gęgotek *, Iwona Jarocka-Karpowicz and Elżbieta Skrzydlewska

Department of Analytical Chemistry, Medical University of Bialystok, 15-089 Bialystok, Poland; iwona.jarocka-karpowicz@umb.edu.pl (I.J.-K.); elzbieta.skrzydlewska@umb.edu.pl (E.S.)
* Correspondence: agnieszka.gegotek@umb.edu.pl; Tel.: +48-857485708

Received: 15 October 2019; Accepted: 29 October 2019; Published: 5 November 2019

Abstract: The combination of ascorbic acid and rutin, often used in oral preparations, due to antioxidant and anti-inflammatory properties, can be used to protect skin cells against the effects of UV radiation from sunlight. Therefore, the aim of this study was to investigate the synergistic effect of rutin and ascorbic acid on the proteomic profile of UVA and UVB irradiated keratinocytes cultured in a three-dimensional (3D) system. Results showed that the combination of rutin and ascorbic acid protects skin cells against UV-induced changes. In particular, alterations were observed in the expression of proteins involved in the antioxidant response, DNA repairing, inflammation, apoptosis, and protein biosynthesis. The combination of rutin and ascorbic acid also showed a stronger cytoprotective effect than when using either compound alone. Significant differences were visible between rutin and ascorbic acid single treatments in the case of protein carboxymethylation/carboxyethylation. Ascorbic acid prevented UV or rutin-induced protein modifications. Therefore, the synergistic effect of rutin and ascorbic acid creates a potentially effective protective system against skin damages caused by UVA and UVB radiation.

Keywords: keratinocytes; rutin; ascorbic acid; UV radiation; proteomics; 3D cell culture

---

1. Introduction

Maintaining the appropriate skin proteomic profile is a critical parameter for proper cell function and maintenance of healthy skin. For this reason, skin cells are characterized by the well-developed cytoprotective system, comprised of antioxidant, anti-inflammatory, and anti-apoptotic proteins. This protection system also involves active cytoprotective transcription factors which are responsible for protein biosynthesis [1]. However, despite these cytoprotective mechanisms, frequent exposure of skin cells to ultraviolet radiation (UV), mainly UVB (electromagnetic radiation with wavelength from 280 nm to 315 nm) and UVA (from 315 nm to 380 nm) contained in sunlight, stimulates pro-oxidant enzyme activity and impairs the action of antioxidants, resulting in oxidative stress [2,3]. As a consequence, UV enhances ROS-dependent modifications in skin cells covering all cellular components including nucleic acids, proteins, and lipids. This is particularly seen in lipid modification which causes an increase in the level of lipid peroxidation products, including 4-hydroxynonenal (4-HNE), which functions as an important signaling molecule [4]. In addition, the resulting highly reactive electrophilic aldehydes can interact with proteins and significantly modify their structure [5]. Also, UV-induced DNA damage activate proteins responsible for nucleic acids repairing [6]. On the other hand, in the case of keratinocytes, UVB radiation also enhances protein modifications such as advanced glycation end products (AGEs), including lysine carboxymethylation (CML) [7]. Generation of CML leads to dysfunction of the proteins biological functions, their translocation into the cell membrane, and activation of G-protein [8]. As a result, the expression of G-protein coupled receptors is often increased.

As a result, changes in the proteomic profile, including modifications of protein activity, are observed [9,10]. Previous studies have shown that skin cells exposure to UVA and UVB radiation leads to the activation of many factors involved in mitogen-activated protein (MAP)-dependent signaling kinases, including ERK1/2 and transcription factors dependent on redox potential, e.g., Nrf2 [11,12]. As a consequence, the described action of UV radiation leads to metabolic disorders in the skin which can significantly deteriorate the condition of the skin [13].

Due to the hazards of UV exposure, there is a constant need for natural, effective, daily use skin protection compounds. Examples of such compounds are rutin and ascorbic acid, not only due to their potent antioxidant properties resulting from their chemical structure (Figure 1), but also due to their synergistic cytoprotective activity [14]. It has been shown that rutin affects cellular metabolism not only via its antioxidant activity, but also by affecting the biological activity of proteins in varying signaling pathways. Rutin activates cytoprotective transcription factor Nrf2, and also inhibits the activity of cyclooxygenases and lipoxygenases, thereby reducing pro-inflammatory processes [15,16]. Additionally, rutin exerts cytoprotective effects on cells exposed to different types of radiation by substantially increasing their viability [17]. The use of rutin in in vitro cultures may be limited by its toxicity in a high concentration, however, different literature data provide different results regarding the range of its toxicity. In addition, this value depends on the type of cells as well as the length of treatment; long-term rutin treatment of cancer cells shows that the toxic concentration of rutin is already in the range 125–250 μM [18,19]. However, in the case of HaCaT keratinocytes, rutin toxicity is visible only above a concentration of 1 mM [20]. However, the action of rutin is limited by cell membrane permeability, which is significantly increased by UV radiation as a result of activation of membrane transporter, bilitranslocase [10]. Ascorbic acid also enhances rutin membrane permeability [21]. On the other hand, ascorbic acid applied to the skin exhibits a number of cytoprotective activities, such as supporting the antioxidant effect of vitamin E [22] and normalizing respiratory chain action by stabilizing mitochondrial membrane polarization in UV-irradiated human skin fibroblasts [23], as well as providing significant protection against inflammation and sunburn [24]. Treatment with ascorbic acid induces well-organized multilayers in 3D keratinocyte cultures, however, exceeding a concentration of 1 mM promotes too fast and uncontrolled cell differentiation [25,26].

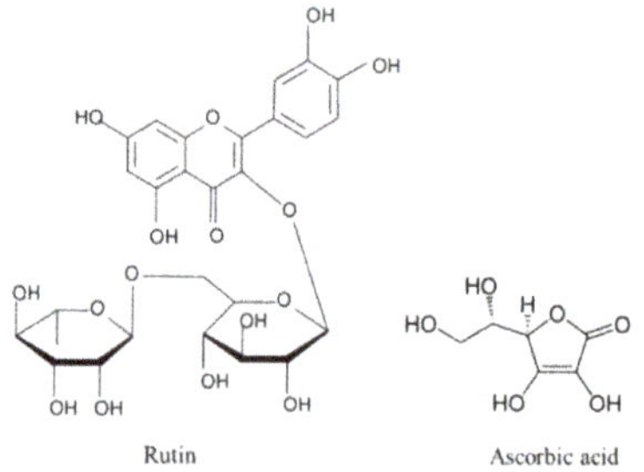

Figure 1. The comparison of chemical structure of rutin and ascorbic acid.

To date, the interaction of orally-administered rutin with ascorbic acid has been demonstrated in terms of anti-inflammatory and vascular sealing actions [27]. Moreover, recent work has shown that combining rutin with ascorbic acid results in a synergistic, cytoprotective effect from UV radiation in vitro [14]. The aim of this study was to examine the combined effect of rutin and ascorbic acid on the proteomic profile in UVA and UVB irradiated keratinocytes grown in a three-dimensional (3D) epidermal-like system.

2. Materials and Methods

2.1. Cell Culture and Treatment

Human cell keratinocytes line CDD 1102 KERTr were obtained from American Type Culture Collection (ATCC, Manassas, VA, USA) and cultured in a humidified atmosphere of 5% CO_2 at 37 °C.

According to the cell culture protocol provided by ATCC, the growth medium for keratinocytes was keratinocyte-SFM medium supplemented with 1% bovine pituitary extract (BPE), human recombinant epidermal growth factor (hEGF), 50 µg/mL streptomycin, and 50 U/mL penicillin. Sterile and cell culture reagents were obtained from Gibco (Grand Island, NY, USA). 3D culture was carried out in AlgiMatrix plates (Life Technologies, Carlsbad, CA, USA).

Keratinocytes, following four days of 3D gel culturing, were irradiated with the following UV doses: UVA (365 nm)—30 J/cm^2 and UVB (312 nm)—60 mJ/cm^2 (Bio-Link Crosslinker BLX 312/365; Vilber Lourmat, Germany). To observe the effect of ascorbic acid and rutin on UV radiated cells following exposure to UV radiation, cells were incubated for 24 h under standard conditions in medium containing 100 µM ascorbic acid (19.81 mg/L) or/and 25 µM rutin (15.25 mg/L) in 0.1% DMSO. The concentration of ascorbic acid was selected according to the suggested concentration of this compound obtained in the body through a balanced diet, while the rutin concentration was chosen as the highest non-cytotoxic dose [14,28]. In parallel, cells were cultured without irradiation in medium containing the above supplements. In order to maintain the same conditions for all experimental and control groups, all media contained 0.1% DMSO.

Following incubation, keratinocytes were collected from 3D gel with AlgiMatrix™ dissolving buffer (Life Technologies, Carlsbad, CA, USA), lysed by sonication on ice, and centrifuged (15 min, 12,000× *g*). The total protein content in supernatant was measured using a Bradford assay [29]. Figure 2 shows the diagram of the experiment scheme.

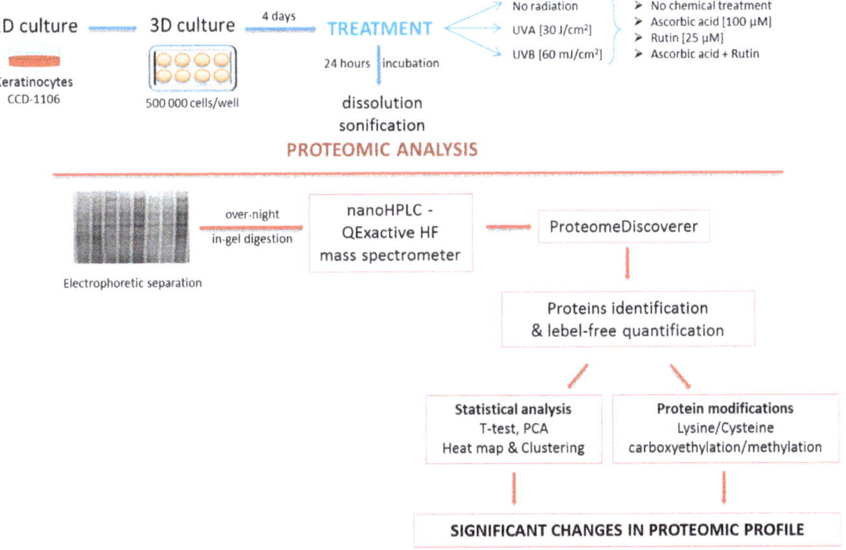

Figure 2. The scheme of the experiment including cells treatment, sample processing, and data statistical analysis.

2.2. Proteomic Analysis

The supernatant was mixed 1:1 with Laemmle buffer (supplemented with 5% 2-mercaptoethanol) and heated at 95 °C for 10 min for protein denaturation. Samples were separated on 10% Tris-Glycine SDS-PAGE gels and stained with Coomassie brilliant blue R-250. Complete lanes were cut out of the gel, sliced into 12 sections and in-gel digested overnight with trypsin (Promega, Madison, WI, USA). The resulting peptide mixture was extracted from the gel, dried, and dissolved in 5% ACN + 0.1% formic acid (FA) and separated using an Ultimate 3000 (Dionex, Idstein, Germany) onto a 150 mm × 75 mm PepMap RSLC capillary analytical C18 column (Dionex, LC Packings). The peptides eluted from the column were analyzed using a QExactive HF mass spectrometer with an electrospray ionization source (ESI) (Thermo Fisher Scientific, Bremen, Germany).

2.3. Protein Identification, Grouping, and Label-Free Quantification

Processing of the raw data generated from LC-MS/MS analysis was carried out using Proteome Discoverer 2.0 (Thermo Fisher Scientific, Bremen, Germany) and Sequest HT (SEQUEST HT algorithm, license Thermo Scientific, registered trademark University of Washington, USA). Input data were searched against the UniProtKB-SwissProt database (taxonomy: *Homo sapiens*, release 04/2018). For protein identification the following search parameters were used: peptide mass tolerance set to 10 ppm, MS/MS mass tolerance set to 0.02 Da, up to two missed cleavages allowed. Dynamic modifications of lysine or cysteine carboxyethylation (CEL/CEC) and carboxymethylation (CML/CMC) were set [30].

2.4. Statistical Analysis

Analysis of each sample were performed in three independent experiments. Results from individual protein label-free quantification were normalized by the sample sum, log transformed and analyzed using the standard statistical analysis methods, including T-test, principal component analysis (PCA), heat map and dendrogram creation with free available MetaboAnalyst 4.0 software (http://www.metaboanalyst.ca).

3. Results

The results of this study showed that rutin and ascorbic acid treatments lead to significant changes in protein expression in UVA or UVB irradiated keratinocytes. For all samples 862 proteins with at least two unique peptides were identified and measured (Table S1). As is shown in Figure 3, there was a strong differentiation between all treatment conditions following principal component analysis (PCA) (component 1–42.8%; component 2–8.9%). The top 20 proteins from PCA component 1, along with their VIP scores, are presented in Figure 2. Moreover, these results were confirmed by the hierarchical clustering of the top 100 proteins with a significant *p*-value (Figure 4). Comparing the *p*-values allowed us to identify which proteins had the highest differences in expression following rutin and ascorbic acid treatment (Figure 5). The main changes observed were in the case of proteins involved in antioxidant response, including peroxiredoxin 1, glutathione reductase, glutathione S-transferase, protein disulfide-isomerase, superoxide dismutase (Cu–Zn), thioredoxin reductase 1, thioredoxin-dependent peroxide reductase, and the cytoprotective transcription factor, Nrf2. The expression of these proteins changed following rutin and ascorbic acid treatment, which counteracted UV-induced alterations. In addition, the UV induced increase in DNA repairing proteins level, such as PARP-1 and FEN1 (Poly(ADP–ribose) polymerase 1, Flap endonuclease 1), was partially decreased by the protective rutin and ascorbic acid action. Simultaneously, rutin and ascorbic acid decreased the expression of some proinflammatory (NFκB, TNFα) and proapoptotic (p53, cell cycle/apoptosis regulator protein 2, caspase 3) proteins. In all, cases rutin treatment, following UV radiation, showed slightly stronger cytoprotective effects than ascorbic acid, however, the use of these two antioxidants in the cases of proapoptotic and proinflammatory proteins allowed the UV irradiated cells to restore the level of mentioned proteins following cell stress.

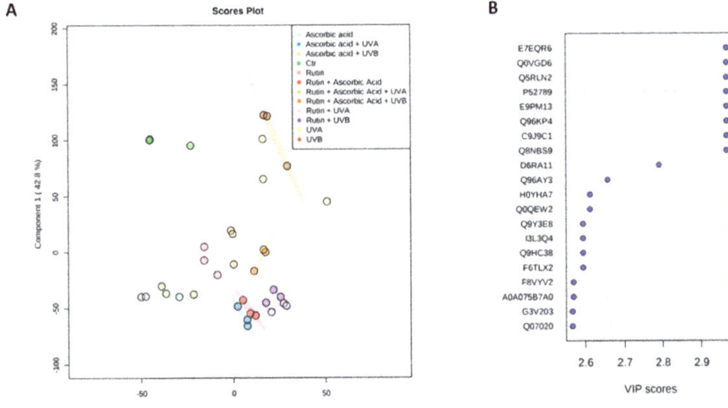

Figure 3. Principal component analysis (PCA) (**A**) and top 20 of component 1 VIP scores (**B**) for proteins from the 3D cultured keratinocytes exposed to UVA (30 J/cm^2) or UVB irradiation (60 mJ/cm^2) and treated with rutin (25 μM) or/and ascorbic acid (100 μM). Proteins: E7EQR6—T-complex protein 1 subunit α; Q0VGD6—heterogeneous nuclear ribonucleoprotein R protein; P52789—hexokinase-2; E9PM13—heat shock cognate 71 kDa protein; Q96KP4—cytosolic non-specific dipeptidase; C9J9C1—serine/threonine-protein phosphatase 2A; Q8NBS9—thioredoxin domain-containing protein 5; D6RA11—ubiquitin-conjugating enzyme E2 D3; Q96AY3—peptidyl-prolyl cis-trans isomerase; H0YHA7, Q0QEW2, F8VYV2, A0A075B7A0, G3V203, Q07020—ribosomal proteins; Q9Y3E8—CGI-150 protein; I3L3Q4, Q9HC38, F6TLX2—glyoxalase domain-containing proteins.

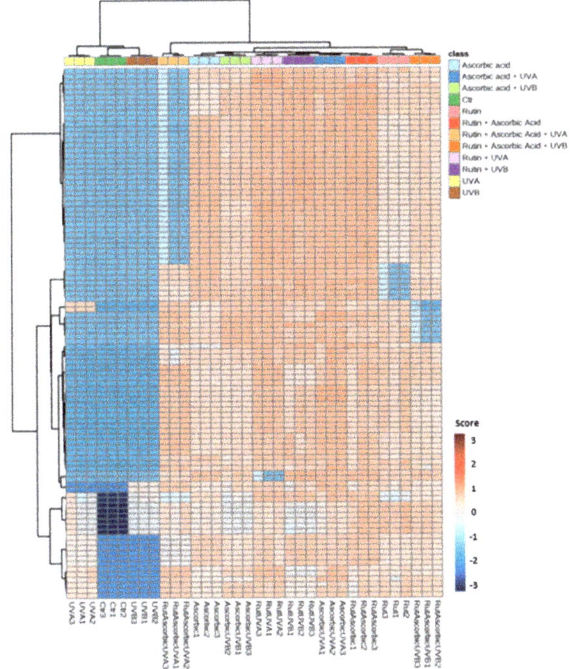

Figure 4. Heat map and clustering for the top 100 proteins from the 3D cultured keratinocytes exposed to UVA (30 J/cm^2) or UVB irradiation (60 mJ/cm^2) and treated with treated with rutin (25 μM) or/and ascorbic acid (100 μM). Protein expression levels (log transformed) were scaled to the row mean.

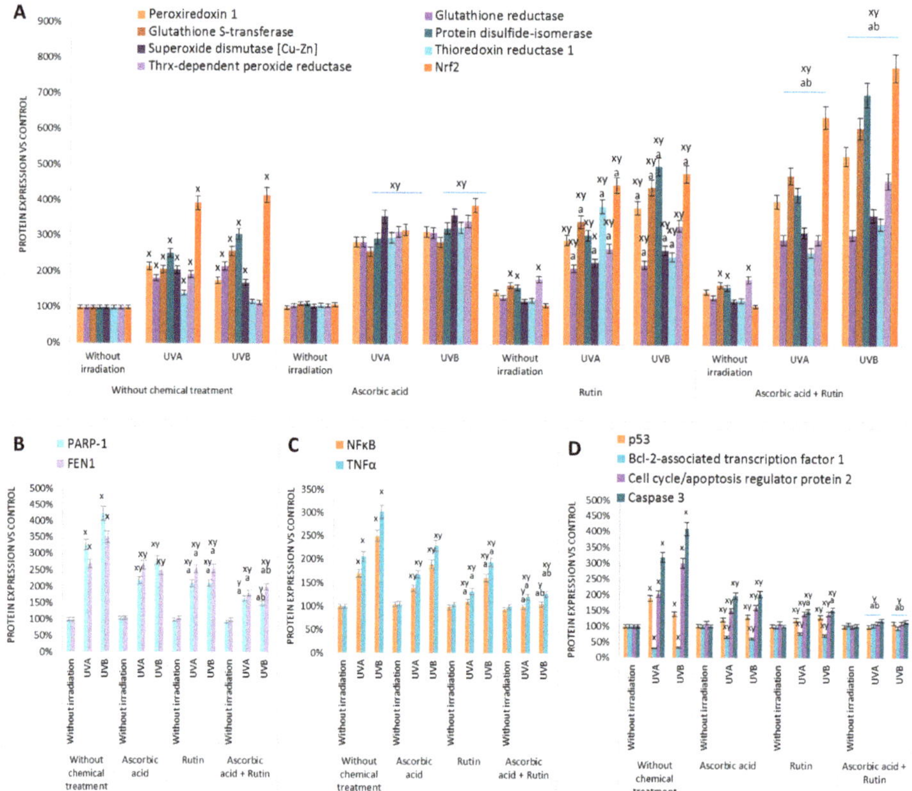

Figure 5. The expression of antioxidant (**A**), DNA repairing (**B**), proinflammatory (**C**), and pro/anti-apoptotic (**D**) proteins of which levels were changed in 3D cultured keratinocytes exposed to UVA (30 J/cm^2) or UVB irradiation (60 mJ/cm^2) and treated with treated with rutin (25 µM) or/and ascorbic acid (100 µM). Data obtained from label-free analysis. Mean values ± SD of three independent experiments are presented. [x] statistically significant differences vs. non-treated group, $p < 0.05$; [y] statistically significant differences vs. respectively group without chemical treatment, $p < 0.05$; [a] statistically significant differences vs. ascorbic acid treated group, $p < 0.05$; [b] statistically significant differences vs. rutin treated group, $p < 0.05$.

UV-induced oxidative stress also significantly increased post-translational protein modifications (PTMs), including lysine/cysteine carboxyethylation and carboxymethylation (Figure 6). While ascorbic acid significantly reduced the level of these modifications, rutin favored their creation under both standard conditions and following UV-induced stress. However, combined use of rutin and ascorbic acid decreased the level of analyzed PTMs in keratinocytes exposed to UV radiations.

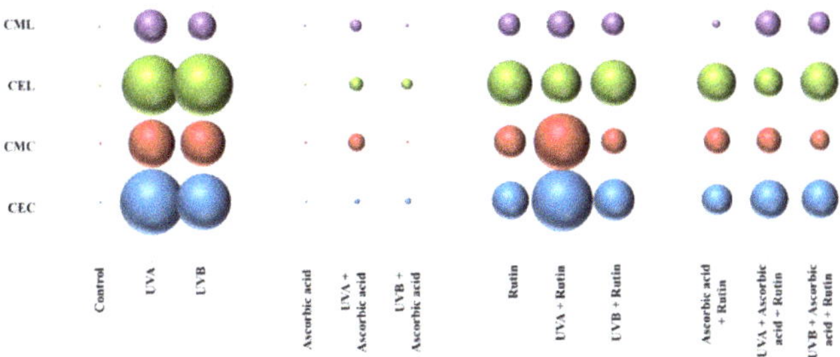

Figure 6. The comparison of the lysine or cysteine modifications (carboxyethylation (CEL/CEC) and carboxymethylation (CML/CMC)) level in 3D cultured keratinocytes exposed to UVA (30 J/cm^2) or UVB irradiation (60 mJ/cm^2) and treated with treated with rutin (25 µM) or/and ascorbic acid (100 µM). Data obtained from label-free analysis.

4. Discussion

Permanent exposure of skin cells to solar radiation induces disturbances in cell metabolism via UV radiation. This creates the need for research and development of natural methods of protection against harmful effects of UV radiation. UV-induced modifications of both bioactive and structural proteins are some of the most dangerous changes in the cellular proteome. In order to maximize the effective protection of skin cells, single treatment therapies are now being replaced by combination therapies [14]. In addition, the testing of harmful effects of UV radiation and cytoprotection of selected compounds in the traditional two-dimensional (2D) culture system does not allow full understanding of the cellular phenotype and cell–cell interaction [31]. Therefore, we assessed the combined effect of ascorbic acid and rutin on the proteomic profile in keratinocytes grown in a three-dimensional (3D) system in an attempt to replicate the skin microenvironment.

This study combined the use of rutin and ascorbic acid, widely known as oral pharmacological agents for the prevention of flu and colds. However, such formulations, despite cytoprotective properties, are not commonly used together in preparations for skin care and protection. While ascorbic acid is frequently added to topical ointments, its interaction with rutin is still unknown apart from the fact that ascorbic acid enhances membranes penetration by rutin [21]. Moreover, water soluble ascorbic acid, as well as high molecular rutin have low absorption into the skin, which can be additionally compensated by oral supplementation. Literature data shows that oral supplementation of rutin in a standard dose 500 mg/day induces a 100 mg/L concentration of this molecule in the plasma following 5 h after ingestion, and decreases in the next five hours to 10 mg/L [32], which are comparable to the conditions used in this study.

Both of the compounds used in this study, have exhibited in previous studies selective anti-inflammatory effects in UV-irradiated cells. It has been shown that ascorbic acid is able to downregulate IL-1β mRNA expression in UVA-irradiated and IL-8 mRNA in UVB-irradiated keratinocytes [33,34]. Additionally, rutin may influence UV-induced proinflammatory signaling via inhibition of the enzymes responsible for metabolism of fatty acids and reducing the generation of proinflammatory signaling molecules [10,16]. However, the combined action of rutin and ascorbic acid, as observed in this study and previous data from 2D culture keratinocytes [14], led to a reduction in the biological activity of NFκB, resulting in a significant reduction in TNFα levels. Therefore, the functional activity of rutin and ascorbic acid in UV irradiated keratinocytes leads to inhibition of inflammatory processes. Presented proteomic data are consistent with previous data obtained by Western blot analysis for 2D cultured keratinocytes [14].

UV radiation induced inflammation was partially reduced by the combined action of rutin and ascorbic acid. Both rutin and ascorbic acid support the cellular antioxidant system, not only as scavengers of free radicals, but also by stimulating the biosynthesis of antioxidant proteins. Data obtained in this study showed that the most UV-sensitive proteins were combined with a glutathione-based system (GSH) (glutathione reductase, glutathione S-transferase) or thioredoxin-based system (thioredoxin reductase, peroxiredoxin, thioredoxin-dependent peroxide reductase, protein disulfide-isomerase). GSH is a cofactor for GSH peroxidase, which is also responsible for protecting lipids from peroxidation [35]. Therefore, the appropriate level of GSH, delivered by the high activity of glutathione reductase and glutathione S-transferase, supports the antioxidant system of keratinocytes during oxidative stress. Alternatively, the thioredoxin-dependent antioxidant system can be activated under oxidative conditions, which is particularly important in the repair of UV-induced protein modifications [36]. However, it has also been shown that many antioxidant compounds with the ability to thioredoxin associated activate enzymes can act as cytoprotectors against UV radiation in keratinocytes [37–39].

The reaction of keratinocytes to UV-induced oxidative stress is also associated with the activation of intracellular and extracellular signaling pathways. These modifications can lead to changes in Nrf2 transcription factor activity. Nrf2 is a transcription factor bound to the cytoplasm by Keap1 that is responsible for the expression of genes encoding antioxidant and anti-apoptotic proteins [40,41]. This reaction was further intensified by rutin and ascorbic acid, especially when used in combination, which prepared the cells for oxidative stress [42]. Similar data have been presented previously for the 2D cultured keratinocytes, where rutin and ascorbic acid additionally increased Nrf2 level also by enhancing its activator KAP1 and decreasing nuclear inhibitor Bach1 [14]. On the other hand, rutin, by forming adducts with Keap1, changes the conformation of this molecule and additionally leads to dissociation of the Keap1–Nrf2 complex [15].

The mentioned antioxidant activity of ascorbic acid and rutin may also be an explanation for the reduction in DNA repairing protein expression following cells exposure to UVA and UVB radiation. It is known that UV irradiation causes direct nucleic acid damage or leads to indirect genetic material destruction also by ROS dependent DNA oxidation [43]. These disturbances induce the DNA repairing system that also includes an increase in the expression of protein involved in DNA repair [6]. Examples of such proteins are PARP-1 and FEN1, of which increased expression following cell UV irradiation has been found previously [44,45]. However, because of the activation of the antioxidant system and reduction of the oxidative UV radiation potential, ascorbic acid and rutin partially prevented the UV-induced increase in the expression of these proteins.

UV radiation induced apoptosis also reduces cell viability [46]. This has been confirmed as associated with increased levels of pro-apoptotic proteins, such as cytochrome c or caspase 3 [47]. As shown in this study, combination of rutin and ascorbic acid protects keratinocytes from UV-induced increased expression of caspase 3, which was previously confirmed by Western blot analysis for 2D cultured keratinocytes [14]. As of now, ascorbic acid has been shown to modify the process of apoptosis. It also has been shown to stimulate apoptosis in cancer cells [48] and prevent apoptosis of unchanged cancer cells subjected to a single stress signal [49]. In the case of keratinocytes, ascorbic acid is important for the initiation of keratinization, and thus is also required for cell survival and differentiation [50]. However, rutin is known as an inhibitor of caspase 9, and may be linked to a lack of caspase 3 activation [51]. This was observed in the present study as a decreased expression of the protein. Similar effects of rutin have been previously observed in the case of protein p53 in the mouse kidney [52]. However, in the case of UV irradiated keratinocytes, decreased levels of p53 and cell cycle/apoptosis proteins prevented cell cycle arrest and promoted proper keratinization and natural keratosis of the epidermis that previously was not visible in the case of cells cultured in monolayer [14].

As is shown in the PCA analysis, the proteins most sensitive to UV radiation and rutin/ascorbic acid treatment were also the molecules involved in the protein biosynthesis, including ribonucleoproteins and ribosomal proteins, as well as molecules involved in protein folding and maturation (heat shock proteins, chaperon: T-complex protein (1)). The data presented here show that rutin alone was able to restore the level of these proteins after stress more than ascorbic acid alone, in agreement with previous work [53]. Another important cell metabolism protein affected by UV radiation, and protected by use of antioxidants, is serine/threonine-protein phosphatase 2A (PP2A), a major phosphatase for microtubule-associated proteins [54]. PP2A modulates the activity of a number of kinases (mitogen-stimulated S6 kinase, MAP-2 kinase, casein kinase 2, or Raf1) and provides cells with fluent signal transmission [55]. Moreover, the combined effect of rutin and ascorbic acid, restores PP2A levels in UV irradiated keratinocytes and may protect cells against apoptosis via PP2A dependent de-phosphorylation of p53 [56]. Further, UV induced changes in expression of glyoxalase domain-containing proteins may lead to disturbances in intracellular cadherin binding and inhibition of cell growth [57]. Therefore, the rutin and ascorbic acid induced decrease in protein levels contributed to the protection of UV-irradiated keratinocytes against apoptosis.

Alternatively, rutin and ascorbic acid can also influence glucose metabolism by modifying hexokinase 2 levels. While ascorbic acid restored the UV-enhanced level of this enzyme, rutin significantly increased hexokinase 2 levels in both control and UV irradiated keratinocytes. Hexokinase phosphorylates glucose to produce glucose-6-phosphate (G6P), which is the first step in energetic metabolism [58]. Flavonoids, including rutin are known to stimulate this process [59]. As a result of increased sugar metabolism, the enhanced levels of AGEs are observed [60]. On the other hand, medium supplemented with rutin was characterized by enhanced pH that additionally favors protein carboxymethylation/carboxyethylation [61]. Therefore, rutin may be a stimulator of this process. Such modifications lead to interruption of protein–protein interactions, polysaccharide–protein complex formation that are considered markers of arteriosclerosis, diabetes mellitus, and aging [62], and also have been shown to operate as a potent antitumor factors [63]. As a consequence, rutin may affect intracellular signaling pathways. However, rutin combined with ascorbic acid partially reduced this effect.

The results shown here indicate that rutin and ascorbic acid, could modify UV induced dysregulation of cellular metabolism. However, the combination of rutin and ascorbic acid showed a stronger cytoprotective effect than when using either of compound alone. These results suggest that this synergistic effect may result from mutual support of penetration through biological membranes. Therefore, the cooperation of rutin and ascorbic acid in the cytoprotective effect on keratinocytes exposed to UVA and UVB radiation makes them a potentially effective protective system against skin damage caused by UV radiation.

Supplementary Materials: The following are available online at http://www.mdpi.com/2072-6643/11/11/2672/s1, Table S1: The list of proteins with at least 2 unique peptides identified in 3D cultured keratinocytes exposed to UVA (30 J/cm^2) or UVB irradiation (60 mJ/cm^2) and treated with treated with rutin (25 µM) or/and ascorbic acid (100 µM).

Author Contributions: Conceptualization, E.S.; formal analysis, A.G. and I.J.-K.; investigation, A.G. and I.J.-K.; methodology, A.G.; project administration, A.G.; supervision, E.S.; validation, A.G. and I.J.-K.; visualization, A.G. and I.J.-K.; writing—original draft, A.G.; writing—review and editing, E.S.

Funding: This study was financed by the National Science Centre Poland (NCN) grant no. 2017/25/N/NZ7/00863. A.G., co-author of the work, was supported by the Foundation for Polish Science (FNP).

Conflicts of Interest: The authors have no conflict of interest to declare.

References

1. Bickers, D.R.; Athar, M. Oxidative stress in the pathogenesis of skin disease. *J. Invest. Dermatol.* **2006**, *126*, 2565–2575. [CrossRef]

2. Bernerd, F.; Marionnet, C.; Duval, C. Solar ultraviolet radiation induces biological alterations in human skin in vitro: Relevance of a well-balanced UVA/UVB protection. *Indian J. Dermatol. Venereol. Leprol.* **2012**, *78*, 15–23. [CrossRef]

3. Gęgotek, A.; Bielawska, K.; Biernacki, M.; Zaręba, I.; Surażyński, A.; Skrzydlewska, E. Comparison of protective effect of ascorbic acid on redox and endocannabinoid systems interactions in in vitro cultured human skin fibroblasts exposed to UV radiation and hydrogen peroxide. *Arch. Dermatol. Res.* **2017**, *309*, 285–303. [CrossRef] [PubMed]

4. Łuczaj, W.; Gęgotek, A.; Skrzydlewska, E. Antioxidants and HNE in redox homeostasis. *Free Radic. Biol. Med.* **2017**, *111*, 87–101. [CrossRef] [PubMed]

5. Møller, I.M.; Sweetlove, L.J. ROS signaling-specificity is required. *Trends Plant Sci.* **2010**, *15*, 370–374. [CrossRef]

6. Sinha, R.P.; Häder, D.P. UV-induced DNA damage and repair: A review. *Photochem. Photobiol. Sci.* **2002**, *1*, 225–236. [CrossRef] [PubMed]

7. Mori, Y.; Aki, K.; Kuge, K.; Tajima, S.; Yamanaka, N.; Kaji, Y.; Yamamoto, N.; Nagai, R.; Yoshii, H.; Fujii, N.; et al. UVB-irradiation enhances the racemization and isomerizaiton of aspartyl residues and production of Nε-carboxymethyl lysine (CML) in keratin of skin. *J. Chromatogr. B Analyt. Technol. Biomed. Life Sci.* **2011**, *879*, 3303–3309. [CrossRef] [PubMed]

8. Grabarczyk, D.B.; Chappell, P.E.; Eisel, B.; Johnson, S.; Lea, S.M.; Berks, B.C. Mechanism of thiosulfate oxidation in the SoxA family of cysteine-ligated cytochromes. *J. Biol. Chem.* **2015**, *290*, 9209–9221. [CrossRef] [PubMed]

9. Roduit, R.; Schorderet, D.F. MAP kinase pathways in UV-induced apoptosis of retinal pigment epithelium ARPE19 cells. *Apoptosis* **2008**, *13*, 343–353. [CrossRef]

10. Gęgotek, A.; Rybałtowska-Kawałko, P.; Skrzydlewska, E. Rutin as a Mediator of Lipid Metabolism and Cellular Signaling Pathways Interactions in Fibroblasts Altered by UVA and UVB Radiation. *Oxid. Med. Cell. Longev.* **2017**, *2017*, 4721352. [CrossRef]

11. Krutmann, J.; Sondenheimer, K.; Grether-Beck, S.; Haarmann-Stemmann, T. Combined, Simultaneous Exposure to Radiation within and Beyond the UV Spectrum: A Novel Approach to Better Understand Skin Damage by Natural Sunlight. In *Environment and Skin*; Springer: Berlin, Germany, 2018; pp. 11–16.

12. Ikehata, H.; Yamamoto, M. Roles of the KEAP1-NRF2 system in mammalian skin exposed to UV radiation. *Toxicol. Appl. Pharmacol.* **2018**, *360*, 69–77. [CrossRef] [PubMed]

13. Deshmukh, J.; Pofahl, R.; Haase, I. Epidermal Rac1 regulates the DNA damage response and protects from UV-light-induced keratinocyte apoptosis and skin carcinogenesis. *Cell Death Dis.* **2017**, *8*, e2664. [CrossRef] [PubMed]

14. Gęgotek, A.; Ambrożewicz, E.; Jastrząb, A.; Jarocka-Karpowicz, I.; Skrzydlewska, E. Rutin and ascorbic acid cooperation in antioxidant and antiapoptotic effect on human skin keratinocytes and fibroblasts exposed to UVA and UVB radiation. *Arch. Dermatol Res.* **2019**, *311*, 203–219. [CrossRef] [PubMed]

15. Gęgotek, A.; Domingues, P.; Skrzydlewska, E. Proteins involved in the antioxidant and inflammatory response in rutin-treated human skin fibroblasts exposed to UVA or UVB irradiation. *J. Dermatol. Sci.* **2018**, *90*, 241–252. [CrossRef]

16. Bouriche, H.; Miles, E.A.; Selloum, L.; Calder, P.C. Effect of *Cleome arabica* leaf extract, rutin and quercetin on soybean lipoxygenase activity and on generation of inflammatory eicosanoids by human neutrophils. *Prostaglandins Leukot. Essent. Fatty Acids* **2005**, *72*, 195–201. [CrossRef]

17. Sunada, S.; Fujisawa, H.; Cartwright, I.M.; Maeda, J.; Brents, C.A.; Mizuno, K.; Aizawa, Y.; Kato, T.A.; Uesaka, M. Monoglucosyl rutin as a potential radioprotector in mammalian cells. *Mol. Med. Rep.* **2014**, *10*, 10–14. [CrossRef]

18. Alonso-Castro, A.J.; Domínguez, F.; García-Carrancá, A. Rutin exerts antitumor effects on nude mice bearing SW480 tumor. *Arch. Med. Res.* **2013**, *44*, 346–351. [CrossRef]

19. Ben Sghaier, M.; Pagano, A.; Mousslim, M.; Ammari, Y.; Kovacic, H.; Luis, J. Rutin inhibits proliferation, attenuates superoxide production and decreases adhesion and migration of human cancerous cells. *Biomed. Pharmacother.* **2016**, *84*, 1972–1978. [CrossRef]

20. Si, Y.X.; Yin, S.J.; Oh, S.; Wang, Z.J.; Ye, S.; Yan, L.; Yang, J.M.; Park, Y.D.; Lee, J.; Qian, G.Y. An integrated study of tyrosinase inhibition by rutin: Progress using a computational simulation. *J. Biomol. Struct. Dyn.* **2012**, *29*, 999–1012. [CrossRef]

21. Guo, R.; Wei, P.; Liu, W. Combined antioxidant effects of rutin and vitamin C in Triton X-100 micelles. *J. Pharm. Biomed. Anal.* **2007**, *43*, 1580–1586. [CrossRef]

22. Pullar, J.M.; Carr, A.C.; Vissers, M. The roles of vitamin C in skin health. *Nutrients* **2017**, *9*, 866. [CrossRef] [PubMed]

23. Sagun, K.C.; Cárcamo, J.M.; Golde, D.W. Vitamin C enters mitochondria via facilitative glucose transporter 1 (Glut1) and confers mitochondrial protection against oxidative injury. *FASEB J.* **2005**, *19*, 1657–1667.

24. Al-Niaimi, F.; Chiang, N.Y.Z. Topical Vitamin C and the Skin: Mechanisms of Action and Clinical Applications. *J. Clin. Aesthet. Dermatol.* **2017**, *10*, 14–17. [PubMed]

25. Seo, A.; Kitagawa, N.; Matsuura, T.; Sato, H.; Inai, T. Formation of keratinocyte multilayers on filters under airlifted or submerged culture conditions in medium containing calcium, ascorbic acid, and keratinocyte growth factor. *Histochem. Cell Biol.* **2016**, *146*, 585–597. [CrossRef]

26. Savini, I.; Rossi, A.; Duranti, G.; Avigliano, L.; Catani, M.V.; Melino, G. Characterization of keratinocyte differentiation induced by ascorbic acid: Protein kinase C involvement and vitamin C homeostasis. *J. Invest. Dermatol.* **2002**, *118*, 372–379. [CrossRef]

27. Milde, J.; Elstner, E.F.; Grassmann, J. Synergistic inhibition of low-density lipoprotein oxidation by rutin, γ-terpinene, and ascorbic acid. *Phytomedicine* **2004**, *11*, 105–113. [CrossRef]

28. Jacob, R.A.; Sotoudeh, G. Vitamin C function and status in chronic disease. *Nutr. Clin. Care* **2002**, *5*, 66–74. [CrossRef]

29. Bradford, M.M. A rapid and sensitive method for the quantitation of microgram quantities of protein utilizing the principle of protein-dye binding. *Anal. Biochem.* **1976**, *72*, 248–254. [CrossRef]

30. Thornalley, P.J.; Battah, S.; Ahmed, N.; Karachalias, N.; Agalou, S.; Babaei-Jadidi, R.; Dawnay, A. Quantitative screening of advanced glycation endproducts in cellular and extracellular proteins by tandem mass spectrometry. *Biochem. J.* **2003**, *375*, 581–592. [CrossRef]

31. Bissell, M.J. Architecture Is the Message: The role of extracellular matrix and 3-D structure in tissue-specific gene expression and breast cancer. *Pezcoller Found. J.* **2007**, *16*, 2–17.

32. Ishii, K.; Furuta, T.; Kasuya, Y. Determination of rutin in human plasma by high-performance liquid chromatography utilizing solid-phase extraction and ultraviolet detection. *J. Chromatogr. B Biomed. Sci. Appl.* **2001**, *759*, 161–168. [CrossRef]

33. Serrano, G.; Almudéver, P.; Serrano, J.M.; Milara, J.; Torrens, A.; Expósito, I.; Cortijo, J. Phosphatidylcholine liposomes as carriers to improve topical ascorbic acid treatment of skin disorders. *Clin. Cosmet. Invest. Dermatol.* **2015**, *8*, 591–599.

34. Kang, J.S.; Kim, H.N.; Kim, J.E.; Mun, G.H.; Kim, Y.S.; Cho, D.; Shin, D.H.; Hwang, Y.I.; Lee, W.J. Regulation of UVB-induced IL-8 and MCP-1 production in skin keratinocytes by increasing vitamin C uptake via the redistribution of SVCT-1 from the cytosol to the membrane. *J. Invest. Dermatol.* **2007**, *127*, 698–706. [CrossRef] [PubMed]

35. Njälsson, R.; Norgren, S. Physiological and pathological aspects of GSH metabolism. *Acta Paediatr.* **2005**, *94*, 132–137. [CrossRef]

36. Lu, J.; Holmgren, A. The thioredoxin antioxidant system. *Free Radic. Biol. Med.* **2014**, *66*, 75–87. [CrossRef]

37. Ono, R.; Masaki, T.; Dien, S.; Yu, X.; Fukunaga, A.; Yodoi, J.; Nishigori, C. Suppressive effect of recombinant human thioredoxin on ultraviolet light-induced inflammation and apoptosis in murine skin. *J. Dermatol.* **2012**, *39*, 843–851. [CrossRef]

38. Telorack, M.; Meyer, M.; Ingold, I.; Conrad, M.; Bloch, W.; Werner, S. A glutathione-Nrf2-thioredoxin cross-talk ensures keratinocyte survival and efficient wound repair. *PLoS Genet.* **2016**, *12*, e1005800. [CrossRef]

39. Thongrakard, V.; Ruangrungsi, N.; Ekkapongpisit, M.; Isidoro, C.; Tencomnao, T. Protection from UVB toxicity in human keratinocytes by Thailand native herbs extracts. *Photochem. Photobiol.* **2014**, *90*, 214–224. [CrossRef]

40. Konstantinopoulos, P.A.; Spentzos, D.; Fountzilas, E.; Francoeur, N.; Sanisetty, S.; Grammatikos, A.P.; Hecht, J.L.; Cannistra, S.A. Keap1 mutations and Nrf2 pathway activation in epithelial ovarian cancer. *Cancer Res.* **2011**, *71*, 5081–5089. [CrossRef]

41. Zhang, D.D. Mechanistic studies of the Nrf2-Keap1 signaling pathway. *Drug Metab. Rev.* **2006**, *38*, 769–789. [CrossRef]

42. Tarumoto, T.; Nagai, T.; Ohmine, K.; Miyoshi, T.; Nakamura, M.; Kondo, T.; Mitsugi, K.; Nakano, S.; Muroi, K.; Komatsu, N.; et al. Ascorbic acid restores sensitivity to imatinib via suppression of Nrf2-dependent gene expression in the imatinib-resistant cell line. *Exp. Hematol.* **2004**, *32*, 375–381. [CrossRef] [PubMed]

43. Boeing, S.; Williamson, L.; Encheva, V.; Gori, I.; Saunders, R.E.; Instrell, R.; Aygün, O.; Rodriguez-Martinez, M.; Weems, J.C.; Kelly, G.P.; et al. Multiomic analysis of the UV-induced DNA damage response. *Cell Rep.* **2016**, *15*, 1597–1610. [CrossRef] [PubMed]

44. Purohit, N.K.; Robu, M.; Shah, R.G.; Geacintov, N.E.; Shah, G.M. Characterization of the interactions of PARP-1 with UV-damaged DNA in vivo and in vitro. *Sci. Rep.* **2016**, *6*, 19020. [CrossRef]

45. Yoon, J.H.; Swiderski, P.M.; Kaplan, B.E.; Takao, M.; Yasui, A.; Shen, B.; Pfeifer, G.P. Processing of UV damage in vitro by FEN-1 proteins as part of an alternative DNA excision repair pathway. *Biochemistry* **1999**, *38*, 4809–4817. [CrossRef] [PubMed]

46. Gęgotek, A.; Biernacki, M.; Ambrożewicz, E.; Surażyński, A.; Wroński, A.; Skrzydlewska, E. The cross-talk between electrophiles, antioxidant defence and the endocannabinoid system in fibroblasts and keratinocytes after UVA and UVB irradiation. *J. Dermatol. Sci.* **2016**, *81*, 107–117. [CrossRef]

47. Assefa, Z.; Van Laethem, A.; Garmyn, M.; Agostinis, P. Ultraviolet radiation-induced apoptosis in keratinocytes: On the role of cytosolic factors. *Biochim. Biophys. Acta* **2005**, *1755*, 90–106. [CrossRef]

48. Shinozaki, K.; Hosokawa, Y.; Hazawa, M.; Kashiwakura, I.; Okumura, K.; Kaku, T.; Nakayama, E. Ascorbic acid enhances radiation-induced apoptosis in an HL60 human leukemia cell line. *J. Radiat. Res.* **2011**, *52*, 229–237. [CrossRef]

49. Singh, S.; Rana, S.V.S. Ascorbic acid improves mitochondrial function in liver of arsenic-treated rat. *Toxicol. Ind. Health* **2010**, *26*, 265–272. [CrossRef]

50. Catani, M.V.; Savini, I.; Rossi, A.; Melino, G.; Avigliano, L. Biological role of vitamin C in keratinocytes. *Nutr. Rev.* **2005**, *63*, 81–90. [CrossRef]

51. Perk, A.A.; Shatynska-Mytsyk, I.; Gerçek, Y.C.; Boztaş, K.; Yazgan, M.; Fayyaz, S.; Farooqi, A.A. Rutin mediated targeting of signaling machinery in cancer cells. *Cancer Cell Int.* **2014**, *14*, 124–129. [CrossRef]

52. Ma, J.Q.; Liu, C.M.; Yang, W. Protective effect of rutin against carbon tetrachloride-induced oxidative stress, inflammation and apoptosis in mouse kidney associated with the ceramide, MAPKs, p53 and calpain activities. *Chem. Biol. Interact.* **2018**, *286*, 26–33. [CrossRef] [PubMed]

53. Ghobadi, E.; Moloudizargari, M.; Asghari, M.H.; Abdollahi, M. The mechanisms of cyclophosphamide-induced testicular toxicity and the protective agents. *Expert Opin. Drug Metab. Toxicol.* **2017**, *13*, 525–536. [CrossRef]

54. Golden, T.; Swingle, M.; Honkanen, R.E. The role of serine/threonine protein phosphatase type 5 (PP5) in the regulation of stress-induced signaling networks and cancer. *Cancer Metastasis Rev.* **2008**, *27*, 169–178. [CrossRef] [PubMed]

55. País, S.M.; Téllez-Iñón, M.T.; Capiati, D.A. Serine/threonine protein phosphatases type 2A and their roles in stress signaling. *Plant. Signal. Behav.* **2009**, *4*, 1013–1015. [CrossRef] [PubMed]

56. Qin, J.; Chen, H.G.; Yan, Q.; Deng, M.; Liu, J.; Doerge, S.; Ma, W.; Dong, Z.; Li, D.W. Protein phosphatase-2A is a target of epigallocatechin-3-gallate and modulates p53-Bak apoptotic pathway. *Cancer Res.* **2008**, *68*, 4150–4162. [CrossRef] [PubMed]

57. Kaur, C.; Sharma, S.; Hasan, M.; Pareek, A.; Singla-Pareek, S.; Sopory, S. Characteristic variations and similarities in biochemical, molecular, and functional properties of glyoxalases across prokaryotes and eukaryotes. *Int. J. Mol. Sci.* **2017**, *18*, 250. [CrossRef] [PubMed]

58. Kriegel, T.M.; Kettner, K.; Rödel, G.; Sträter, N. Regulatory function of hexokinase 2 in glucose signaling in *Saccharomyces cerevisiae*. *J. Biol. Chem.* **2016**, *291*, 16477. [CrossRef]

59. Boesten, D.M.; von Ungern-Sternberg, S.N.; den Hartog, G.J.; Bast, A. Protective Pleiotropic Effect of Flavonoids on NAD. *Oxid. Med. Cell Longev.* **2015**, *2015*, 894597. [CrossRef]

60. Singh, V.P.; Bali, A.; Singh, N.; Jaggi, A.S. Advanced glycation end products and diabetic complications. *Korean J. Physiol. Pharmacol.* **2014**, *18*, 1–14. [CrossRef]

61. Chen, S.; Zou, Y.; Yan, Z.; Shen, W.; Shi, S.; Zhang, X.; Wang, H. Carboxymethylated-bacterial cellulose for copper and lead ion removal. *J. Hazard. Mater.* **2009**, *161*, 1355–1359. [CrossRef]

62. Yu, J.; Wei, W.; Danner, E.; Ashley, R.K.; Israelachvili, J.N.; Waite, J.H. Mussel protein adhesion depends on interprotein thiol-mediated redox modulation. *Nat. Chem. Biol.* **2011**, *7*, 588–590. [CrossRef] [PubMed]

63. Tao, Y.; Zhang, Y.; Zhang, L. Chemical modification and antitumor activities of two polysaccharide-protein complexes from *Pleurotus* tuber-regium. *Int. J. Biol. Macromol.* **2009**, *45*, 109–115. [CrossRef] [PubMed]

Review

A Review on Antifungal Efficiency of Plant Extracts Entrenched Polysaccharide-Based Nanohydrogels

Navkiranjeet Kaur [1], Aarti Bains [2], Ravinder Kaushik [3], Sanju B. Dhull [4], Fogarasi Melinda [5,*] and Prince Chawla [1,*]

1   Department of Food Technology and Nutrition, Lovely Professional University, Phagwara 144411, Punjab, India; navikooner1998@gmail.com
2   Department of Biotechnology, Chandigarh Group of Colleges Landran, Mohali 140307, Punjab, India; aarti05888@gmail.com
3   School of Health Sciences, University of Petroleum and Energy Studies, Dehradun 248007, Uttrakhand, India; ravinder_foodtech2007@rediffmail.com
4   Department of Food Science and Technology, Chaudhary Devi Lal University, Sirsa 125055, Haryana, India; sanjudhull@gmail.com
5   Department of Food Engineering, University of Agricultural Sciences and Veterinary Medicine of Cluj-Napoca, Calea Mănăstur 3–5, 400372 Cluj-Napoca, Romania
*   Correspondence: melinda.fogarasi@usamvcluj.ro (F.M.); princefoodtech@gmail.com (P.C.)

Abstract: Human skin acts as a physical barrier; however, sometimes the skin gets infected by fungi, which becomes more severe if the infection occurs on the third layer of the skin. Azole derivative-based antifungal creams, liquids, or sprays are available to treat fungal infections; however, these formulations show various side effects on the application site. Over the past few years, herbal extracts and various essential oils have shown effective antifungal activity. Additionally, autoxidation and epimerization are significant problems with the direct use of herbal extracts. Hence, to overcome these obstacles, polysaccharide-based nanohydrogels embedded with natural plant extracts and oils have become the primary choice of pharmaceutical scientists. These gels protect plant-based bioactive compounds and are effective delivery agents because they release multiple bioactive compounds in the targeted area. Nanohydrogels can be applied to infected areas, and due to their contagious nature and penetration power, they get directly absorbed through the skin, quickly reaching the skin's third layer and effectively reducing the fungal infection. In this review, we explain various skin fungal infections, possible treatments, and the effective utilization of plant extract and oil-embedded polysaccharide-based nanohydrogels.

Keywords: fungal infections; nanohydrogel; skin; polysaccharide; essential oils

Citation: Kaur, N.; Bains, A.; Kaushik, R.; Dhull, S.B.; Melinda, F.; Chawla, P. A Review on Antifungal Efficiency of Plant Extracts Entrenched Polysaccharide-Based Nanohydrogels. *Nutrients* **2021**, *13*, 2055. https://doi.org/10.3390/nu13062055

Academic Editors: Jean Christopher Chamcheu, Anthony L. Walker and Felicite Noubissi-Kamdem

Received: 23 April 2021
Accepted: 9 June 2021
Published: 15 June 2021

Publisher's Note: MDPI stays neutral with regard to jurisdictional claims in published maps and institutional affiliations.

1. Introduction

Skin acts as a protector of the internal organs by shielding against external agents, sunburn, and by regulating body temperature; however, sometimes pathogens invade the body and disturb the skin protective properties, leading to skin diseases or infections [1]. Bacteria, viruses, parasites, and fungi can cause skin diseases. Fungal infections are more severe because they occur on the third layer of the skin [2]. Fungi act on keratin tissue such as skin, nails, and hair [3]. In the skin, fungi lead to subcutaneous infections, and over the past years, the cases of fungal skin infections have been increasing rapidly, especially in immune-compromised individuals [4]. Several well-known severe skin infections (Table 1) such as Tinea corporis (ringworm), Tinea pedis, Tinea faciei, Tinea manuum, Tinea cruris (Jock-itch), and Tinea barbae are caused mainly by *Trichophyton* species [5,6].

Table 1. The occurrence of Tinea infection in various body parts.

Tinea Infection	Affected Locations	References
Tinea capitis	Scalp	[7]
Tinea corporis	Trunk	[8]
Tinea faciei	Face	[9]
Tinea manuum	Hands	[10]
Tinea pedis	Feet	[11]
Tinea unguium	Nails	[12]

Fungal infections are typically recognized by symptoms such as itchy red color patches, hair loss, and crusted patches [13]. Some common conditions leading to fungal infection are wearing tight-fitting clothes or sharing a locker room, clothes, or furniture with an infected person [14]. Antifungal drugs, primarily topical, oral, and intravenous, are used to treat various types of fungal infections; however, oral antifungal drugs are more toxic to the human body as compared to topical antifungal drugs. Additionally, commonly used antifungal drugs contain different types of broad categories of components such as azole, echinocandin, and polyenes [15]. Azoles inhibit the oxidative enzymes present in the fungal cell membrane, which prevents the cell wall of the fungus from forming sterol (ergosterol), and due to incomplete synthesis, cells become permeable. On the other hand, echinocandins inhibit the synthesis of important polysaccharides (1,3-β-glucan) responsible for developing the cell wall, whereas polyenes directly bind to the ergosterol and move inside the cell through the cell membrane by creating pores, and through these pores, cellular organelles come out that cause the death of the cell [16]. While the topical antifungal drugs act on the different sites to target the molecules for the treatment of fungal infections, they show various side effects (Figure 1) on the application site, such as burning, redness, and some allergic reactions [17].

Due to the immediate release of the drug, treatments for an extended period are sometimes needed due to low penetration. Additionally, these drugs may not reach the target location, which could lead to incomplete clearance of the infection. To overcome this problem, the use of natural plant extracts and oils as antifungal agents could be a practical approach [18]. Several plants such as *Bucida buceras* (black olive tree), *Breonadia salicina*, *Harpephyllum caffrum*, *Olinia ventosa*, *Vangueria infausta*, and *Xylotheca kraussiana* have been explored for their antifungal efficacy [19]. On the other hand, due to effective antifungal activity, cinnamon, anise, clove, citronella, peppermint, pepper, camphor essential oils have been used in the formulation of antimycotic drugs [20]. However, the synthesis of an antimycotic drug involves a variety of processes, including high-heat and high-temperature treatments, and as a result of these treatments, the structure of the phytochemicals in the herbal extract is disrupted, leading to the epimerization process. Several studies have found that combining high temperatures and an alkaline state causes structural changes in polyphenolic components. [21–23]. As drug carriers, various approaches such as liposomes, ethosomes, trans-ferosomes, niosomes, spanlistics, nanoemulsions, and nanohydrogels are used to overcome this problem. Among all, polysaccharide-based nanohydrogel is an emerging technology, as it shows the same flexibility as natural tissue, with the lowest chance of rejection, minimal side effects, and maximum advantages; for instance, it can target the desired site with controlled release of the antifungal component inside the tissue due to its high penetration power [24]. It can be defined as a three-dimensional nano-sized porous structure with several unique properties, such as high stability, solubility, biodegradability, and biocompatibility with bioactive compounds [5,25]. These gels are macromolecules that can hold a large amount of water and swell up without dissolving due to crosslinking between polymers, which increases their surface area [26]. Therefore, polysaccharide-based nanohydrogels embedded with natural plant extracts and oils have become the primary choice of pharma scientists. These gels protect plant-based bioactive compounds and are effective delivery agents, as they release multiple bioactive compounds in the targeted area. Nanohydrogels can be applied to infected areas, and due to their spreadable nature and penetration power, they get directly absorbed through the skin and reach up to the third layer quickly, thereby effectively reducing the fungal infection. Therefore, in this review, we explain in detail various skin fungal infections, possible treatments, and the effective utilization of plant extract and oil-embedded polysaccharide-based nanohydrogels.

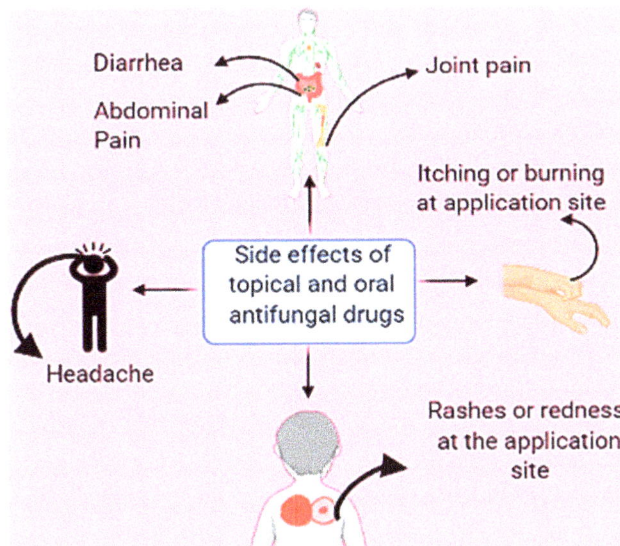

Figure 1. Side effects of synthetic antifungal drugs. These side effects are due to uncontrolled drug release and can lead to prolonged treatment due to low penetration; consequently, these drugs may not reach the target location and could lead to the incomplete clearance of infection.

2. Skin Fungal Infections and Causative Agents

Fungal exposure causes tissue necrosis of epidermal layers and can lead to superficial, systemic, and subcutaneous mycosis [27]. Except for the production of protease enzymes by yeast and non-dermatophytes, dermatophytes, yeast, and non-dermatophytes share the same pathologic pathway. They always act upon the skin, nails, and hair keratin of human beings and animals, as they utilize keratin as a nutrient source, which ultimately leads to tissue necrosis [28]. According to the natural epidemiological standpoint, dermatophytes can be classified into three major categories: anthropophilic, zoophillic, and geophilic (Table 2).

Table 2. Most relevant species of fungal infection according to their natural habitat.

Dermatophytes Based on Their Habitat	Fungal Species Belonging to Different Dermatophyte Group	Infection Site	References
Anthropophillic	*Microsporum audouinii*	Scalp	[29]
	Trichophyton concenricum	Body	
	Microsporum ferrugineum	Scalp	
	Trichophyton interdigitale	Foot, groin, nails	
	Trichophyton megninii	Scalp, beard	
	Trichophyton rubrum	Foot, nails, body	
	Trichophyton schoenleinii	Scalp	
	Trichophyton soudanense	Scalp	
	Trichophyton tonsurans	Scalp, body	
	Trichophyton violaceum	Scalp, body, nails	

Table 2. *Cont.*

Dermatophytes Based on Their Habitat	Fungal Species Belonging to Different Dermatophyte Group	Infection Site	References
Zoophillic	*Microsporum canis*	Scalp, body	[29,30]
	Microsporum distortum	Scalp	
	Trichophyton equinum	Scalp	
	Microsporum nanum	Scalp, body	
	Trichophyton verrucosum	Exposed areas	
Geophilic	*Microsporum fulvum*	Scalp, body	[29]
	Microsporum gypseum	Scalp, body	

According to etiology, these can be classified into three genera, viz. *Epidermophyton*, *Microsporum*, and *Trichophyton* [29,31].

Anthropophillicorganisms are a group of dermatophytes predominantly parasitic to humans; they are transmitted from animals and cause dermatophytosis. Zoophilic organisms are parasites to lower animals and are transmitted to humans through direct or indirect contact, whereas geophilic organisms are dermatophytes that are mainly present in the soil as saprophytes; exposure can lead to skin infection or, if severe, can cause systematic infections [27]. Epidermophyton is the species whose microscopic analysis shows between 1 and 9 thin to thick smooth-walled septa; known species are *Epidermophyton floccosum* and *Epidermophyton stockdaleae*. *Microsporum* can cause skin and hair fungal infection; however, it cannot cause nail fungal infection. Due to the presence of keratin tissue, *Trichophyton* can cause infection in all three areas: nails, hair, and skin. [30]. Moreover, genus *Candida* can cause severe fungal infections such as candidiasis [32]. Twenty species of *Candida* are responsible for human skin infections, and the most common species is *Candida albicans* [33]. These species can cause superficial and systematic types of fungal infections, while dermatophytes can cause superficial types of fungal infections. Primarily, they are found alive on the human skin surface; however, overgrowth of these species can cause severe oral thrush and genital yeast infection [34]. Furthermore, several *Candida* species implicated in human infections are classified as standard, less common, and rare species. *Candida albicans*, *Candida glabrata*, *Candida tropicalis*, *Candida parapsilosis*, *Candida krusei*, *Candida guilliermondii*, *Candida lusitaniae*, and *Candida kefyr* are the examples of common species of *Candida* that are associated with infections in humans. Hyperkeratosis and epidermal hypertrophy persistent with inflammation are diagnostic symptoms for the fungal infection caused by candida species [35].

Sporotrichosis is a fungal infection caused by the fungus *Sporothrix*, which is also known as "Rose Garden Disease" [36]. This fungus takes growth in soil, plant matter, and skin. Minor cuts or wounds that come in contact with fungus-enriched soil and plant matter can lead to severe infection. Domestic animal bites such as bites from cats and dogs are also the cause of infection [37]. This type of fungus can cause three types of sporotrichosis, such as lymphocutaneous, fixed, and multifocal or disseminated cutaneous sporotrichosis on the mammal's skin. Lymphocutaneous sporotrichosis is a common type of cutaneous sporotrichosis in which a small nodule is present at the infectious site; after that, a string of this nodule passes towards the proximal lymphatics with different morphology and surfaces such as with a smooth, watery surface like ulcers [38]. In lymphocutaneous sporotrichosis, fixed cutaneous sporotrichosis is uncommon. In this type of fungal infection, nodule formation at injury sites is common but cannot spread; it remains and develops at the origin. On the other hand, dermatophytes can cause several and severe types of fungal infections because they can cause infection on the innermost layer of human skin (Figure 2).

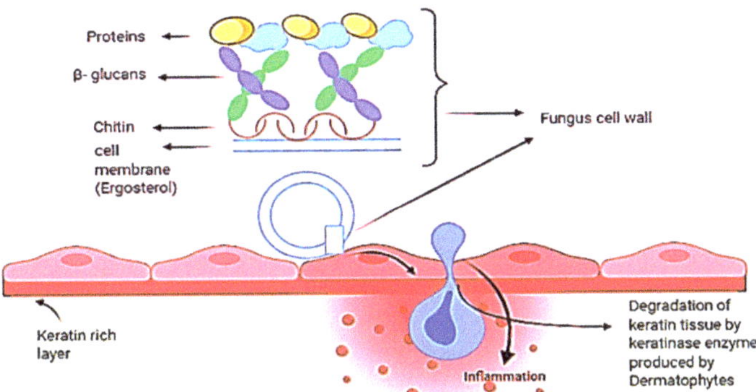

Figure 2. A possible mechanism for dermatophytosis. Dermatophytes are the fungi that need keratin for their growth; therefore, when a fungus cell invades the skin, it produces keratinase, an enzyme that feeds on the skin layer keratin, due to which keratin tissue degradation occurs and causes skin inflammation.

Tinea capitis, a fungal infection commonly found in preadolescent children, is caused by the dermatophyte species of *Microsporum* and *Trichophyton* [39]. *Epidermophyton floccosum* and *Trichophyton concentricum* are two species that are unable to cause this fungal infection. Tinea capitis born due to direct contact with an infectious person or by sharing utensils with infected people [40]. When sebum production is low, it reduces fatty acids and increases the pH of the scalp, which enhances colonization and results in fungal disease by dermatophytes (tinea capitis koreon) [41]. It is characterized by primary symptoms such as blemishes, blisters, or nodules on the scalp. If not treated well at this stage, it can lead to secondary infections such as superficial reddening of the skin and scaling (dry skin). Fungi enter the hair shaft through the outer root sheath of the hair follicle [42]. The hair shaft comprises three layers, the cuticle (which is the outermost layer of hair that protects the inner structure of hair), the cortex (which is the middle layer that provides color, strength, and texture to the hairs), and the medulla (which is the innermost layer, also known as the marrow of the hair) [43]. When the fungus invades the hair shaft, all these above activities are inhibited and result in Tinea capitis infection; however, sometimes it is non-inflammatory (not permanent hair loss), while sometimes infection is inflammatory (painful nodules with pus and permanent hair loss). It typically affects children of 3 to 7 years of age [44]. Treatment of Tinea capitis uses griseofulvin, terbinafine, itraconazole, and fluconazole. *Penicillium* produces griseofulvin, which can be divided into two micronized and ultra-micronized antifungal drugs for better absorption. The FDA approved intake for the 2-year-old children is 10/mg/kg/day, its micronized type is 20–25 mg/kg/day and ultra-micronized is 10–15 mg/kg/day for 6–12 weeks [45]. It is very effective against *Microsporum* species and shows resistance to *Trichophyton tonsurans*. Griseofulvin reaches the infectious site through sweat glands, and it shows some side effects such as headaches and gastrointestinal upset. Furthermore, terbinafine is an allylamine derivative with fungicidal properties, and it inhibits the enzyme epoxidase, which produces ergosterol. [46]. It is lipophilic; therefore, it can reach the hair follicle, skin, and nails. When taken orally, it binds to the plasma protein and takes 2 h to attain peak plasma value [47]. Itraconazole belongs to the azole class, having 3 nitrogens and one cyclic, heterocyclic ring, which shows fungicidal properties by inhibiting the production of ergosterol; it interferes with the 14-α-demethylase enzyme that converts lanosterol to 14-dimethyl lanosterol [48]. It is a lipophilic drug and can reach the keratinous tissue; however, the time taken to reach the tissue depends upon the dosage, and the optimum time is 4 h. The recommended dose varies in a period for different species; for instance, it is 5 mg/kg/day for 2–4 weeks for *Trichophyton tonsurans* and 4–6 weeks for *Microsporum canis*, and infants use this dose for 3–6 weeks. Stomach pain, diarrhea, and rashes are the side effects of this drug [49].

Fluconazole is an antifungal drug with low molecular weight and high solubility in water that does not require plasma protein to travel to the plasma, which it reaches within 1–2 h after oral consumption. Fluconazole's efficiency in treating disease is the same as griseofulvin. Gastrointestinal distress and headache are the side effects of it [50].

Tinea corporis is commonly known as ringworm, and the responsible species for this infection are *Trichophyton rubrum*, *Trichophyton tonsurans*, and *Trichophyton mentagrophytes* [51]. In this infection, dermatophytes penetrate the hair follicle and produce scaly ring-shaped itchy rashes on the skin. These ring vesicles are brown in the center and red around the edges, and they mostly appear on the exposed skin. If not treated in its initial stage, kerion-like secondary lesions appear in which infected rings merge and multiply and puss-filled scars develop near the rings. It is spread by conidia, which are spread through hair loss or direct contact with an infected person [52]. Tinea pedis, which is also known as athlete's foot, occurs by the growth of fungus on the feet in warm and moist conditions. Tinea pedis forms a fluid-rich layer and cause more severe infections if the fluid releases. In moist areas, the fungus spreads by direct contact with infected skin. Tinea pedis is mainly found in people who wear shoes for an extended time and among those who live in industrial areas [53]. *Trichophyton interdigitale*, *Trichophyton tonsurans*, *Trichophyton rubrum*, and *Epidermophyton floccosum* are the main causative agents of tinea pedis which is the fungal infection that occurs in interdigital form and mainly consists of pain-causing scales and ulcers, found between the third and fourth toes [54,55]. Antifungal creams are used for four weeks to treat this fungal infection. However, interdigital tinea pedis requires only one week. Topical antifungals used to treat are azoles, allylamines, butenafine, coclopirox, tolnaflate, and econazole nitrate [56]. At the same time, terbinafine or neftifine(allylamines) show a more effective peak than the azoles. For *Trichophyton* species treatment, a luliconazole formulation is used twice a day for 1–2 weeks [57]. Tinea barbae, also known as barber's itch and can affect the face's beard area in two ways: deep and superficial. Zoophile dermatophytes, *Trichophyton mentagrophytes*, and *Trichophyton verrucosum* are the main causative agents of the deep type of this fungal infection which is more severe than the superficial type that is primarily found in farmers. In contrast, crusted patches caused by *Trichophyton rubrum* and *Trichophyton violaceum* species are used to diagnose superficial infection [58]. Tinea faciei is a fungal infection that causes pustules and scales in the non-beard region of the face, causing itching and rashes when exposed to sunlight. Zoophilic dermatophytes such as *Microsporum canis* can cause more severe infection than anthropophilic dermatophytes by causing a reaction under the stratum corneum by producing keratinases, or keratin-degrading enzymes [59]. Tinea manuum is somehow similar to tinea pedis and is diagnosed by dry, scaly patches on the palms and in interdigital areas of the hand. Mainly, *Trichophyton rubrum*, *Trichophyton mentagrophytes*, and *Epidermophytes* can cause this fungal infection [60]. Tinea cruris, also known as jock-itch, is characterized by the patches produced by *Trichopyton rubrum* species that infect hair follicles of the inguinal and perineal area by causing irritation and burning [61]. Treatment of tinea cruris usually lasts 2–4 weeks; however, treatment length can vary depending on the severity of the condition [62]. Tinea versicolor, a superficial infection of the skin, causes hyperpigmentation of the neck, also known as pityriasis versicolor, which is caused by the fungus *Malassezia* species, commonly found on human skin [63]. There are fourteen species found, with *Malassezia furfur* and *Malassezia globasa* being the main causatives [64]. Both topical and oral medications containing specific and nonspecific antifungal agents are used for treatment, with topical treatment serving as first-line therapy and oral treatment serving as second-line therapy for severe infections [65]. Nonspecific agents including sulfur plus salicylic acid, selenium sulfide 2.5%, and zinc pyrithone help remove the dead tissue of fungi from the skin to stop the further invasion of the fungus [66]. In contrast, antifungal agents such as imidazole and allylamine are used to stop the fungus' mechanism and cure the infection. Ketoconazole is the most effective imidazole topical form of treatment and is available as a cream or a foaming solution [67]. Itraconazole and fluconazole are also used; however, terbinafine is not a viable treatment option for pityriasis versicolor [68].

3. Fungal Infections of Nails (Onychomycosis)

Onychomycosis is known as tinea unguium and is a fungal infection of nails in which foot toenails are more affected than fingernails. Onychomycosismainly occurs after tinea pedis, through which the transmission of the fungus takes place [69]. The microbiology of onychomycosis (Table 3) is that dermatophytes, yeast, and non-dermatophytes can cause nail fungal infection; however, dermatophytes participate more and cause nail dysfunction and pain [70].

Table 3. Most common fungal agents for the treatment of onychomycosis.

Onychomycosis Microbiology	Name of Species Cause Onychomycosis
Dermatophytes	*Epidermophyton floccosum*
	Microsporum species
	Trichophyton interdigital
	Trichophyton mentagrophytes
	Trichophyton rubrum
	Trichophyton tonsurans
Nondermatophyte	*Acremonium species*
	Alternaria species
	Aspergillus species
	Cladosporium carrionii
	Fusarium species
	Geotrichum cadidum
	Lasiodiplodia theobromae
	Onychocola species
	Scopulariopsiss pecies
	Scytalidium species
Yeast	*Candida albicans*
	Candida parapsilosis

Patients suffering from peripheral vascular immunologic disorder and diabetes are more prone to onychomycosis. On the other hand, others such as smokers and persons sharing bathing facilities can also face this infection [71].

4. Oral and Topical Antifungals and Their Use

According to Table 4, oral or topical antifungal drugs or a combination of both can be used to treat fungal infections caused by dermatophytes [72]. All are aimed at degrading the cell wall of the fungus to inhibit its mechanism of infection and to cause cell death. Topical antifungals are commonly thought of as a first-line treatment for dermatomycosis because as creams, liquids, or sprays they are treatments that can be directly applied to the skin, nails, hair, or even to the mouth. Topical antifungals are more effective than systemic treatment, and their method of administration gives them the advantage of curing skin diseases by their direct application at the site of infection [73]. Candida infections, pityriasis versicolor, tinea barbae, tinea capitis, tinea corporis, tinea cruris, tinea faciei, tinea manuum, tinea nigra, and tinea pedis are examples of fungal infections that are treated with both first- and second-line topical antifungal drugs, depending on the severity of the infection [74,75].

Table 4. Classification of antifungal components and their effectiveness against various fungal infections.

Class of Antifungal Agents	Antifungal Agents	Chemical Structure of Different Antifungal Agents	Fungal Infections	Desired Treatment Duration	References
Imidazoles	Clotrimazole (1%)		Tinea corporis Tinea cruris Tinea pedis	4–6 weeks	[10]
	Econazole (1%)		Tinea corporis Tinea cruris Tinea pedis	4–6 weeks	[11]
	Miconazole (1%)		Tinea corporis Tinea cruris Tinea pedis	4–6 weeks	[73]
	Oxiconazole (2%)		Tinea corporis Tinea cruris Tinea pedis	4 weeks	[57]
	Sertaconazole (2%)		Tinea corporis Tinea cruris Tinea pedis	4 weeks	[11]
	Luliconazole (1%)		Tinea corporis Tinea cruris Tinea pedis	2 weeks	[54]
	Eberconazole (1%)		Tinea corporis Tinea cruris Tinea pedis	2–4 weeks	[10]

Table 4. Cont.

Class of Antifungal Agents	Antifungal Agents	Chemical Structure of Different Antifungal Agents	Fungal Infections	Desired Treatment Duration	References
Triazoles	Efinaconazole (10%)		Tinea pedis	52 weeks	[12]
Allylamines	Terbinafine		Tinea corporis Tinea cruris Tinea manuum Tinea pedis	2 weeks 2 weeks 4 weeks 4 weeks	[76]
	Naftifine (1%)		Tinea corporis Tinea cruris Tinea pedis	Used 2 weeks beyond the resolution of symptoms	[77]
	Butenafine (1%)		Tinea corporis Tinea cruris Tinea pedis	2–4 weeks	[78]
Others	Amorolfine (0.25%)		Tinea corporis	4 weeks	[79]
	Amphotericin B (0.1%)		Tinea corporis	4 weeks	[80]

The primary class includes azoles, polyenes, and benzylamine. On the other hand, in addition to drugs in the primary class—such as clotrimazole, econazole, ketoconazole, miconazole, and trioconazole—the drugs terbinafine and amorolfine are also used for the treatment of fungal infection. Ketoconazole shampoo is used to treat fungal infections on the scalp [81]. Antifungal injections are also available, such as amphotericin, flucytosine, itraconazole, voriconazole, anidulafungin, caspofungin, and micafungin, which are also used depending on the type of fungal infection. Antifungals are different from antibiotics because antibiotics can only kill bacteria [82].

Imidazoles and triazoles are the two types of azole antifungals. Both work with the same mechanism, which is by stopping the conversion of lanosterol to ergosterol by inhibiting the work of enzyme lanosterol 14-alpha-demethylase, thus causing porousness in the fungal cell wall [74]. However, both categories show their activity at the different points of the spectrum and have structural differences in the number of nitrogen atoms: in imidazole, it is two, whereas triazole has three nitrogen atoms in it [83]. The polyene antifungals are nystatin, natamycin, and amphotericin B; they are effective against common fungal infections such as candidiasis, aspergillosis, mucormycosis, and cryptococcosis. Polyene antifungals bind to ergosterol, which is the main sterol present in the cell membrane, and they form a polyene–ergosterol complex that creates pores that increase cell permeability [84,85]. On the other hand, amphotericin B shows fungicidal activity for *Candida* species, *Histoplasma capsulatum*, *Cryptococcus neoformans*, *Blastomyces*, and *coccidioides immitis*, although effective treatment depends on parameters such as drug amount and pH (6.0 to 7.5) [86]. Nystatin belongs to the polyene antifungal group and is effective against the mucosal and cutaneous infections caused by candida species, although it is less effective against dermatophytes [87]. Butenafine and allylamines are types of benzylamine drugs and are used topically for the treatment of dermatophytosis. They disturb the cycle of ergosterol synthesis by inhibiting squalene epoxidase enzyme synthesis. Allylamines are considered less effective antifungal agents; however, they have an advantage in the treatment of tinea pedis [88]. When the infection reaches a more severe level, second-line treatment with oral antifungal drugs such as griseofulvin, itraconazole, fluconazole, and terbinafine are among the medications mainly used, with allitraconazole being the most effective [89,90].

5. Candidiasis Treatment by Both Oral and Topical Antifungal Treatments

Antifungal medications such as polyenes, azoles, and echinocandins are used to treat candidiasis in both topical and oral forms, depending on the severity of the infection. Nystatin and amphotericin B from polyene; miconazole, clotrimazole, itraconazole, ketoconazole, fluconazole, voriconazole, and econazole from azoles; and caspofungin, micafungin, and anidulafungin from echinocandins are used [91]. These are cyclic hexapeptides with an N-acyl aliphatic or aryl side chain that aid in the treatment of fungal infections caused by Candida and Aspergillus species by disrupting the fungus' cell wall structure [92]. Echinocandins named caspofungin, micafungin, and anidulafungin are mainly used and these are lipopeptides that act as an inhibitor of the β-d-glucan enzyme (the main component of cell walls) [93]. Therefore, the inhibition leads to cell death and helps to prevent the infection. The pathogenicity of candidiasis depends upon the host strength or factors by which yeast can multiply [94]. Upon candida infection, most microorganisms gather and form a three-dimensional structure on the surface, which is called biofilm [95]. These biofilms are resistant to amphotericin B and fluconazole antifungal drugs. However, the biofilm of *Candida albicans* is more pathogenic than all other species. For a long time, contaminations caused by *Candida* species were treated with azoles, the most common class of antifungal medications. As of late, protection from azoles has expanded in Candida species, both in clinical settings and in vitro. Azoles can treat infection by interfering with the catalyst lanosterol 14-α-demethylase [96]. This catalyst engages with the biosynthesis of a critical component, ergosterol [97]. Azoles mainly target components such as chitin and glucan, which are not present in human skin due to the structural difference between ergosterol and cholesterol [98]. Previously azoles and echinocandins were considered effective drugs against *Candida* species. However, nowadays, these species have become resistant to these drugs [99].

6. Resistance to Antifungal Drugs

Fungi have developed a variety of adaptation mechanisms, including cell reactions, drug resistance, and stress reactions. [100]. They can cause a change in osmolarity and drug exposure, interfere with the cell wall, and interrupt the drug-acting pathway. Due to these mechanisms, antifungal drugs show resistance to some fungal species and cannot cure the disease [101]. To withstand cellular pressures such as those generated by drug exposure, fungi must be able to detect and respond to changes in the environment. They have evolved a variety of adaption processes, including complicated circuitries involving cellular responses, such as signaling molecules' interactions, stress responses, and drug resistance. In this context, Michaelides et al. (1961) considered griseofulvin resistant against dermatophytes. In the 1980s, azoles were reported as antifungal drugs resistant to candidiasis. Mukherjee et al. (2003) *demonstrated the resistance of Trichophyton rubrum* species to terbinafine in onychomycosis-infected patients [102].

6.1. Mechanism of Drug Resistance

Drug Efflux

Efflux cell membrane transporters are proteins that attach to a variety of substances, including antifungal medicines, and then expel them from the cell. Treatment failure is linked to the development of these molecular pumps, which operate as a protective mechanism against the drug's cytotoxic effects by limiting the accumulation of hazardous chemicals [103]. By reducing the expected effect in vivo, efflux cell activity causes a gradual and non-specific rise in drug resistance. As a result, increased drug export is one of the most important strategies employed by fungi to develop resistance to antifungal treatments. The pathogenicity of fungal resistance is aided by the mutation of efflux transporters, which favors the emergence of dermatophytosis instances by giving the fungus a colonization advantage [104]. MDR is caused in part by the overexpression of genes from the ATP-binding cassette superfamily (ABC), which has been found in both prokaryotes and eukaryotes throughout evolution. These proteins have two different regions: a nucleotide-binding domain that is extremely conserved and a transmembrane domain that is extremely changeable. ATP-binding cassette transporters actively convey a wide variety of structurally and chemically unrelated substances across membranes by binding and hydrolyzing ATP, decreasing drug accumulation even in cancer cells [105]. MDR, MDR-associated protein (MRP), and the pleiotropic drug resistance (PDR) families are three well-studied families involved in the efflux of hazardous chemicals. The genome sequences of many dermatophytes have revealed a homogeneous group of these transporters with very little genetic variability and only a few distinctive genes in each species studied. Martinez-Rossi et al. (2018) studied about seven species; despite the morphological and adaptive diversity of each species, these dermatophyte genomes discovered a substantial number of encoded ABC transporter domains, with many of the related genes having analogs in all analyzed species, implying that ABC transporter genes work similarly. In the *Trichopyton interdigitale* H6 strain, evidence implicating drug efflux as a mechanism of resistance has been thoroughly examined (previously identified as *Trichophyton rubrum*). Due to reported resistance to both GRS and tioconazole in vitro, the occurrence of a drug efflux phenomenon was initially suggested. Following that, the mdr1 and mdr2 genes were discovered to have elevated transcription levels after exposure to several antifungal drugs. The deletion of the mdr2 gene made a mutant more susceptible to terbinafine (TRB) but not to other medications tested, ruling out the possibility of the mdr2 gene playing a modulatory role in drug sensitivity. When challenged with antifungal medicines, a gene expression study revealed that the mdr4 gene is somewhat dependent on the MDR2 transporter, implying the existence of network connection about MDR activity as well as the reliance of distinct ABC transporters in drug efflux. Different ABC transporters with overlapping substrate profiles may have a wide range of substrate affinity and transport capacity, as well as behave differentially to cellular drug concentrations, affecting the substrates' actual fate. For example, transcriptional profiling of the pdr1, mdr2, and mdr4 genes in four Trichophyton species—*Trichophyton equinum*, *Trichophyton interdigitale*, *Trichophyton rubrum*, and *Trichophyton tonsurans*—did not reflect the intrinsic phylogenetic relationship among these fungi, nor did it reveal a functional correlation between species and efflux modulation under the tested conditions. The evaluated genes, on the other hand, appear to work in concert, leading to the conclusion that one mdr gene may be compensated by others in terms of extrusion activity [106].

6.2. Mutations Affecting Drug Target Genes

In this context, Robbin et al. (2017) demonstrate that mutations in the drug target gene that impairs drug binding and efficacy are one of the most common methods by which microorganisms evolve resistance to drugs. This is a common method by which azole resistance emerges in *Candida albicans*. In clinical isolates of *Candida albicans*, many mutations in the azole target gene ERG11 have been discovered, and they are frequently found in hot-spot regions near the enzyme active site. As *Candida albicans* is a diploid organism, mutations in ERG11 frequently result in the loss of heterozygosity, conferring higher levels of azole resistance via increased dosage of the resistance allele. Mutations in the drug target gene are frequently linked to azole resistance in *Candida neoformans* and *Aspergillus fumigatus*. Resistance to echinocandins is generally caused by mutations in the drug target gene, which are frequently concentrated in two different areas. In *Candida albicans*, changes in FKS1 are frequently followed by a loss of heterozygosity, resulting in two mutant alleles with reduced affinity for the antifungal agent [107]. On the other hand, Khurana et al. (2019) studied Erg11, a gene that codes for 14-lanosterol demethylase. In yeasts, overexpression of gene products and mutations in Erg11 have been documented as mechanisms of azole resistance, but these have yet to be described in dermatophytes. As a result, some mutations affect the hydrogen bond between the medication and the target protein, lowering the affinity of short-tailed triazoles but not long-tailed triazoles, according to a recent study [108].

6.2.1. Decreased Concentration of Drug within Fungi

Increasing the drug efflux mechanism helps to reduce drug buildup within the fungal cell. Membrane proteins called multidrug efflux transporters are found in all living species. These proteins bind to a wide range of structurally and chemically different substances and actively expel them from cells. The medication accumulates less in the cell when the genes encoding these efflux pumps are mutated (upregulated or overexpressed). Antifungal medicines are affected by a variety of efflux mechanisms. The ATP-binding cassette (ABC) superfamily and the main facilitator superfamily are two drug efflux mechanisms that modulate azole resistance. In the *Trichopyton rubrum*, the genes TruMDR1 and TruMDR2 are overexpressed [105,109].

6.2.2. Drug Detoxification

When fungi are exposed to cytotoxic medicines at sub-inhibitory doses, a large number of genes involved in cellular detoxification are activated, contributing to increased drug tolerance. RNA-sequencing (RNA-Seq) examination of the effect of acriflavine (ACR) on *Trichophyton rubrum* revealed that genes involved in cellular detoxification, such as those encoding catalases that protect cells from oxidative stress and reactive oxygen species, were considerably upregulated. Increased production of catalase genes could be a compensatory mechanism to keep the intracellular level of this enzyme stable, sparing cells from the drug's apoptotic effects. Acriflavine, on the other hand, inhibits catalase activity in vitro. Furthermore, the RNASeq study of the differential gene expression of T. rubrum challenged with UDA indicated that several antioxidant enzyme genes were upregulated. In addition to catalases, UDA stimulates superoxide dismutase, peroxidases, glutathione transferases, and glutathione peroxidases, resulting in enzymatic oxidative detoxification of the body. A previous report showed that Candida resistance to amphotericin B (AMB) could be linked to these findings. Increased catalase activity is also a factor.

Several azole-resistant species have overexpression of target enzymes, which is a compensation mechanism for ergosterol depletion. In *Candida albicans*, overexpression of erg11 is caused by the duplication of the left arm of chromosome 5, which contains the erg11 gene. The development of chromosome 1 disomies allows *Cryptococcus neoformans* to adapt to high FLC concentrations. This resistance is caused by the duplication of two genes on this chromosome: ERG11, which is the target of FLC, and AFR1, which is an azole transporter in *Cryptococcus neoformans*. Extra copies of the *Aspergillus fumigatus* squalene epoxidase gene also give terbinafine resistance. The resistant phenotype was reversed after the strain was cultivated for several generations without terbinafine, resulting in the loss of plasmids encoding the salA gene, confirming that extra copies of this gene caused resistance. Furthermore, the original strain without any plasmid increased the expression of the native salA gene in response to the terbinafine challenge. Unlike bacteria, which can develop antibiotic resistance by the breakdown of these substances, fungi can develop resistance to antifungal drugs by inactivation or destruction [107]. Given that fungi secrete a vast number of enzymes; similar pathways leading to antifungal resistance or tolerance are likely common. Enzymes that fulfill this role in fungi have been discovered, which as a result might be beneficial in the production of enzyme inhibitors and these inhibitors can be used alone or in combination with traditional therapeutic methods.

6.2.3. Changes in Metabolism to Counteract the Drug's Effect on De Novo Synthesis of Pyrimidines

The antifungal medicine 5FCcompetes with typical pyrimidine intermediate metabolites for nucleic acid incorporation. *Candida glabrata* has developed 5FC resistance as a result of a de novo increase in pyrimidine synthesis [110].

6.2.4. Variation in Plasma Membrane Composition

A decrease or complete absence of ergosterol in the plasma membrane caused by mutations in non-essential ergosterol genes is a rare mechanism of resistance among polyene drugs, such as the ERG3 mutation in clinical isolates of *Candida albicans* and the ERG6 mutation in *Candida glabrata*.

6.2.5. Biofilms

Biofilms are sessile microbial populations encased in extracellular polymeric compounds that are resistant to antimicrobials and host defenses. Biofilms can be produced by both *Trichophyton rubrum* and *Trichophyton mentagrophytes*.

6.2.6. Modifications in the Biosynthesis of Ergosterol

Azole medicines' antifungal efficacy is based on the removal of ergosterol from the fungal membrane and the formation of the toxic result 14-methyl-3,6-diol, which causes growth arrest. Inactivation of the ERG3 gene in the late stages of the ergosterol biosynthetic pathway can result in the entire inactivation of C5 sterol desaturase, as well as cross-resistance to a variety of ergosterols.

7. Nanohydrogels

Hydrogels are a polymeric network that has a functional group to absorb water, swell up, and not dissolve again into the water due to the presence of binders or crosslinkers. Capillary, osmotic, and hydration forces are responsible for the collaboration between the polymeric chain system and organic liquids, which contribute to the chain network's balance and expansion [111]. Nanohydrogels are an application that reaches into wide areas, such as drug delivery systems, 3D networking systems, tissue engineering, and biosensors. Historically, hydrogels have been classified into three generations: first-generation, second-generation, and third-generation [112]. Nanohydrogels' use in drug delivery is very effective due to their extracellular material, which is somehow similar to the extracellular material of human beings. Nanohydrogels can be synthesized in a variety of ways, including one-step and multi-step procedures. They are classified either as copolymers, homopolymers, interpenetrating networks, and semi-interpenetrating networks based on the method of their synthesis. [113]. The copolymer type requires two types of monomers to bind with the crosslinkers. Monomers such as methacrylic acid, PEG-PEGMA, carboxymethyl cellulose, and polyvinylpyrrolidone with crosslinker tetradimethacrylate are used for the synthesis of copolymer hydrogel by following the free radical photopolymerization mechanism that is used for drug delivery and wound dressing material [114–116].

In a homopolymer, one type of monomer is used, wherein the choice of crosslinker depends upon the monomer and the nature of the polymerization reaction. Their use also depends upon the nature of the crosslinker, as (Cretu et al. 2004; Das in 2013) showed that the synthesis of hydrogels from monomers such as poly(2-hydroxyethyl methacrylate), 2-hydroxyethyl methacrylate, and polyethylene glycol with crosslinker polyethylene glycol dimethacrylate for drug delivery systems, contact lenses, scaffolds for protein recombination and gels having crosslinker Triethylene glycol dimethacrylate can be used in wound healing and the production of functional tissues [114,117]. Interpenetrating network hydrogels have a pre-prepared hydrogel in which monomers and initiators are dissolved in a solution and are primarily used for drug delivery due to advantages such as pore size and surface properties that can be modified to control the drug release mechanism. On the other hand, semi-interpenetrating network hydrogels have slow drug-releasing properties with advantages similar to interpenetrating network hydrogels, and they also deliver drugs effectively [113].Polymerization is an essential step in the synthesis of hydrogels and can be accomplished in a single step or multiple steps involving different mechanisms, such as in the presence of benzoin isobutyl ether as the UV-sensitive initiator, template copolymerization, or free-radical photopolymerization. Their products are also classified based on the materials used in their manufacture, the types of crosslinkers used, how they appear physically, the net electrical charge, and how they are affected by both physical and chemical changes. [118]. Physical nanohydrogels are referred to as reversible nanohydrogels due to conformational changes, whereas chemically crosslinked hydrogels are referred to as irreversible hydrogels due to configurational changes. [119]. Reversible and irreversible hydrogels can be created using bulk polymerization, which involves dissolving one or more types of monomers and a solvent such as water, a water-ethanol mixture, or benzyl alcohol to form a transparent polymeric network that swells up in the water and acquires flexibility [26]. Crosslinking polymerization is a thermal process that involves the use of ionic and neutral monomers as well as crosslinking agents. A grafting technique is used to strengthen the weak bulk of the polymerized structure. Sometimes unsaturated compounds such as gamma rays and electron beams are used as initiators for nanohydrogel synthesis [120]. In these types of nanohydrogels, first, macro radicals are produced, which further combine with other polymer chains by covalent bonding, and a crosslinked network structure is formed [26]. Stimuli-sensitive hydrogels are hydrogels that are sensitive to certain environmental changes and respond by changing their form or volume when exposed to certain conditions. The physical stimuli to which they respond include light, pressure, temperature, electric field, magnetic field, and ultrasound; chemical stimuli include pH, redox, ionic strength, CO_2, glucose; and biological stimuli include enzymes, antigens, glutathione, and DNA. Based on their source at the time of application to the hydrogels in vivo, these stimuli can also be classified as internal or exterior stimuli. Except for the temperature, which can be considered an external or an internal stimulus, chemical, and biological stimuli fall into the first group, whereas physical stimuli fall into the second. These hydrogels have been termed "smart" or "intelligent" in the sense that they perceive a stimulus and respond by changing their physical and/or chemical behavior, allowing the contained medication to be released.

7.1. Mechanisms of Different Stimuli-Responsive Hydrogels

(a)   Physical stimuli

    1.   Temperature Temperature changes fluctuate the polymer-polymer and polymer–water interactions that are responsible for swelling and drug release.

    2.   Pressure Increased pressure causes swelling, and vice versa. This is because the lower critical solution temperature (LCST) of hydrogels rises with pressure. The temperature below which negative thermoresponsive hydrogels swell is known as the LCST.

    3.   Light The hydrogel is reversibly changed from a flowable to a non-flowable state when exposed to light (UV and visible light).

4. Electric field Swelling–deswelling is caused by changes in the electrical charge distribution within the hydrogel matrix when an electric field is applied, and this is responsible for on-demand drug release.
5. Magnetic field The application of a magnetic field causes pores in the gel to expand, resulting in drug release.
6. Ultrasound irradiation The drug is released when the ionic crosslinks in the hydrogels are briefly broken by ultrasound waves, but the crosslinks are repaired when the ultrasound waves are turned off. This allows for on-demand medicine delivery.

(b) Chemical stimuli

1. pH The charge on the polymer chains changes when the pH changes, causing swelling and drug release.
2. Ionic strength Change in ion concentration also causes swelling and drug release
3. CO_2 A pH-sensitive hydrogel disc comes into touch with a bicarbonate solution in CO2 sensors. When exposed to CO2, the pH of the solution changes, causing the hydrogel to swell or de-swell, causing a change in pressure, which is a measure of CO2 partial pressure.
4. Glucose In reaction to an increase in glucose concentration, hydrogels swell. The combination generated by glucose and phenylboronic acid causes the hydrogels to enlarge, resulting in insulin release.
5. Redox In a reductive environment (high glutathione concentration = 0.5–10 mM), disulfide links in reduction-sensitive hydrogels cleave in the intracellular matrix, releasing bioactive molecules/drugs.

(c) Biological stimuli

1. Enzymes Enzymes are responsible for hydrogel decomposition and, as a result, drug release. This is termed a chemically regulated drug release mechanism.
2. Antigen When hydrogels detect free antigens, they swell and release the molecule.
3. DNA In the presence of ssDNA, single-stranded (ss) DNA grafted hydrogel probes swell.

Hydrogels show unique physical properties, such as high porosity that allows a gel to be loaded or released with highly active components. As a result, the application of hydrogels in drug delivery systems is of great interest [121] (Table 5).

Table 5. Clinical trials of hydrogels in the treatment of skin disorders.

Type of Hydrogel	Clinical Study	Agent	Skin Disorder	References
Clindamycin/Tretinoin Hydrogel	Clindamycin/Tretinoin Hydrogel	Combination Of Clindamycin (1%) and Tretinoin (0.025%)	Acne vulgaris	[79]
Liposomal Methylene Blue Hydrogel	Randomized and comparative study of 35 patients (21 men and 14 women) with varying degrees of acne vulgaris on the back	Methylene Blue	Acne vulgaris (Truncal)	[80]
Carboxymethylcellulosebased Hydrogel	Single-blind study on 20 patients (12 men and 8 women)	Resveratrol	Acne vulgaris (Facial)	[122]
Hydrogel Patch	Men and women with plaque-type psoriasislesions	Mometasone Furoate	Psoriasis	[123]
Hydrogel Micropatch	100 psoriatic patients (75 men and 25 women) and 100 healthy volunteers	Mometasone Furoate	Psoriasis	[124]

However, sometimes chemically synthesized or drug-loaded nanohydrogels show some side effects such as itching or rashes at the site of application. As a result, antifungal drug-loaded gels based on polysaccharides can be a more highly effective combination of natural and chemical antifungal compounds over chemically synthesized nanohydrogels [125]. Several polysaccharide-based nanohydrogels containing a wide range of active compound formulations have been reported in various studies, and all the developed formulations revealed potential against various types of skin fungal infection-causing strains (Table 6).

Table 6. Existing antifungal drug-loaded polysaccharide-based formulations.

Polysaccharide	Active Compounds	References
Galan gum/cyclodextrin	Fluconazole	[126]
Galan gum	Terbinafine HCL	[127]
Chitosan/carbopol/natrosol	Terbinafine HCL	[128]
Galan gum/carrageenan	Econazole	[129]
Galan gum/carbopol934P hydroxyl propyl methyl cellulose E50LV	Clotrimazole	[130]
Galan gum/glycerol	Fluconazole	[131]
Galan gum	Natamycin	[132]

Various reports revealed that some strains of *Aspergillus* species are becoming resistant to azoles and nanohydrogels consisting of azole components that showed less activity against the fungal strains. Therefore, to overcome this problem, various plant-based extracts and essential oils have shown promising results against such pathogenic strains of skin fungal infection. However, environmental stress, high temperature, and oxygen are significant factors that lead to detrimental effects on plant-based extracts and essential oils. Therefore, to overcome this problem, scientists have turned to nanohydrogels as their first choice for transporting and protecting bioactive constituents (Table 7).

Table 7. Natural extract entrenched antifungal formulations.

Formulation	Natural Extract	Active Ingredients of the Natural Extract	Effective against	References
Methylcellulose hydrogel	*Melissa officinalis*	citronellal (50%), citronellol (10%), and geraniol (14%)	Candidiasis	[133]
Polyherbal gel	*Piper betal* and *Piper nigrum* leaf extract	methanolic hydro extracts	*Candida albicans*	[134]
Copper chitosan nanocomposite hydrogel	*Thymus vulgaris*	p-cymene, thymol, and 1,8-cineole	*Aspergillus flavus*	[134]
Hydroxypropylmethylcellulose hydrogel	*Melaleuca alternifolia* (Tea tree oil)	terpinene-4-ol	Oral candidiasis	[135]
Carbopol hydrogel	*Cymbopogon citratus* (Lemongrass oil)	geraniol, geranylacetate, and monoterpene olefins	*Candida albicans*	[136]

Plants have been utilized therapeutically for thousands of years everywhere in the world. Medicinal plants incorporate herbs, herbal materials, and items that contain various pieces of plants or other plant materials that have traditionally been used to combat health disorders [137]. Essential oils and herbal extracts have antifungal properties because their phenolic groups act as the primary antimicrobial bioactive compound [138–142]. Phenolic groups are complex, volatile, aromatic compounds with different chemical structures and are stored in various parts of the plant, in particular tissue such as glandular hairs, oil cells, and oil ducts [143]. They are now well known due to their antimicrobial, germ-killing, anti-inflammatory, and antioxidant properties [144]. Many essential oils and their components, such as eucalyptus oil, clove oil, thyme oil, bitter almond oil, cinnamon oil, tea tree oil, and lemongrass oil, are commonly used and have antifungal activity. However, on the other hand, Tabassum and Hamdani (2014) observed that several plants can be used to treat skin diseases, such as aloe vera (*Barbadosaloe*), the gel of which has properties that act against bacteria and fungi (Table 8). In addition, *Bauhinia variegates, Beta vulgaris, Brassica oleraceae, Calendula officinalis, Camellia sinesis, Cannabis sativus* can aid in the treatment of viral, bacterial, and fungal infections.

Table 8. Plant extracts having antifungal properties.

Antifungal Plant Extracts	Effective against	References
Leaves of *Piperregnellii*	*Trichophyton rubrum, Trichophyton mentagrophytes, Microsporum canis*	[145]
Roots of *Rubiatinctorum*	*Asperegilus niger, Alternaria lternaria, Penicillium verrucosum, Mucor mucedo*	[146]
Tithoniadiversifolia	*Microbotryum violaceum, Chlorella fusca*	[147]
Daturametel	*Candida albicans, Candida tropicalis*	[148]
Alliumcepa and *Alliumsativum*	*Malassezia furfur, Candida albicans,* and other *Candida* species	[18]

Phytochemical compounds are responsible for the antifungal activity of essential oils. Other factors, such as concentration, dosage, length of treatment, and the primary type of fungal species, have an impact on an oil's antifungal action [149]. The vulnerability of various pathogens to essential oil dosage is determined by the morphological and physiological characteristics of the hyphae. Sometimes fungal species produce enzymes that oxidize and inactivate oils, and when oils come in contact with high temperatures that heat the phytochemicals, epimerization [22] may occur. As a result, to inhibit these types of problems, their use in nanohydrogel formulations allows them to reach an infectious site, such as the third layer of skin, due to their high penetration power, which is why they are gaining more attention in pharmaceutical medical science. Natural polymeric nanohydrogels have numerous advantages due to the popularity of natural materials in the therapeutic area. In addition, natural polymeric nanohydrogels have various applications—such as use as sensors or in drug delivery—because of their ability to concentrate on controlled change during the delivery of bioactive compounds, even when environmental conditions change. Moreover, natural polymeric nanohydrogels gained a preferred position as compared to engineered materials because they show biodegradability, biocompatibility, inexhaustibility, nontoxicity, low-cost, and expanding application [150]. Hosseini and Nabid studied a basil seed mucilage-based hydrogel for drug delivery in 2020. Active compounds used in disease treatment can be incorporated into nanohydrogels using various methods such as (1) physical entrapment; (2) covalent conjugation; and (3) controlled self-assembly. Nowadays, the controlled self-assembly method is widely known for encapsulating active components or drugs inside the gel [151,152].

7.2. Physical Entrapment Method for Incorporation

Noncovalent interactions such as ionic, lipophilic, and hydrogen bonding can be used to entrap drugs within nanohydrogels [153]. The self-assembly of cholesterol-containing hyaluronic acid into nanohydrogel for protein delivery is one example. Injection of this gel containing recombinant human growth hormone (rhGH) into rats resulted in a week of sustained release [154]. Hydrogen bonding is demonstrated in the encapsulation of curcumin into chitin nanohydrogels for skin cancer treatment. Chitin is commonly used in nanohydrogel synthesis due to its high biocompatibility, biodegradability, skin non-irritability, ease of availability, and cost-effectiveness. The cationic charge of chitin and the lipophilic nature of both chitin and curcumin help in skin penetration. On the other hand, nanohydrogels are hydrophilic, so the hydrophilic-lipophilic balance of chitin–curcumin nanohydrogels is advantageous. Curcumin interacts through its terminal-OH group with-NHCOCH$_3$ of chitin [155].

7.3. Covalent Conjugation Method for Incorporation

Nanosystems offer a convenient drug delivery platform. This is because their inherent functional groups play a role in determining the structure and properties of nanoparticles. The drug is covalently conjugated to the crosslinked nanohydrogels, which gives the encapsulated drug additional stability [156]. Polysaccharides contain hydroxyl groups that easily interact with hydroxyl groups formed by forming ester bonds with the drug's carboxyl groups. Such a drug would be released prematurely in this case due to the cleavage of functional groups by enzymes such as esterase.

7.4. Controlled Self-Assembly Method for Incorporation

The term "self-assembly" refers to the autonomous organization of components that results in a good structure. Many molecules self-assemble, which is characterized by diffusion, and the subsequent precise binding of molecules occurs by non-covalent interactions, hydrophobic associations, or by including electrostatic interactions. Due to these a lot of interactions show weaknesses and dominates the assembly's structure and conformational behavior. In the presence of oppositely charged solutes, polyelectrolyte-based nanohydrogels tend to self-assemble. In an aqueous environment, amphoteric molecules self-assemble into nanoparticles, allowing for improved drug interaction and release from the nanohydrogels. The hydrophilic part of the drug molecule should be exposed to polar or aqueous media, while the hydrophobic part should be fixed in the component's core. The concentration of polymer above which chains aggregate is referred to as critical micelle concentration or critical aggregate concentration [156,157].

8. Antifungal Mechanism of Essential Oil and Plant Extract-Based Nanohydrogel

Essential oils and plant extracts exhibit effective antifungal activity; however, several limitations inhibit the utilization of essential oils and plant extracts [100]. Scientists already revealed that nanohydrogels could pass through the three tissue layers of skin and could thus reach the dermis layer because of their properties (Figure 3). The pH of nanohydrogels is similar to that of human skin; however, when infection occurs, it causes changes at the infectious site. As a result of these changes—and according to the nature, concentration, and hydrophilicity of the polymer used to synthesize the nanohydrogel—a swollen nanohydrogel can receive stimuli to release its antifungal agent through three mechanisms: (1) a diffusion-controlled mechanism; (2) a chemical-controlled mechanism; and (3) a swelling-controlled mechanism [121].

Mainly the nanosized gels release active compounds with the help of diffusion. In a diffusion-controlled mechanism, the gel comes in contact with a high temperature, which leads to an increase in the hydrophobicity of the structure and causes a slow and controlled release of the antifungal agent from the gel, which moves across the concentration gradient, such as from high to low concentration, as calculated by the Fick's first law of diffusion [158]:

$$dQ/dt = ADK_d(Co - C)/h \tag{1}$$

where dQ/dt is the rate of mass transfer, A is the surface area of film, Kd is the partition coefficient, h is the thickness of the hydrogel, and Co and C both are the concentrations of the drug or active compound at nanohydrogel and infectious site [125]. If the molecular dimensions of the drug molecules are significantly smaller than the pore size of the porous hydrogels, the porosity of the hydrogels is related to the diffusion coefficient of the hydrogels. The release of the drug molecules is hampered by the crosslinked polymer chains when the pore size in the hydrogels and the size of the drug molecules are comparable. The diffusion coefficient decreases as a result. If the rate of drug release exceeds the rate of swelling, the release of the drug is regulated by the swelling-controlled mechanism. In a chemical-controlled mechanism, the active component is released via matrix reactions; this can be an enzymatic, hydrolytic, reversible, or irreversible reaction between the releasing component and the matrix; the releasing action is determined by the rate of crosslinking structure degradation. On the other hand, in the swelling-controlled mechanism, the nanohydrogel starts to swell more when solvent molecules attract towards it, and after this interaction temperature decreases and it leads to the release of the desired component from the polymeric structure. This swelling-controlled mechanism is mainly the function of temperature, in which the transition from swelling to shrinkage occurs [159]. It is also classified as a (1) kinetically controlled release mechanism and (2) reaction diffusion-controlled release mechanism. After the release of the active component, the active component acts against the fungal infection by following different mechanisms, such as an oil that inhibits the growth of a fungus by inserting emulsions in the fungus cell wall where the oil emulsions can cause respiration inhibition and can destroy the cell wall of the fungus hyphae, resulting in cell death, while the oil helps to cure the infection [160]. On the other hand, volatile essential oils inhibit spore formation, such as in a study by Fajinmi et al. (2019) showed the effect of 40 μL *Agathosma betulina* essential oil against the *Trichophyton rubrum* species. These species can produce hyphae that can penetrate the host's skin and cause an infection inside the skin. The research team subcultured the essential oil-treated fungus on a separate petri plate and observed no fungus growth during and after seven days of an incubation period [143]. Orange oil also acts as an essential oil because of the high content of limonene, myrcene, linalool, and citral components; these show fungicidal effects against a broad spectrum of organisms, such as *Candida albicans, Candida krusei, Candida tropical, Aspergillus niger, Aspergillus fumigatus*, and *Penicillium chrysogenum* [161]. Eucalyptus oil acts as both an antibacterial and an antifungal oil [8]. Monoterpenes are the active compounds in eucalyptus oil and work against dermatophytes such as *Microsporum canis* and *Microsporum gypseum* [9].

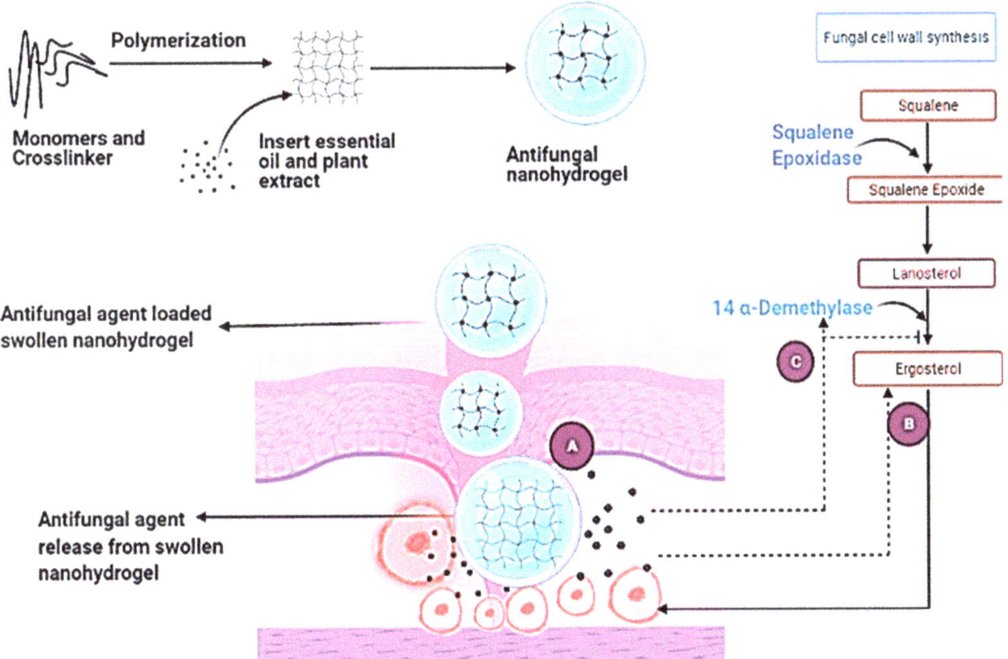

Figure 3. Mechanism action of nanohydrogel and antifungal components. **A:** Swollen nanohydrogel contacts stimuli in the environment and releases antifungal agents by following three mechanisms; **B:** antifungal agent binds to a vital component of the fungus cell wall, create spores, and leads to the death of the cell or causes respiration inhibition of the fungus cell; **C:** antifungal agent interferes with the catalyst and inhibits the ergosterol synthesis from lanosterol, causing cell death.

In this context, Wang and Vonrecum (2011) stated that sometimes nanohydrogels show minimal drug loading efficiency [162]. On the other hand, Vinogradov et al. (2002) revealed that the hydrophilicity of nanohydrogels decreases due to the strong interaction between drug and polymer and causes structural changes, also showing adverse effects due to the presence of surfactant or monomers [67,163]. Hydrogels show significantly less potential for the delivery of hydrophobic drugs. Some of the main disadvantages of the hydrogel are their low mechanical strength, fast dissolution, and low tensile strength [6].

9. Conclusions and Future Perspective

Several types of fungal skin infections and their impact on human skin were discussed in detail. The various advantages and applications of antifungal nanocarriers outlined above definitely hold the promise that nanocarriers could potentially revolutionize the pharmaceutical and food sectors. Among all nanocarriers, the scope of nanohydrogels is broad, and formulating functional nanohydrogels for antifungal applications from nature-derived polymers is widely conceived as a sustainable approach that is safer for humans. An array of plant-based biopolymers and essential oils can be incorporated into nanohydrogels for practical components for various skin fungal diseases. A potential mechanism, as well as examples of the successful use of nanohydrogels in the treatment of fungal skin infections, was presented and discussed. Moreover, in the future, concerns over the safety of the polysaccharide-based nanohydrogels will be directly affected by public acceptance of such emerging technologies. In the case of biopolymeric nanohydrogels, the assurance of acceptance comes from the fact that all of the materials discussed here are polymers that, unlike allopathic products, can be quickly degraded in the body. Hence, toxicological studies with nanoformulations are necessary to ensure the effective utilization of nanohydrogels.

Author Contributions: Conceptualization, P.C.; investigation N.K., A.B. and S.B.D.; writing—original draft preparation, N.K.; writing—review and editing, N.K. and R.K.; supervision, P.C. and F.M. All authors have read and agreed to the published version of the manuscript.

Funding: This work was supported by a grant from the Romanian Ministry of Education and Research, CNCS-UEFISCDI, project number PN-III-P1-1.1-PD-2019-0475, within PNCDI III.

Institutional Review Board Statement: Not applicable.

Informed Consent Statement: Not applicable.

Data Availability Statement: Not applicable.

Acknowledgments: We are grateful for the administrative and financial support offered by the University of Agricultural Sciences and Veterinary Medicine Cluj-Napoca, Romania, and the support of library access from Lovely Professional University, Phagwara, Punjab.

Conflicts of Interest: The authors declare no conflict of interest.

References

1. Imtiaz, N.; Niazi, M.B.; Fasim, F.; Khan, B.A.; Bano, S.A.; Shah, G.; Badshah, M.; Menaa, F.; Uzair, B. Fabrication of an Original Transparent PVA/Gelatin Hydrogel: In Vitro Antimicrobial Activity against Skin Pathogens. *Int. J. Polym. Sci.* **2019**, *2019*, 1–11. [CrossRef]

2. Shields, B.E.; Rosenbach, M.; Brown-Joel, Z.; Berger, A.P.; Ford, B.A.; Wanat, K.A. Angioinvasive fungal infections impacting the skin: Background, epidemiology, and clinical presentation. *JAAD* **2019**, *80*, 869–880. [CrossRef] [PubMed]

3. Rai, M.; Ingle, A.P.; Pandit, R.; Paralikar, P.; Gupta, I.; Anasane, N.; Dolenc-Voljč, M. *Nanotechnology for the Treatment of Fungal Infections on Human Skin. Clinical Microbiology Diagnosis, Treatment and prophylaxis of Infections the Microbiology of Skin, Soft Tissue, Bone and Joint Infections*; Kon, K., Rai, M., Eds.; Elsevier: Amsterdam, The Netherlands, 2017; Volume 2, pp. 169–184.

4. Bongomin, F.; Gago, S.; Oladele, R.O.; Denning, D.W. Global and multi-national prevalence of fungal diseases—Estimate preci-sion. *J. Fungi* **2017**, *3*, 57. [CrossRef] [PubMed]

5. Wijesiri, N.; Yu, Z.; Tang, H.; Zhang, P. Antifungal photodynamic inactivation against dermatophyte Trichophyton rubrum us-ing nanoparticle-based hybrid photosensitizers. *Photodiagnosis Photodyn. Ther.* **2018**, *23*, 202–208. [CrossRef]

6. Gulyuz, U.; Okay, O. Self-healing poly (acrylic acid) hydrogels with shape memory behaviour of high mechanical strength. *Macromolecules* **2014**, *47*, 6889–6899. [CrossRef]

7.   Black, A.T. Dermatological Drugs, Topical Agents, and Cosmetics. *Side Eff. Drugs Annu.* **2015**, *37*, 175–184.

8.   Baptista, E.B.; Zimmermann-Franco, D.C.; Lataliza, A.A.; Raposo, N.R. Chemical composition and antifungal activity of essential oil from Eucalyptus smithii against dermatophytes. *Rev. Soc. Bras. Med. Trop.* **2015**, *48*, 746–752. [CrossRef]

9.   Elaissi, A.; Rouis, Z.; Salem, N.A.; Mabrouk, S.; Ben Salem, Y.; Salah, K.B.; Aouni, M.; Farhat, F.; Chemli, R.; Harzallah-Skhiri, F.; et al. Chemical composition of 8 eucalyptus species' essential oils and the evaluation of their antibacterial, antifungal and antiviral activities. *BMC Complement. Altern Med.* **2012**, *12*, 1–5. [CrossRef]

10.   Gupta, A.K.; Foley, K.A.; Versteeg, S.G. New antifungal agents and new formulations against dermatophytes. *Mycopathologia* **2017**, *182*, 127–141. [CrossRef]

11.   Lakshmi, C.V.; Bengalorkar, G.M.; Kumar, V.S. Clinical efficacy of topical terbinafine versus topical luliconazole in treatment of tinea corporis/tinea cruris patients. *J. Pharm. Res. Int.* **2013**, *24*, 1001–1014. [CrossRef]

12.   Glynn, M.; Jo, W.; Minowa, K.; Sanada, H.; Nejishima, H.; Matsuuchi, H.; Okamura, H.; Pillai, R.; Mutter, L. Efinaconazole: Developmental and reproductive toxicity potential of a novel antifungal azole. *Reprod. Toxicol.* **2015**, *52*, 18–25. [CrossRef]

13.   Abd Elaziz, D.; Abd El-Ghany, M.; Meshaal, S.; El Hawary, R.; Lotfy, S.; Galal, N.; Ouf, S.A.; Elmarsafy, A. Fungal infections in primary immunodeficiency diseases. *J. Clin. Immunol.* **2020**, *219*, 108553. [CrossRef]

14.   Jain, A.; Jain, S.; Rawat, S. Emerging fungal infections among children: A review on its clinical manifestations, diagnosis, and prevention. *J. Pharm. Bioallied. Sci.* **2010**, *2*, 314–320. [CrossRef]

15.   Nami, S.; Aghebati-Maleki, A.; Morovati, H.; Aghebati-Maleki, L. Current antifungal drugs and immunotherapeutic ap-proaches as promising strategies to treatment of fungal diseases. *Biomed. Pharmacother.* **2019**, *110*, 857–868. [CrossRef]

16.   Marek, C.L.; Timmons, S.R. Antimicrobials in Pediatric Dentistry. *Pediatr. Dent.* **2019**, 128–141.e1. [CrossRef]

17.   Girois, S.B.; Chapuis, F.; Decullier, E.; Revol, B.G. Adverse effects of antifungal therapies in invasive fungal infections: Review and meta-analysis. *Eur. J. Clin. Microbiol.* **2006**, *25*, 138–149. [CrossRef]

18.   Tabassum, N.; Hamdani, M. Plants used to treat skin diseases. *Pharmacogn. Rev.* **2014**, *8*, 52. [CrossRef]

19.   Mahlo, S.M.; Chauke, H.R.; McGaw, L.; Eloff, J. Antioxidant and antifungal activity of selected medicinal plant extracts against phytopathogenic fungi. *Afr. J. Tradit. Complement. Altern. Med.* **2016**, *13*, 216–222. [CrossRef]

20.   Hu, F.; Tu, X.F.; Thakur, K.; Hu, F.; Li, X.L.; Zhang, Y.S.; Zhang, J.G.; Wei, Z.J. Comparison of antifungal activity of essential oils from different plants against three fungi. *Food Chem. Toxicol.* **2019**, *134*, 110821. [CrossRef]

21.   Kothe, L.; Zimmermann, B.F.; Galensa, R. Temperature influences epimerization and composition of flavanol monomers, di-mers and trimers during cocoa bean roasting. *Food Chem.* **2013**, *141*, 3656–3663. [CrossRef]

22.   Dai, J.; Mumper, R.J. Plant phenolics: Extraction, analysis and their antioxidant and anticancer properties. *Molecules* **2010**, *15*, 7313–7352. [CrossRef]

23.   Andrés-Bello, A.; Barreto-Palacios, V.I.; García-Segovia, P.; Mir-Bel, J.; Martínez-Monzó, J. Effect of pH on color and texture of food products. *Food Eng. Rev.* **2013**, *5*, 158–170. [CrossRef]

24.   Gonçalves, C.; Pereira, P.; Gama, M. Self-assembled hydrogel nanoparticles for drug delivery applications. *Materials* **2010**, *3*, 1420–1460. [CrossRef]

25.   Martínez-Martínez, M.; Rodríguez-Berna, G.; Gonzalez-Alvarez, I.; Hernández, M.A.; Corma, A.; Bermejo, M.; Merino, V.; Gon-zalez-Alvarez, M. Ionic hydrogel based on chitosan cross-linked with 6-phosphogluconic trisodium salt as a drug delivery system. *Biomacromolecules* **2018**, *19*, 1294–1304. [CrossRef] [PubMed]

26.   Ahmed, E.M. Hydrogel: Preparation, characterization, and applications: A review. *J. Adv. Res.* **2015**, *6*, 105–121. [CrossRef]

27.   Boddy, L. Interactions with Humans and Other Animals. *Fungi* **2016**, 293–336. [CrossRef]

28.   Natarajan, V.; Nath, A.K.; Thappa, D.M.; Singh, R.; Verma, S.K. Coexistence of onychomycosis in psoriatic nails: A descriptive study. *Indian J. Dermatol.* **2010**, *76*, 723.

29.   Ajello, L. A taxonomic review of the dermatophytes and related species. *Sabouraudia* **1968**, *6*, 147. [CrossRef]

30.   Sulaiman, I.M.; Jacobs, E.; Simpson, S.; Kerdahi, K. Genetic characterization of fungi isolated from the environmental swabs collected from a compounding center known to cause multistate meningitis outbreak in united states using ITS sequencing. *Pathogens* **2014**, *3*, 732–742. [CrossRef]

31.   Kalita, J.M.; Sharma, A.; Bhardwaj, A.; Nag, V.L. Dermatophytoses and spectrum of dermatophytes in patients attending a teaching hospital in Western Rajasthan, India. *J. Family Med. Prim. Care* **2019**, *8*, 1418.

32.   Sawant, B.; Khan, T. Recent advances in delivery of antifungal agents for therapeutic management of candidiasis. *Biomed. Pharmacother.* **2017**, *96*, 1478–1490. [CrossRef] [PubMed]

33.   AbouSamra, M.M.; Basha, M.; Awad, G.E.; Mansy, S.S. A promising nystatin nanocapsular hydrogel as an antifungal polymer-ic carrier for the treatment of topical candidiasis. *J. Drug Deliv. Sci. Technol.* **2019**, *49*, 365–374. [CrossRef]

34.   Nobile, C.J.; Johnson, A.D. Candida albicans biofilms and human disease. *Annu. Rev. Microbiol.* **2015**, *69*, 71–92. [CrossRef] [PubMed]

35.   Dabas, P.S. An approach to etiology, diagnosis and management of different types of candidiasis. *J. Yeast Fungal Res.* **2013**, *4*, 63–74.

36.   Tovikkai, D.; Maitrisathit, W.; Srisuttiyakorn, C.; Vanichanan, J.; Thammahong, A.; Suankratay, C.S. The case series in Thai-land and literature review in Southeast Asia. *Med. Mycol. Case Rep.* **2020**, *27*, 59–63. [CrossRef]

37.   De Lima Barros, M.B.; de Almeida Paes, R.; Schubach, A.O. Sporothrixschenckii and Sporotrichosis. *Clin. Microbiol. Rev.* **2011**, *24*, 633–654. [CrossRef]

38. Mahajan, V.K. Sporotrichosis: An Overview and Therapeutic Options. *Dermatol. Res. Pract.* **2014**, *2014*, 1–13. [CrossRef]
39. Chokoeva, A.A.; Zisova, L.; Sotiriou, E.; Miteva-Katrandzhieva, T. Tinea capitis: A retrospective epidemiological comparative study. *Wien. Med. Wochenschr.* **2017**, *167*, 51–57. [CrossRef]
40. Baumgardner, D.J. Fungal infections from human and animal contact. *J. Patient Cent. Res. Rev.* **2017**, *4*, 78. [CrossRef]
41. White, T.C.; Findley, K.; Dawson, T.L.; Scheynius, A.; Boekhout, T.; Cuomo, C.A.; Xu, J.; Saunders, C.W. Fungi on the skin: Dermatophytes and Malassezia. *Cold Spring Harb. Perspect. Med.* **2014**, *4*, 019802. [CrossRef]
42. Hay, R.J. Tinea capitis: Current status. *Mycopathologia* **2017**, *182*, 87–93. [CrossRef]
43. Brown, T.M.; Krishnamurthy, K. *Histology, Hair and Follicle*; StatPearls Publishing: Treasure Island, FL, USA, 2020. Available online: http://creativecommons.org/licenses/by/4.0/Last (accessed on 10 May 2021).
44. Al Aboud, A.M.; Crane, J.S. *Tinea Capitis*; StatPearls Publishing: Treasure Island, FL, USA, 2020; Bookshelf ID: NBK536909.
45. Alk, T.M.; Krishnamurthy, A.; Cantrell, W.; Elewski, B. Treatment of tinea capitis. *Skin Appendage Disord.* **2019**, *5*, 201–210.
46. Newland, J.G.; Abdel-Rahman, S.M. Update on terbinafine with a focus on dermatophytoses. *Clin. Cosmet. Investing. Dermatol.* **2009**, *2*, 49.
47. Bennassar, A.; Grimalt, R. Management of tinea capitis in childhood. *Clin. Cosmet. Investing. Dermatol.* **2010**, *3*, 89.
48. Trösken, E.R. Toxicological Evaluation of Azole Fungicides in Agriculture and Food Chemistry. Ph.D. Thesis, Julius-Maximilians-Universität Würzburg, Würzburg, Germany, 2005.
49. Michaels, B.D.; Del Rosso, J.Q. Tinea capitis in infants: Recognition, evaluation, and management suggestions. *J. Clin. Aesthet. Dermatol.* **2012**, *5*, 49.
50. Basha, B.N.; Prakasam, K.; Goli, D. Formulation and evaluation of gel containing fluconazole-antifungal agent. *Int. J. Drug Dev. Res.* **2011**, *3*, 119–127.
51. Sahoo, A.K.; Mahajan, R. Management of tinea corporis, tinea cruris, and tinea pedis: A comprehensive review. *Indian Dermatol. Online J.* **2016**, *7*, 77.
52. Shy, R. Tinea corporis and tinea capitis. *Pediatr. Rev.* **2007**, *28*, 164–174. [CrossRef]
53. Diongue, K.; Ndiaye, M.; Diallo, M.A.; Seck, M.C.; Badiane, A.S.; Diop, A.; Ndiaye, Y.D.; Déme, A.; Ndiaye, T.; Ndir, O.; et al. Fungal interdigital tinea pedis in Dakar (Senegal). *J. Mycol. Med.* **2016**, *26*, 312–316. [CrossRef]
54. Salehi, Z.; Fatahi, N.; Taran, M.; Izadi, A.; Badali, H.; Hashemi, S.J.; Rezaie, S.; Ghazvini, R.D.; Ghaffari, M.; Aala, F.; et al. Com-parison of in vitro antifungal activity of novel triazoles with available antifungal agents against dermatophyte species caused tinea pedis. *J. Mycol. Med.* **2020**, *30*, 100935. [CrossRef]
55. Ilkit, M.; Durdu, M. Tinea pedis: The etiology and global epidemiology of a common fungal infection. *Crit. Rev. Microbiol.* **2015**, *41*, 374–388. [CrossRef]
56. Rotta, I.; Ziegelmann, P.K.; Otuki, M.F.; Riveros, B.S.; Bernardo, N.L.; Correr, C.J. Efficacy of topical antifungals in the treatment of dermatophytosis: A mixed-treatment comparison meta-analysis involving 14 treatments. *JAMA Dermatol.* **2013**, *149*, 341–349. [CrossRef]
57. Khanna, D.; Bharti, S. Luliconazole for the treatment of fungal infections: An evidence-based review. *Core Evid.* **2014**, *9*, 113. [CrossRef]
58. Raugi, G.; Nguyen, T.U. Superficial Dermatophyte infections of the skin. In *Netter's Infectious Diseases*; Elsevier: Amsterdam, The Netherlands, 2012; pp. 102–109.
59. Banki, A.; Castiglione, F.M. Infections of the Facial Skin and Scalp. In *Head, Neck and Orofacial Infections: An Interdisciplinary Approach E-Book*; Elsevier: St. Louis, MO, USA, 2015; p. 318, ISBN 978-0-323-3907-2.
60. Chamorro, M.J.; House, S.A. *Tinea Manuum*; StatPearls Publishing: Treasure Island, FL, USA, 2020. Available online: http://creativecommons.org/licenses/by/4.0/ (accessed on 10 August 2020).
61. Reddy, K.R. Fungal Infections (Mycoses): Dermatophytoses (Tinea, Ringworm). *J. GMC-N.* **2017**, *10*. [CrossRef]
62. Noble, S.L.; Forbes, R.C.; Stamm, P.L. Diagnosis and management of common tinea infections. *Am. Fam. Physician.* **1998**, *58*, 163.
63. Ferry, M.; Shedlofsky, L.; Newman, A.; Mengesha, Y.; Blumetti, B. Tinea in Versicolor: A rare distribution of a common erup-tion. *Cureus* **2020**, *12*. [CrossRef]
64. Gaitanis, G.; Magiatis, P.; Hantschke, M.; Bassukas, I.D.; Velegraki, A. The Malassezia genus in skin and systemic diseases. *Clin. Microbial. Rev.* **2012**, *25*, 106–141. [CrossRef]
65. Gupta, M.; Sharma, V.; Chauhan, N.S. Promising novel nanopharmaceuticals for improving topical antifungal drug delivery. In *Nano-and Microscale Drug Delivery Systems*; Elsevier: Amsterdam, The Netherlands, 2017; pp. 197–228.
66. Karray, M.; McKinney, W.P. *Tinea (Pityriasis) Versicolor*; StatPearls Publishing: Treasure Island, FL, USA, 2020. Available online: http://creativecommons.org/licenses/by/4.0/ (accessed on 10 August 2020).
67. Garg, S.; Chandra, A.; Mazumder, A.; Mazumder, R. Green synthesis of silver nanoparticles using Arnebia nobilis root extract and wound healing potential of its hydrogel. *Asian J. Pharm. Sci.* **2014**, *8*. [CrossRef]
68. Gupta, A.K.; Foley, K.A. Antifungal treatment for pityriasis versicolor. *J. Fungi.* **2015**, *1*, 13–29. [CrossRef] [PubMed]
69. Klafke, G.B.; Silva, R.A.; Pellegrin, K.T.; Xavier, M.O. Analysis of the role of nail polish in the transmission of onychomycosis. *Ann. Bras. Dermatol.* **2018**, *93*, 930–931. [CrossRef] [PubMed]
70. Sharma, P.; Sharma, S. Non-dermatophytes emerging as predominant cause of onychomycosis in a tertiary care centre in rural part of Punjab, India. *JACM* **2016**, *18*, 36. [CrossRef]
71. Westerberg, D.P.; Voyack, M.J. Onychomycosis: Current trends in diagnosis and treatment. *Am. Fam. Phys.* **2013**, *88*, 762–770.

72. Sahni, K.; Singh, S.; Dogra, S. Newer topical treatments in skin and nail dermatophyte infections. *Indian Dermatol. Online J.* **2018**, *9*, 149.

73. Hay, R. Therapy of Skin, Hair and Nail Fungal Infections. *J. Fungi* **2018**, *4*, 99. [CrossRef]

74. McKeny, P.T.; Nessel, T.A.; Zito, P.M. *Antifungal Antibiotics*; StatPearls Publishing: Treasure Island, FL, USA, 2020. Available online: http://creativecommons.org/licenses/by/4.0/ (accessed on 4 May 2021).

75. Sahni, T.; Sharma, S.; Arora, G.; Verma, D. Synthesis, Characterization and Antifungal Activity of a Substituted Coumarin and its Derivatives. *Pestic. Res. J.* **2020**, *32*, 39–48. [CrossRef]

76. Danielli, L.J.; Pippi, B.; Duarte, J.A.; Maciel, A.J.; Machado, M.M.; Oliveira, L.F.; Vainstein, M.H.; Teixeira, M.L.; Bordignon, S.A.; Fuentefria, A.M. Antifungal mechanism of action of Schinuslentiscifolius Marchand essential oil and its synergistic effect in vitro with terbinafine and ciclopirox against dermatophytes. *J. Pharm. Pharmacol.* **2018**, *70*, 1216–1227. [CrossRef]

77. Pârvu, M.; Moţ, C.A.; Pârvu, A.E.; Mircea, C.; Stoeber, L.; Roşca-Casian, O.; Ţigu, A.B. Allium sativum extract chemical composition, antioxidant activity and antifungal effect against Meyerozymaguilliermondii and Rhodotorulamucilaginosa causing onychomycosis. *Molecules* **2019**, *21*, 3958. [CrossRef]

78. Zhang, A.Y.; Camp, W.L.; Elewski, B.E. Advances in topical and systemic antifungals. *Clin. Dermatol.* **2007**, *25*, 165–183. [CrossRef]

79. Jachak, G.R.; Ramesh, R.; Sant, D.G.; Jorwekar, S.U.; Jadhav, M.R.; Tupe, S.G.; Deshpande, M.V.; Reddy, D.S. Silicon incorporated morpholine antifungals: Design, synthesis, and biological evaluation. *ACS Med. Chem. Lett.* **2015**, *6*, 1111–1116. [CrossRef]

80. Tutaj, K.; Szlazak, R.; Szalapata, K.; Starzyk, J.; Luchowski, R.; Grudzinski, W.; Osinska-Jaroszuk, M.; Jarosz-Wilkolazka, A.; Szuster-Ciesielska, A.; Gruszecki, W.I. Amphotericin B-silver hybrid nanoparticles: Synthesis, properties and antifungal activity. *Nanomed. Nanotechnol. Biol.* **2016**, *12*, 1095–1103. [CrossRef]

81. Chowdhry, S.; Gupta, S.; D'souza, P. Topical antifungals used for treatment of seborrheic dermatitis. *J. Bacteriol. Mycol. Open Access.* **2017**, *4*, 1–7.

82. Nankervis, H.; Thomas, K.S.; Delamere, F.M.; Barbarot, S.; Rogers, N.K.; Williams, H.C. Scoping systematic review of treatments for eczema. *Program. Grants Appl. Res.* **2016**, *4*, 1–480. [CrossRef] [PubMed]

83. Shalini, K.; Kumar, N.; Drabu, S.; Sharma, P.K. Advances in synthetic approach to and antifungal activity of triazoles. *Beilstein J. Org. Chem.* **2011**, *7*, 668–677. [CrossRef] [PubMed]

84. Ashbee, H.R.; Barnes, R.A.; Johnson, E.M.; Richardson, M.D.; Gorton, R.; Hope, W.W. Therapeutic drug monitoring (TDM) of antifungal agents: Guidelines from the British Society for Medical Mycology. *J. Antimicrob. Chemother.* **2014**, *69*, 1162–1176. [CrossRef] [PubMed]

85. Prasad, R.; Shah, A.H.; Rawal, M.K. Antifungals: Mechanism of action and drug resistance. In *Yeast Membrane Transport*; Springer: Berlin/Heidelberg, Germany, 2016; pp. 327–349.

86. Dowd, F.J.; Johnson, B.; Mariotti, A. *Pharmacology and Therapeutics for Dentistry-E-Book*, 7th ed.; Elsevier Health Sciences: St. Louis, MO, USA, 2016; ISBN 978-0-323-39307-2.

87. Goldstein, A.O.; Goldstein, B.G. *Dermatophyte (Tinea) Infections*; UpToDate: Walthman, MA, USA, 2017. Available online: https://www.uptodate.com/ (accessed on 27 May 2021).

88. Waller, D.G.; Sampson, T. *Medical Pharmacology and Therapeutics E-Book*, 5th ed.; Elsevier Health Sciences: St. Louis, MO, USA, 2017.

89. Kaul, S.; Yadav, S.; Dogra, S. Treatment of dermatophytosis in elderly, children, and pregnant women. *Indian Dermatol. Online J.* **2017**, *8*, 310.

90. Rengasamy, M.; Chellam, J.; Ganapati, S. Systemic therapy of dermatophytosis: Practical and systematic approach. *Clin. Dermatol. Rev.* **2017**, *1*, 19. [CrossRef]

91. Bondaryk, M.; KurząTkowski, W.; Staniszewska, M. Antifungal agents commonly used in the superficial and mucosal candidi-asis treatment: Mode of action and resistance development. *Postepy Dermatol. Alergol.* **2013**, *30*, 293. [CrossRef]

92. Mroczyńska, M.; Brillowska-Dąbrowska, A. Review on Current Status of Echinocandins Use. *Antibiotics* **2020**, *9*, 227. [CrossRef]

93. Kofla, G.; Ruhnke, M. Pharmacology and metabolism of anidulafungin, caspofungin and micafungin in the treatment of in-vasivecandidosis—Review of the literature. *Eur. J. Med. Res.* **2011**, *16*, 159–166. [CrossRef] [PubMed]

94. Marak, M.B.; Dhanashree, B. Antifungal Susceptibility and Biofilm Production of Candida spp. Isolated from Clinical Samples. *Int. J. Microbiol.* **2018**, *2018*, 1–5. [CrossRef] [PubMed]

95. Harriott, M.M.; Lilly, E.A.; Rodriguez, T.E.; Fidel, P.L., Jr.; Noverr, M.C. Candida albicans forms biofilms on the vaginal mucosa. *Microbiology* **2010**, *156*, 3635. [CrossRef]

96. Spampinato, C.; Leonardi, D. CandidaInfections, Causes, Targets, and Resistance Mechanisms: Traditional and Alternative Antifungal Agents. *BioMed. Res. Int.* **2013**, *2013*, 1–13. [CrossRef]

97. Chong, P.P.; Chin, V.K.; Wong, W.F.; Madhavan, P.; Yong, V.C.; Looi, C.Y. Transcriptomic and genomic approaches for unrav-elling Candida albicans biofilm formation and drug resistance—An update. *Genes* **2018**, *9*, 540. [CrossRef]

98. Long, N.; Xu, X.; Zeng, Q.; Sang, H.; Lu, L. Erg4A and Erg4B Are Required for Conidiation and Azole Resistance via Regulation of Ergosterol Biosynthesis in Aspergillus fumigatus. *Appl. Environ. Microbiol.* **2016**, *83*, e02924-16. [CrossRef]

99. Pristov, K.E.; Ghannoum, M.A. Resistance of Candida to azoles and echinocandins worldwide. *Clin. Microbiol Infect.* **2019**, *25*, 792–798. [CrossRef]

100. Cowen, L.E. The evolution of fungal drug resistance: Modulating the trajectory from genotype to phenotype. *Nat. Rev. Microbiol.* **2008**, *6*, 187–198. [CrossRef]

101. Hayes, B.M.; Anderson, M.A.; Traven, A.; van der Weerden, N.L.; Bleackley, M.R. Activation of stress signalling pathways en-hances tolerance of fungi to chemical fungicides and antifungal proteins. *Cell. Mol. Life Sci.* **2014**, *71*, 2651–2666. [CrossRef]
102. Pai, V.; Ganavalli, A.; Kikkeri, N.N. Antifungal resistance in dermatology. *Indian J. Dermatol.* **2018**, *63*, 361. [CrossRef]
103. El-Awady, R.; Saleh, E.; Hashim, A.; Soliman, N.; Dallah, A.; Elrasheed, A.; Elakraa, G. The Role of Eukaryotic and Prokaryotic ABC Transporter Family in Failure of Chemotherapy. *Front. Pharmacol.* **2017**, *7*, 535. [CrossRef]
104. Coleman, J.J.; Mylonakis, E. Efflux in fungi: La piece de resistance. *PLoS Pathog.* **2009**, *5*, e1000486. [CrossRef]
105. Martinez-Rossi, N.M.; Peres, N.T.A.; Rossi, A. Antifungal Resistance Mechanisms in Dermatophytes. *Mycopathology* **2008**, *166*, 369–383. [CrossRef]
106. Martinez-Rossi, N.M.; Bitencourt, T.A.; Peres, N.T.; Lang, E.A.; Gomes, E.V.; Quaresemin, N.R.; Martins, M.P.; Lopes, L.; Rossi, A. Derma-tophyte resistance to antifungal drugs: Mechanisms and prospectus. *Front. Microbiol.* **2018**, *29*, 1108. [CrossRef]
107. Robbins, N.; Caplan, T.; Cowen, L.E. Molecular Evolution of Antifungal Drug Resistance. *Annu. Rev. Microbiol.* **2017**, *71*, 753–775. [CrossRef]
108. Sagatova, A.; Keniya, M.V.; Wilson, R.K.; Sabherwal, M.; Tyndall, J.D.A.; Monk, B.C. Triazole resistance mediated by mutations of a conserved active site tyrosine in fungal lanosterol 14α-demethylase. *Sci. Rep.* **2016**, *6*, 26213. [CrossRef] [PubMed]
109. Perlin, D.S.; Shor, E.; Zhao, Y. Update on Antifungal Drug Resistance. *Curr. Clin. Microbiol. Rep.* **2015**, *2*, 84–95. [CrossRef] [PubMed]
110. Vandeputte, P.; Ferrari, S.; Coste, A.T. Antifungal Resistance and New Strategies to Control Fungal Infections. *Int. J. Microbiol.* **2011**, *2012*, 1–26. [CrossRef] [PubMed]
111. Varaprasad, K.; Raghavendra, G.M.; Jayaramudu, T.; Yallapu, M.M.; Sadiku, R. A mini review on hydrogels classification and recent developments in miscellaneous applications. *Mater. Sci. Eng.* **2017**, *79*, 958–971. [CrossRef] [PubMed]
112. Laftah, W.A.; Hashim, S.; Ibrahim, A.N. Polymer hydrogels: A review. *Polym. Plast. Technol. Eng.* **2011**, *50*, 1475–1486. [CrossRef]
113. Bahram, M.; Mohseni, N.; Moghtader, M. *An Introduction to Hydrogels and Some Recent Applications. Emerging Concepts in Analysis and Applications of Hydrogels*; IntechOpen: London, UK, 2016. [CrossRef]
114. Das, N. Preparation methods and properties of hydrogel: A review. *Int. J. Pharm. Sci.* **2013**, *5*, 112–117.
115. Kim, B.; Peppas, N.A. Poly (ethylene glycol)-containing hydrogels for oral protein delivery applications. Biomed. *Microdevices* **2003**, *5*, 333–341. [CrossRef]
116. Richter, A.; Paschew, G.; Klatt, S.; Lienig, J.; Arndt, K.F.; Adler, H.J. Review on hydrogel-based pH sensors and microsensors. *Sensors* **2008**, *8*, 561–581. [CrossRef]
117. Cretu, A.; Gattin, R.; Brachais, L.; Barbier-Baudry, D. Synthesis and degradation of poly (2-hydroxyethyl methacrylate)-graft-poly (ε-caprolactone) copolymers. *Polym. Degrad. Stab.* **2004**, *83*, 399–404. [CrossRef]
118. Mondal, A.; Gebeyehu, A.; Miranda, M.; Bahadur, D.; Patel, N.; Ramakrishnan, S.; Rishi, A.K.; Singh, M. Characterization and printability of sodium alginate-gelatin hydrogel for bioprinting NSCLC co-culture. *Sci. Rep.* **2019**, *9*, 1–12. [CrossRef]
119. Raghavendra, G.M.; Jayaramudu, T.; Varaprasad, K.; Reddy, G.S.; Raju, K.M. Antibacterial nanocomposite hydrogels for superi-or biomedical applications: A Facile eco-friendly approach. *RSC Adv.* **2015**, *5*, 14351–14358. [CrossRef]
120. Ajji, Z.; Mirjalili, G.; Alkhatab, A.; Dada, H. Use of electron beam for the production of hydrogel dressings. *Radiat. Phys. Chem.* **2008**, *77*, 200–202. [CrossRef]
121. Caló, E.; Khutoryanskiy, V.V. Biomedical applications of hydrogels: A review of patents and commercial products. *Eur. Poly. J.* **2015**, *65*, 252–267. [CrossRef]
122. Leyden, J.J.; Krochmal, L.; Yaroshinsky, A. Two randomized, double-blind, controlled trials of 2219 subjects to compare the combination clindamycin/tretinoin hydrogel with each agent alone and vehicle for the treatment of acne vulgaris. *JAAD* **2006**, *54*, 73–81. [CrossRef]
123. Moftah, N.H.; Ibrahim, S.M.; Wahba, N.H. Intense pulsed light versus photodynamic therapy using liposomal methylene blue gel for the treatment of truncal acne vulgaris: A comparative randomized split body study. *Arch. Dermatol. Res.* **2016**, *308*, 263–268. [CrossRef]
124. Fabbrocini, G.; Staibano, S.; De Rosa, G.; Battimiello, V.; Fardella, N.; Ilardi, G.; La Rotonda, M.I.; Longobardi, A.; Mazzella, M.; Siano, M.; et al. Resveratrol-Containing Gel for the Treatment of Acne Vulgaris. *Am. J. Clin. Dermatol.* **2011**, *12*, 133–141. [CrossRef]
125. Dalwadi, C.; Patel, G. Application of nanohydrogels in drug delivery systems: Recent patents review. *Recent Pat. Nanotech.* **2015**, *9*, 17–25. [CrossRef]
126. Fernández-Ferreiro, A.; Bargiela, N.F.; Varela, M.S.; Martínez, M.G.; Pardo, M.; Ces, A.P.; Méndez, J.B.; Barcia, M.G.; Lamas, M.J.; Otero-Espinar, F.J. Cyclodextrin–polysaccharide-based, in situ-gelled system for ocular antifungal delivery. *Beilstein J. Org. Chem.* **2014**, *10*, 2903–2911. [CrossRef] [PubMed]
127. Tayel, S.A.; El-Nabarawi, M.A.; Tadros, M.I.; Abd-Elsalam, W.H. Promising ion-sensitive in situ ocular nanoemulsion gels of terbinafine hydrochloride: Design, in vitro characterization and in vivo estimation of the ocular irritation and drug pharmacokinetics in the aqueous humor of rabbits. *Int. J. Pharm.* **2013**, *443*, 293–305. [CrossRef] [PubMed]
128. Çelebi, N.; Ermiş, S.; Özkan, S. Development of topical hydrogels of terbinafine hydrochloride and evaluation of their anti-fungal activity. *Drug Dev. Pharm.* **2015**, *41*, 631–639. [CrossRef] [PubMed]

129. Díaz-Tomé, V.; Luaces-Rodríguez, A.; Silva-Rodríguez, J.; Blanco-Dorado, S.; García-Quintanilla, L.; Llovo-Taboada, J.; Blanco-Méndez, J.; García-Otero, X.; Varela-Fernández, R.; Herranz, M.; et al. Ophthalmic econazole hydrogels for the treatment of fungal keratitis. *J. Pharm. Sci.* **2018**, *107*, 1342–1351. [CrossRef]
130. Harish, N.M.; Prabhu, P.; Charyulu, R.N.; Gulzar, M.A.; Subrahmanyam, E.V. Formulation and evaluation of in situ gels con-taining clotrimazole for oral candidiasis. *Indian J. Pharm. Sci.* **2009**, *71*, 421. [CrossRef]
131. Paolicelli, P.; Petralito, S.; Varani, G.; Nardoni, M.; Pacelli, S.; Di Muzio, L.; Tirillò, J.; Bartuli, C.; Cesa, S.; Casadei, M.A.; et al. Effect of glycerol on the physical and mechanical properties of thin gellan gum films for oral drug delivery. *Int. J. Pharm.* **2018**, *547*, 226–234. [CrossRef]
132. Janga, K.Y.; Tatke, A.; Balguri, S.P.; Lamichanne, S.P.; Ibrahim, M.M.; Maria, D.N.; Jablonski, M.M.; Majumdar, S. Ion-sensitive in situ hydrogels of natamycin bilosomes for enhanced and prolonged ocular pharmacotherapy: In vitro permeability, cyto-toxicity and in vivo evaluation. *Artifi. Cells Nanomed. Biotechnol.* **2018**, *46*, 1039–1050. [CrossRef]
133. Serra, E.; Saubade, F.; Ligorio, C.; Whitehead, K.; Sloan, A.; Williams, D.W.; Hidalgo-Bastida, A.; Verran, J.; Malic, S. Methylcellu-lose hydrogel with Melissa officinalis essential oil as a potential treatment for oral candidiasis. *Microorganisms* **2020**, *8*, 215. [CrossRef]
134. Umadevi, A.; Kumari, C.; Kumar, P.A.; Am, H.S.; Divya, K.; Hisana, P.V. Development and Evaluation of Polyherbal GEL FOR Antifungal Activity. *Int. J. Curr. Pharm. Res.* **2018**, *10*, 40–43.
135. Khan, S.; Gowda, B.H.; Deveswaran, R.; Mishra, S.; Sharma, A. Comparison of fluconazole and tea tree oil hydrogels designed for oral candidiasis: An invitro study. In *AIP Conference Proceedings 2020*; AIP Publishing LLC: Melville, NY, USA, 2020; Volume 2274, p. 050004.
136. Aldawsari, H.M.; Badr-Eldin, S.M.; Labib, G.S.; El-Kamel, A.H. Design and formulation of a topical hydrogel integrating lemongrass-loaded nanosponges with an enhanced antifungal effect: In Vitro/In Vivo evaluation. *Int. J. Nanomed.* **2015**, *10*, 893.
137. Petrovska, B.B. Historical review of medicinal plants' usage. *Pharmacogn. Rev.* **2012**, *6*, 1. [CrossRef]
138. Natu, K.N.; Tatke, P.A. Essential oils—Prospective candidates for antifungal treatment? *J. Essent. Oil Res.* **2019**, *31*, 347–360. [CrossRef]
139. Fogarasi, M.; Socaci, S.A.; Fogarasi, S.; Jimborean, M.; Pop, C.; Tofană, M.; Rotar, A.; Tibulca, D.; Salagean, D.; Salanta, L. Evaluation of biochemical and microbiological changes occurring in fresh cheese with essential oils during storage time. *Stud. Univ. Babeș-Bolyai Chem.* **2019**, *64*, 527–537. [CrossRef]
140. Semeniuc, C.A.; Socaciu, M.-I.; Socaci, S.A.; Mureșan, V.; Fogarasi, M.; Rotar, A.M. Chemometric comparison and Classification of some essential oils extracted from plants belonging to apiaceae and lamiaceae families based on their chemical composition and biological activities. *Molecules* **2018**, *23*, 2261. [CrossRef]
141. Semeniuc, C.A.; Pop, C.; Rotar, A.M. Antibacterial activity and interactions of plant essential oil combinations against Gram-positive and Gram-negative bacteria. *J. Food Drug Anal.* **2017**, *25*, 403–408. [CrossRef]
142. Socaciu, M.-I.; Fogarasi, M.; Semeniuc, C.A.; Socaci, S.A.; Rotar, M.A.; Mureşan, V.; Pop, O.L.; Vodnar, D.C. Formulation and Characterization of Antimicrobial Edible Films Based on Whey Protein Isolate and Tarragon Essential Oil. *Polymer* **2020**, *12*, 1748. [CrossRef]
143. Fajinmi, O.O.; Kulkarni, M.G.; Benická, S.; Zeljković, S.Ć.; Doležal, K.; Tarkowski, P.; Finnie, J.F.; Van Staden, J. Antifungal activity of the volatiles of Agathosmabetulina and Coleonema album commercial essential oil and their effect on the morphology of fungal strains Trichophyton rubrum and T. mentagrophytes. *S. Afr. J. Bot.* **2019**, *122*, 492–497. [CrossRef]
144. Swamy, M.K.; Akhtar, M.S.; Sinniah, U.R. Antimicrobial Properties of Plant Essential Oils against Human Pathogens and Their Mode of Action: An Updated Review. *Evid. Based Complement. Altern. Med.* **2016**, *2016*, 1–21. [CrossRef]
145. Koroishi, A.M.; Foss, S.R.; Cortez, D.A.; Ueda-Nakamura, T.; Nakamura, C.V.; Dias Filho, B.P. In vitro antifungal activity of extracts and neolignans from Piper regnellii against dermatophytes. *J. Ethnopharmacol.* **2008**, *117*, 270–277. [CrossRef]
146. Manojlovic, N.T.; Solujic, S.; Sukdolak, S.; Milosev, M. Antifungal activity of Rubia tinctorum, Rhamnus frangula and Caloplacac-erina. *Fitoterapia* **2005**, *76*, 244–246. [CrossRef]
147. Yemele, B.M.; Krohn, K.; Hussain, H.; Dongo, E.; Schulz, B.; Hu, Q. Tithoniamarin and tithoniamide: A structurally unique iso-coumarin dimer and a new ceramide from Tithonia diversifolia. *Nat. Prod. Res.* **2006**, *20*, 842–849. [CrossRef] [PubMed]
148. Dabur, R.; Chhillar, A.K.; Yadav, V.; Kamal, P.K.; Gupta, J.; Sharma, G.L. In vitro antifungal activity of 2-(3, 4-dimethyl-2, 5-dihydro-1H-pyrrol-2-yl)-1-methylethyl pentanoate, a dihydropyrrole derivative. *J. Med. Microbiol.* **2005**, *54*, 549–552. [CrossRef] [PubMed]
149. Moghaddam, M.; Miran, S.N.; Pirbalouti, A.G.; Mehdizadeh, L.; Ghaderi, Y. Variation in essential oil composition and antioxi-dant activity of cumin (Cuminum cyminum L.) fruits during stages of maturity. *Ind. Crops Prod.* **2015**, *70*, 163–169. [CrossRef]
150. Akram, M.; Hussain, R. Nanohydrogels: History, development, and applications in drug delivery. In *Nanocellulose and Nanohydro-gel Matrices: Biotechnological and Biomedical Applications*; Wiley: Hoboken, NJ, USA, 2017; pp. 297–330.
151. Neamtu, I.; Rusu, A.G.; Diaconu, A.; Nita, L.E.; Chiriac, A.P. Basic concepts and recent advances in nanogels as carriers for medical applications. *Drug Deliv.* **2017**, *24*, 539–557. [CrossRef]
152. Lombardo, D.; Kiselev, M.A.; Caccamo, M.T. Smart Nanoparticles for Drug Delivery Application: Development of Versatile Nanocarrier Platforms in Biotechnology and Nanomedicine. *J. Nanomater.* **2019**, *2019*, 1–26. [CrossRef]
153. Zhang, X.; Malhotra, S.; Molina, M.; Haag, R. Micro-and nanogels with labile crosslinks–from synthesis to biomedical applications. *Chem. Soc. Rev.* **2015**, *44*, 1948–1973. [CrossRef]

154. Nakai, T.; Hirakura, T.; Sakurai, Y.; Shimoboji, T.; Ishigai, M.; Akiyoshi, K. Injectable Hydrogel for Sustained Protein Release by Salt-Induced Association of Hyaluronic Acid Nanogel. *Macromol. Biosci.* **2012**, *12*, 475–483. [CrossRef]
155. Mangalathillam, S.; Rejinold, N.S.; Nair, A.; Lakshmanan, V.K.; Nair, S.V.; Jayakumar, R. Curcumin loaded chitin nanogels for skin cancer treatment via the transdermal route. *Nanoscale* **2012**, *4*, 239–250. [CrossRef]
156. Kabanov, A.V.; Vinogradov, S.V. ChemInform Abstract: Nanogels as Pharmaceutical Carriers: Finite Networks of Infinite Capabilities. *Chemin* **2009**, *40*, 5418. [CrossRef]
157. Herbal Nanogel Formulation: A Novel Approch. *J. Sci. Technol.* **2020**, *5*, 138–146. [CrossRef]
158. Missirlis, D.; Kawamura, R.; Tirelli, N.; Hubbell, J.A. Doxorubicin encapsulation and diffusional release from stable, polymer-ic, hydrogel nanoparticles. *Eur. J. Pharm. Sci.* **2006**, *29*, 120–129. [CrossRef]
159. Wang, C.; Mallela, J.; Garapati, U.S.; Ravi, S.; Chinnasamy, V.; Girard, Y.; Howell, M.; Mohapatra, S. A chitosan-modified gra-phene nanogel for noninvasive controlled drug release. *Nanomed. Nanotechnol. Biol. Med.* **2013**, *9*, 903–911. [CrossRef]
160. Nazzaro, F.; Fratianni, F.; Coppola, R.; Feo, V.D. Essential oils and antifungal activity. *Pharmaceuticals* **2017**, *10*, 86. [CrossRef]
161. Hussein, A.; Abdel-Mottaleb, M.M.; El-assal, M.; Sammour, O. Novel biocompatible essential oil-based lipid nanocapsules with antifungal properties. *J. Drug Deliv. Sci. Technol.* **2020**, *56*, 101605. [CrossRef]
162. Wang, N.X.; von Recum, H.A. Affinity-based drug delivery. *Macromol. Biosci.* **2011**, *11*, 321–332. [CrossRef]
163. Vinogradov, S.V.; Bronich, T.K.; Kabanov, A.V. Nanosized cationic hydrogels for drug delivery: Preparation, properties and interactions with cells. *Adv. Drug Deliv. Rev.* **2002**, *54*, 135–147. [CrossRef]

Review

Bioactive Compounds for Skin Health: A Review

Monika Michalak [1,*], **Monika Pierzak** [2], **Beata Kręcisz** [1] **and Edyta Suliga** [2]

[1] Department of Dermatology, Cosmetology and Aesthetic Surgery, The Institute of Medical Sciences, Medical College, Jan Kochanowski University, Żeromskiego 5, 25-369 Kielce, Poland; beata.krecisz@ujk.edu.pl

[2] Department of Nutrition and Dietetics, The Institute of Health Sciences, Medical College, Jan Kochanowski University, Żeromskiego 5, 25-369 Kielce, Poland; monika.pierzak@ujk.edu.pl (M.P.); edyta.suliga@ujk.edu.pl (E.S.)

* Correspondence: monika.michalak@ujk.edu.pl; Tel.: +48-41-349-69-70

Abstract: Human skin is continually changing. The condition of the skin largely depends on the individual's overall state of health. A balanced diet plays an important role in the proper functioning of the human body, including the skin. The present study draws attention to bioactive substances, i.e., vitamins, minerals, fatty acids, polyphenols, and carotenoids, with a particular focus on their effects on the condition of the skin. The aim of the study was to review the literature on the effects of bioactive substances on skin parameters such as elasticity, firmness, wrinkles, senile dryness, hydration and color, and to define their role in the process of skin ageing.

Keywords: skin care; skin health; bioactive substances; phytonutrients; antioxidants; nutraceuticals

Citation: Michalak, M.; Pierzak, M.; Kręcisz, B.; Suliga, E. Bioactive Compounds for Skin Health: A Review. *Nutrients* **2021**, *14*, 203. https://doi.org/10.3390/nu13010203

Received: 11 December 2020
Accepted: 9 January 2021
Published: 12 January 2021

Publisher's Note: MDPI stays neutral with regard to jurisdictional claims in published maps and institutional affiliations.

1. Introduction

The skin is the largest organ of the human body. It is composed of the epidermis, which consists of epithelial tissue, and the dermis, which consists of connective tissue. Under the dermis, there is a layer of subcutaneous tissue called the hypodermis (Figure 1). The epidermis comprises a horny layer (stratum corneum), a clear layer (stratum lucidum), a granular layer (stratum granulosum), a spinous layer (stratum spinosum) and a basal layer (stratum basale). Apart from keratinocytes—cells involved in keratinization—the five-layer epidermis also contains pigment cells and melanocytes, as well as Langerhans cells, mastocytes, and Merkel cells. It is closely connected to the dermis underneath by the basement membrane. The dermis, which comprises a papillary layer (primarily loose connective tissue) and a reticular layer (dense connective tissue), contains fibroblasts responsible for the production of collagen, elastin, and glycosaminoglycans (GAGs), as well as numerous blood vessels, nerve endings, and appendages, such as hair follicles and sweat and sebaceous glands. The subcutaneous tissue consists of loose connective tissue containing fat cells (adipocytes) forming fat lobules [1–3].

The skin performs a wide variety of complex functions. It provides an optimal environment for deeper tissues by separating them from the external environment and, at the same time, ensures contact with it through the exchange of substances and reception of stimuli. The skin protects against biological agents (potentially pathogenic microbes), chemical agents (corrosive, irritating and allergenic substances) and physical factors (sunlight, ionizing radiation, infrared radiation, and mechanical and thermal factors). It performs important functions in water and electrolyte balance (epidermal barrier and sweat glands), thermoregulation (thermoreceptors) and the immune response (skin-associated lymphoid tissues (SALT)). It is also an important sensory organ (with free nerve endings, Pacinian corpuscles, Meissner corpuscles, Ruffini corpuscles, Krause end bulbs and Merkel cells). In addition, it is involved in metabolism and homeostasis and is responsible for the elimination, selective absorption and storage of substances [1,4].

One of the most common dermatological and cosmetic concerns is skin ageing: a natural, complex process influenced by two mechanisms—intrinsic (genetic, chronological) ageing resulting from the passage of time, and extrinsic ageing (photoaging), caused by environmental factors (including UV radiation, environmental pollution and cigarette smoke) [5–8]. The two processes overlap and are closely linked to increased reactive oxygen species (ROS) and oxidative stress in the skin [9]. Both the intrinsic and extrinsic processes are associated with biochemical disturbances (e.g., the excessive formation of oxygen radicals, leading to protein and DNA damage, amino acid racemization, and non-enzymatic glycosylation, leading to the abnormal cross-linking of collagen fibers and other structural proteins), as well as changes in the physical, morphological and physiological properties of the epidermis and dermis. These include disturbances in the function of the epidermal barrier, the flattening of the dermal–epidermal junction, a reduced number and activity of fibroblasts, the accumulation of abnormal elastin fibers (elastosis), and the impaired functioning of Langerhans cells [6,10–13]. Features characteristic of mature skin include wrinkles, a loss of elasticity, changes in color, uneven pigmentation and discoloring, dryness, foci of abnormal epidermal keratosis, telangiectasias, susceptibility to irritation, and slower skin regeneration and healing [5,8]. With age, the degradation of blood vessels leads to insufficient blood supply and, thus, the inadequate oxygenation and nourishment of the skin [8].

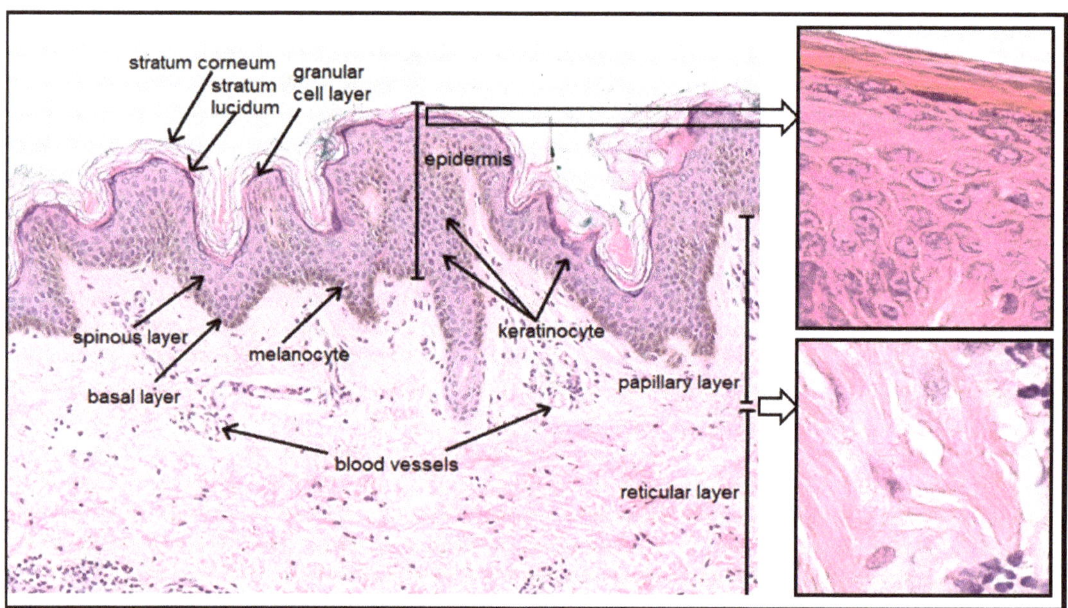

Figure 1. Skin structure (own work, photo: Department of Clinical and Experimental Pathology, Medical College, Jan Kochanowski University).

Many authors stress the relationship between a suitably balanced diet and the condition of the human body, including the appearance and functioning of the skin [14–17] (Table 1). The intake of essential nutrients in the daily diet is extremely important for the biological processes taking place in both young and ageing skin [15,18]. The skin is a tissue with high proliferative potential, which is why an adequate intake of proteins, carbohydrates and fats, which are essential for cellular generation, is so important [18–20]. The overall condition of the skin—its surface texture, color, and physiological properties—results from factors such as hydration, i.e., the presence of an adequate amount of water in the stratum corneum, sebum content and surface acidity. Natural moisturizing factor (NMF), consisting mainly of amino acids, plays an important role in hydration and acidity [15,21]. Specific fatty acids are also important for maintaining the function of the skin barrier and the integrity of the stratum corneum [22]. Research increasingly suggests that a well-balanced diet significantly affects the skin ageing process. Functional anti-ageing ingredients in food include substances involved in the synthesis and metabolism of skin components (e.g., protein peptides and essential fatty acids) and those that inhibit the degradation of skin components and maintain its structural integrity (e.g., substances regulating the expression of enzymes such as matrix metalloproteinases (MMPs) and activating protein 1 (AP-1)) [18]. Due to their ability to protect the skin from harmful UV-induced effects through their antimutagenic, antioxidant and free-radical-scavenging properties, some dietary botanicals could be useful supplements for mature skin care [23,24]. These include carotenoids and polyphenols, such as apigenin (a flavonoid occurring in numerous herbs, fruits and vegetables), quercetin (a flavonol found in onion skin and apple peel), curcumin (obtained from the turmeric rhizome), silymarin (a standardized extract of flavonolignans from milk thistle), genistein (an isoflavone from soybeans), proanthocyanidins (from the seeds of grapes), and resveratrol (a polyphenol found in grapes, peanuts, fruits, red wine and mulberries). Other key elements of an anti-ageing diet are vitamins and minerals with antioxidant properties [15,25].

The present study is a review of the literature on the effects of nutrition on the condition of the skin, as well as an attempt to establish whether a healthy diet supplemented with vitamins and minerals has photoprotective and anti-ageing effects. The aim of the study was to present the effects of bioactives on skin parameters such as a wrinkled appearance, elasticity, firmness, senile dryness, hydration, and color, as well as their role in the skin ageing process.

2. Selected Substances of Importance for Skin Function

The present study focused on substances that combat skin ageing and contribute to a younger-looking face with a healthy appearance; nutrients and dietary supplements reported to have the potential to prevent skin ageing are discussed. Particular focus is placed on substances that protect the skin against oxidative and UV-induced damage, dehydration and a loss of elasticity: vitamins A, C and E; selenium; zinc; copper; silicon; polyphenols; carotenoids; and essential polyunsaturated omega-3 and omega-6 fatty acids.

2.1. Vitamins

Vitamins are a group of compounds with diverse chemical structures that are essential for the normal functioning of the human body. They do not play an active role in supplying energy and are not building blocks for tissues, but they are essential for normal growth and development. They are biological catalysts and building blocks for the prosthetic groups of various enzymes, and thus, they enable numerous biochemical reactions at various levels [26].

2.1.1. Vitamin A

Vitamin A is one of the fat-soluble vitamins. The name refers to polyene compounds comprising a beta-ionone ring and a polyene side chain containing a functional group: an alcohol group, retinol; an aldehyde group, retinal; an acid group, retinoic acid; or an ester group, retinyl ester. Derivatives of vitamin A, called retinoids, include both natural and many synthetic retinol derivatives with activity similar to that of vitamin A [27–31] (Figure 2).

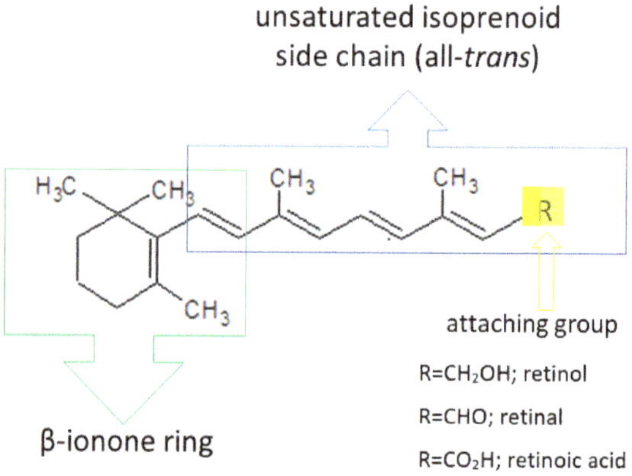

Figure 2. Basic structure of retinoids (own work based on [32]).

Vitamin A is obtained from the diet. Retinol and its various chemical forms are derived from foods of animal and plant origin (Table 1). Vitamin A from animal sources is mainly acquired in the form of retinyl esters (palmitate, propionate and acetate). Plants contain provitamin A (β-carotene), which is converted to vitamin A in the skin. Beta-carotene, one of the most important carotenoids, is a precursor of vitamin A in the human body [27,29,31,33,34] (Scheme 1).

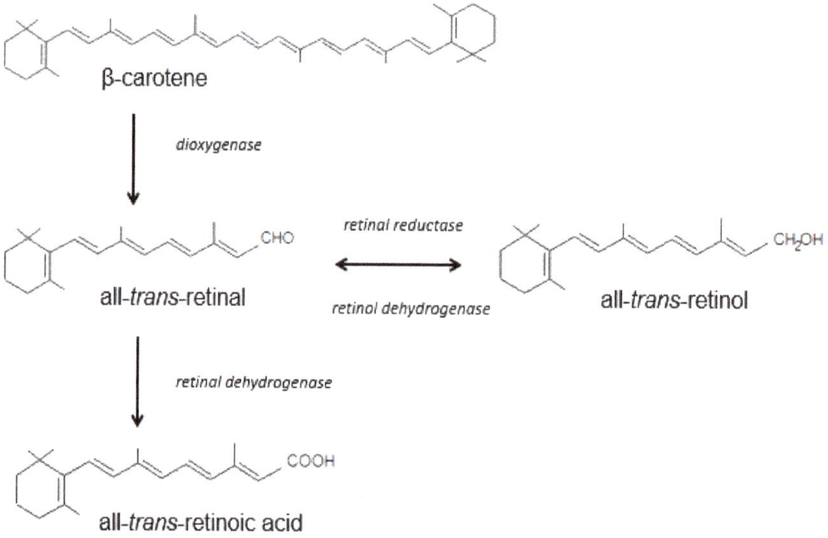

Scheme 1. Metabolism of retinoids (own work based on [34,35]).

Beta-carotene is metabolized in the small intestine, leading to the formation of the aldehyde retinal, which can be reduced to retinol through enzymatic conversion. This process is reversible, and the alcohol can be converted as needed into retinal in an oxidation process catalyzed by retinol dehydrogenase. Further oxidation is irreversible and produces tretinoin. Tretinoin is thus the main metabolite of retinol, but it can be synthesized in certain tissues (e.g., the epidermal cells) without retinol, from retinal generated via the degradation of β-carotene. Retinyl esters in the enterocytes are converted to retinol, after which they are modified again and stored as retinyl esters in chylomicrons, before being released into the blood and supplied to cells and tissues associated with their carrier protein. The transport and activity of retinoids require specific proteins occurring in both the plasma (RBP—retinoid-binding proteins) and the cytoplasm (CRBP—cellular retinoid-binding proteins). In cell nuclei, retinoic acid receptors (RARs) and retinoid X receptors (RXRs, for all retinoids) have been detected [27,29,31,33,34]. In the blood and tissues, the two dominant endogenous retinoids are retinol and its esters. Retinol and retinyl esters account for more than 99% of all the retinoids present in the skin [36].

Vitamin A and all of its derivatives, applied systemically or topically, have a significant influence on skin health. Retinoids are a group of chemical compounds that inhibit cell division during excessive proliferation and activate it when the process is too slow. They play an active role in protein production, cellular metabolism and cell division [28,34,37]. Retinoids also affect the thickness and color of the skin, regulate the function of the sebaceous glands, and limit sebum production, as well as being responsible for hair and nail growth, and also influencing the distribution of melanin in the skin [27,30,35].

The anti-ageing effects of retinoids are mediated by three major types of skin cells: epidermal keratinocytes, dermal endothelial cells and fibroblasts [38]. Topically applied retinoids penetrate the keratinized epidermis and, to a lesser extent, the dermis and subcutaneous tissue [27,30,35]. A study in mice by Törmä et al. showed that retinol esters accounted for 90% of the vitamin A in the epidermis, with retinol making up the remaining 10% [39]. After penetrating the epidermis, retinoids play a role in the release of transcription and growth factors. They are responsible for the renewal of the epidermis through their active role in the exfoliation of dead cells in the stratum corneum and the proliferation of live cells in the epidermal layers [28,34,35]. Moreover, vitamin A promotes the compaction of the stratum corneum and deposition of GAGs in the stratum corneum and intercellular spaces in the epidermis [40]. A study conducted by Kang et al. showed that the application of all-trans-retinol on normal human skin induced epidermal thickening and increased the mRNA expression of cellular retinoic acid binding protein and cellular retinol binding protein [41]. Retinoids also reduce transepidermal water loss (TEWL) by strengthening the epidermal barrier [28,34,35]. After penetrating the dermis, retinoids increase the synthesis of elastin and collagen by stimulating the fibroblasts [42]. Topical retinol improves the dermal extracellular matrix (ECM) microenvironment (type I collagen, fibronectin, and elastin) and activates ECM-producing cells in aged skin [32,38]. The activation of fibroblast production is stimulated through the TGF-β/CTGF pathway, the major regulator of ECM homeostasis [38]. Retinoids and all biologically active forms of vitamin A also promote the remodeling of reticulin fibers and synthesis of a new capillary network in the dermis. This process influences the condition of the connective tissue of the dermis, improving its firmness, hydration and elasticity [35,42]. Retinoids protect collagen against the destruction induced by MMPs, as well as enhancing the synthesis of tissue inhibitors of metalloproteinases (TIMPs) [32,42]. Varani et al. observed that topical vitamin A (retinol) reduced MMP expression and increased fibroblast growth and collagen synthesis in naturally aged, as it did in photoaged, skin [42]. Pierard-Franchimont et al. showed that a retinol formulation improved the physical properties of ageing facial skin, including tensile properties and contours, after 12 weeks of treatment [43].

Retinol and retinyl esters exhibit a strong capacity to absorb ultraviolet radiation in the range of 300–350 nm [39]. Antille et al. conducted a study to assess the ability of retinyl palmitate to protect the skin, prevent DNA damage and protect against the induction of erythema by a single exposure to strong UVB radiation. The results clearly indicate that retinyl esters present in the epidermis have strong antiphotocarcinogenic properties and protect DNA against damage by UV radiation [44].

Topically applied retinoids reduce the discoloration of the skin and its pigmentation by about 60%. Retinoids influence the function of melanocytes and the distribution of melanin in the skin, as well as blocking the transport of melanin to epidermal cells. A decrease in the number of melanocytes is related to the inhibition of melanogenesis [32].

Retinoids have been shown to improve the clinical features of aged appearance by a reduction in fine wrinkling, increased smoothness and decreased hyperpigmentation [45]. Vitamin A and its biologically active forms, retinoids, are used in the care of ageing skin and to eliminate signs of photoaging, but also in the treatment of numerous dermatological conditions, which may be seborrheic (e.g., acne vulgaris and rosacea), viral (flat warts, molluscum contagiosum or genital warts induced by the human papilloma virus (HPV)), proliferative, cancerous or precancerous (keratoacanthoma, cutaneous T-cell lymphoma, leukoplakia of the oral mucosa, actinic keratosis and xeroderma pigmentosum), autoimmune (lupus or lichen sclerosus), or papulosquamous dermatoses (psoriasis, pityriasis rubra pilaris, or lichen planus), as well as genodermatoses with keratosis disorders (congenital and hereditary ichthyosis or Darier disease) [27,28,44].

2.1.2. Vitamin C

Vitamin C (L-ascorbic acid) is a highly water-soluble, sugar-like alpha-ketolactone produced from D-glucose (Figure 3).

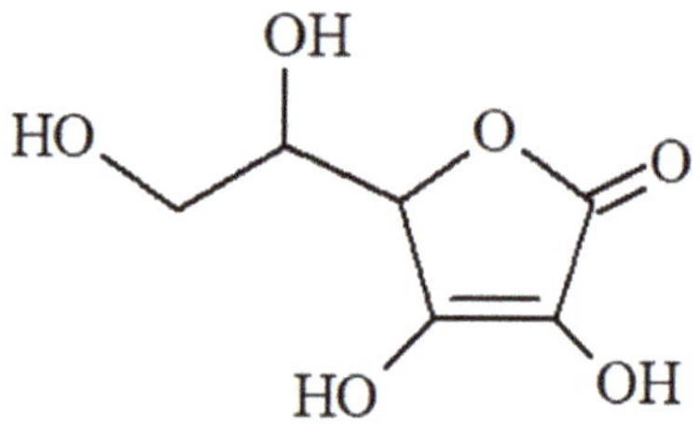

Figure 3. Chemical structure of ascorbic acid (own work).

As humans lack L-gulonolactone oxidase, the enzyme enabling its production, this vitamin must be supplied in the diet (Table 1). About 70–80% of it is absorbed in the duodenum and the proximal part of the small intestine. The absorption and bioavailability of vitamin C are affected by the current metabolic state of the body, as well as by age and sex. Digestive and absorption disorders, vomiting, diarrhea, smoking, and certain medicines, such as aspirin, significantly affect its concentration in the body [46–48]. Reserves of vitamin C are stored in organs with high metabolic activity, such as the liver, pancreas, lungs, brain and adrenal glands [49]. The vitamin C concentration in the skin is higher than in other human tissues, at 6–64 mg/100 g wet weight in the epidermis and 3–13 mg/100 g in the dermis [50–52]. There are two transport mechanisms for ascorbic acid, and these depend on the sodium–ascorbate cotransporters (SVCTs) present in various tissues and organs. In the skin, sodium–ascorbate cotransporter-1 (SVCT1) is responsible for the transport of epidermal vitamin C to the keratinocytes, while sodium–ascorbate cotransporter-2 (SVCT2) is responsible for intradermal transport [53,54].

Vitamin C promotes the formation of the epidermal barrier and collagen in the dermis, protects against skin oxidation, helps to counteract skin ageing, and plays a role in the signaling pathways of cell growth and differentiation, which are linked to the occurrence of various skin diseases [55,56].

The role of vitamin C in differentiating keratinocytes is being researched. The vitamin has been shown to enhance keratinocyte differentiation, reduce differentiation-dependent oxidative stress and maintain the integrity of the skin barrier, which in turn helps to prevent water loss from the skin [55,57]. Pasonen-Seppanen et al. showed that vitamin C enhanced the differentiation of rat epidermal keratinocytes cell line in an organotypic culture model, resulting in a stronger stratum corneum, which indicates that it has a skin-protective function [58]. Savini et al., in an in vitro study on normal human epidermal keratinocytes, showed that ascorbate improves the barrier functions of the epidermis and promotes the formation of the cornified cell envelope, a specialized structure that protects the body against harmful chemical and physical factors [57]. Research by Cosgrove et al. indicates that a higher vitamin C intake is associated with a reduced likelihood of a wrinkled appearance and senile dryness [59].

The biological function of vitamin C in the skin is its active role in collagen synthesis. It is responsible for the biosynthesis of collagen through its role in the hydroxylation of proline and lysine residues to hydroxyproline and hydroxylysine. It donates electrons to enzymes involved in hydroxylation, resulting in the conversion of procollagen to collagen [46,49]. It functions as a cofactor of proline and lysine hydroxylases, which are responsible for the tertiary structure of collagen; promotes the expression of collagen genes; and activates the production of collagen mRNA by fibroblasts [50]. Vitamin C exhibits anti-wrinkle activity. There is scientific evidence that ascorbic acid plays an active role in the proliferation and migration of skin fibroblasts and stimulates the production of collagen and elastin in cultures of fibroblasts [46,49]. Supplementation with vitamin C enhances the production of GAGs, promotes the expression of genes coding for antioxidant enzymes and enzymes involved in DNA repair, and inhibits the production of pro-inflammatory cytokines and apoptosis induced by UV radiation or other harmful environmental factors [49].

The antioxidant properties of vitamin C protect the skin, especially the epidermis, against oxidants generated by ultraviolet radiation and other environmental factors. Although the direct antioxidant protection provided by vitamin C is limited to aqueous compartments, vitamin C significantly curtails lipid oxidation by regenerating fat-soluble vitamin E; it regenerates α-tocopherol from α-tocopherol radicals found in cell membranes and lipoproteins [60]. Topical vitamin C supplementation can also counteract UVA-induced oxidative stress [61]; Offord et al. assessed the photoprotective potential of vitamin C and other antioxidants, confirming that vitamin C plays a significant role in protecting the skin against UVA radiation [62]. The photoprotective efficacy of vitamin C is enhanced by combining it with vitamin E [63]. Placzek et al., in a study with human volunteers, showed that the oral administration of ascorbic acid and D-alpha-tocopherol for three months significantly reduced the sunburn reaction to UVB irradiation. The results of that study also suggest that supplementation with a combination of the two antioxidants protects against DNA damage [64]. Eberlein-König and Ring have also shown that vitamin C used in combination with vitamin E and, to a lesser extent, with other photoprotective compounds dramatically increases their photoprotective effects as compared to monotherapies [65].

Vitamin C is also thought to be involved in inhibiting melanogenesis. Melanin, a natural skin pigment resulting from complex biochemical conversions in melanocytes catalysed by tyrosinase, plays an important role in photoprotection. The overproduction of melanin, however, may contribute to hyperpigmented diseases as well as the initiation of melanomas [66,67]. A review of the literature indicates that vitamin C inhibits melanogenesis with low cytotoxicity, although some studies suggest that the role of vitamin C in inhibiting melanogenesis is very minor [67–69]. Furthermore, a combination of vitamin C and other vitamins, including vitamin E, is more effective at reducing melanin than vitamin C alone and can be used for treating skin discoloration, e.g., age spots or melasma [50,68].

Numerous scientific papers emphasize the potential of vitamin C for the treatment of such diseases as porphyria cutanea tarda, atopic dermatitis, malignant melanoma, herpes zoster and postherpetic neuralgia, as well as the clinical application of this vitamin in the treatment of other skin conditions, such as acne, acne scars, allergic contact dermatitis, psoriatic progressive pigmented purpuric dermatosis (PPPD), genital herpes and vitiligo [56,70–75].

2.1.3. Vitamin E

Vitamin E is a group of lipophilic compounds that includes four tocopherols (α-, β-, γ- and δ-tocopherol) and four tocotrienols (α-, β-, γ- and δ-tocotrienol) [76,77]. The chemical structures of all the compounds referred to as vitamin E have a two-ring 6-hydroxychroman skeleton and a side chain composed of three isoprenoid units. Tocopherols have a chromanol ring and a saturated phytyl tail, while the chemical structures of tocotrienols have an unsaturated tail [76–78] (Figure 4).

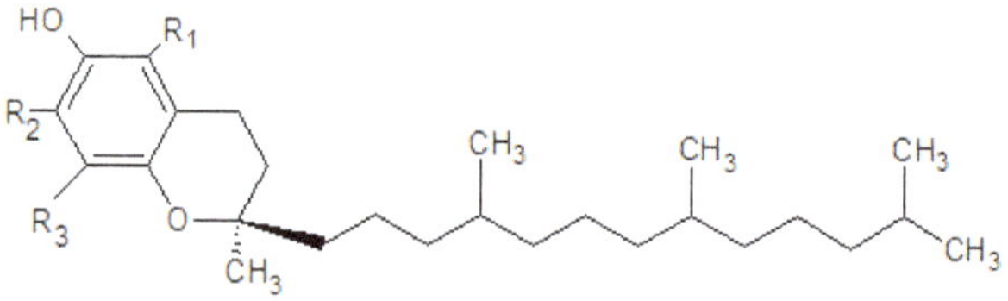

(a)

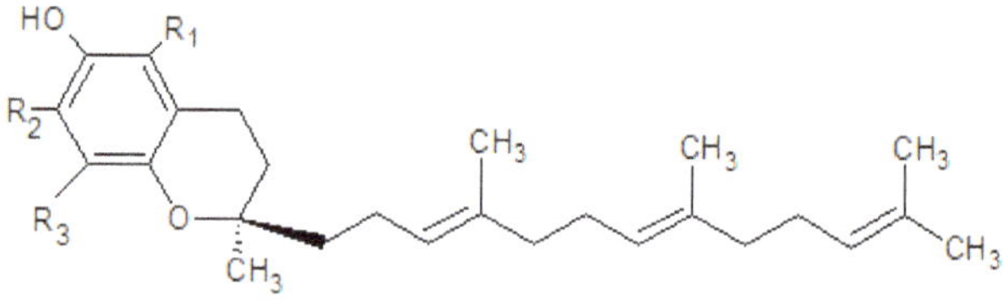

(b)

Figure 4. Basic structure of (**a**) tocopherols and (**b**) tocotrienols (own work).

Vitamin E is exclusively synthesized by plants [76], and all forms are supplied to the human body by food (Table 1). Alpha-tocopherol is the most important form, showing an affinity for the specialized protein alpha-TTP, which binds and transports this form of the vitamin only. The remaining dietary forms are metabolized in the liver and eliminated from the body with bile [79,80].

Vitamin E plays an important role in maintaining skin health, and it has been used for over 50 years in dermatology. The most probable physiologic function of epidermal vitamin E is contributing to the antioxidant defences of the skin and protecting the epidermis and dermis against oxidative stress induced by environmental factors. Vitamin E is the major lipid-soluble antioxidant in humans [40]. Owing to the antioxidant properties of vitamin E and its ability to scavenge free radicals and become part of lipid structures, it protects against lipid peroxidation and slows skin ageing [80]. Alpha-tocopherol has been shown to decrease the amount of 8-hydroxydeoxyguanosine produced indirectly by reactive oxygen species, so it may reduce ROS-induced DNA damage and thereby help to retard the development of skin cancer [81].

Alpha-tocopherol supplementation has been shown to improve facial hyperpigmentation. Ichihashi et al., in an in vitro study using cultured human melanoma cells and normal human melanocytes, showed that alpha-tocopherol inhibits tyrosine hydroxylase activity and suppresses melanogenesis [81]. Thus, vitamin E may be a candidate whitening agent for the treatment of hyperpigmentation, including age-related conditions or those arising from sun exposure.

Some research also indicates that vitamin E displays strong photoprotective, firming, hydrating and anti-ageing properties, as well as improving the elasticity, structure and softness of the epidermis and dermis [40,82]. It is thought that vitamin E, incorporated into the intercellular cement and lipid structures, protects the skin against solar UVB radiation and, thus, against redness and swelling. Although the topical application of alpha-tocopherol shows promising photoprotective effects, especially when combined with systemic and topical antioxidant substances such as vitamin C or carotenoids, controlled studies in humans are needed before vitamin E can be recommended as an effective cosmeceutical anti-ageing agent [40,83].

The available literature concerning the efficacy of the systemic and topical use of vitamin E is extensive, but the results are often contradictory and range from an improvement in skin appearance to no effect at all [40]. Vitamin E plays a role in the healing of wounds of varying aetiology; in the treatment of dermatological conditions such as subcorneal pustular dermatoses, cutaneous amyloidosis, atopic dermatitis, epidermolysis bullosa, psoriasis, acne vulgaris and scleroderma; in skin cancer prevention; and in the treatment of Hailey–Hailey disease [82]. Tsoureli-Nikita et al. conducted a study to evaluate the effect of oral vitamin E supplementation on the improvement and remission of atopic dermatitis. The results, which indicated relationships between vitamin E, the IgE level and the clinical symptoms of atopy, suggest that vitamin E can be used to treat atopic dermatitis [84]. Butt et al. evaluated the usefulness of vitamin E in protecting skin against thermal trauma. The results suggest that the clinical transplantation of keratinocytes preconditioned with vitamin E, alone or in combination with skin fibroblasts in skin substitutes, could be used to treat thermally damaged skin [85]. There are conditions for which vitamin E has no confirmed effects, including keratosis follicularis, chronic cutaneous lupus erythematosus, pseudoxanthoma elasticum, and porphyria cutanea tarda [82]. Other studies have also failed to confirm that the topical application of vitamin E improves the cosmetic appearance of scars [86].

2.2. Minerals

Minerals, alongside vitamins, are essential micronutrients that cannot be synthesized in humans and therefore must be obtained via the diet. A healthy diet should ensure an adequate intake of both macroelements (calcium, phosphorus, potassium, sodium and magnesium) and microelements (iodine, sulphur, zinc, iron, chlorine, cobalt, copper, manganese, molybdenum and selenium). Minerals are responsible for the functioning of the skeletal, circulatory, nervous and endocrine systems. They also have numerous health benefits as cofactors and coenzymes in various enzyme systems, aiding the regulation and coordination of biochemical and physiological functions. A deficiency of microelements has an adverse effect on human development and health, including the functioning and appearance of the skin [26,87]. In the context of skin ageing, attention is also paid to the role of minerals such as selenium, zinc, copper and silicon.

2.2.1. Selenium

Selenium (Se) is one of the most important trace elements without which the body cannot function properly. In terms of its chemical structure, selenium is one of the chalcogens. It occurs in two forms: inorganic (selenates and selenites) and organic (selenomethionine and selenocysteine) [88]. Selenium should be supplied to the human body with foods of plant and animal origin [26,88] (Table 1).

Selenium plays a role as part of the structure of enzymatic proteins. It is also an element of three families of enzymes: glutathione peroxidases (GPx), thioredoxin reductases (TrxR) and iodothyronionic deionidases (DIOs). Twelve single selenoproteins occur in the human body, including selenoprotein P, responsible for controlling redox potential in cells and the transport of selenium to peripheral tissues. Selenium exhibits strong antioxidant properties and protects against DNA damage [88,89].

The mineral is also important for skin function, having a protective effect and the ability to scavenge free radicals [89]. By stimulating the activity of selenium-dependent antioxidant enzymes, such as glutathione peroxidase and thioredoxin reductase, selenium protects the skin against the oxidative stress induced by UV radiation [87]. Zhu et al. draw attention to the role of selenium in protecting skin cells against the ageing induced by UVB radiation [90]. This was confirmed in a study by Jobeili et al., which showed that selenium delays skin ageing by protecting keratinocyte stem cells [91]. Favrot et al. showed that low doses of Se (30 nM) provide potent protection against UVA-induced cytotoxicity in young keratinocytes (from 20–30-year-old donors), while higher concentrations (240 nM) were required for protective efficacy in old keratinocytes (from donors 60–70 years old) [92].

Selenium supplementation is an important strategy for inhibiting wrinkles, because it can reverse ultraviolet light damage and play an anti-ageing role in the skin [89,92]. Kim et al. showed that a selenium-rich tuna heart extract enhanced collagen synthesis and promoted the proliferation of skin fibroblasts, thus having anti-ageing and anti-wrinkle effects [93]. These studies indicate that Se supplementation could be a new strategy for combating skin ageing and photoaging.

Due to the anti-inflammatory and antioxidant properties of selenium, some research has focused on its role in skin cancer prevention. Whether selenium can prevent skin cancer or not is controversial. Research by Pols et al. indicates that a high serum concentration of selenium is associated with an approximately 60% decrease in the incidence of basal cell carcinoma (BCC) and squamous cell carcinoma (SCC) of the skin [94]. However, some researchers have demonstrated that supplementation with Se does not significantly affect the incidence of BCC or is ineffective in preventing it [95,96].

2.2.2. Zinc

Zinc (Zn) is an essential microelement with an important role in many physiological processes [97]. A deficiency can cause a number of health problems, so it is essential to ensure an adequate intake, mainly through a balanced diet containing both animal and plant products, but also by addressing factors that can impede the absorption of this element. Zinc can be supplied in food products (Table 1) or in the form of dietary supplements [26,97,98]. The element has been shown to be present in all human tissues and body fluids. The average content of this element in an adult is 2–3 g [97,99]. Zinc is present in the muscles, bones, skin and liver, as well as in the brain, kidneys, spleen and pancreas [97,99–101]. The skin is the third most Zn-abundant tissue in the body, with greater amounts in the epidermis than in the dermis [99,102]. In the epidermis, zinc is more abundant in the stratum spinosum than in the other layers of keratinocytes, while in the dermis, the zinc concentration is higher in the upper parts than in the deeper areas [102,103].

Zinc performs catalytic, structural and regulatory functions [97,98], being important for the regulation of lipid, protein and nucleic acid metabolism, as well as gene transcription [98]. The molecular mechanisms of action of zinc are linked to its involvement in the structure and function of over 300 enzymes, including oxidoreductases, transferases, hydrolases, lyases, isomerases and ligases [97,104]. Zinc is also a component of numerous DNA transcription factors (zinc fingers) [100,105].

Zinc ions are involved in maintaining the balance between oxidation and reduction. As an antioxidant, zinc protects the cells of the body against the harmful effects of free radicals. An adequate level of this element is required for the activity of antioxidant enzymes, including cooper zinc superoxide dismutase (Cu/Zn-SOD), which is responsible for neutralizing the superoxide radical [97].

Zn is essential for the division and differentiation of new cells, and it plays a role in apoptosis and the ageing of the body [99,100]. The element influences the structure and proper functioning of the skin and mucous membranes [98,100]. It stabilizes skin cell membranes and participates in basal cell mitosis and differentiation [106], while also playing crucial roles in the survival of keratinocytes. The ZIP2 protein, a zinc transporter, is essential for the differentiation of keratinocytes [103].

Zinc also affects the immune function of the skin, modulating macrophage and neutrophil functions, phagocytic activity and various inflammatory cytokines [98,99]. It plays an important role in preventing the damage induced by UV radiation, influences collagen metabolism, exhibits antiandrogen properties by modulating the activity of type 1 and 2 5α-reductase, and also promotes lipogenesis and glucose transport through insulin-like effects on 3T3-L1 fibroblasts and adipocytes [98,107,108]. The oral and/or topical administration of Zn has long been used for its effect on skin regeneration and healing [99,109]. Topical preparations such as calamine or zinc pyrithione have been used as soothing agents or active ingredients in anti-dandruff shampoos [98]. Zinc preparations in the form of pastes or ointments cleanse the skin of excess sebum, restore its natural pH, and have astringent, anti-inflammatory and anti-acne effects. Zinc oxide, due to its ability to reflect and disperse UV rays, is used as a physical filter in sunscreens [97–99,107,110].

Zinc is used to treat numerous dermatological conditions, such as infections (e.g., warts), inflammatory dermatoses (acne vulgaris, rosacea, atopic dermatitis and alopecia areata) and pigmentary disorders (melasma) [97–99,107,110]. Zinc administered orally or topically has been shown to have therapeutic applications in skin ageing (a 0.1% copper–zinc malonate cream applied topically for 6 weeks significantly reduced wrinkles) [111]; melasma (a 10% zinc sulphate solution applied topically twice daily for 2 months significantly reduced MASI scores) [112]; actinic keratoses (a 25% zinc sulphate solution applied topically twice daily for 12 weeks was safe and effective, especially in patients with multiple actinic keratosis lesions) [113]; xeroderma pigmentosum (a 20% topical application of a zinc sulphate solution for 4 months to 2 years improved all types of skin lesions, softened the skin, lightened the skin color, and cleared the skin of solar keratosis and small malignancies) [114]; eczema (a 0.05% Clobetasol + 2.5% zinc sulphate cream applied topically was effective in hand eczemas) [115]; rosacea (100 mg of oral zinc sulphate three times per day was effective after 3 months of therapy) [116]; and alopecia areata (5 mg/kg/day, in three divided doses, of oral zinc sulphate induced significant hair growth after 6 months) [117].

2.2.3. Copper

Copper (Cu) is naturally found in many food sources [26,118] (Table 1). Dietary cupric copper (Cu^{2+}) is reduced by several reductases to cuprous copper (Cu^+) and taken up by copper transporter 1 (CTR1) at the plasma membrane [119]. In the human body, copper has been found in the bones, muscles, skin, bone marrow, liver and brain [120]. More than 30 proteins essential for living organisms (e.g., lysyl oxidase, dopamine β-hydroxylase, cytochrome *c* oxidase, superoxide dismutase and tyrosinase) contain copper [121], so Cu is involved in numerous physiological and metabolic processes important for the functioning of the human body [122].

Copper also exhibits a wide range of antimicrobial activity, including that against Gram-positive (e.g., *Staphylococcus aureus*, *Enterococcus faecalis*, *Clostridium difficile* and *Listeria monocytogenes*) and Gram-negative (e.g., *Escherichia coli*, *Pseudomonas aeruginosa* and *Klebsiella pneumoniae*) bacteria, fungi (*Candida albicans* and *Aspergillus brasiliensis*) and viruses *(adenovirus, norovirus* and *poliovirus)* [123–125]. Copper and copper compounds have been used for centuries by many civilizations for general hygiene and to treat various conditions, such as headaches, burns, intestinal worms, and ear infections, as well as skin diseases [118,126]. Pulverized malachite (basic cupric carbonate) was used by the Sumerians for generic medical purposes and by the ancient Egyptians to prevent and cure eye infections. The ancient Chinese and ancient Romans also used various copper compounds to treat eye and skin diseases [118]. Preparations containing copper have been used throughout history in cases of chronic inflammation, skin eczemas, tuberculosis, syphilis and lupus [126]. Copper sulphate is widely used in Africa to heal wounds. Copper ions reduce the risk of the fungal and bacterial infection of minor wounds and cuts, as well as enhancing wound healing [118].

Copper is also important for the overall condition of the skin. Cu stimulates the proliferation of dermal fibroblasts and is involved in the synthesis and stabilization of extracellular matrix skin proteins and angiogenesis. It is a cofactor of superoxide dismutase (SOD), an enzyme involved in protecting the skin against the harmful effects of free radicals, and also prevents oxidative damage to cell membranes and lipid peroxidation [118,126]. Copper deficiency reduces the activity of the enzyme Cu/Zn-SOD and ceruloplasmin, as well as copper-independent enzymes, including catalase and glutathione peroxidase. Copper also promotes the function of free radical scavengers such as metallothionein and glutathione. A deficiency of copper ions in the plasma, liver, erythrocytes and heart increases the concentration of the lipid peroxidation product malondialdehyde (MDA) [127].

Furthermore, Cu is a cofactor of tyrosinase, the main enzyme involved in synthesis of the skin pigment melanin [128]. In the active center of the enzyme, in the N-terminal domain, there are two copper atoms, which are coordinated by three histidine residues. Three types of tyrosinase are involved in melanogenesis: oxytyrosinase (E_{oxy}), mettyrosinase (E_{met}) and deoxytyrosinase (E_{deoxy}) [129].

Copper ions are commonly included in cosmetics. Cu is used as an active component of face creams, including those for the care of mature, tired, dull or oily skin [118,130]. The effectiveness of the active ingredients is known to depend on their ability to penetrate the skin [130,131]. A study by Mazurowka and Mojski showed that the peptides glycyl-histydyl-lysine (GHK) and γ-glutamyl-cysteinyl-glycine (GSH) influence the permeation of copper ions through the horny layer of the epidermis [131]. The copper tripeptide complex (GHK-Cu and GSH-Cu) plays an important role in the protection and regeneration of skin tissue, increases the synthesis of collagen and elastin, and improves the condition of ageing skin. For this reason, copper peptides are often used as cosmetic ingredients [130]. In comparison with other metal compounds, copper has a low allergenic potential, and the risk of adverse side effects due to skin contact with the metal is small [132,133].

2.2.4. Silicon

Silicon (Si) is the second most abundant element on Earth, present in water, plants and animals [134] and it has semi-metallic properties. In nature, it is not normally found in free form but is usually present as a chemical compound of silicon dioxide, a complex compound, or silicon silicate [135,136]. The silicon present in food is solubilized in the acidic environment of the stomach, becoming easily absorbed orthosilicic acid (OSA) [134]. Si is mainly absorbed from the diet [135,137] (Table 1). Another important source of silicon used in dietary supplements is horsetail (*Equisetum arvense*) [134].

Silicon is present in all healthy tissues of the human body, including the connective tissues, bones, liver, heart, muscle, kidneys and lungs [137]. Si is also present in the skin, hair and nails [138]. The amount of silicon in tissues decreases with age, most likely because the organ responsible for silicon absorption is the thymus, which atrophies with age [137].

Organic silicon plays an important role in skin structure, promotes neocollagenesis, strengthens connective tissue, and reduces the risk of alopecia [139]. Orthosilicic acid (OSA) stimulates fibroblasts to secrete collagen type I [140]. Si also promotes the synthesis of elastin and GAGs, helps to preserve blood vessel elasticity, and increases the resistance and thickness of nail and hair fibers [135]. Barel et al. demonstrated that the oral intake of choline-stabilized orthosilicic acid for 20 weeks had a significant positive effect on the surface and mechanical properties of skin. Treatment with silicon may also improve keratin structure in the hair and nails and reduce brittleness [138]. Supplementation with silicon in a highly bioavailable form can be used for skin rejuvenation [141]. Kalil et al. described the positive effect of orthosilicic acid stabilized with hydrolyzed marine collagen at 600 mg/day on skin texture, firmness and hydration [141]. Ferreira et al. investigated the oral intake of two forms of Si—maltodextrin-stabilized orthosilicic acid (M-OSA) and monomethylsilanetriol (MMST)—and their effects on the nails, skin and hair. They were found to improve skin parameters, increase eyelash length, reduce facial wrinkles and UV spots, and act as aluminum detoxification agents [135]. Barel et al. studied the effect of supplements containing choline-stabilized orthosilicic acid (ch-OSA) on the skin, hair and nails. In a group of 50 women with photodamaged facial skin taking 10 mg of Si/day orally for 20 weeks, significant improvements were observed in the surface characteristics and mechanical properties of the skin, as was a positive effect on the brittleness of the hair and nails [138].

Silicones are included in cosmetic formulations such as hair conditioners and shampoos to give hair a silky appearance and shine, as well as in facial creams to protect and improve the skin's strength and elasticity [134,136]. Silicone products applied to the skin impart a pleasant smoothing sensation. Silicone elastomeric particles can absorb various liquids, including emollients and sebum, and can thus be used as an agent to carry active ingredients to the skin or to control sebum deposition in the skin [136,142].

Table 1. Sources and roles of vitamins and minerals of importance for skin health and function.

Micronutrient	Source	Functions in Skin	Hypovitaminosis and Skin Disorders	References
Vitamin A (retinol)	liver, fish liver oils, dairy products, butter, cheese, egg yolk, meat products, certain saltwater fish, plant products (β-carotene)	-stimulates synthesis of epidermal proteins -stimulates proliferation of epidermal cells and regulates exfoliation of keratinocytes -stimulates fibroblasts to synthesize collagen and elastin, protects collagen against MMP-induced degradation -takes part in angiogenesis in the dermis -influences skin thickness -influences melanin degradation and skin color -protects the skin against UV radiation -reduces sebum and limits the development of *Propionibacterium acnes*	-abnormal keratinization of the epidermis -excessive epidermal exfoliation -excessive skin dryness -reduced sebum secretion -follicular keratosis	[27,34,35,143–145]
Vitamin C (ascorbic acid)	citrus fruits, seaberry, dog-rose, blackcurrant, strawberry, raspberry, kiwi, hawthorn, rowan, cruciferous vegetables (Brussels sprouts, kohlrabi, broccoli, cabbage, cauliflower), spinach, chicory, green bean, red pepper, chives, parsley, nettle, five-flavor berry, oregano	-takes part in differentiation of keratinocytes -stimulates ceramide synthesis -influences skin hydration -takes part in collagen biosynthesis and the formation of the extracellular matrix, increases synthesis of GAGs -inhibits melanogenesis -protects against UV radiation, photoaging	-skin fragility -impaired collagen biosynthesis -thickening of the stratum corneum -impaired wound healing -subcutaneous bleeding -scurvy	[48–50,58,62]
Vitamin E (toco-pherol)	wheat germ oil, sunflower oil, safflower oil, soybean oil, maize oil, cottonseed oil, palm oil, cereal products, nuts	-takes part in biosynthesis of collagen, elastin and GAGs -protects lipid structures of the stratum corneum -prevents the development of UV-induced erythema -has strong antioxidant properties	-exfoliative dermatitis -skin inflammation	[76,78,80,82,84,85,146]
Selenium (Se)	Brazil nuts, saltwater fish (yellowfin tuna, halibut, sardines), seafood, sea salt, meats, poultry, eggs, dairy products, cereals, cereal products, broccoli, white cabbage, asparagus, kohlrabi, garlic, onion, legumes, mushrooms	-acts as an antioxidant -is a cofactor for GPx, which removes harmful peroxides -reduces ROS-mediated inflammation and DNA damage -protects the skin from UV-induced oxidative stress -plays roles in fighting ageing and preventing ageing-related diseases	-several skin diseases, including psoriasis, acne vulgaris, and atopic dermatitis	[26,88–90,106,147]

Table 1. *Cont.*

Micronutrient	Source	Functions in Skin	Hypovitaminosis and Skin Disorders	References
Zinc (Zn)	red meat, fish, poultry, seafood (shrimp, oysters), nuts, pumpkin seeds, sunflower seeds, legumes, whole grains, dairy products	-prevents UV-induced skin damage -protects against photodamage -plays role in skin morphogenesis, repair and maintenance -modulates activity of 5α-reductase	-acrodermatitis enteropathica -atopic dermatitis, epidermolysis bullosa	[26,97,98,106,107,109,148,149]
Copper (Cu)	liver, seafood, nuts, oysters, seeds, some whole grains and legumes, chocolate, cocoa	-exhibits antimicrobial activity -stimulates collagen maturation -modulates melanin synthesis	-Menkes kinky hair disease (hypopigmented skin and hair, sparse, short, brittle scalp hair, pale, mottled, doughy skin)	[118,126,128,150]
Silicon (Si)	grains (rice, barley, oat, wheat), grain products (breakfast cereals, bread, pasta), root vegetables (carrots, beetroot, radish, onion, potatoes), beans, maize, fruit (bananas), dried fruit (raisins), nuts, field horsetail	-plays role in synthesis of collagen, elastin and GAGs -improves skin surface and its mechanical properties -improves keratin structure -increases resistance and thickness of nails and hair	-faster appearance of wrinkles -soft and brittle nails -slower wound healing	[135,137,138]

2.3. Fatty Acids

Unsaturated fatty acids include omega-9 (ω-9; n-9) monounsaturated fatty acids and omega-3 (ω-3; n-3) and omega-6 (ω-6; n-6) polyunsaturated fatty acids (PUFAs) [151]. Omega-3 and omega-6 fatty acids, including alpha-linolenic (ALA), linoleic (LA) and gamma-linolenic (GLA) acids, included among the essential fatty acids (EFAs), are of significant dietary and cosmetic importance (Figure 5) [152].

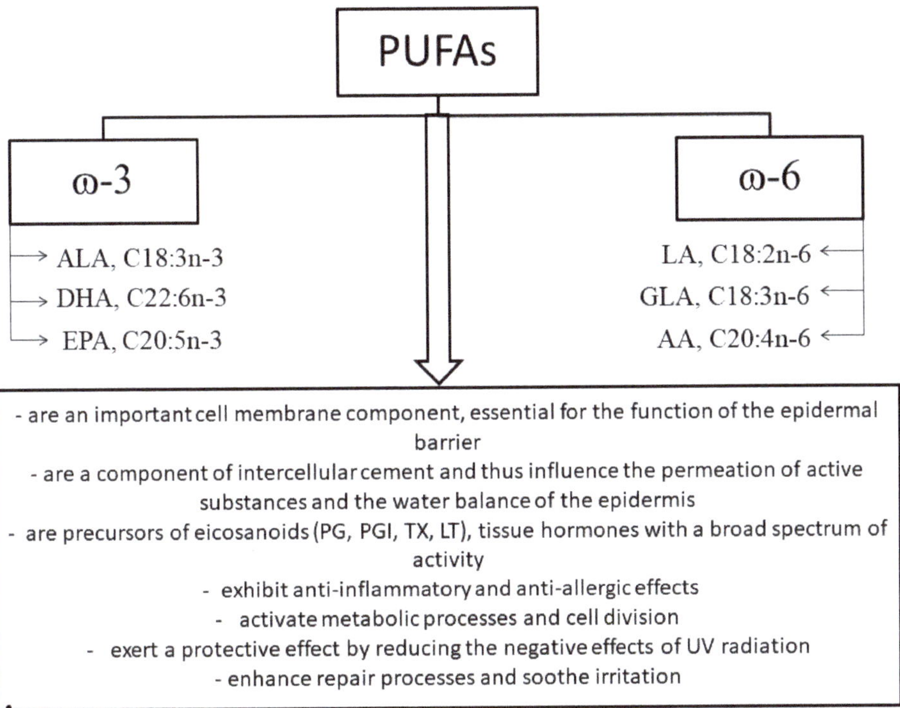

Figure 5. Classification of polyunsaturated fatty acids and the effects of omega-3 and omega-6 fatty acids on physiological processes in the skin; ALA, alpha-linolenic acid; DHA, docosahexaenoic acid; EPA, eicosapentaenoic acid; LA, linoleic acid; GLA, gamma-linolenic acid; AA, arachidonic acid; PG, prostaglandins; PGI, prostacyclins; TXA, thromboxanes; LT, leukotrienes (own work based on [151,153,154]).

Essential fatty acids are ascribed an important role in prophylaxis, especially of cardiovascular disease and allergic or inflammatory conditions [155]. EFAs also play an important role in skin structure and function.

Owing to unsaturated fatty acids, which, alongside ceramides and cholesterol, are components of the intracellular cement, the skin can act as an effective barrier limiting TEWL, thus ensuring adequate hydration and protecting against external factors [153,154,156]. EFAs have therapeutic properties (e.g., anti-inflammatory and anti-allergic) and exert protective effects. The results of recent research on the beneficial effects of GLA on various dermatological conditions are promising and support the hypothesis that it is an essential fatty acid for skin function. GLA applied topically as a cream penetrates to the stratum corneum, while taken orally, it reaches the dermis, enhancing its cohesiveness and preventing excessive TEWL [156,157].

The symptoms of fatty acid deficiency include dryness of the epidermis, peeling, flabby skin, skin inflammation, an increased susceptibility to irritation and slower healing. A deficiency of LA, a component of ceramide 1, which plays an important role in the cohesiveness of the intracellular cement, results in the dysfunction of the skin barrier and symptoms of dry skin. Research by Cosgrove et al. showed that a higher LA intake is associated with a lower likelihood of senile dryness and skin atrophy [59]. Many skin problems, including excessive exfoliation of the epidermis, are also caused by a deficiency of GLA, which is formed from LA via an enzymatic reaction involving delta-6 desaturase [153,154,158,159]. A deficiency of essential fatty acids can also reduce the fluidity of sebum, which leads to the obstruction of the sebaceous glands and the appearance of blackheads and inflammation [152,157].

Unsaturated fatty acids are not synthesized in the human body, so they must be supplied through the diet. An important source of EFAs is vegetable oils obtained from seeds, fruit, nuts and sprouts [156]. The biological value and cosmetic suitability of oils depend on their percentages of fatty acids, both saturated (e.g., stearic and palmitic acid) and unsaturated (e.g., oleic, linoleic, linolenic and arachidonic acid). Vegetable oils rich in LA include wheat germ, soybean, sunflower seed and sesame seed oil [152,160], while oils from the seeds of borage, evening primrose or blackcurrant are important sources of GLA [161,162]. Oils rich in EFAs improve skin hydration, have a regenerative effect on the damaged epidermal lipid barrier, and regulate skin metabolism. Vegetable oils play an important role in the care of dry, sensitive, oily, acne-prone and mature skin [152,157] (Table 2).

Table 2. Selected vegetable oils that are sources of fatty acids of importance for the skin.

Oil Example	Source	Fatty Acids Composition	Reference
Dry, sensitive and atopic skin			
Borage oil	*Borago officinalis*	γ-linolenic (26–38%), linoleic (35–38%), oleic (16–20%), palmitic (10–11%), stearic (3.5–4.5%), eicosenoic (3.5–5.5%), erucic (1.5–3.5%) acid	[163]
Evening primrose oil	*Oenothera biennis*	linoleic (70–75%), γ-linolenic (8–14%) acid	[161,164]
Blackcurrant seed oil	*Ribes nigrum*	linoleic (>40%), linolenic (10–20%), γ-linolenic (13–15%), stearidonic (2–3%) acid	[162,165]
Hemp oil	*Canabis sativa*	α-linolenic (16–19%), linoleic (55–58%), oleic (11–13%), γ-linolenic (4%), palmitic (5–6%), stearic (2.3–2.5%) acid	[166,167]
Oily, combination and acne skin			
Baobab oil	*Adansonia digitata*	linoleic (36%), palmitic (28.8%), oleic (25.1%) acid	[168–170]
Neem oil	*Azadirachta indica*	oleic (40%), stearic (16%), palmitic (14%), linoleic (21%) acid	[171,172]
Black cumin oil	*Nigella sativa*	γ-linolenic (50–60%), oleic (20%), eicosadienoic (3%), dihomo-γ-linolenic (10%), palmitic and stearic (>30%) acid	[173,174]
Coconut oil	*Cocos nucifera*	capric (7%), lauric (49%), myristic (18%), palmitic (9%), oleic (6%), stearic (2%), linoleic (2%) acid	[175]
Mature skin, photoaging, UV protection			
Sea buckthorn oil	*Hippophaë rhamnoides*	linoleic (34–40%), α-linolenic (23–36%), palmitic (26.3%), oleic (15–20%) acid	[158,176,177]
Green coffee oil	*Coffea arabica*	linoleic (44%), palmitic (34%), oleic (9%), stearic (7%), arachidonic (3%), linolenic (1.5%) acid	[178,179]
Wheat germ oil	*Triticum aestivum*	linoleic (58.6%), α-linolenic (9.6%), oleic (11.2%), palmitic (17.9%), stearic (1.2%), palmitoleic (0.2%) acid	[180]
Barbary fig oil	*Opuntia ficus-indica*	linoleic (62–49%), palmitic (10.6–12.8%), oleic (13–23.5%), stearic (3.3–5.4%) acid	[181]
Argan oil	*Argania spinosa*	oleic (43–49%), linoleic (29–36%), palmitic (11–15%), stearic (4–7%) acid	[182–184]

Omega-3 acids obtained from fish oil—eicosapentaenoic acid (EPA) and docosahexaenoic acid (DHA)—also play an important role in skin function. Although these acids are not present in the normal epidermis, their metabolites (epidermal 15-lipoxygenase transforms EPA into

15-hydroxyeicosapentaenoic acid (15-HEPE) and DHA into 17-hydroxydocosahexaenoic acid (17-HDoHE)) accumulate in it after the consumption of fish oil [185]. There are studies indicating that fish oils have a protective role and can reduce the severity of erythema. Rhodes et al., in a study with a group of 15 people given fish oil containing 1.8 g of EPA and 1.2 g of DHA, observed a marked reduction in UVB-erythemal sensitivity after six months of supplementation [186]. In a randomized double-blind trial in 42 healthy patients taking 4 g/day of EPA or monounsaturated oleic acid for three months, an increase in the UV-induced erythematous threshold was observed, as was a decrease in the expression of the protein p53, a marker of DNA damage induced by UV radiation. The results indicate that EPAs have a UV-filtering effect and suggest that longer-term supplementation may reduce the risk of skin cancer in humans [187]. Furthermore, the supplementation of diets with vegetable or fish oils may generate local cutaneous anti-inflammatory metabolites, which could serve as adjuncts in the management of skin inflammatory disorders [185].

2.4. Polyphenols

Polyphenols are compounds widespread in the world of plants, imparting color—ranging from red to yellow to blue—to flowers and fruits [188]. Rich sources of polyphenols include spices and herbs, such as cloves (eugenol), star anise (anethole), peppermint (eriocitrin), oregano (pinocembrin), sage, thyme, spearmint and rosemary (rosmarinic acid), as well as fruits, including berries (black chokeberries, black elderberries, blueberries and blackcurrants), plums, cherries, strawberries, raspberries, grapes and drupe fruits (apples, peaches, apricots, nectarines and pears), some seeds (flaxseeds and soybeans), nuts (chestnuts, walnuts, hazelnuts, pecans and almonds) and vegetables, including black and green olives, artichoke heads, red and green chicory, onions, spinach, broccoli, asparagus and lettuce [189]. Polyphenols are organic chemical compounds containing two or more hydroxyl groups attached to an aromatic ring [190]. Five main groups of phenolic compounds can be distinguished based on chemical structure: flavonoids, phenolic acids, tannins, stilbenes and diferuloylmethane [190–192] (Figure 6).

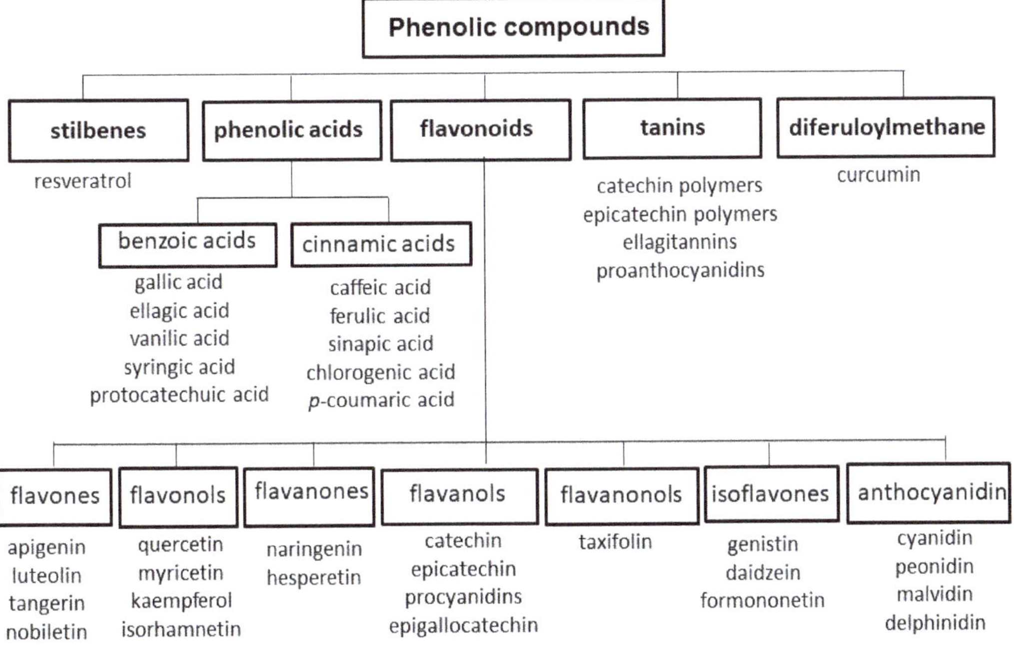

Figure 6. Classification of polyphenols (own work based on [190–193]).

The most thoroughly researched group of polyphenols is the flavonoids [193]. The structure of flavonoids is based on a flavan structure consisting of three rings [193,194] (Figure 7).

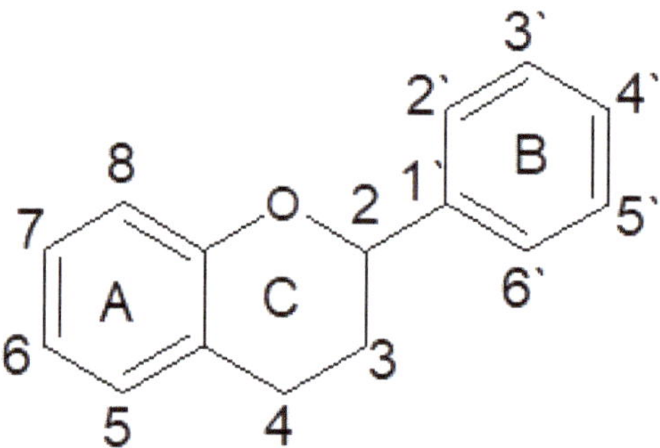

Figure 7. Basic structure and numbering system of flavonoids (own work).

The health-promoting effects of polyphenols taken orally are linked to their bioavailability, which is largely dependent on their chemical structure. The bioavailability depends on the amounts of nutrients that are digested, absorbed and included in metabolic processes [195,196]. The diversity of polyphenols' structures is reflected in their multi-faceted biological activity [190]. They exhibit anti-inflammatory, antibacterial, antifungal, antiviral, antiallergic, anticancer and anticoagulant properties [190–193,197]. Plant polyphenols are considered important substances for skin function, with hydrating, smoothing, softening, soothing and astringent effects [25,190,193,197–199]. Polyphenols inhibit the activity of enzymes present in the skin—collagenase and elastase, which catalyze the hydrolysis of collagen and elastin fibers, and hyaluronidase, which degrades hyaluronic acid. Furthermore, they soothe irritation and reduce the redness of skin, accelerating the natural regeneration of the epidermis, stabilizing the capillaries, improving microcirculation and elasticity in the skin, and protecting against harmful external factors, including UV radiation [188,197,198,200].

The presence of antioxidants has been shown to be linked to a lower frequency of ROS-induced photoaging [201]. The antioxidant and antiradical properties of polyphenols result from the elimination of radicals through direct reactions, scavenging, or the reduction of free radicals (e.g., hydroxyl, superoxide, peroxide and alcoxyl radicals) to less reactive compounds. Polyphenols can also chelate transition metal cations (e.g., Cu^{2+} and Fe^{2+}), thus preventing Haber–Weiss and Fenton reactions (which lead to the generation of the extremely reactive hydroxyl radical •OH) and also inhibiting the activity of many enzymes involved in free radical generation (e.g., xanthine oxidase, protein kinase and lipoxygenase). Their activity also involves the stimulation and protection of other antioxidants, such as ascorbate in the cytosol or tocopherol in biological membranes [194,195,202].

Raw polyphenolic materials exert synergistic effects with other antioxidants in preventing skin ageing. Cho et al. investigated the effect of the oral administration of an antioxidant mixture of pycnogenol, evening primrose oil, vitamin C and vitamin E on UVB-induced wrinkle formation. The study showed that a 10-week administration of the antioxidants to hairless mice irradiated three times a week with UVB radiation significantly prevented the UVB-induced expression of MMPs and mitogen-activated protein (MAP) kinase, as well as the activation of the transcription factor AP-1. It also enhanced the expression of type I procollagen and TGF-β2). The results indicate that the oral administration of the antioxidant mixture can inhibit wrinkle formation by preventing the expression of MMPs and increasing collagen synthesis [203].

Examples of polyphenolic plant materials of value in the prevention of skin ageing and protection of skin cells against UV radiation and photoaging, present in the diet or used as ingredients in dietary supplements, include sylimarin, a complex of the flavonolignans silibinin, isosilibin, silychristin and sylidianin obtained from milk thistle seed coats (*Silybum marianum*) [204]; genistein, an isoflavonoid obtained from soybeans (*Glycine max*) [205]; curcumin, a component of the spice turmeric (*Curcuma longa*) [206]; and resveratrol, a polyphenolic phytoalexin present in grape seeds (*Vitis vinifera*) [207].

Another source of polyphenols (anthocyanins and hydrolysed tannins) with beneficial effects on the condition of the skin is pomegranates (*Punica granatum*) [208]. Research has confirmed the photochemoprotective, antioxidant, anti-inflammatory and antiproliferative properties of pomegranate extract. Pomegranate fruit extract has been shown to reduce UVB-induced oxidative stress and the oxidation of skin proteins [209], and also to improve the color of the skin and restore the glow of skin exposed to UV radiation [210].

Another material rich in polyphenols exhibiting properties that protect skin cells is green tea. The leaves of *Camelia sinensis* contain four major polyphenols: epicatechin (EC), epicatechin gallate (ECG), epigallocatechin (EGC) and epigallocatechin gallate (EGCG), the last of which accounts for 40% of all the polyphenols [211]. Green tea polyphenols (GTPs) are an example of a component of plant-based dietary supplements used to prevent the adverse biological effects of UV radiation, including immunosuppression and photocarcinogenesis [212]. Research also indicates that GTPs can alleviate the symptoms of premature skin ageing induced by UVB radiation [213]. Chiu et al. studied the effect of GTPs on the symptoms of photoaging of the skin in a double-blind trial using a placebo in a group of 40 women. In half of the subjects a 10-percent green tea cream was applied together with 300 mg, twice-daily green tea oral supplementation, while the other half received the placebo. Histological examination showed significant improvements in elastic tissue content, but no significant changes were demonstrated clinically. The authors concluded that a clinically perceptible improvement might require a longer period of GTP administration. GTPs also exhibit anti-inflammatory and antioxidant properties and the ability to scavenge free radicals [214]. In vivo studies indicate an increase (from 2 to 15%) in plasma antioxidant activity following the consumption of tea or tea polyphenols [215]. A study by Vayalil et al. demonstrated the inhibition of oxidative stress induced by UVB radiation in the skin of SKH-1 hairless mice that received GTPs in their drinking water (0.2%, w/v) and were then exposed to multiple doses of UVB light (90 mJ/cm^2) for two months. The inhibition of protein oxidation was observed, as was the inhibition of the UVB-induced expression of matrix-degrading MMPs, such as MMP-2 (67%), MMP-3 (63%), MMP-7 (62%) and MMP-9 (60%) [213]. A study in C57BL/6 mice showed that green tea extract reduced collagen cross-linking and UVB-induced oxidative changes in the structures of proteins [216].

A particularly interesting substance of plant origin that can function as a functional analogue of retinol is bakuchiol. This is a phenolic compound with a monoterpene side chain, obtained from the seeds and leaves of *Psoralea corylifolia*. Chaudhuri and Bojanowski showed that bakuchiol can prevent wrinkles; improve pigmentation, elasticity and firmness; and reduce overall photodamage. The authors stress that this compound can be used as a retinol-like anti-ageing functional compound [37].

The literature includes reports of many clinical trials evaluating the effectiveness of polyphenol-based therapies. Polyphenols used topically and orally may be effective in treating certain dermatological conditions, including anogenital warts, alopecia, acne vulgaris, fungal infections, hyperpigmentation and photoaged skin [217].

2.5. Carotenoids

Carotenoids are polyene pigments, having a conjugated system of double bonds. They can occur in the form of acyclic, monocyclic or bicyclic compounds. They include carotenes and their oxygenated derivatives, xanthophylls [218–220] (Figure 8).

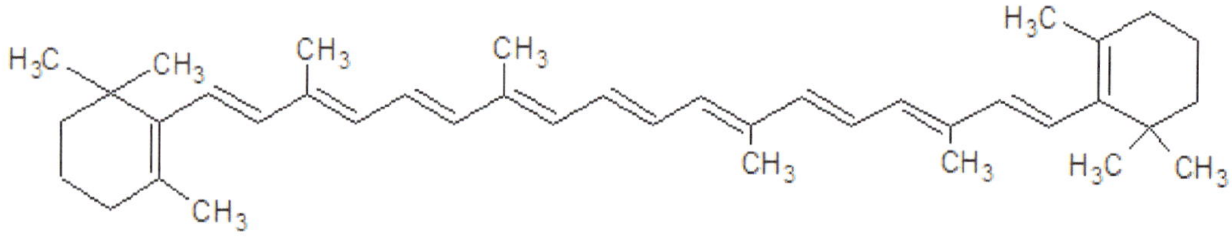

(a)

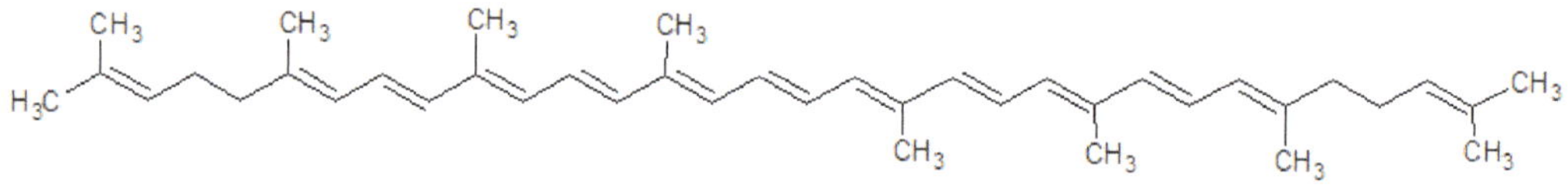

(b)

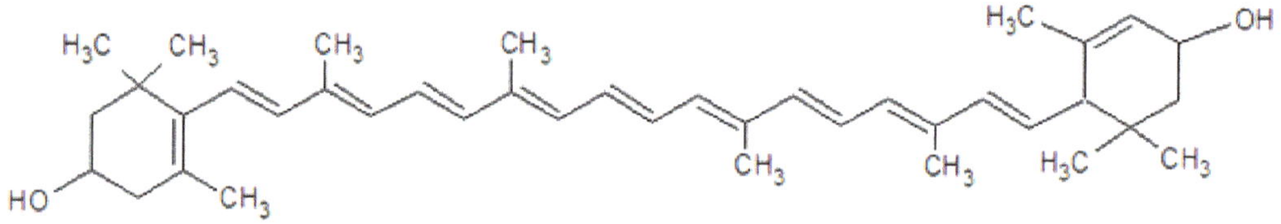

(c)

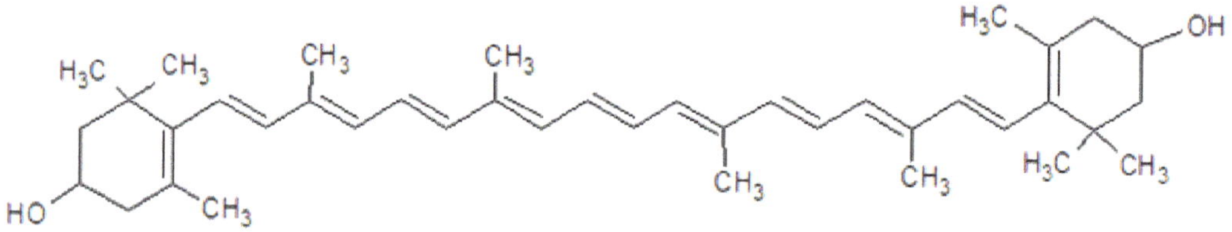

(d)

Figure 8. Structures of the major carotenoids: (**a**) β-carotene; (**b**) lycopene; (**c**) lutein; (**d**) zeaxanthin (own work).

Carotenoids are present in plants—imparting a yellow-to-red color to the flowers, fruits or leaves—and in the marine environment, as a red–orange pigment common in many aquatic animals [220–222] (Table 3).

Table 3. Classification of carotenoids, and their sources, functions and importance in skin health [219–228].

Classification	Example	Source	Function
Carotenes—the hydrocarbon carotenoids	β-carotene	beet root, apricots, cantaloupe, carrots, pumpkin, sweet potato, pink grapefruit, tomatoes, watermelon, mango, papaya, peaches, prunes, oranges, bilberry green fruits and vegetables such as green beans, broccoli, brussels sprouts, cabbage, kale, kiwi, lettuce, peas, spinach, acerola, nuts, oil palm, buriti (*Mauritia vinifera*), sea buckthorn (*Hippophaë rhamnoides*), camu-camu (*Myrciaria dubia*), rose hip (*Rosa canina*), alga *Dunaliella salina*	-conversion to retinoids (provitamin A carotenoids: α- and β-carotene, β-cryptoxanthin) -modulation of the enzymatic activity of lipoxygenases -enhancement of immune system function -antioxidant activity (ability to quench singlet oxygen, free radical scavenging (including superoxide anions and hydroxyl radicals) -reduction of lipid peroxidation in human skin -promotion of secretion of transcription factors and growth factors
	lycopene	tomato and its derivatives, such as juices, soups, sauces, and ketchup, cherry, guava, watermelon, papaya, peaches, grapefruit, asparagus, rose hip	
Xanthophylls—the oxygenated carotenoids	lutein	green and dark green leafy vegetables, like broccoli, brussels sprouts, spinach, parsley, pumpkin, acerola, rose hip, *Tropaeolum majus*, caja (*Spondias lutea*), camu-camu (*Myrciaria dubia*), microalgae *Chlorella vulgaris*, eggs	-suppression of UVA-mediated induction of MMP-1, MMP-3, and MMP-10 -protection of dermal collagen against UV-induced oxidation of proline -promotion of synthesis of procollagen type I -anti-inflammatory properties -protection of cell membranes and tissues against effects of UV light -protection of skin against UVB-induced photoaging and photocarcinogenesis -reduction of sensitivity to UV radiation-induced erythema -protection from sunburn -skin-lightening effects -improvement of skin condition (hydration, skin texture, radiance, elasticity, reduction in furrows and wrinkles)
	zeaxanthin	broccoli, brussels sprouts, spinach, parsley, maize, pequi (*Caryocar villosum*), microalgae *Chlorella saccharophila*	
	astaxanthin	aquatic animals, such as lobster, crab, salmon, trout, krill, shrimp, crayfish and crustacea, microalgae *Phaffia rhodozyma*, *Chlorella vulgaris*, Haematococcus pluvialis, yeast	
	antheraxanthin	many plants, especially maize	
	α-, β-cryptoxanthin	many colored plants as maize and papaya, sea buckthorn	
	capsanthin	peppers, paprika	

Among 800 recognized carotenoids, about 40 are present in the typical human diet, while only 14 have been identified in the blood and tissues [220,224]. In the human body, carotenoids mainly accumulate in lipid tissue cells and the liver. They are also present in the horny layer of the epidermis, as confirmed by Raman spectroscopy. The content of carotenoids in human skin varies, with the highest levels observed in parts of the body with high concentrations of sweat and sebaceous glands (e.g., the forehead and hands). The total content of carotenoids in the skin is influenced by numerous factors, such as the intake of fruits and vegetables, their bioavailability from various foods, supplementation, exposure to UV radiation, air pollution, alcohol consumption, smoking and stress [220,229]. The external application of preparations containing carotenoids in combination with oral supplementation has been shown to increase the concentrations of these compounds in the skin [25,230].

Due to the coloring properties of carotenoids, they are often used in the food, pharmaceutical, cosmetics and feed industries [220,225].

Carotenoids prevent ageing, stimulate fibroblasts to produce collagen and elastin, inhibit the activity of MMPs, and exhibit anti-inflammatory and UV-filtering effects [222,226,231]. They have been shown to improve the elasticity, hydration and texture of skin; reduce TEWL; lighten the skin; reduce discoloration; and delay the signs of photoaging [222,231–233]. Yoon et al. investigated the effect of astaxanthin (2 mg/day) and collagen hydrolysate

(3 g/day) supplementation on moderately photoaged skin in humans. The bioactives significantly improved barrier integrity, reduced TEWL, and increased the expression of procollagen type I mRNA. The results indicate that astaxanthin combined with collagen hydrolysate, due to the beneficial effects of the compounds on skin elasticity and hydration, can be used as an anti-ageing agent for photoaged skin [234]. A study by Juturu et al. showed that supplementation with 10 mg of lutein and 2 mg of zeaxanthin isomers daily for 12 weeks improves the appearance of skin and lightens skin tone [233]. Schwartz et al. evaluated the effect of zeaxanthin on skin parameters such as fine and deep lines, total wrinkles, wrinkle severity, radiance/skin color, discoloration and skin pigment homogeneity. The results indicate that a zeaxanthin-based dietary supplement and topical serum improve hydration and reduce wrinkles in facial skin [235]. Furthermore, carotenoids support healing processes and protect the skin against the oxidative stress resulting from excessive ROS activity, thereby reducing the risk of skin cancer [220,226,232,236,237]. Research confirms the photoprotective effects of carotenoids in the diet and in the form of dietary supplements. Wertz et al. investigated this effect in the case of β-carotene. This compound was found to suppress the UVA induction of MMP-1, MMP-3 and MMP-10, the three main metalloproteinases involved in photoaging. Moreover, β-carotene was shown to block the $^1O_2$-mediated induction of MMP-1 and MMP-10 in a dose-dependent manner [228]. Cesarini et al., in a group of 25 subjects, studied the effect of an orally administered antioxidant complex consisting of β-carotene, lycopene, selenium and α-tocopherol on parameters of the protection of the epidermis against UV-induced skin damage. After 6 weeks of the application of the antioxidants, an increase in protection against UV-induced skin damage was observed, accompanied by a reduction in lipoperoxide levels and sunburn cells (SBCs), as well as an increase in skin pigmentation [238]. Other research has shown a significant reduction in UV-induced erythema and improvement in the hydration and elasticity of skin following the oral application of lutein [237]. Heinrich et al. compared the erythema-protective effect of 24 mg/d of beta-carotene from an algal source to that of 24 mg/d of a carotenoid mixture (beta-carotene, lutein and lycopene, at 8 mg/d each). An increase in carotenoid levels in the serum and skin was observed after 12 weeks of supplementation in both groups, as was a comparable reduction in UV-induced erythema [239]. Aust et al. evaluated the photoprotective effects of synthetic lycopene in comparison with those of a tomato extract and a drink containing solubilized tomato extract. The subjects ingested similar amounts of lycopene (about 10 mg/d) from all three sources. After 12 weeks of supplementation, significant increases in serum lycopene levels and total skin carotenoids were observed in all the groups, as was less severe erythema following exposure to UV radiation. The protective effects were more pronounced in the groups that received the tomato extract or a drink containing solubilized tomato extract [240]. These studies indicate that a photoprotective effect and reduction in the severity of UV-induced erythema can be ascribed to the entire family of carotenoids.

Researchers draw attention to role of carotenoids in the prevention and treatment of photodermatoses, such as erythropoietic protoporphyria (EPP), porphyria cutanea tarda (PCT) and polymorphous light eruption (PMLE) [220]. Some studies indicate that a high dietary intake of lutein and zeaxanthin is associated with a reduced incidence of squamous cell carcinomas (SCC) in subjects with a history of skin cancer at baseline [241]. In other research, treatment with β-carotene did not significantly reduce the occurrence of new skin cancers in persons with previous non-melanoma skin cancer [242].

3. Conclusions

The skin is a sensitive indicator of nutritional deficiencies. The most effective way to improve the condition of the skin is to supply it with essential nutrients, both externally and—importantly—internally, through a varied diet. An increasing body of research suggests that a well-balanced diet significantly affects the skin ageing process. It is worth noting the substances that protect and restore the epidermal barrier, which reduces TEWL, ensuring an appropriate level of skin hydration and protecting against external factors and the damage induced by inflammation (e.g., omega-3 and omega-6 fatty acids). Antioxidants and other phytonutrients that scavenge ROS and alleviate oxidative skin damage also play an important role in the prophylaxis and care of ageing skin, as do substances that protect the skin against the negative effects of ultraviolet radiation (including vitamins A, C and E; selenium; zinc; copper; silicon; polyphenols; and carotenoids). The oral administration of antioxidants can be an effective supplement to chemical and physical UV-filtering agents and can reduce the DNA damage leading to skin ageing and the development of skin cancer. The inclusion of these substances in the daily diet could be a useful approach in anti-ageing interventions. In conclusion, the promotion of healthy dietary habits can benefit the appearance of the skin, delay ageing processes and reduce the risk of skin cancer.

Author Contributions: Conceptualization, M.M. and M.P.; writing—original draft preparation, M.M. and M.P.; critical revision of the manuscript, E.S. and B.K. All authors have read and agreed to the published version of the manuscript.

Funding: The project was supported under the program of the Minister of Science and Higher Education under the name "Regional Initiative of Excellence" in 2019–2022, project number: 024/RID/2018/19, financing amount: 11.999.000,00 PLN.

Institutional Review Board Statement: Not applicable.

Informed Consent Statement: Not applicable.

Data Availability Statement: Data sharing not applicable.

Conflicts of Interest: The authors declare no conflict of interest.

References

1. Ndiaye, M.A.; Nihal, M.; Wood, G.S.; Ahmad, N. Skin, reactive oxygen species, and circadian clocks. *Antioxid. Redox Signal.* **2014**, *20*, 2982–2996. [CrossRef]
2. Tagami, H. Location-related differences in structure and function of the stratum corneum with special emphasis on those of the facial skin. *Int. J. Cosmet. Sci.* **2008**, *30*, 413–434. [CrossRef]
3. Driskell, R.; Jahoda, C.; Chuong, C.M.; Watt, F.; Horsley, V. Defining dermal adipose tissue. *Exp. Dermatol.* **2014**, *23*, 629–631. [CrossRef]
4. Venus, M.; Waterman, J.; McNab, I. Basic physiology of the skin. *Surgery* **2011**, *29*, 471–474.
5. Wölfle, U.; Bauer, G.; Meinke, M.C.; Lademann, J.; Schempp, C.M. Reactive molecule species and antioxidative mechanisms in normal skin and skin aging. *Skin Pharmacol. Physiol.* **2014**, *27*, 316–332. [CrossRef]
6. Poljšak, B.; Dahmane, R.G.; Godić, A. Intrinsic skin aging: The role of oxidative stress. *Acta Dermatovener.* **2012**, *21*, 33–36.
7. Nishigori, C.; Hattori, Y.; Arima, Y.; Miyachi, Y. Photoaging and oxidative stress. *Exp. Dermatol.* **2003**, *12*, 18–21. [CrossRef]
8. Binic, I.; Lazarevic, V.; Ljubenovic, M.; Mojsa, J.; Sokolovic, D. Skin ageing: Natural weapons and strategies. *Evid. Based Complement. Alternat. Med.* **2013**. [CrossRef]
9. Poljšak, B.; Dahmane, R.G. Free radicals and extrinsic skin aging. *Dermatol. Res. Pract.* **2012**, 1–10. [CrossRef]
10. Biesalski, H.K.; Berneburg, M.; Grune, T.; Kerscher, M.; Krutmann, J.; Raab, W.; Reimann, J.; Reuther, T.; Robert, L.; Schwarz, T. Hohenheimer Consensus Talk. Oxidative and premature skin ageing. *Exp. Dermatol.* **2003**, *12*, 3–15. [CrossRef]
11. Pinnel, S.R. Cutaneous photodamage, oxidative stress, and topical antioxidant protection. *J. Am. Acad. Dermatol.* **2003**, *48*, 1–19. [CrossRef] [PubMed]
12. Masaki, H. Role of antioxidants in the skin: Anti-aging effects. *J. Dermatol. Sci.* **2010**, *58*, 85–90. [CrossRef] [PubMed]
13. Fisher, G.J.; Kang, S.; Varani, J.; Bata-Csorgo, Z.; Wan, Y.; Datta, S.; Voorhees, J.J. Mechanisms of photoaging and chronological skin aging. *Arch. Dermatol.* **2002**, *138*, 1462–1470. [CrossRef] [PubMed]
14. Kaimal, S.; Thappa, D.M. Diet in dermatology: Revisited. *IndianJ. Dermatol. Venereol. Leprol.* **2010**, *76*, 103–115.
15. Boelsma, E.; van de Vijver, L.P.L.; Goldbohm, R.A.; Klöpping-Ketelaars, I.A.A.; Hendriks, H.F.J.; Roza, L. Human skin condition and its associations with nutrient concentrations in serum and diet. *Am. J. Clin. Nutr.* **2003**, *77*, 348–355. [CrossRef]
16. Divya, S.A.; Sriharsha, M.; Narotham, R.K.; Krupa, S.N.; Siva, T.R.K. Role of diet in dermatological conditions. *J. Nutr. Food Sci.* **2015**, *5*, 1–7.
17. Rezaković, S.; Pavlić, M.; Navratil, M.; Počanić, L.; Žužul, K.; Kostović, K. The impact of diet on common skin disorders. *J. Nutr. Ther.* **2014**, *3*, 149–155. [CrossRef]
18. Cao, C.; Xiao, Z.; Wu, Y.; Ge, C. Diet and skin aging—from the perspective of food nutrition. *Nutrients* **2020**, *12*, 870. [CrossRef]
19. Strasser, B.; Volaklis, K.; Fuchs, D.; Burtscher, M. Role of dietary protein and muscular fitness on longevity and aging. *Aging Dis.* **2018**, *9*, 119–132. [CrossRef]

20. Balić, A.; Vlašić, D.; Žužul, K.; Marinović, B.; Bukvić Mokos, Z. Omega-3 versus omega-6 polyunsaturated fatty acids in the prevention and treatment of inflammatory skin diseases. *Int. J. Mol. Sci.* **2020**, *21*, 741. [CrossRef]

21. Rawlings, A.V.; Scott, I.R.; Harding, C.R.; Bowser, P.A. Stratum corneum moisturization at the molecular level. *J. Investig. Dermatol.* **1994**, *103*, 731–740. [CrossRef] [PubMed]

22. Schurer, N.Y.; Plewig, G.; Elias, P.W. Stratum corneum lipid function. *Dermatologica* **1991**, *183*, 77–94. [CrossRef] [PubMed]

23. Reuter, J.; Merfort, I.; Schempp, C.M. Botanicals in dermatology: An evidence-Based review. *Am. J. Clin. Dermatol.* **2010**, *11*, 247–267. [CrossRef] [PubMed]

24. Tungmunnithum, D.; Thongboonyou, A.; Pholboon, A.; Yangsabai, A. Flavonoids and other phenolic compounds from medicinal plants for pharmaceutical and medical aspects: An overview. *Medicines* **2018**, *5*, 93. [CrossRef]

25. Anunciato, T.P.; da Rocha Filho, P.A. Carotenoids and polyphenols in nutricosmetics, nutraceuticals, and cosmeceuticals. *J. Cosmet. Dermatol.* **2012**, *11*, 51–54. [CrossRef]

26. Godswill, A.G.; Somtochukwu, I.V.; Ikechukwu, A.O.; Kate, E.C. Health benefits of micronutrients (vitamins and minerals) and their associated deficiency diseases: A systematic review. *Int. J. Food. Sci.* **2020**, *3*, 1–32.

27. Khali, S.; Bardawil, T.; Stephan, C.; Darwiche, N.; Abbas, O.; Kibbi, A.G. Retinoids: A journey from the molecular structures and mechanisms of action to clinical uses in dermatology and adverse effects. *J. Dermatolog. Treat.* **2017**, *8*, 684–696. [CrossRef]

28. Sorg, O.; Kuenzli, S.; Kaya, S.; Saurat, J.H. Proposed mechanisms of action for retinoid derivatives in the treatment of skin aging. *J. Cosmetic. Dermato.* **2005**, *4*, 237–244. [CrossRef]

29. Mukherjee, S.; Date, A.; Patravale, P.; Korting, H.C.; Roeder, A.; Weindel, G. Retinoids in the treatment of skin aging: An overview of clinical efficacy and safety. *Clin. Interv.Aging* **2006**, *1*, 327–348. [CrossRef]

30. Polcz, E.M.; Barbul, A. The Role of Vitamin A in Wound Healing. *Nutr. Clin. Pract.* **2019**, *34*, 695–700. [CrossRef]

31. Bono, M.R.; Tejon, G.; Flores-Santibañez, F.; Fernandez, G.; Rosemblatt, M.; Sauma, D. Retinoic acid as a modulator of T cell immunity. *Nutrients* **2016**, *8*, 349. [CrossRef] [PubMed]

32. Zasada, M.; Budzisz, E. Retinoids: Active molecules influencing skin structure formation in cosmetic and dermatological treatments. *Postepy Dermatol. Alergol.* **2019**, *36*, 392–397. [CrossRef] [PubMed]

33. Erkelens, M.N.; Mebius, R.E. Retinoic acid and immune homeostasis: A balancing act. *Trends Immunol.* **2017**, *38*, 168–180. [CrossRef] [PubMed]

34. Oliveira, L.M.; Teixeira, F.M.E.; Sato, M. N Impact of retinoic acid on immune cells and inflammatory diseases. *Mediat. Inflamm.* **2018**. [CrossRef]

35. Ferreira, R.; Napoli, J.; Enver, T.; Bernardino, L.; Ferrrira, L. Advances and challenges in retinoid delivery systems in regenerative and therapeutic medicine. *Nat. Commun.* **2020**, *1*, 4265. [CrossRef]

36. Sorg, O.; Antille, C.; Kaya, G.; Saurat, J.H. Retinoids in cosmeceuticals. *Dermatol. Ther.* **2006**, *19*, 289–296. [CrossRef]

37. Chaudhuri, R.K.; Bojanowski, K. Bakuchiol: A retinol-like functional compound revealed by gene expression profiling and clinically proven to have anti-aging effects. *Int. J. Cosmet. Sci.* **2014**, *36*, 221–230. [CrossRef]

38. Shao, Y.; Te, H.; Fisher, G.J.; Voorhees, J.J.; Quan, T. Molecular basis of retinol anti-ageing properties in naturally aged human skin in vivo. *Int. J. Cosmet. Sci.* **2017**, *39*, 56–65. [CrossRef]

39. Törmä, H.; Brunnberg, L.; Vahlquist, A. Age-related variations in acyl-CoA: Retinol acyltransferase activity and vitamin A concentration in the liver and epidermis of hairless mice. *Biochim. Biophys. Acta.* **1987**, *921*, 254–258. [CrossRef]

40. Manela-Azulay, M.; Bagatin, E. Cosmeceuticals vitamins. *Clin. Dermatol.* **2009**, *27*, 469–474. [CrossRef]

41. Kang, S.; Duell, E.A.; Fisher, G.J. Application of retinol to human skin in vivo induces epidermal hyperplasia and cellular retinoid binding proteins characteristic of retinoic acid but without measurable retinoic acid levels or irritation. *J. Investig. Dermatol.* **1995**, *105*, 549–556. [CrossRef] [PubMed]

42. Varani, J.; Warner, R.L.; Gharaee-Kermani, M.; Phan, S.H.; Kang, S.; Chung, J.H.; Wang, Z.Q.; Datta, S.C.; Fisher, G.J.; Voorhees, J.J. Vitamin A antagonizes decreased cell growth and elevated collagen degrading matrix metaloproteinases and stimulates collagen accumulation in naturally aged human skin. *J. Investig. Dermatol.* **2000**, *114*, 480–486. [CrossRef] [PubMed]

43. Piérard-Franchimont, C.; Castelli, D.; Cromphaut, I.V. Tensile properties and contours of aging facial skin. A controlled double-blind comparative study of the effects of retinol, melibose-lactose and their association. *Skin Res. Technol.* **1998**, *4*, 237–243. [CrossRef] [PubMed]

44. Antille, C.; Tran, C.; Sorg, O.; Carraux, P.; Didierjean, L.; Saurat, J.H. Vitamin A exerts a photoprotective action in skin by absorbing ultraviolet B radiation. *J. Investig. Dermatol.* **2003**, *121*, 1163–1167. [CrossRef]

45. Fisher, G.J.; Talwar, H.S.; Lin, J.; Voorhees, J.J. Molecular mechanisms of photoaging in human skin in vivo and their prevention by all-trans retinoic acid. *Photochem. Photobiol.* **1999**, *69*, 154–157. [CrossRef]

46. Hacışevki, A. An overview of ascorbic acid biochemistry. *J. Fac. Pharm. Ankara.* **2009**, *38*, 233–255.

47. Lykkesfeldt, J.; Tveden-Nyborg, P. The pharmacokinetics of vitamin C. *Nutrients* **2019**, *11*, 2412. [CrossRef]

48. Janda, K.; Kasprzak, M.; Wolska, J. Vitamin C – structure, properties, occurrence and functions. *Pom. J. Life Sci.* **2015**, *61*, 419–425.

49. Pullar, J.M.; Carr, A.C.; Vissers, M.C.M. The roles of vitamin C in skin health. *Nutrients* **2017**, *9*, 866–892.

50. Rhie, G.; Shin, M.H.; Seo, J.Y.; Choi, W.W.; Cho, K.H.; Kim, K.H.; Park, K.C.; Eun, H.C.; Chung, J.H. Aging-and photoaging-dependent changes of enzymic and nonenzymic antioxidants in the epidermis and dermis of human skin in vivo. *J. Investig. Dermatol.* **2001**, *117*, 1212–1217. [CrossRef]

51. Shindo, Y.; Witt, E.; Han, D.; Epstein, W.; Packer, L. Enzymic and non-enzymic antioxidants in epidermis and dermis of human skin. *J. Investig. Dermatol.* **1994**, *102*, 122–124. [CrossRef] [PubMed]

52. McArdle, F.; Rhodes, L.E.; Parslew, R.; Jack, C.I.; Friedmann, P.S.; Jackson, M.J. UVR-induced oxidative stress in human skin in vivo: Effects of oral vitamin C supplementation. *Free Radic. Biol. Med.* **2002**, *33*, 1355–1362. [CrossRef]

53. Steiling, H.; Longet, K.; Moodycliffe, A.; Mansourian, R.; Bertschy, E.; Smola, H.; Mauch, C.; Williamson, G. Sodium-dependent vitamin C transporter isoforms in skin: Distribution, kinetics, and effect of UVB-induced oxidative stress. *Free Radic. Biol. Med.* **2007**, *43*, 752–762. [CrossRef] [PubMed]

54. Butler, J.D.; Bergsten, P.; Welch, R.W.; Levine, M. Ascorbic acid accumulation in human skin fibroblasts. *Am. Soc. Clin. Nutr.* **1991**, *54*, 1144–1146. [CrossRef]

55. Ponec, M.; Weerheim, A.; Kempenaar, J.; Mulder, A.; Gooris, G.S.; Bouwstra, J.; Mommaas, A.M. The formation of competent barrier lipids in reconstructed human epidermis requires the presence of vitamin C. *J. Investig. Dermatol.* **1997**, *109*, 348–355. [CrossRef]

56. Wang, K.; Jiang, H.; Li, W.; Qiang, M.; Li, H. Role of Vitamin C in Skin Diseases. *Front. Physiol.* **2018**, *9*, 819. [CrossRef]

57. Savini, I.; Valeria, M.; Duranti, G.; Avigliano, L.; Catani, M.V.; Melino, G. Characterization of keratinocyte differentiation induced by ascorbic acid: Protein kinase C involvement and vitamin C homeostasis. *J. Investig. Derm.* **2002**, *118*, 372–379. [CrossRef]

58. Pasonen-Seppanen, S.; Suhonen, T.M.; Kirjavainen, M.; Suihko, E.; Urtti, A.; Miettinen, M.; Hyttinen, M.; Tammi, M.; Tammi, R. Vitamin C enhances differentiation of a continuous keratinocyte cell line (REK) into epidermis with normal stratum corneum ultrastructure and functional permeability barrier. *Histochem. Cell Biol.* **2001**, *116*, 287–297. [CrossRef]

59. Cosgrove, M.C.; Franco, O.H.; Granger, S.P.; Murray, P.G.; Mayes, A.E. Dietary nutrient intakes and skin-aging appearance among middle-aged American women. *Am. J. Clin. Nutr.* **2007**, *86*, 1225–1231. [CrossRef]

60. Jones, D.P.; Kagan, V.E.; Aust, S.D.; Reed, D.J.; Omaye, S.T. Impact of nutrients on cellular lipid peroxidation and antioxidant defense system. *Fundam. Appl. Toxicol.* **1995**, *26*, 1–7. [CrossRef]

61. Ou-Yang, H.; Stamatas, G.; Saliou, C.; Kollias, N. A chemiluminescence study of UVA-induced oxidative stress in human skin in vivo. *J. Investig. Dermatol.* **2004**, *122*, 1020–1029. [CrossRef] [PubMed]

62. Offord, E.A.; Gautier, J.C.; Avanti, O.; Scaletta, C.; Runge, F.; Krämer, K.; Applegate, L.A. Photoprotective potential of lycopene, beta-carotene, vitamin E, vitamin C and carnosic acid in UVA-irradiated human skin fibroblasts. *Free Radic. Biol. Med.* **2002**, *32*, 1293–1303. [CrossRef]

63. Darr, D.; Dunston, H.; Faust, H.; Pinnell, S. Effectiveness of antioxidants (vitamin C and E) with and without sunscreens as topical photoprotectants. *Acta Derm. Venereol.* **1996**, *76*, 264–268. [PubMed]

64. Placzek, M.; Gaube, S.; Kerkmann, U.; Gilbertz, K.P.; Herzinger, T.; Haen, E.; Przybilla, B. Ultraviolet B-induced DNA damage in human epidermis is modified by the antioxidants ascorbic acid and D-alpha-tocopherol. *J. Investig. Dermatol.* **2005**, *124*, 304–307. [CrossRef]

65. Eberlein-König, B.; Ring, J. Relevance of vitamins C and E in cutaneous photoprotection. *J. Cosmet. Dermatol.* **2005**, *4*, 4–9. [CrossRef]

66. d'Ischia, M.; Wakamatsu, K.; Cicoira, F.; Di, M.E.; Garcia-Borron, J.C.; Commo, S.; Galván, I.; Ghanem, G.; Kenzo, K.; Meredith, P.; et al. Melanins and melanogenesis: From pigment cells to human health and technological applications. *Pigment Cell Melanoma Res.* **2015**, *28*, 520–544. [CrossRef]

67. Panich, U.; Tangsupa-a-nan, V.; Onkoksoong, T.; Kongtaphan, K.; Kasetsinsombat, K.; Akarasereenont, P.; Wongkajornsilp, A. Inhibition of UVA-mediated Melanogenesis by Ascorbic Acid through Modulation of Antioxidant Defense and Nitric Oxide System. *Arch. Pharm. Res.* **2011**, *34*, 811–820. [CrossRef]

68. Choi, Y.K.; Rho, Y.K.; Yoo, K.H.; Lim, Y.Y.; Li, K.; Kim, B.J.; Seo, S.J.; Kim, M.N.; Hong, C.K.; Kim, D.S. Effects of vitamin C vs. multivitamin on melanogenesis: Comparative study in vitro and in vivo. *Int. J. Dermatol.* **2010**, *49*, 218–226. [CrossRef]

69. Shimada, Y.; Tai, H.; Tanaka, A.; Ikezawa-Suzuki, I.; Takagi, K.; Yoshida, Y.; Yoshie, H. Effects of ascorbic acid on gingival melanin pigmentation in vitro and in vivo. *J. Periodontol.* **2009**, *80*, 317–323. [CrossRef]

70. Chawla, S. Split face comparative study of microneedling with PRP versus microneedling with vitamin C in treating atrophic post acne scars. *J. Cutan. Aesthet. Surg.* **2014**, *7*, 209–212. [CrossRef]

71. Basketter, D.A.; White, I.R.; Kullavanijaya, P.; Tresukosol, P.; Wichaidit, M.; McFadden, J.P. Influence of vitamin C on the elicitation of allergiccontact dermatitis top-phenylenediamine. *Contact Dermat.* **2016**, *74*, 368–372. [CrossRef] [PubMed]

72. Soodgupta, D.; Kaul, D.; Kanwar, A.J.; Parsad, D. Modulation of LXR-αand the effector genes by Ascorbic acid and Statins in psoriatic keratinocytes. *Mol. Cell. Biochem.* **2014**, *397*, 1–6. [CrossRef] [PubMed]

73. Schober, S.M.; Peitsch, W.K.; Bonsmann, G.; Metze, D.; Thomas, K.; Goerge, T.; Luger, T.A.; Schneider, S.W. Early treatment with rutoside and ascorbic acid is highly effectivefor progressive pigmented purpuric dermatosis. *J. Dtsch. Dermatol. Ges.* **2016**, *12*, 1112–1119.

74. Zervoudis, S.; Iatrakis, G.; Peitsidis, P.; Peitsidou, A.; Papandonopolos, L.; Nikolopoulou, M.K.; Papadopoulos, L.; Vladareanu, R. Complementary treatment with oralpidotimod plus vitamin C after laser vaporization for female genital warts: Aprospective study. *J. Med. Life* **2010**, *3*, 286–288.

75. Don, P.; Iuga, A.; Dacko, A.; Hardick, K. Treatment of vitiligo with broadband ultraviolet B and vitamins. *Int. J. Dermatol.* **2006**, *45*, 63–65. [CrossRef]

76. Khadangi, F.; Azzi, A. Vitamin-The Next 100 Years. *IUBMB Life* **2019**, *71*, 411–415. [CrossRef]

77. Abraham, A.; Kattoor, A.J.; Saldeen, T.; Mehta, J.L. Vitamin E and its anticancer effects. *Crit. Rev. Food. Sci. Nutr.* **2019**, *59*, 2831–2838. [CrossRef]
78. Traber, M.G.; Atkinson, J. Vitamin E, antioxidant and nothing more. *Free. Rad. Biol. Med.* **2007**, *43*, 4–15. [CrossRef]
79. Mustacich, D.J.; Bruno, R.S.; Traber, M.G. Vitamin E. *Vitam. Horm.* **2007**, *76*, 1–21.
80. Lee, G.Y.; Han, S.N. The role of Vitamin E in Immunity. *Nutrients* **2018**, *10*, 1614. [CrossRef]
81. Ichihashi, M.; Funasaka, Y.; Ohashi, A.; Chacraborty, A.; Ahmed, N.U.; Ueda, M.; Osawa, T. The inhibitory effect of DL-alpha-tocopheryl ferulate in lecithin on melanogenesis. *Anticancer Res.* **1999**, *19*, 3769–3774. [PubMed]
82. Keen, M.A.; Hassan, I. Vitamin E in dermatology. *Indian. Dermatol. Online J.* **2016**, *7*, 311–315. [CrossRef] [PubMed]
83. Dahiaya, A.; Romano, J.F. Cosmeceuticals: A review of their use for aging and photoaged skin. *Cosmet. Dermatol.* **2006**, *19*, 479–484.
84. Tsoureli-Nikita, E.; Hercogova, J.; Lotti, T.; Menchini, G. Evaluation of dietary intake of vitamin E in the treatment of atopic dermatitis: A study of the clinical course and evaluation of the immunoglobulin E serum levels. *Int. J. Dermatol.* **2002**, *41*, 146–150. [CrossRef]
85. Butt, H.; Mehmood, A.; Ali, M.; Tasneem, S.; Tarar, M.N.; Riazuddin, S. Vitamin E preconditioning alleviates in vitro thermal stress in cultured human epidermal keratinocytes. *Life Sci.* **2019**, *239*, 116–972. [CrossRef]
86. Baumann, L.S.; Spencer, J. The Effects of Topical Vitamin E on the Cosmetic Appearance of Scars. *Dermatol. Surg.* **1999**, *25*, 311–315. [CrossRef]
87. Park, K. Role of micronutrients in skin health and function. *Biomol. Ther.* **2015**, *23*, 207–217. [CrossRef]
88. Fairweather-Tait, S.J.; Bao, Y.; Broadley, M.R.; Collings, R.; Ford, D.; Hesketh, J.H.; Hurst, R. Selenium in human health and disease. *Antioxid. Redox Signal.* **2011**, *14*, 1337–1383. [CrossRef]
89. Cai, Z.; Zhang, J.; Li, H. Selenium, aging and aging-related diseases. *Aging Clin. Exp. Res.* **2019**, *31*, 1035–1047. [CrossRef]
90. Zhu, X.; Jiang, M.; Song, E.; Jiang, X.; Song, Y. Selenium deficiency sensitizes the skin for UVB-induced oxidative damage and inflammation which involved the activation of p38 MAPK signaling. *Food Chem. Toxicol.* **2015**, *75*, 139–145. [CrossRef]
91. Jobeili, L.; Rousselle, P.; Béal, D.; Blouin, E.; Roussel, AM.; Damour, O.; Rachidi, W. Selenium preserves keratinocyte stemness and delays senescence by maintaining epidermal adhesion. *Aging* **2017**, *9*, 2302–2315. [CrossRef] [PubMed]
92. Favrot, C.; Beal, D.; Blouin, E.; Leccia, M.E.; Roussel, A.M. Age-dependent protective effect of selenium against UVA irradiation in primary human keratinocytes and the associated DNA repair signature. *Oxid. Med. Cell. Longev.* **2018**. [CrossRef] [PubMed]
93. Kim, Y.M.; Jung, H.J.; Choi, J.S.; Nam, T.J. Anti-wrinkle effects of a tuna heart H_2O fraction on Hs27 human fibroblasts. *Int. J. Mol. Med.* **2016**, *37*, 92–98. [CrossRef] [PubMed]
94. van der Pols, J.C.; Heinen, M.M.; Hughes, M.C.; Ibiebele, T.I.; Marks, G.C.; Green, A.C. Serum antioxidants and skin cancer risk: An 8-year community-based follow-up study. *Cancer Epidemiol. Biomark. Prev.* **2009**, *18*, 1167–1173. [CrossRef]
95. Combos, G.F.; Clark, L.C.; Turnbull, B.W. Reduction of cancer mortality and incidence by selenium supplementation. *Med Klin* **1997**, *92*, 42–45. [CrossRef]
96. Duffield-Lillico, A.J.; Slate, E.H.; Reid, M.E.; Turnbull, B.W.; Wilkins, P.A.; Jr, G.F.C.; Park, H.K.; Gross, E.G.; Graham, G.F.; Stratton, M.S.; et al. Selenium supplementation and secondary prevention of nonmelanoma skin cancer in a randomized trial. *J. Natl. Cancer. Inst.* **2003**, *95*, 1477–1481. [CrossRef]
97. Devi, C.B.; Nandakishore, Th.; Sangeeta, N.; Basar, G.; Deni, N.O.; Sungdirenla, J.; Singh, M.A. Zinc in human health. *IOSR-JDMS* **2014**, *13*, 18–23. [CrossRef]
98. Gupta, M.; Mahajan, V.K.; Mehta, K.S.; Chauhan, P.S. Zinc therapy in dermatology: A review. *Dermatol. Res. Pract.* 2014. [CrossRef]
99. Ogawa, Y.; Kinoshita, M.; Shimada, S.; Kawamura, T. Zinc and skin disorders. *Nutrients* **2018**, *10*, 199. [CrossRef]
100. Manet, W. Zinc biochemistry: From a single zinc enzyme to a key element of life. *Adv. Nutr.* **2013**, *4*, 82–91.
101. Fraga, C.G. Relevance, essentiality and toxicity of trace elements in human health. *Mol. Aspects Med.* **2005**, *26*, 235–244. [CrossRef] [PubMed]
102. Michaelsson, G.; Ljunghall, K.; Danielson, B.G. Zinc in epidermis and dermis in healthy subjects. *Acta Derm. Venereol.* **1980**, *60*, 295–299. [PubMed]
103. Inoue, Y.; Hasegawa, S.; Ban, S.; Yamada, T.; Date, Y. ZIP2 protein, a zinc transporter, is associated with keratinocyte differentiation. *J. Biol. Chem.* **2014**, *289*, 21451–21462. [CrossRef] [PubMed]
104. Vallee, B.L.; Galdes, A. The metallobiochemistry of zinc enzymes. *Adv. Enzymol. Relat. Areas Mol. Biol.* **1984**, *56*, 283–430.
105. Klug, A. The discovery of zinc fingers and their applications in gene regulation and genome manipulation. *Annu. Rev. Biochem.* **2010**, *79*, 213–231. [CrossRef]
106. Vollmer, D.L.; West, V.A.; Lephart, E.D. Enhancing skin health: By oral administration of natural compounds and minerals with implications to the dermal microbiome. *Int. J. Mol. Sci.* **2018**, *19*, 3059. [CrossRef]
107. Nitzan, Y.B.; Cohen, A.D. Zinc in skin pathology and care. *J. Dermatolog. Treat.* **2006**, *17*, 205–210. [CrossRef]
108. Tang, X.; Shay, N.F. Zinc has an insulin-like effect on glucose transport mediated by phosphoinositol-3-kinaseand Akt in 3T3-L1 fibroblasts and adipocytes. *J. Nutr.* **2001**, *131*, 1414–1420. [CrossRef]
109. Lin, P.H.; Sermersheim, M.; Li, H.; Lee, P.H.U.; Steinberg, S.M.; Ma, J. Zinc in wound healing modulation. *Nutrients* **2018**, *10*, 16. [CrossRef]

110. Kim, J.E.; Yoo, S.R.; Jeong, M.G.; Ko, J.Y.; Ro, Y.S. Hair zinc levels and the efficacy of oral zinc supplementation in patients with atopic dermatitis. *Acta Derm. Venereol.* **2014**, *94*, 558–562. [CrossRef]
111. Mahoney, M.G.; Brennan, D.; Starcher, B.; Faryniarz, J.; Ramirez, J.; Parr, L.; Uitto, J. Extracellular matrix in cutaneous ageing: The effects of 0.1% copper-zincmalonate-containing cream on elastin biosynthesis. *Exp. Dermatol.* **2009**, *18*, 205–211. [CrossRef] [PubMed]
112. Sharquie, K.E.; Al-Mashhadani, S.A.; Salman, H.A. Topical 10% zinc sulfate solution for treatment of melisma. *Dermatol. Surg.* **2008**, *34*, 1346–1349. [PubMed]
113. Sharquie, K.E.; Al-Mashhadani, S.A.; Noaimi, A.A.; Hasan, A.A. Topical zinc sulphate (25%) solution: A new therapy for actinic keratosis. *J. Cutan. Aesthet. Surg.* **2012**, *5*, 53–56. [CrossRef] [PubMed]
114. Sharquie, K.E. Topical therapy of xeroderma pigmentosa with 20% zinc sulfate solution. *IASJ* **2008**, *7*, 231–236.
115. Faghihi, G.; Iraji, F.; Shahingohar, A.; Saidat, A.H. The efficacy of 0.05% Clobetasol + 2.5% zinc sulphate cream versus 0.05% Clobetasol alone cream in the treatment of the chronich and eczema: A double-blind study. *J. Eur. Acad. Dermatol. Venereol.* **2008**, *22*, 531–536. [CrossRef]
116. Sharquie, K.E.; Najim, R.A.; Al-Salman, H.N. Oral zinc sulfate in the treatment of rosacea: A double-blind, placebo-controlled study. *Int. J. Dermatol.* **2006**, *45*, 857–861. [CrossRef]
117. Sharquie, K.E.; Noaimi, A.A.; Shwail, E.R. Oral zinc sulphate in treatment of alopecia areata (Double Blind; Cross-Over Study). *J. Clin. Exp. Dermatol. Res.* **2012**, *3*, 150. [CrossRef]
118. Borkow, G. Using Copper to Improve the Well-Being of the Skin. *Curr. Chem. Biol.* **2014**, *8*, 89–102. [CrossRef]
119. Nishito, Y.; Kambe, T. Absorption Mechanisms of Iron, Copper, and Zinc: An Overview. *J. Nutr. Sci. Vitaminol.* **2018**, *64*, 1–7. [CrossRef]
120. Linder, M.C.; Wooten, L.; Cerveza, P.; Cotton, S.; Shulze, R.; Lomeli, N. Copper transport. *Am. J. Clin. Nutr.* **1998**, *67*, 965–971. [CrossRef]
121. Karlin, K.D. Metalloenzymes, structural motifs, and inorganic models. *Science* **1993**, *261*, 701–708. [CrossRef] [PubMed]
122. Uauy, R.; Olivares, M.; Gonzalez, M. Essentiality of copper in humans. *Am. J. Clin. Nutr.* **1998**, *67*, 952–959. [CrossRef] [PubMed]
123. Borkow, G.; Gabbay, J. Copper as a biocidal tool. *Curr. Med. Chem.* **2005**, *12*, 2163–2175. [CrossRef] [PubMed]
124. Santo, C.E.; Quaranta, D.; Grass, G. Antimicrobial metallic copper surface kill *Staphylococcus haemolyticus* via membrane demage. *Microbiol. Open* **2012**, *1*, 46–52. [CrossRef] [PubMed]
125. Grass, G.; Rensing, C.; Solioz, M. Metallic copper as an antimicrobial surface. *Appl. Environ. Microbiol.* **2011**, *77*, 1541–1547. [CrossRef] [PubMed]
126. Philips, N.; Hwang, H.; Chauhan, S.; Leonardi, D.; Gonzalez, S. Stimulation of cell proliferation and expression of matrix metalloproteinase-1 and interluekin-8 genes in dermal fibroblasts by copper. *Connect. Tissue Res.* **2010**, *51*, 224–229. [CrossRef]
127. Wołonciej, M.; Milewska, E.; Roszkowska-Jakimiec, W. Trace elements as an activator of antioxidant enzymes. *Postepy. Hig. Med. Dosw.* **2016**, *70*, 1483–1498. [CrossRef]
128. Olivares, C.; Solano, F. New insights into the active site structure and catalytic mechanism of tyrosinase and its related proteins. *Pigment Cell Melanoma Res.* **2009**, *22*, 750–760. [CrossRef]
129. Chang, T.S. An updated review of tyrosinase inhibitors. *Int. J. Mol. Sci.* **2009**, *10*, 2440–2475. [CrossRef]
130. Pickart, L.; Margolina, A. Skin regenerative and anti-cancer actions of copper peptides. *Cosmetics* **2018**, *5*, 29. [CrossRef]
131. Mazurowska, L.; Mojski, M. Biological activities of selected peptides: Skin penetration ability of copper complexes with peptides. *J. Cosmet. Sci.* **2008**, *59*, 59–69. [PubMed]
132. Hostynek, J.J.; Maibach, H.I. Copper hypersensitivity: Dermatologic aspects-an overview. *Rev. Environ. Health* **2003**, *18*, 153–183. [CrossRef] [PubMed]
133. Fage, S.W.; Faurschou, A.; Thyssen, J.P. Copper hypersensitivity. *Contact Dermat.* **2014**, *71*, 191–201. [CrossRef] [PubMed]
134. De Araújo, L.A.; Addor, F.; Campos, P.M.B.G.M. Use of silicon for skin and hair care: An approach of chemical forms available and efficacy. *An. Bras. Dermatol.* **2016**, *91*, 331–335. [CrossRef] [PubMed]
135. Ferreira, A.O.; Freire, É.S.; Polonini, H.C.; da Silva, P.J.L.C.; Brandão, M.A.F.; Raposo, N.R.B. Anti-aging effects of monomethylsilanetriol and maltodextrin-stabilized orthosilicic acid on nails, skin and hair. *Cosmetics* **2018**, *5*, 41. [CrossRef]
136. Harlim, A.; Aisah, S.; Sihombing, R. Silicon level in skin tissues of normal female individuals. *J. Pakistan Assoc. Dermatologists.* **2018**, *28*, 134–138.
137. Boguszewska-Czubara, A.; Pasternak, K. Silicon in medicine and therapy. *J. Elem.* **2011**, *3*, 489–497. [CrossRef]
138. Barel, A.; Calomme, M.; Timchenko, A.; de Paepe, K.; Demeester, N.; Rogiers, V.; Clarys, P.; Vandenberghe, D. Effect of oral intake of choline-stabilized orthosilicic acid on skin, nails and hair in women with photodamaged skin. *Arch. Dermatol. Res.* **2005**, *297*, 147–153. [CrossRef]
139. Mancinella, A. Silicon, a trace element essential for living organisms. Recent knowledge on its preventive role in atherosclerotic process, aging and neoplasms. *Clin. Ter.* **1991**, *137*, 343–350.
140. Reffitt, D.M.; Ogston, N.; Jugdaohsingh, R.; Cheung, H.F.; Evans, B.A.; Thompson, R.P.; Powell, J.J.; Hampson, G.N. Orthosilicic acid stimulates collagen type 1 synthesis and osteoblastic differentiation in human osteoblast-like cells in vitro. *Bone* **2003**, *32*, 127–135. [CrossRef]

141. Kalil, C.L.P.V.; Campos, V.; Cignachi, S.; Izidoro, J.V.; Reinehr, C.P.H.; Chaves, C. Evaluation of cutaneous rejuvenation associated with the use of ortho-silicic acid stabilized by hydrolyzed marine collagen. *J. Cosmet. Dermatol.* **2017**, *17*, 814–820. [CrossRef] [PubMed]

142. Nair, A.; Jacob, S.; Al-Dhublab, B.; Attimarad, M.; Harsha, S. Basic considerations in the dermatokinetics of topical formulations. *Bras. J. Pharmaceut. Sci.* **2013**, *49*, 423–434. [CrossRef]

143. Penniston, K.L.; Tanumihardjo, S.A. The acute and chronic toxic effects of vitamin A. *Am. J. Clin. Nutr.* **2006**, *83*, 191–201. [CrossRef] [PubMed]

144. Tanumihardjo, S.A.; Russell, R.M.; Stephensen, C.H.B.; Gannon, B.M.; Craft, N.E.; Haskell, M.J.; Lietz, G.; Schulze, K.; Raiten, D.J. Biomarkers of nutrition for development (BOND)-vitamin A overview. *J. Nutr.* **2016**, *146*, 1816–1848. [CrossRef] [PubMed]

145. Arscott, S.A.; Howe, J.A.; Davis, C.R.; Tanumihardjo, S.A. Carotenoid profiles in provitamin A-containing fruits and vegetables affect the bioefficacy in Mongolian gerbils. *Exp. Biol. Med.* **2010**, *235*, 839–848. [CrossRef] [PubMed]

146. Szymańska, R.; Nowicka, B.; Kruk, J. Vitamin E-occurrence, biosynthesis by plants and functions in human nutrition. *Mini. Rev. Med. Chem.* **2017**, *17*, 1039–1052. [CrossRef]

147. Lv, J.; Ai, P.; Lei, S.; Zhou, F.; Chen, S.; Zhang, Y. Selenium levels and skin diseases: Systematic review and meta-analysis. *J. Trace. Elem. Med. Biol.* **2020**, *62*, 126548. [CrossRef]

148. Fine, J.D.; Tamura, T.; Johnson, L. Blood vitamin and trace metal levels in epidermolysis bullosa. *Arch. Dermatol.* **1989**, *125*, 374–379. [CrossRef]

149. Ewing, C.I.; Gibbs, A.C.; Ashcroft, C.; David, T.J. Failure of oral zinc supplementation in atopic eczema. *Eur. J. Clin. Nutr.* **1991**, *45*, 507–510.

150. Choudhary, S.V.; Gadegone, R.W.; Koley, S. Menkes Kinky hair disease. *Indian J. Dermatol.* **2012**, *57*, 407–409. [CrossRef]

151. Hornych, A.; Oravec, S.; Girault, F.; Forette, B.; Horrobin, D.F. The effect of gamma-linolenic acid on plasma and membranelipids and renal prostaglandin synthesis in older subjects. *Bratisl. Lek. Listy.* **2002**, *103*, 101–107. [PubMed]

152. Michalak, M.; Glinka, R. Plant oils in cosmetology and dermatology. *Pol. J. Cosmetol.* **2018**, *21*, 2–9.

153. Correa, M.C.; Mao, G.; Saad, P.; Flach, C.R.; Mendelsohn, R.; Walters, R.M. Molecular interactions of plant oil components with stratum corneum lipids correlate with clinical measures of skin barrier function. *Exp. Dermatol.* **2014**, *23*, 39–44. [CrossRef] [PubMed]

154. Feingold, K.R.; Elias, P.M. Role of lipids in the formation and maintenance of the cutaneous permeability barier. *Biochim. Biophys. Acta.* **2014**, *1841*, 280–294. [CrossRef]

155. Kapoor, R.; Huang, Y.-S. Gamma linolenic acid: An antiinflammatory omega-6 fatty acid. *Curr. Pharm. Biotechno.* **2006**, *7*, 531–534. [CrossRef] [PubMed]

156. Michalak, M.; Kiełtyka-Dadasiewicz, A. Oils from fruit seeds and their dietetic and cosmetic significance. *Herba Pol.* **2018**, *64*, 63–70. [CrossRef]

157. Burris, J.; Rietkerk, W.; Woolf, K. Acne: The role of medical nutrition therapy. *J. Acad. Nutr. Diet.* **2013**, *113*, 416–430. [CrossRef]

158. Cakir, A. Essential oil and fatty acid composition of the fruits of Hippophae rhamnoides L. (Sea Buckthorn) and Myrtus communis L. from Turkey. *Biochem. Syst. Ecol.* **2004**, *32*, 809–816. [CrossRef]

159. Van Smeden, J.; Janssens, M.; Gooris, G.S.; Bouwstra, J.A. The important role of stratum corneum lipids for the cutaneous barrier function. *Biochim. Biophys. Acta.* **2014**, *1841*, 295–313. [CrossRef]

160. Sabikhi, L.; Kumar, M.H.S. Fatty acid profile of unconventional oil seeds. *Adv. Food. Nutr. Res.* **2012**, *67*, 141–184.

161. Simon, D.; Eng, P.A.; Borelli, S.; Kägi, R.; Zimmermann, C.; Zahner, C.; Drewe, J.; Hess, L.; Ferrari, G.; Lautenschlager, S.; et al. Gamma-linolenic acid levels correlate with clinical efficacy of evening primrose oil in patients with atopic dermatitis. *Adv. Ther.* **2014**, *31*, 180–188. [CrossRef] [PubMed]

162. Larmo, P.; Ulvinen, T.; Määttä, P.; Judin, V.-P. CO_2-extracted blackcurrant seed oil for well-being of the skin. *Innov. Food Technol.* **2011**, 68–69.

163. Asadi-Samani, M.; Bahmani, M.; Rafieian-Kopaei, M. The chemical composition, botanical characteristic and biological activities of *Borago officinalis*: A review. *Asian. Pac. J. Trop. Med.* **2014**, *7*, 22–28. [CrossRef]

164. Muggli, R. Systemic evening primrose oil improves the biophysical skin parameters of healthy adults. *Int. J. Cosmet. Sci.* **2005**, *27*, 243–249. [CrossRef] [PubMed]

165. Linnamaa, P.; Savolainen, J.; Koulu, L.; Tuomasjukka, S.; Kallio, H.; Yang, B.; Vahlberg, T.; Tahvonen, R. Blackcurrant seed oil for prevention of atopic dermatitis in newborns: A randomized, double-blind, placebo-controlled trial. *Clin. Exp. Allergy* **2010**, *40*, 1247–1255. [CrossRef] [PubMed]

166. Oomah, B.D.; Busson, M.; Godfrey, D.V.; Drover, J.C.G. Characteristics of hemp (Cannabis sativa L.) seed oil. *Food Chem.* **2002**, *76*, 33–43. [CrossRef]

167. Montserrat-de la Paz, S.; Marín-Aguilar, F.; García-Giménez, M.D. Fernández-Arche, M.A. Hemp (Cannabis sativa L.) seed oil: Analytical and phytochemical characterization of the unsaponifiable fraction. *J. Agric. Food Chem.* **2014**, *62*, 1105–1110. [CrossRef]

168. Razafimamonjison, G.; Tsy, J.M.L.P.; Randriamiarinarivo, M.; Ramanoelina, P.; Rasoarahona, J.; Fawbush, F.; Danthu, P. Fatty acid composition of baobab seed and its relationship with the genus *Adansonia* taxonomy. *Chem. Biodiversity* **2017**, *14*. [CrossRef]

169. Osman, M.A. Chemical and nutrient analysis of baobab (Adansonia digitata) fruit and seed protein solubility. Plant Foods Hum. Nutrients **2004**, *59*, 29–33.

170. Komane, B.M.; Vermaak, I.; Kamatou, G.P.P.; Summers, B.; Viljoen, A.M. Beauty in baobab: A pilot study of the safety and efficacy of Adansonia digitata seed oil. *Rev. Bras. Farmacogn.* **2017**, *27*, 1–8. [CrossRef]
171. Momchilova, S.; Antonova, D.; Marekov, I.; Kuleva, L.; Nikolova-Damyanova, B. Fatty acids, triacylglycerols, and sterols in neem oil (Azadirachta indica A. Juss) as determined by a combination of chromatographic and spectral techniques. *J. Liq. Chromatogr. Relat. Technol.* **2007**, *30*, 11–25. [CrossRef]
172. Campos, E.V.R.; de Oliveira, J.L.; Pascoli, M.; de Lima, R.; Fraceto, L.F. Neem oil and crop protection: From now to the future. *Front. Plant Sci.* **2016**, *7*, 1494. [CrossRef] [PubMed]
173. Ahmad, A.; Husain, A.; Mujeeb, M.; Khan, S.A.; Najmi, A.K.; Siddique, N.A.; Damanhouri, Z.A.; Anwar, F. A review on therapeutic potential of *Nigella sativa*: A miracle herb. *Asian Pac. J. Trop. Biomed.* **2013**, *3*, 337–352. [CrossRef]
174. Ghorbanibirgani, A.; Khalili, A.; Rokhafrooz, D. Comparing *Nigella sativa* oil and fish oil in treatment of vitiligo. *Iran Red. Crescent. Med. J.* **2014**, *16*, 4515. [CrossRef]
175. DebMandal, M.; Mandal, S. Coconut (*Cocos nucifera* L.: Arecaceae): In health promotion and disease prevention. *Asian Pac. J. Trop. Med.* **2011**, *4*, 241–247. [CrossRef]
176. Yang, B.; Kalimo, K.O.; Mattila, L.M.; Kallio, S.E.; Katajisto, J.K.; Peltola, O.J.; Kallio, H.P. Effects of dietary supplementation with sea buckthorn (Hippophae¨rhamnoides) seed and pulp oils on atopic dermatitis. *J. Nutr. Biochem.* **1999**, *10*, 622–630. [CrossRef]
177. Christaki, E. Hippophae rhamnoides L. (Sea Buckthorn): A potential source of nutraceuticals. *Food Pub. Health* **2012**, *2*, 69–72. [CrossRef]
178. Wagemaker, T.A.; Rijo, P.; Rodrigues, L.M.; Campos, P.M.B.G.M.; Fernandes, A.S.; Rosado, C. Integrated approach in the assessment of skin compatibility of cosmetic formulations with green coffee oil. *Int. J. Cosmet. Sci.* **2015**, *37*, 506–510. [CrossRef]
179. Oliveira, L.S.; Franca, A.S.; Mendonça, J.C.F.; Barros-Júnior, M.C. Proximate composition and fatty acids profile of green and roasted defective coffee beans. *Food Sci. Technol.* **2006**, *39*, 235–239. [CrossRef]
180. Attia, R.S.; Abou-Gharbia, H.A. Evaluation and stabilization of wheat germ and its oil characteristics. *Alex. J. Fd. Sci. Technol.* **2011**, *8*, 31–39.
181. Matthäus, B.; Özcan, M.M. Habitat effects on yield, fatty acid composition and tocopherol contents of prickly pear (Opuntiaficus-indica L.) seed oils. *Sci. Horticulturae* **2011**, *131*, 95–98. [CrossRef]
182. Boucetta, K.Q.; Charrouf, Z.; Aguenaou, H.; Derouiche, A.; Bensouda, Y. Does argan oil have a moisturizing effect on the skin of postmenopausal women? *Skin Res. Technol.* **2013**, *19*, 356–357. [CrossRef]
183. Guillaume, D.; Charrouf, Z. Argan oil and other argan products: Use in dermocosmetology. *Eur. J. Lipid Sci. Technol.* **2011**, *113*, 403–408. [CrossRef]
184. Boucetta, K.Q.; Charrouf, Z.; Aguenaou, H.; Derouiche, A.; Bensouda, Y. The effect of dietary and/or cosmetic argan oil on postmenopausal skin elasticity. *Clin. Interv. Aging* **2015**, *30*, 339–349.
185. Ziboh, V.A. The significance of polyunsaturated fatty acids in cutaneous biology. *Lipids* **1996**, *31*, 249–253. [CrossRef] [PubMed]
186. Rhodes, L.E.; O'Farrell, S.; Jackson, M.J.; Friedmann, P.S. Dietary fish-oil supplementation in humans reduces UVB-erythemal sensitivity but increases epidermal lipid peroxidation. *J. Investig. Derm.* **1994**, *103*, 151–154. [CrossRef] [PubMed]
187. Rhodes, L.E.; Shahbakhti, H.; Azurdia, M.R.; Moison, R.M.W.; Steenwinkel, M.-J.S.T.; Homburg, M.I.; Dean, M.P.; McArdle, F.; van Henegouwen, G.M.J.B.; Epe, B.; et al. Effect of eicosapentaenoic acid, an omega-3 polyunsaturated fatty acid, on UVR-related cancer risk in humans. An assessment of early genotoxic markers. *Carcinogenesis* **2003**, *24*, 919–925. [CrossRef]
188. Michalak, M.; Glinka, R. Sources of vegetable dyes and their use in cosmetology. *Pol. J. Cosmetol.* **2017**, *20*, 196–205.
189. Pérez-Jiménez, J.; Neveu, V.; Vos, F.; Scalbert, A. Identification of the 100 richest dietary sources of polyphenols: An application of the Phenol-Explorer database. *Eur. J. Clin. Nutr.* **2010**, *64*, 112–120. [CrossRef]
190. Han, X.; Shen, T.; Lou, H. Dietary polyphenols and their biological significance. *Int. J. Mol. Sci.* **2007**, *8*, 950–988. [CrossRef]
191. Tsao, R. Chemistry and biochemistry of dietary polyphenols. *Nutrients* **2010**, *2*, 1231–1246. [CrossRef] [PubMed]
192. Shahidi, F.; Ambigaipalan, P. Phenolics and polyphenolics in foods, beverages and spices: Antioxidant activity and health effects—A review. *J. Funct. Foods* **2015**, *18*, 820–897. [CrossRef]
193. Jain, P.K.; Kharya, M.D.; Gajbhiye, A.; Sara, U.V.S.; Sharma, V.K. Flavonoids as nutraceuticals. A review. *Herba Pol.* **2010**, *56*, 105–117.
194. Rice-Evans, C.A.; Miller, N.J.; Paganga, G. Structure-antioxidant activity relationships of flavonoids and phenolic acids. *Free Radic. Biol. Med.* **1996**, *20*, 933–956. [CrossRef]
195. Bravo, L. Polyphenols: Chemistry, dietary sources, metabolism, and nutritional significance. *Nutr. Rev.* **1998**, *56*, 317–333. [CrossRef]
196. Epstein, H. Cosmeceuticals and polyphenols. *Clin. Dermatol.* **2009**, *27*, 475–478. [CrossRef]
197. Nichols, J.A.; Katiyar, S.K. Skin photoprotection by natural polyphenols: Anti-inflammatory, antioxidant and DNA repair mechanisms. *Arch. Dermatol. Res.* **2010**, *302*, 71–83. [CrossRef]
198. Zillich, O.V.; Schweiggert-Weisz, U.; Eisner, P.; Kerscher, M. Polyphenols as active ingredients for cosmetic products. *Int. J. Cosmet. Sci.* **2015**, *37*, 455–464. [CrossRef]
199. Chuarienthong, P.; Lourith, N.; Leelapornpisid, P. Clinical efficacy comparison of anti-wrinkle cosmetics containing herbal flavonoids. *Int. J. Cosmet. Sci.* **2010**, *32*, 99–106. [CrossRef]
200. Chen, L.; Hu, J.Y.; Wang, S.Q. The role of antioxidants in photoprotection: A critical review. *J. Am. Acad. Dermatol.* **2012**, *67*, 1013–1024. [CrossRef]

201. Afag, F.; Mukhtar, H. Botanical antioxidants in the prevention of photocarcinogenesis and photoaging. *Exp. Derm.* **2006**, *15*, 678–684. [CrossRef] [PubMed]
202. Gupta, V.K.; Kumria, R.; Garg, M.; Gupta, M. Recent updates on free radicals scavenging flavonoids: An overview. *Asian. J. Plant Sci.* **2010**, *9*, 108–117. [CrossRef]
203. Cho, H.S.; Lee, M.H.; Lee, J.W.; No, K.O.; Park, S.K.; Lee, H.S. Anti-wrinkling effects of the mixture of vitamin C, vitamin E, pycnogenol and evening primrose oil, and molecular mechanisms on hairless mouse skin caused by chronic ultraviolet B irradiation. *Photodermatol. Photoimmunol. Photomed* **2007**, *23*, 155–162. [CrossRef] [PubMed]
204. Svobodová, A.; Zdarilová, A.; Malisková, J.; Mikulková, H.; Walterová, D.; Vostalová, J. Attenuation of UVA-induced damage to human keratinocytes by silymarin. *J. Dermatol. Sci.* **2007**, *46*, 21–30. [CrossRef]
205. Moore, J.O.; Yongyin, W.; Stebbins, W.G.; Gao, D.; Zhou, X.; Phelps, R.; Lebwohl, M.; Wei, H. Photoprotective effect of isoflavone genistein on ultraviolet B-induced pyrimidine dimer formation and PCNA expression in human reconstituted skin and its implications in dermatology and prevention of cutaneous carcinogenesis. *Carcinogenesis* **2006**, *27*, 1627–1635. [CrossRef]
206. Lima, C.F.; Pereira-Wilson, C.; Rattan, S.I. Curcumin induces heme oxygenase-1 in normal human skin fibroblasts through redox signaling: Relevance for anti-aging intervention. *Mol. Nutr. Food Res.* **2011**, *55*, 430–442. [CrossRef]
207. Afaq, F.; Adhami, V.M.; Ahmad, N. Prevention of short-term ultraviolet B radiation-mediated damages by resveratrol in SKH-1 hairless mice. *Toxicol. Appl. Pharmacol.* **2003**, *186*, 28–37. [CrossRef]
208. Afaq, F.; Saleem, M.; Krueger, C.G.; Reed, J.D.; Mukhtar, H. Anthocyanin- and hydrolyzable tannin-rich pomegranate fruit extract modulates MAPK and NF-kappa B pathways and inhibits skin tumorigenesis in CD-1 mice. *Int. J. Cancer* **2005**, *113*, 423–433. [CrossRef]
209. Khan, N.; Syed, D.N.; Pal, C.H.; Mukhtar, H.; Afag, F. Pomegranate fruit extract inhibits UVB-induced inflammation and proliferation by modulating NF-κB and MAPK signaling pathways in mouse skin. *Photochem. Photobiol.* **2012**, *88*, 1126–1134. [CrossRef]
210. Kasai, K.; Yoshimura, M.; Koga, T.; Arii, M.; Kawasaki, S. Effects of oral administration of ellagic acid-rich pomegranate extract on ultraviolet-induced pigmentation in the human skin. *J. Nutr. Sci. Vitaminol.* **2006**, *52*, 383–388. [CrossRef]
211. Henning, S.M.T.; Niua, Y.; Liua, Y.; Leea, N.H.; Harac, Y.; Thamesa, G.D.; Minuttia, R.R.; Carpentera, C.L.; Wang, H.; Heber, D. Bioavailability and antioxidant effect of epigallocatechin gallate administered in purified form versus as green tea extract in healthy individuals. *J. Nutr. Biochem.* **2005**, *16*, 610–616. [CrossRef] [PubMed]
212. Katiyar, S.K.; Afaq, F.; Perez, A.; Mukhtar, H. Green tea polyphenol (-)-epigallocatechin-3-gallate treatment of human skin inhibits ultraviolet radiation-induced oxidative stress. *Carcinogenesis* **2001**, *22*, 287–294. [CrossRef] [PubMed]
213. Vayalil, P.K.; Mittal, A.; Hara, Y.; Elmets, C.E.; Katiyar, S.K. Green tea polyphenols prevent ultraviolet light-induced oxidative damage and matrix metalloproteinases expression in mouse skin. *J. Investig. Dermatol.* **2004**, *122*, 1480–1487. [CrossRef] [PubMed]
214. Chiu, A.E.; Chan, J.L.; Kern, D.G.; Kohler, S.; Rehmus, W.E.; Kimball, A.B. Double-blinded, placebo-controlled trial of green tea extracts in the clinical and histologic appearance of photoaging skin. *Dermatol. Surg.* **2005**, *31*, 855–860. [CrossRef]
215. Higdon, J.V.; Frei, B. Tea catechins and polyphenols: Health effects, metabolism, and antioxidant functions. *Crit. Rev. Food Sci. Nutr.* **2003**, *43*, 89–143. [CrossRef]
216. Rutter, K.; David, R.S.; Nalani, F.; Obrenovich, M.; Zito, M.; Starke-Reed, P.; Monnier, V.M. Green tea extract suppresses the age-related increase in collagen crosslinking and fluorescent products in C57BL/6Mice. *Int. J. Vitam. Nutr. Res.* **2003**, *73*, 453–460. [CrossRef] [PubMed]
217. Tuong, W.; Walker, L.; Sivamani, R.K. Polyphenols as novel treatment options for dermatological diseases: A systematic review of clinical trials. *J. Dermatolog. Treat.* **2015**, *26*, 381–388. [CrossRef]
218. Beutner, S.; Bloedorn, B.; Frixel, S.; Hermandez, B.; Hoffmann, T.; Martin, H.; Mayer, B.; Noack, P.; Ruck, C.; Schmidt, M.; et al. Quantitative assessment of antioxidant properties of natural carotene in antioxidant functions. *J. Sci. Food Argic.* **2001**, *81*, 559–568. [CrossRef]
219. Delgado-Vargas, F.; Jiménez, A.R.; Paredes-López, O. Natural pigments: Carotenoids, anthocyanins, and betalains-characteristics, biosynthesis, processing, and stability. *Crit. Rev. Food Sci. Nutr.* **2000**, *40*, 173–289. [CrossRef] [PubMed]
220. Balić, A.; Mokos, M. Do we utilize our knowledge of the skin protective effects of carotenoids enough? *Antioxidants* **2019**, *8*, 259. [CrossRef] [PubMed]
221. Eldahshan, O.A.; Singab, A.N.B. Carotenoids. *J. Pharmacogn. Phytochem.* **2013**, *2*, 225–234.
222. Davinelli, S.; Nielsen, M.E.; Scapagnini, G. Astaxanthin in skin health, repair, and disease: A comprehensive review. *Nutrients* **2018**, *10*, 522. [CrossRef] [PubMed]
223. Mathews-Roth, M.M. Plasma concentration of carotenoids after large doses of beta-carotene. *Am. J. Clin. Nutr.* **1990**, *52*, 500–501. [CrossRef]
224. Bendich, A. Biological functions of dietary carotenoids. *Ann. N. Y. Acad. Sci.* **1993**, *691*, 61–67. [CrossRef] [PubMed]
225. Mezzomo, N.; Ferreira, S.R.S. Carotenoids functionality, sources, and processing by supercritical technology: A review. *J. Chem.* **2016**, *2016*, 1–16. [CrossRef]
226. Bayerl, C. Beta-carotene in dermatology-does it help? *Acta. Dermatovenerol. Alp. Pannonica. Adriat.* **2008**, *17*, 160–166. [PubMed]
227. Astner, S.; Wu, A.; Chen, J.; Philips, N.; Rius-Diaz, F.; Parrado, C.; Mihm, M.C.; Goukassian, D.A.; Pathak, M.A.; González, S. Dietary lutein/zeaxanthin partially reduces photoaging and photocarcinogenesis in chronically UVB-irradiated Skh-1 hairless mice. *Skin. Pharmacol. Physiol.* **2007**, *20*, 283–291. [CrossRef]

228. Wertz, K.; Seifert, N.; Hunziker, P.B.; Riss, G.; Wyss, A.; Lankin, C.; Goralczyk, R. β-carotene inhibits UVA-induced matrix metalloprotease 1 and 10 expression in keratinocytes by a singlet oxygen-dependent mechanism. *Free. Radic. Biolol. Med.* **2004**, *37*, 654–670. [CrossRef]

229. Darvin, M.E.; Sterry, W.; Lademann, J.; Vergou, T. The role of carotenoids in human skin. *Molecules* **2011**, *16*, 10491–10506. [CrossRef]

230. Darvin, M.E.; Fluhr, J.W.; Schanzer, S.; Richter, H.; Patzelt, A.; Meinke, M.C.; Zastrow, L.; Golz, K.; Doucet, O.; Sterry, W. Dermal carotenoid level and kinetics after topical and systemic administration of antioxidants: Enrichment strategies in a controlled in vivo study. *J. Dermatol. Sci.* **2011**, *64*, 53–58. [CrossRef]

231. Tominaga, K.; Hongo, N.; Karato, M.; Yamashita, E. Cosmetic benefits of astaxanthin on human subjects. *Acta. Biochim. Pol.* **2012**, *59*, 43–47. [CrossRef] [PubMed]

232. Yamashita, E. Cosmetic benefit of dietary supplements including astaxanthin and tocotrienol on human skin. *Food Style* **2002**, *216*, 112–117.

233. Yoon, H.-S.; Cho, H.H.; Cho, S.; Lee, S.-R.; Shin, M.-H.; Chung, J.H. Supplementing with dietary astaxanthin combined with collagen hydrolysate improves facial elasticity and decreases matrix metalloproteinase-1and-12 expression: A comparative study with placebo. *J. Med. Food* **2014**, *17*, 810–816. [CrossRef] [PubMed]

234. Juturu, V.; Bowman, J.P.; Deshpande, J. Overall skin tone and skin-lightening-improving effects with oral supplementation of lutein and zeaxanthin isomers: A double-blind, placebo-controlled clinical trial. *Clin. Cosmet. Investig. Dermatol.* **2016**, *9*, 325–332. [CrossRef]

235. Schwartz, S.; Frank, E.; Gierhart, D.; Simpson, P.; Frumento, R. Zeaxanthin-based dietary supplement and topical serum improve hydration and reduce wrinkle count in female subjects. *J. Cosmet. Dermatol.* **2016**, *15*, 13–20. [CrossRef]

236. Meephansan, J.; Rungjang, A.; Yingmema, W.; Deenonpoe, R.; Ponnikorn, S. Effect of astaxanthin on cutaneous wound healing. *Clin. Cosmet. Investig. Dermatol.* **2017**, *10*, 259–265. [CrossRef]

237. Palombo, P.; Fabrizi, G.; Ruocco, V.; Ruocco, E.; Fluhr, J.; Roberts, R.; Morganti, P. Beneficial long-term effects of combined oral/topical antioxidant treatment with the carotenoids lutein and zeaxanthin on human skin: A double-blind, placebo-controlled study. *Skin Pharmacol. Physiol.* **2007**, *20*, 199–210. [CrossRef]

238. Césarini, J.P.; Michel, L.; Maurette, J.M.; Adhoute, H.; Béjot, M. Immediate effects of UV radiation on the skin: Modification by an antioxidant complex containing carotenoids. *Photodermatol. Photoimmunol. Photomed.* **2003**, *19*, 182–189. [CrossRef]

239. Heinrich, U.; Gärtner, C.; Wiebusch, M.; Eichler, O.; Sies, H.; Tronnier, H.; Stahl, W. Supplementation with beta-carotene or a similar amount of mixed carotenoids protects humans from UV-induced erythema. *J. Nutr.* **2003**, *133*, 98–101. [CrossRef]

240. Aust, O.; Stahl, W.; Sies, H.; Tronnier, H.; Heinrich, U. Supplementation with tomato-based products increases lycopene, phytofluene, and phytoene levels in human serum and protects against UV-light-induced erythema. *Int. J. Vitam. Nutr. Res.* **2005**, *75*, 54–60. [CrossRef]

241. Heinen, M.M.; Hughes, M.C.; Ibiebele, T.I.; Marks, G.C.; Green, A.C.; van der Pols, J.C. Intake of antioxidant nutrients and the risk of skin cancer. *Eur. J. Cancer.* **2007**, *43*, 2707–2716. [CrossRef] [PubMed]

242. Greenberg, E.R.; Baron, J.A.; Stukel, T.A.; Stevens, M.M.; Mandei, J.S.; Spencer, S.K. A clinical trial of β-carotene to prevent basal-cell and squamous-cell cancers of the skin. *N. Engl. J. Med.* **1990**, *323*, 789–795. [CrossRef] [PubMed]

MDPI

St. Alban-Anlage 66

4052 Basel

Switzerland

Tel. +41 61 683 77 34

Fax +41 61 302 89 18

www.mdpi.com

Nutrients Editorial Office

E-mail: nutrients@mdpi.com

www.mdpi.com/journal/nutrients